Merlin Carl
Ordinal Computability

De Gruyter Series in Logic and Its Applications

Volume 9

Merlin Carl

Ordinal Computability

An Introduction to Infinitary Machines

DE GRUYTER

Mathematics Subject Classification 2010
Primary: 03D60, 03D65, 03Exx, 03-XX; Secondary: 03E15, 03Dxx

Author
Dr. Merlin Carl
Europa-Universität Flensburg
Institut für mathematische,
naturwissenschaftliche
und technische Bildung
Auf dem Campus 1b
24943 Flensburg
Germany
and
Universität Konstanz
Fachbereich Mathematik und Statistik
78457 Konstanz
Germany
merlin.carl@uni-flensburg.de

ISBN 978-3-11-049562-1
e-ISBN (PDF) 978-3-11-049615-4
e-ISBN (EPUB) 978-3-11-049291-0
ISSN 1438-1893

Library of Congress Control Number: 2019946003

Bibliographic information published by the Deutsche Nationalbibliothek
The Deutsche Nationalbibliothek lists this publication in the Deutsche Nationalbibliografie; detailed bibliographic data are available on the Internet at http://dnb.dnb.de.

Typesetting: VTeX UAB, Lithuania
Printing and binding: CPI books GmbH, Leck

www.degruyter.com

Contents

1 Introduction

1.1 About this book

Ordinal computability is by now a well established and flowering area of mathematical logic, somewhere at the border of set theory and recursion theory. Up to summer 2018, there are well over 60 research papers by over 20 authors on the subject, some of them in the top journals in mathematical logic, and moreover, at least five Ph. D. theses and a hardly surveyable number of Bachelor, Master's and diploma theses. As a relatively new field with many open questions and many new directions to explore, it is a fruitful research topic for anyone interested in set theory and recursion theory.

Unfortunately, the area is not easy to access for a beginner. On the one hand, there is not, like Turing computability in the finite case, a single notion of computation, but a variety of models. On the other hand, the field is spread over a large number of research papers, and unless one has a supervisor who already knows her or his way around, it is hard to see where to start and how to proceed. Finally, the required background varies strongly: while some research papers on ordinal computability require only basic set-theoretical and recursion-theoretical knowledge, others use fine structure, descriptive set theory, classical recursion theory, forcing, inner model theory, etc.

What thus seems to be missing – in spite of some good overview articles like Welch's [190] or [189] is a systematical exposition that requires only little background knowledge on the part of the reader that leads up to the current research frontier and treats the area comprehensively, so that one can get a good grasp on the fundamentals while at the same time getting at least an outlook of the whole field. Ideally, such a treatment would also show that the "zoo" of machines models is actually a well-structured hierarchy with a compelling amount of stability at many levels, resembling more the realm of complexity classes known from classical computability than a proliferation of "Turing machine fan fiction," and thereby contribute to an increased interaction between the researchers investigating these models.

This is what this book attempts to achieve. In large parts, it originated from my 2016 habilitation thesis "Infinite Time Machines: A Computational Approach to Transfinite Mathematics" at the University of Konstanz, and my initial intention was merely to supplement the results obtained there with the necessary prerequisites and amend them with expositions of results by others. However, working on it offered the rare opportunity to—one might also say: forced me to—rework the area from its origins, and in particular to review older open questions in the light of the recently developed tools. In this way, working on this book turned out to be surprisingly productive: It was, for example, the impulse behind determining the computational strength of (α,β)-Turing machines in [34], the remarks on variants of machine models in Chapter 2, various new results on transfinite register machines in Chapter 3 such as Theorem 3.4.18 or

https://doi.org/10.1515/9783110496154-001

Corollary 3.4.13, Theorem 7.2.21 on the connection between time and space complexity for ITTMs answering an old question on the topic and a bunch of other new results in the remaining chapters that appear here for the first time.

For the reader's part, an elementary knowledge of set theory suffices for a good part of the book, roughly what one will find in the first three chapters of [48]. For Chapter 3, a background on Kripke–Platek set theory and Gödel's constructible universe L is required; even though this is briefly recalled, the reader who has never heard of this should have some additional references at hand, such as [92, 10, 135] or [40]. This will be enough for the bulk of Chapter 4, except the section using inner models for Woodin cardinals. Chapter 5 crucially makes use of forcing, which is only very briefly repeated and otherwise deferred to the references. No further background knowledge is necessary, though Chapters 6–9 will depend on the same prerequisites and on each other. The following graph on the chapter numbers roughly suggests various possibilities to either read the book or use it in a lecture or a seminar.

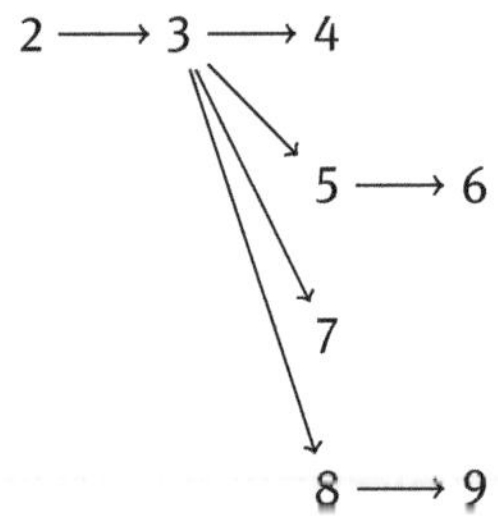

1.2 What happened so far?

In the past two decades, there has been considerable activity in the field of infinitary computability. We give a quick overview of some aspects, without striving for completeness.

The first model of infinitary computability are the Infinite Time Turing Machines (ITTMs) which were invented by Joel Hamkins and Jeffrey Kidder and first published by Joel Hamkins and Andrew Lewis in [81], the paper which marks the beginning of ordinal computability. Peter Koepke introduced stronger versions of ITTMs that have a proper class tape (Ordinal Turing Machines, OTMs) and characterized Gödel's constructible hierarchy in terms of OTM-computability [104, 105]. Independently, this machine type was defined and studied by Barnaby Dawson in his (unpublished) Ph. D. thesis [45]. Koepke also introduced α-Turing machines and characterized, together with Benjamin Seyfferth, α-recursiveness in terms of computability by these machines [112]. Due to him are also further models obtained by generalizing register machines rather than Turing machines into the transfinite, like weak Infinite Time Register Machines [108], Infinite Time Register Machines (introduced with Russell Miller in [110] and further studied in [109]), Ordinal Register Machines (jointly with Ryan Siders, studied in [114, 115, 116]) and infinite time Blum–Shub–Smale-machines [113]. Seyfferth further studied transfinite versions of λ-calculus (jointly with Tim Fischbach, see [54, 55] or [159]) and, together with Philipp Schlicht, gave connections between ordinal computability and descriptive set theory [160].

Joel Hamkins and others worked, e. g., on the degree theory for ITTMs (with Andrew Lewis, [77]) and applied ITTMs to model theory (jointly with Russell Miller, Daniel Seabold and Steve Warner, [79]) and also proposed ITTM-reducibility as a stronger alternative to Borel reducibility (with Sam Coskey, [42, 43]).

Benedikt Löwe found a connection between ITTMs and the revision theory of truth [128].

Robert Lubarsky considered Feedback-ITTMs, which are roughly ITTMs equipped with an oracle x that contains as much information as possible on the halting problem for ITTMs working in that same oracle x and studied their behavior in [133] and went on to consider connections with Σ^0_3-determinacy. Later on, the same idea was applied to classical Turing machines by Lubarsky, jointly with Nathanael Ackerman and Cameron Freer, which led to nice characterizations of hyperarithmetic computability as the feedback version of Turing computability and Turing computability as the feedback version of primitive recursion (see [3]).

Philip Welch gave precise characterizations of ITTM-writability, eventual writability and accidental writability and worked out a considerable amount of ITTM-degree theory [180, 184, 179]. He also connected ITRMs to reverse mathematics (jointly with P. Koepke, [117]) and to determinacy principles [187]. Together with Sy Friedman, he introduced Σ_n-ITTMs, a model considerably stronger than ITTMs [62].

Ralf Schindler considered analogues of classical complexity classes for ITTMs and proved that, for ITTMs, $P \neq \mathrm{NP}$ [155], which led to the study of analogues of various other complexity classes for ITTMs (see, e. g., [46]). Löwe then developed measures for space complexity of ITTMs [129], which were considered further by Joost Winter in [196].

Arnold Beckmann, together with Samuel Buss and Sy Friedman, introduced safe recursive set recursions and showed that the safe recursive set functions on infinite binary strings are exactly those that an ITTM can compute in fewer than ω^ω many steps [12].

A collection of various approaches to an effective mathematics of the transfinite was given in [71].

On the philosophical side, it was argued in [29] that ordinal Turing machines model an intuitive notion of transfinite computability; partially on the basis of [29], Benjamin Rin argued in [151] that transfinite recursion is a fundamental principle in set theory and determined its strength for various versions.

Recently, the topics of recognizability and algorithmic randomness for infinitary machines have been extensively studied; for details, we refer to our summaries of Chapters 4 and 5 below.

1.3 Summary

In Chapter 2, we will introduce the models of computation that we will be concerned within this book: Infinite Time Register Machines (ITRMs), Infinite Time Turing Machines (ITTMs), α-Turing Machines (α-TMs) and Ordinal Turing Machines (OTMs) and Ordinal Register Machines (ORMs). We make some elementary observations on their behavior and their power, without requiring a stronger set-theoretical or model-theoretical background.

In Chapter 3, we will determine the computability strength of each of the machine models introduced in Chapter 2, i. e., we will characterize the realm of objects that the respective machine types can compute. We will use standard results and notions from the Levy hierarchy, constructible set theory and admissibility that will be briefly recalled at the beginning of the chapter. The reader that is totally unfamiliar with theses subjects is advised to consult the numerous references given there.

Chapter 4 will introduce and study the concept of recognizability: Roughly, an object x (usually a real number or a set of ordinals) is recognizable if and only if there is a program that takes some input y and, after working for a while, halts and returns the answer whether $x = y$. Some parts of this chapter only require a basic knowledge of set theory, but others require forcing and the final section on recognizability for parameter-OTM assumes a firm knowledge of inner models for Woodin cardinals on the reader's part.

Chapter 5 is concerned with degree theory. Relativizing the notion of computability, one may consider the equivalence relation between two objects x and y that holds when x is computable from y and vice versa. The set of objects that are equivalent to a given x in this sense is the "degree" of x with respect to the relevant notion of computability, and the study of degrees and their structure plays are large role in classical computability theory. Particularly, prominent points are the existence of degrees that lie strictly between the degree of computable objects and the degree of the halting problem, and the existence of incomparable degrees known as Post's problem. We will see that these questions have positive solutions for some models of infinitary computability and negative solutions for others. Other results on degrees, such as Welch's results on minimality in the ITTM-degrees, will only be mentioned in passing.

Another big topic in classical computability theory is algorithmic randomness, which we consider in Chapter 6: Roughly, we say that an object is "random" if it has no rare properties that can be effectively approached. Similar notions are defined for infinitary machines and studied. We particularly emphasize analogues of Sack's theorem and van Lambalgen's theorem: Sack's theorem says that computability relative to all elements of positive Lebesgue measure implies computability, while van Lambalgen's theorem (roughly) states that, for two random real numbers x and y, if x has no information about y, then y has no information about x.

Chapter 7 deals with complexity theory for infinitary machines. We give an overview of the results of Schindler, Hamkins and Deolalikar on the analogues of the classes P and NP for ITTMs and the approach and Löwe and Winter to measuring space complexity for ITTMs. We then turn to OTMs, showing, e. g., that an infinitary analogue of the satisfiability problem, SAT^{∞}, is NP-complete, that NP-complete problems are undecidable and that there are NP-intermediate problems in this setting.

Chapter 8 concerns applications of ordinal computability in other areas. For example, we explain Koepke's proof of the continuum hypothesis in L using OTMs, the "computational" proof of Shoenfield's absoluteness theorem by Schlicht and Seyfferth, the approach of Coskey and Hamkins to replace Borel reducibility with ITTM-reducibility and an ordinal analogue of Weihrauch reducibility applicable to set-theoretical Π_2-formulas.

Finally, Chapter 9 is about connections between ordinal computability and philosophy. On the one hand, we offer a conceptual analysis showing that there is an intuitive (or at least "heuristical") concept of infinitary computability and that OTMs are an adequate formalization thereof. Then we given examples of the use of idealized agency in the foundations and the philosophy of mathematics and use OTMs to obtain formal counterparts to justifications and criticism of set-theoretical axioms made on the basis of this approach. Finally, we explain Löwe's discovery of a strong correspondence between ITTM-computability and the revision theory of truth.

Fortunately, ordinal computability is both already too extensively developed and develops so fast that no book project can keep up with all of its developments and branches. Moreover, the choice of topic is naturally influenced by the author's interests and familiarity with the respective areas. Therefore, some topics, like ITTM-degree theory for sets of real numbers, have only been touched briefly, while others, like recently discovered connections between ordinal computability and differential equations (see Bournez and Ouazzani [16]) are not covered at all; also, a wealth of new results on the computational strength of parameter-free α-ITTMs [35] and randomness with ITTMs [142] has been obtained while this book was written. Naturally, we hope that this happy tendency will continue.

A topic that some readers may find of interest, but that is not touched in this book is whether transfinite computations are physically possible or even potentially practically relevant. In our view, the appeal of transfinite computability is sufficiently justified by its potential to accommodate heuristic and intuitive notions of generalized effectivity that occur in various areas in mathematics (see Chapter 8), the possibility of an algorithmic access to various areas of mathematics and set theory (which is heuristically as well as didactically relevant), its connections to the philosophy of mathematics (see Chapter 9), its potential impact on classical recursion theory, its ability to foster interdisciplinary research and its aesthetic appeal. Nevertheless, the question of physical realizability of hypercomputations is an interesting one and treated, e. g., in Welch [183, 182] or Syropoulos [170].

As a general maxim, since we are proposing ordinal computability as a new approach to and perspective on transfinite mathematics, we have usually attempted to give the most "algorithmic" proofs of theorems. We have only deviated from this paradigm when a "nonalgorithmic" proof is either much shorter or yields much more information than the best known "algorithmic" proof. As an example, though Welch's original analysis of the computational strength of ITTMs is much more "computational" than the later approach by Welch and Friedman via the Σ_2-machine, it is the latter one that allows for a complete analysis of the computational strength of (α, β)-ITTMs, so that we opted for the latter. In many of these cases, the more computational variants are developed in the exercises.

In general, we have tried to give the original source of a statement, definition etc. as reference. However, some statements and proofs have become folklore; others arise very natural in a variety of considerations. In such cases, references may merely lead to survey articles or are sometimes missing. This is also the case when a statement merely serves as an auxiliary towards a central result that carries a reference. A missing reference does therefore not imply that a statement is due to the author or original to this book. Except for some observations on the strength of α-(w)ITRMs in Chapter 3 and on the space complexity for ITTMs in Chapter 7, most material in this book is not new.

The exercises serve a two-fold purpose: Many of them emphasize a certain proof technique or stress a certain point about a definition; and some of them yield additional results that we did not prove fully either due to space constraints or because we regarded the proofs are sufficiently similar to ones already explained.

1.4 Acknowledgments

First of all, I want to thank Peter Koepke who introduced me to ordinal computability and invented many of the machine models discussed in this book. Thanks are also due to Joel Hamkins who, together with Jeffrey Kidder and Andrew Lewis, started the area of ordinal computability with the invention and publication of Infinite Time Turing Machines and also initiated many of the methods and directions that shape the field.

Thanks are also due to the University of Konstanz for offering me a position, along with a pleasurable working atmosphere, that allowed me to carry out research on ordinal computability, in particular, my supervisor, Salma Kuhlmann, for giving me a lot of freedom in the choice of my research topics.

Further institutions that I thank for hosting me temporarily during research stays are the Laboratoire d'Informatique, de Robotique et de Microelectronique de Montpellier (LIRMM) in Montpellier and the Newton Institute in Cambridge. I should also mention that many of the ideas for my research originated at the annual "Computability in

Europe" (CiE) conference, where I first learned about many subjects, including algorithmic randomness and Weihrauch reducibility, that were then considered in transfinite time in our research.

I gratefully thank my coauthors Bruno Durand (Montpellier), Gregory Lafitte (Montpellier), Benedikt Löwe (Hamburg), Sabrina Ouazzani (Paris), Benjamin Rin (Utrecht), Philipp Schlicht (Bonn) and Philip Welch (Bristol) for the creative and successful cooperation; special thanks are due to Philipp Schlicht for also taking the time to discuss proofs, spotting mistakes and occasionally suggesting simplifications, which greatly helped in gaining confidence in and improving the exposition of several of my results, e. g., in Chapters 5 and 7.

Large parts of Chapters 4–8 consist of portions and adaptations of portions from several of my papers on the subject, and I thank both my coauthors and the publishers for the kind permission to reuse these text parts here. More specifically:

Philipp Schlicht and Philip Welch and Elsevier for [39], Elsevier for [19], Philipp Schlicht and Springer for [37], my coauthors Benjamin Rin and Benedikt Löwe as well as Springer for [33], Springer for [24, 22, 20, 27], Sabrina Ouazzani, Philip Welch and Springer for [34], IOS Press for [25] and [23], Philipp Schlicht and Duke University Press for [36], Cambridge University Press for [21] and the editors Stefan Geschke, Benedikt Löwe and Philipp Schlicht for [30].

Finally, I am grateful to the following people for proofreading portions of the text and providing valuable feedback: Matteo Bianchetti for the introduction and parts of Chapter 1, Philipp Schlicht for Sections 4.4 and 5.2, Dominik Klein for Chapter 9, Benedikt Löwe for Section 9.5 and Olivier Bournez for Chapter 2 and Chapter 3.

2 Machine models of transfinite computability

2.1 Introduction

In this chapter, we introduce the machine models that we will consider in this book. The most extensively studied models are Infinite Time Turing machines (ITTMs), followed by Ordinal Turing/Register Machines (OTMs, ORMs) and weak and strong Infinite Time α-Register Machines (α-wITRMs and α-ITRMs). α-Turing machines (α-TMs and α-ITTMs) have so far not received a comparable amount of attention. We also mention a few other models like the hypermachines of Friedman and Welch and Infinite Time Blum–Shub–Smale-Machines (ITBSSMs).

We will explain all of these models, but deviate slightly from the established literature. The reason is the following: ITTMs and ITRMs were originally introduced to work with tape length and register contents bounded by ω. Only later on, after these models had been successfully studied, generalizations to longer tapes and larger registers were considered. It is basically a piece of folklore knowledge in the area of ordinal computability that many of the results and arguments from the ω-case carry over to these more general machines with few modifications, yet these arguments have never been worked out, thus creating a somewhat shaky ground for someone entering the area. We thus mostly work with more general machines with tape length and register contents bounded by some appropriate ordinal α (closure under ordinal multiplication is usually more than enough), thus hopefully helping to fill this gap in the literature. The more established models, such as ITTMs and ITRMs, will be contained as special cases, and we will point out when their theory actually differs from the more general theory.

The models appear in order with their computational strength which will be determined in the next chapter and which is quite different from the historical order in which they appeared. Chronologically, ITTMs – which were invented by Hamkins and Kidder and are also known as Hamkins–Kidder-machines – came first with the seminal Hamkins–Lewis paper [81], followed by Koepke's Ordinal Turing Machines [104], Ordinal Register Machines [107, 114, 115], weak ITRMs [106], α-Turing machines [112] and Infinite Time Register Machines [110].

This chapter is organized as follows: In the first part, we introduce classical register machines; by allowing transfinite running time, we obtain weak and strong Infinite Time Register Machines (ITRM). Allowing infinite ordinals as register contents, we get α-register machines. Two ways are considered to deal with the event of a "register overflow," which occurs when the limit rules would lead to a register content beyond the allowed limit α, namely breaking the computation off and resetting the content to 0; these lead us to the distinction between weak α-ITRMs and strong α-ITRMs. We investigate some of their basic properties and give algorithms for, e. g., testing the well-foundedness of a partial ordering encoded by a subset of α, which will be of great

https://doi.org/10.1515/9783110496154-002

importance in the following chapters. Dropping the upper bound on the register contents altogether and allowing the registers to contain arbitrary ordinals, we finally get Ordinal Register Machines.

We then turn to "tape models," i. e., infinitary generalizations of Turing machines. Here, we start with the α-ITTMs already mentioned above, which are Turing machines with transfinite running time and tape length α. Bounding the running time of an α-ITTM by an ordinal β, we get α-β-Turing machines. Using the whole class of ordinals both as the tape length and the running time, we arrive at Ordinal Turing Machines.

It is important to observe that some very elementary properties taken for granted in the theory of Turing machines do not necessarily work for machines with a fixed time bound. For example, classically, one can, given two halting programs P and Q, obtain a new halting program R that first runs P and then Q. However, when time is limited, R may fail to halt, as running Q after P may violate the time bound. In order to avoid such difficulties which seem to make the theory more complicated without adding any interesting phenomena,[1] we will usually assume in the following that the ordinal time bounds, where they appear, are closed under ordinal arithmetic, i. e., under ordinal addition, multiplication and exponentiation.

Definition 2.1.1. Let α be an ordinal. Then α is additively/multiplicatively/exponentially closed if and only if, for all $\beta, \gamma < \alpha$, we have $\beta + \gamma < \alpha$ / $\beta \cdot \gamma < \alpha$ / $\beta^{\gamma} < \alpha$.

As a general convention, the words "machine" and "program" in statements like "there is a program with the following properties" or "there is a machine that solves the following problem" will be used as synonyms. Furthermore, computations for machine models without a fixed time bound will be regarded as having length On; i. e., when a program halts, we regard the computation as continuing, repeating the halting state forever.

For coding issues, it will be important to have Cantor's pairing function available, both for natural numbers and for ordinals.

Definition 2.1.2. For $m, n \in \omega$, we let $p(m,n) = \frac{(m+n)(m+n+1)}{2} + n$. $p(m,n)$ is the *Cantor code* for the pair (m,n).

More generally, the ordering $<^*$ on pairs of ordinals is defined by letting $(\alpha,\beta) <^* (\gamma,\delta)$ if and only if

$\max(\alpha,\beta) < \max(\gamma,\delta) \vee [\max(\alpha,\beta) = \max(\gamma,\delta) \wedge \alpha < \gamma] \vee [\max(\alpha,\beta) = \max(\gamma,\delta) \wedge \alpha = \gamma \wedge \beta < \delta]$.

Now, for $(\alpha,\beta) \in \mathrm{On} \times \mathrm{On}$, the ordinal $p(\alpha,\beta)$ coding (α,β) in the sense of Cantor's ordinal pairing function is the order type of the set $\{(\gamma,\delta) : (\gamma,\delta) <^* (\alpha,\beta)$ under the ordering $<^*\}$.

1 The critical reader is very much invited to refute this claim by providing counterexamples.

We can encode finite sequences of natural numbers and ordinals by iteration: For $\alpha, \beta \in \mathrm{On}$, we let $p_2(\alpha,\beta) = p(\alpha,\beta)$ and $p_{n+1}(\alpha_0,\dots,\alpha_n) = p(p_n(\alpha_0,\alpha_1,\dots,\alpha_{n-1}),\alpha_n)$ for $n \geq 2$.

Remark 2.1.3.
(1) The slight clash of notation arising from using p both for pairing on natural numbers and on ordinals is harmless, as the two functions coincide when $\alpha, \beta \in \omega$.
(2) It is easy to check that $p : \mathrm{On} \times \mathrm{On} \to \mathrm{On}$ is indeed a bijective (class) function and that the same holds on $\omega \times \omega$. This is also true for p_k, for every $k \in \omega$.
(3) When $\gamma \in \mathrm{On}$ is closed under ordinal multiplication, it is also closed under p.

Exercise 2.1.4. Prove the statements (1)–(3) from the last remark.

Proposition 2.1.5. *For every $\alpha \geq 2$, α^ω is closed under the pairing function.*

Proof. Let $\beta, \gamma < \alpha^\omega$. Then there is $k \in \omega$ such that $\beta, \gamma < \alpha^k$. Clearly, we have $p(\beta,\gamma) < p(\alpha^k,\alpha^k)$. Now, if $(\alpha_0,\alpha_1) \leq^* (\alpha^k,\alpha^k)$, then $\max\{\alpha_0,\alpha_1\} \leq \alpha^k$. The order type of this set of pairs in the Cantor ordering is at most $(\alpha^k+1)^k < (\alpha+1)^{2k} < \alpha^\omega$. So $p(\alpha^k,\alpha^k) < \alpha^\omega$. □

Exercise 2.1.6. For $\alpha \geq 2$, let $\alpha_0 = \alpha$, $\alpha_{i+1} = \alpha^{\alpha_i}$ and $\bar{\alpha} = \sup\{\alpha_i : i \in \omega\}$. Show that $\bar{\alpha}$ is exponentially closed.

We will frequently use $\oplus$ to denote the join of two sets of ordinals. That is, when $x, y \subseteq \mathrm{On}$ are sets, then $x \oplus y := \{2\alpha : \alpha \in x\} \cup \{2\beta+1 : \beta \in y\}$. When more than two sets are involved, $\bigoplus_{i=1}^n x_i$ denotes $x_1 \oplus x_2$ when $n = 2$ and, recursively, $x_1 \oplus \bigoplus_{i=2}^n x_i$ when $n > 2$.

Definition 2.1.7. The lexicographical ordering $<_{\mathrm{lex}}$ on finite sets $A \neq B$ of ordinals is defined by $A \leq_{\mathrm{lex}} B$ if and only if the smallest element of $B \Delta A$ belongs to B. This is clearly a well-ordering of the finite sets of ordinals.

We will also use $<_{\mathrm{lex}}$ on the finite subsets of any well-ordered set, which will have its obvious meaning.

Definition 2.1.8. If α is an ordinal and $x \subseteq \alpha$, then $\chi_x : \alpha \to \{0,1\}$ denotes the characteristic function of x, i. e., $\chi_x(\iota) = 1$ if and only if $\iota \in x$ and otherwise $\chi_x(\iota) = 0$. We will occasionally identify sets of ordinals with their characteristic functions.

2.1.1 Simulation of machine models

To every model of computation belongs a range of computable objects and functions. If we have two models of computation, and every object and function computable in the sense of the first model is also computable in the sense of the second model, we consider the second model to be "stronger." Information on the strength of models of computation is helpful, as it allows us to carry over results from one model to another;

for example, by the results below, it will suffice to show that a function $f : \omega \to \omega$ is wITRM-computable in order to know that it is computable by α-ITRMs, α-ITTMs, α-TMs with $\alpha \geq \omega$, etc.

Results on the relative strength of models of computability are usually proved by showing something stronger, namely that computations of programs P according to the one model of computability can be "simulated" by computations of programs Q according to another. Roughly, this means that Q not only computes the same function as P, but also the "behavior" of P can be read off from the "behavior" of Q. We will not give a formal definition, but restrict ourselves to the following explanation, which should be enough to understand claims and proofs involving the term of a "simulation" below.

Let M and M' be machine models[2] and let P be an M-program and Q a M'-program. When a program is carried out on a machine, the machine is at every point of the computation in a certain "configuration," which, e. g., describes the content of the memory, however structured, the active program line, the head positions in the case of a tape model, etc. Let $C(M)$ and $C(M')$ be the sets of possible configurations for M-programs and M'-configurations. Also, we suppose that there is a certain reserved bit b in the memory of the machine carrying out P (which can, e. g., be a register or a tape cell in the case of register and tape models) which can take the values 0 and 1. Then we say that P simulates Q if and only if there is an injective function $F : C(M) \to C(M')$ (which can be a proper class function) such that the following holds: If $(c_\iota^P : \iota \in \mathrm{On})$ and $(c_\iota^Q : \iota \in \mathrm{On})$ are the sequences of configurations in the computations of P and Q, and ι_τ is the τth value of ι for which b has the value 1 in c_ι^P, then $F(c_{\iota_\tau}^P) = c_\tau^Q$. Intuitively, b indicates whether a configuration in the computation of P represents a configuration in the computation of Q or whether it is mere intermediate "scratch work" during the simulation. In addition, we require that F is "simple," i. e., it should be straightforward to read off the M'-configuration represented by a M-configuration M.

In actual proofs of simulation claims below, we will not be too strict about b; however, it should in each case be clear that the construction could be modified in order to satisfy this requirement.

For machine models M and M', we say that M can simulate M' if and only if there is a recursive function f mapping M'-programs to M programs such that, for each M'-program P, $f(P)$ is a M-program that simulates P. If P uses parameters $\vec{y}$, then $f(P)$ in the parameter $\vec{y}$ should simulate $P(\vec{y})$.

2 We will not attempt a general definition of "machine model"; an abundance of examples will be defined below.

2.1.2 Halting problems

A notion that will be relevant for every model of computability in this book is that of the halting problem. For every machine model M to be considered below, we will have a countable set $\{P_i : i \in \omega\}$ of programs, enumerated in some natural way, and often also a set of parameters, which will usually be some ordinal α, as well as an oracle x, which will be a set of ordinals. Then the halting problem for M relative to x is the set $x'_M := \{(i, \iota) \in \omega \times \alpha$:'$P_i$ halts in the parameter ι and the oracle x'$\}$. Turing's well-known diagonalization argument also shows very generally that x'_M is never M-computable in the oracle x; we refer the reader to Chapter 3 for a brief discussion.

2.2 Infinitary analogues of register machines

Classical register machines are a frequently used model of computation, the definition of which can, e. g., be found in Cutland [44], on which our exposition is based (and where they are called Unlimited Register Machines, i. e. URMs). A register machine has finitely many registers $R_0, \dots, R_n$, each of which can store a single natural number. A program for a register machines is a sequence $(l_1, \dots, l_m)$ of program lines, indexed with natural numbers, each of which contains one of the following commands:

- $R_i = 0$ with $i \in \{0, \dots, n\}$ replaces the content of the ith register with 0
- $R_i + 1$ with $i \in \{0, \dots, n\}$ increases the content of the ith register by 1
- m IF $R_i = R_j$ THEN GOTO k with $i, j \in \{0, \dots, n\}$ changes the active program line to l_k if R_i and R_j have the same content; otherwise, one continues with l_{m+1}
- COPY(i, j) with $i, j \in \{0, \dots, n\}$ replaces the content of R_j with the content of R_i.
  If one wants to work relative to oracles (which, in the case of URMs, are real numbers), one also uses the following:
- ORACLE(i) works as follows: Say R_i contains the natural number r_i. Then the content of R_i is replaced with the r_ith bit of the oracle.

From now on, we fix a certain natural enumeration $(P_i : i \in \omega)$ of the URM-programs, of which we will make frequent use. If k distinct register indices appear in the program P, we say that P "mentions" k registers. We will usually tacitly assume that these are $R_0, \dots, R_k$.

A URM-program is then executed in the obvious way: Suppose that P is a URM-program that has m program lines and mentions k registers, which we assume without loss of generality to be $R_0, \dots, R_k$. A **configuration** or a **snapshot** of P is a $(k+2)$-tuple $(l, r_0, \dots, r_k) \in \mathbb{N}^{k+2}$. Here, l denotes the active program line and r_i denotes the content of R_i. We will from now on always use r_i as denoting the content of the register R_i in any context in which it makes sense, i. e., in which we talk about such a register.

If $c = (l, r_0, \dots, r_k)$ is a configuration of P, then its **successor configuration** c^+ is defined in the obvious way as follows, where z_l denotes the lth line of P:

- If z_l = '$R_i = 0$,' then c^+ is obtained from c by replacing r_i with 0 and l with $l+1$, leaving the rest unchanged.
- If z_l = '$R_i + 1$', then c^+ is obtained from c by replacing r_i with $r_i + 1$ and l with $l+1$, leaving the rest unchanged.
- If z_l = 'IF $R_i = R_j$ THEN GOTO k', then c^+ is obtained from c by replacing l with k if $r_i = r_j$ and otherwise replacing l with $l+1$, leaving the rest unchanged.
- If z_l = 'COPY(i,j), then c^+ is obtained from c by replacing r_j with r_i, leaving the rest unchanged.

A **partial computation** of P on input $(a_0,\dots,a_k)$ is now a (possibly infinite) sequence $s = (c_0,\dots,c_n)$ such that $c_0 = (1,a_0,\dots,a_k)$ and $c_{i+1} = c_i^+$ for $i \in \{0,2,\dots,n-1\}$. If $c_n = (l,r_0,\dots,r_n)$ and $l > m$, i. e., the first entry of c_n is not the index of a program line of P, the computation cannot continue, as c_n^+ is not defined. In this case, we say that P "halts" in n steps on the input $(a_0,\dots,a_n)$ and call c_n the **final configuration** of P on input $(a_0,\dots,a_k)$. We then call s the computation of P on input $(a_0,\dots,a_k)$; if $c_n = (l,b_0,\dots,b_k)$, we call $(b_0,\dots,b_k)$ the output of P on input $(a_0,\dots,a_k)$. Depending on the context, the formulation output of P may also refer merely to parts of this information such as b_0 or (b_0,b_1), etc. Thus, when we say, e. g., that P halts with output 1, we mean that P halts with a final configuration in which $r_0 = 1$. The computation of P on the input $(a_0,\dots,a_k)$ in the oracle x will be denoted by $P^x(a_0,\dots,a_k)$. If we write $P^x(a_0,\dots,a_i)$ with $i < k$, we mean the computation of P on input $(a_0,\dots,a_k)$ where $a_j = 0$ for $j > i$; when $x = \emptyset$, the oracle is dropped. If there is no $n \in \omega$ such that P halts in step n on the input $(a_0,\dots,a_k)$, we say that P "diverges" or "does not halt" on this input; otherwise, we say that P halts on this input. We write $P^x(a_0,\dots,a_k)\downarrow$ to indicate that P halts on the input $(a_0,\dots,a_k)$ in the oracle x, and $P^x(a_0,\dots,a_k)\uparrow$ to indicate that it does not halt; to say that $P^x(a_0,\dots,a_k)$ halts with output $(b_0,\dots,b_i)$, we write $P^x(a_0,\dots,a_k)\downarrow = (b_0,\dots,b_i)$. When no input is specified, $P\downarrow$ and $P\uparrow$ mean that P halts or does not halt on the **empty input**, i. e., the input $(0,0,\dots,0)$, respectively.

If $i,j \leq k$, $x \subseteq \omega$ and $f : \mathbb{N}^{i+1} \to \mathbb{N}^{j+1}$ is a function, then we say that f is URM-computable in the oracle x if and only if there is a URM-program P such that, for every $(a_0,\dots,a_i) \in \mathbb{N}^{i+1}$, $P^x(a_0,\dots,a_i)\downarrow = f((a_0,\dots,a_i))$.

It is matter of routine exercises to see that all standard arithmetical operations are URM-computable, a fact that we will frequently make use of. We offer the following exercise and otherwise refer the reader to [44].

Exercise 2.2.1 (Cf. [44]). Show that the following functions are URM-computable:
(a) $n \mapsto (n-1)$ when $n > 0$ and $n \mapsto 0$ when $n = 0$
(b) $(a,b) \mapsto a + b$
(c) $(a,b) \mapsto a \cdot b$
(d) $(a,b) \mapsto \lfloor \frac{a}{b} \rfloor$
(e) $(a,b) \mapsto a^b$
(f) $(a,b) \mapsto \lfloor a^{\frac{1}{b}} \rfloor$

An important feature of classical models of computability is the existence of **universal programs** that, given the index i of another program P_i as input, will compute the same function as P_i. In particular, there are universal URM-programs.

Theorem 2.2.2. *There is a URM-program P_u with the following property: If $i \in \omega$ and P_i computes the partial function $f : \omega \to \omega$, then $P_u(j, i)$ will halt if and only if $P_i(j)$ halts and if it halts, it will do so with output $f(j)$. In fact, there is such a program P_u that uses only three registers.*

Proof. See, e. g., [44]. □

Our goal is to consider infinitary variants of this machine model. There are various options to pursue: The working time, currently restricted to ω, could be extended to an arbitrary ordinal α or even the whole class On of ordinals. Also, the registers, which currently can only contain ordinals below ω, could be extended to contain ordinals below a certain fixed ordinal α or even arbitrary ordinals. Finally, these choices can be combined.

All of these options have been worked out as machine models by P. Koepke in a number of papers [106, 110, 114, 112], though not all have received the same attention since. We will consider two variants with the time axis extended to On and the register contents restricted to a multiplicatively closed ordinal α (which will differ in their behavior at limit times) and the variant in which both time and the register contents are arbitrary ordinals.

Clearly, in defining a machine model that works along an ordinal time axis which contains limit points, one needs to determine the machine configuration at limit times in addition to the rules given above for determining the configuration at successor times. A rather natural choice is to choose the configuration c_λ at a limit time λ as a kind of "limit" of the sequence of configurations $(c_\iota : \iota < \lambda)$. This is indeed the idea behind all machine models of infinitary computations.

The first and weakest infinite variant of URMs we will consider are the **weak Infinite Time α-Register Machines**, or α-wITRMs also called "unresetting" Infinite Time α-Register Machines, which were introduced and studied in the case $\alpha = \omega$ by Koepke in [106]. They already exhibit a considerable portion of the abilities that are typical for infinitary machine models. On the other hand, we will see that all the other models that receive closer attention in this book are stronger than ω-wITRMs in the sense that they are able to "simulate their behavior," as we will see later in this chapter. We will thus take some time in the next chapter to prove theorems about the abilities of ω-wITRMs; by the simulation results, these statements then automatically hold for all the other models, and we will frequently make use of this fact.

Definition 2.2.3. Let $s = (\alpha_\iota : \iota < \lambda)$ be a sequence of ordinals, where λ is a limit ordinal. The supremum sup(s) of s is the smallest ordinal β such that there is $\zeta < \lambda$ with $\beta \geq \alpha_\iota$ for all $\iota > \zeta$. The infimum inf(s) of s is the largest ordinal β such that there is $\zeta < \lambda$ with $\beta \leq \alpha_\iota$ for all $\iota > \zeta$.[3] The inferior limit liminf(s) is the infimum of the sequence $(\sup(\alpha_\iota : \gamma < \iota < \lambda) : \gamma < \lambda)$ consisting of the suprema of the final segments of s.

Definition 2.2.4 (Koepke, cf. [106]). Let $\alpha \in \mathrm{On}$ be a limit ordinal.

Any URM-program is also an α-wITRM-program and vice versa. If P is an α-wITRM-program in which all registers have indices $\leq k$, then an α-configuration of P is a tuple $(l, r_0^\iota, \ldots, r_k^\iota) \in \omega \times \alpha^{(k+1)}$. Here, the first component is the active program line, while the others represent the register contents. The prefix α will be dropped when it is clear from the context.

A **partial α-wITRM-computation** of the α-wITRM-program P using k registers on the input $(\alpha_0, \ldots, \alpha_k) \in \alpha^{(k+1)}$ is a sequence $(c_\iota : \iota < \delta)$ with $\delta \in \mathrm{On}$ of α-configurations $c_\iota = (l^\iota, r_0^\iota, \ldots, r_k^\iota)$ of P such that $c_0 = (1, \alpha_0, \ldots, \alpha_k)$ and the following holds:

(1) For $\iota + 1 < \delta$, we have $c_{\iota+1} = c_\iota^+$, where c^+ is defined as for URMs, the successor instruction now taking the effect of replacing a register content with its ordinal successor.[4] Moreover, oracles for α-wITRMs are subsets of α, and the oracle command is carried out accordingly.

(2) If $\lambda < \delta$ is a limit ordinal, then l_λ is the inferior limit $\liminf(l_\iota : \iota < \lambda)$ of the sequence $(l_\iota : \iota < \lambda)$. As the set of potential values of l_ι is a finite set of natural numbers, this limit is always finite.

(3) If $\lambda < \delta$ is a limit ordinal and $\liminf(r_i^\iota : \iota < \lambda)$ is $< \alpha$ for all $\iota < \lambda$, $i \in \{1, 2, \ldots, k\}$, then $r_i^\lambda = \liminf(r_i^\iota : \iota < \lambda)$ for all $i \in \{0, 2, \ldots, k\}$. However, if $\liminf(r_i^\iota : \iota < \lambda)$ is $\geq \alpha$ for some $i \in \{0, 2, \ldots, k\}$, then r_λ^i is undefined. In this case, we say that the computation breaks off.

If $\beta = \gamma + 1$ is a successor ordinal and l_γ is not the index of a program line of P, then we say that the computation of P on input $(\alpha_0, \ldots, \alpha_k)$ halts at time β; in this case, we call $(c_\iota : \iota < \beta)$ the computation of P on input $(\alpha_0, \ldots, \alpha_k)$. If there is no β such that the computation halts at time β, we say that P diverges on the input $(\alpha_0, \ldots, \alpha_k)$; in this case, the computation of P on the input $(\alpha_0, \ldots, \alpha_k)$ is the class-length sequence $(c_\iota : \iota \in \mathrm{On})$ of (α-)configurations of P.

When P runs on input $(\alpha_0, \ldots, \alpha_k)$ in the oracle x, we will also write that $P^x(\alpha_0, \ldots, \alpha_k)$ halts with a certain output, diverges, etc.

A function $f : \alpha \to \alpha$ is α-wITRM-computable relative to $x \subseteq \alpha$ if and only if there are an α-wITRM-program P and an ordinal $\rho < \alpha$ such that, for every $\iota < \alpha$, $P^x(\iota, \rho)$ halts with $f(\iota)$ in the register R_0. If $\rho = \emptyset$, then f is α-wITRM-computable without parameters. A subset $y \subseteq \alpha$ is α-wITRM-computable if and only if its characteristic function is. In this case, we also say that x is α-wITRM-reducible to y and write $x \leq_{\mathrm{wITRM}} y$. When $P(\iota)$

3 You may want to take a minute to show that the infimum always exists.

4 Note that the successor instruction will never lead outside of α, since α is a limit ordinal by assumption.

runs for $< \beta$ many steps for each $\iota < \alpha$ and P computes f, we say that f is computable in β many steps.

When β is an ordinal, an (α, β)-wITRM is defined like an α-wITRM with the modification that computations of length $\geq \beta$ count as diverging. Thus, $f : \alpha \to \alpha$ is (α, β)-wITRM-computable if and only if there are a program P and an ordinal $\gamma < \alpha$ such that, for every $\iota < \alpha$, $P(\iota, \gamma)$ terminates in $< \beta$ many steps with output $f(\iota)$.

When $\alpha = \omega$, the prefix α is dropped and we merely speak of wITRMs. Instead of (α, α)-wITRMs, we also speak of α-register machines (α-RMs).

Proposition 2.2.5. *Suppose that α is a limit ordinal and $f : \alpha \to \alpha$ is (α, β)-wITRM-computable for some $\beta \in \mathrm{On}$. Then f is α-wITRM-computable. In particular, every total α-RM-computable function $f : \alpha \to \alpha$ is also α-wITRM-computable.*

Proof. Trivial. □

Clearly, every URM-computable function is also ω-wITRM-computable. However, much more is possible.

Proposition 2.2.6. *There is a wITRM-program H such that, for all $i \in \omega$, $H(i)$ halts with output* 1 *if and only if the URM-program P_i halts on the empty input* $(0, 0, \ldots, 0)$ *and otherwise, $H(i)$ halts with output* 0. *Thus, wITRMs can solve the halting problem for Turing machines.*

Proof. Consider the following wITRM-program H: Given the input $i \in \omega$, run the universal URM $P_u(0, i)$, which uses 3 registers, say R_1, R_2 and R_3. We use three additional auxiliary registers R_4, R_5, R_6 and reserve two extra registers R_7 and R_8; initially, we let $r_7 = 1$, $r_8 = 0$. Whenever a step in the computation of $P_u(0, i)$ is carried out, the contents of R_7 and R_8 are switched, the contents of R_1, R_2, R_3 are copied to R_4, R_5, R_6, respectively, and r_1, r_2, r_3 are set to 0. Then the contents of R_4, R_5, R_6 are copied back to R_1, R_2, R_3 and r_4, r_5, r_6 are set to 0, after which the computation of $P_u(0, i)$ is continued. (We shall presently see the point of this seemingly pointless business.) When $P_u(0, i)$ halts, output 1 and halt. When $r_7 = r_8 = 0$, output 0 and halt. Otherwise, the computation will continue through all natural numbers and we need to look at the configuration at time ω (if it exists). It does indeed exist, as our practice of replacing the register contents with 0 between any two steps of the computation of $P_u(0, i)$ guarantees that all inferior limits exist and are in fact equal to 0. So the ωth configuration will be the first configuration for which $r_7 = r_8 = 0$, in which case the computation halts with output 0. Thus, H indeed works as desired. □

Remark 2.2.7. This proof exhibits some typical features of programming infinitary machines: First, the use of the "flag" registers r_7, r_8 to determine when the computation reaches a limit time. Second, the program often needs to be arranged in such a way that certain effects are determined to happen at limit times. The reader can get some practice on this kind of thinking by doing the following exercise.

Exercise 2.2.8. (a) Write a wITRM-program that runs for exactly $\omega 2+3$ many steps.

(b) Write a wITRM-program that runs for exactly ω^2+3 many steps.

(c) Write a wITRM-program for computing the second jump, i. e., the halting problem for URMs that have the set $\{i : P_i \text{ halts}\}$ as their oracle.

Exercise 2.2.9. Let $\alpha > \omega$ be a limit ordinal.

(a) Show that there is an α-wITRM-program P that runs for exactly $\omega+1$ many steps and terminates with ω in the register R_1.

(b) Show that, if $f : \omega \to \omega$ is wITRM-computable and $\alpha \geq \omega$, then f is also α-wITRM-computable.

(c) More generally, show that, if $\beta < \gamma$ are limit ordinals and $f : \beta \to \beta$ is β-wITRM-computable, then f is also γ-wITRM-computable, using β as a parameter.

2.2.1 wITRM-computations

An important feature of finitary models of computation such as URMs is the following: If a configuration occurs more than once during a URM-computation, the computation will "loop," i. e., it will repeatedly carry out the same steps from that point on, returning to the configuration over and over again without ever halting.

This is false already for wITRMs, and models of infinitary computations in general. To see this, consider the following wITRM-program P:

1. $R_0 + 1$
2. IF $R_0 = R_1$ THEN GOTO 8
3. $R_1 + 1$
4. $R_0 = 0$
5. $R_1 = 0$
6. $R_0 + 1$
7. IF $R_2 = R_3$ THEN GOTO 2

The computation of this program looks as follows:

Time	Program line	Register 0	Register 1	Register 2	Register 3
0	1	0	0	0	0
1	**2**	**1**	**0**	**0**	**0**
2	3	1	0	0	0
3	4	1	1	0	0
4	5	0	1	0	0
5	6	0	0	0	0
6	7	1	0	0	0
7	**2**	**1**	**0**	**0**	**0**
...	...	...	...	...	...
ω	2	0	0	0	0
$\omega+1$	8 (HALT)	0	0	0	0

Thus, in the first ω many steps, P repeats, e. g., the configuration $(2,1,0,0,0)$ infinitely many times. However, at time ω, the configuration is $(2,0,0,0,0)$, so the jump condition of line 2 is satisfied and the program stops in the next step.

Nevertheless, we do have a looping criterion for α-wITRMs, which will in fact in some variant apply to all models of infinitary computability that we will consider (and which we will thus not restate and reprove in each case to avoid annoying redundancies).

Lemma 2.2.10 (Koepke, [106]). *If a snapshot appears at two times $\tau_0 < \tau_1$ during an α-wITRM-computation and no register content nor the active line index ever drops below the value, it has at time τ_0 and at time τ_1 between these times, then the machine is looping.*

Proof. This can be seen inductively as follows: Let $\tau_0 + \gamma = \tau_1$. For an ordinal $\delta \geq \tau_1$, write $\delta = \tau_1 + \gamma\eta + \rho$, $\rho < \gamma$. Assume by induction that, for $\delta' := \tau_1 + \gamma\eta' + \rho' < \delta$, the snapshot at time δ' is the same as at time $\tau_0 + \rho'$. If $\rho \neq 0$, the sequence of snapshots between $\tau_1 + \gamma\eta$ and δ is the same as that between τ_0 and $\tau_0 + \rho$, hence the snapshots at times δ and $\tau_0 + \rho$ will also be equal. If, on the other hand, $\rho = 0$, we further distinguish whether η is a successor or a limit ordinal. If $\eta = \xi + 1$ is a successor, then by induction, the snapshot sequence between times $\tau_1 + \gamma\xi$ and $\tau_1 + \gamma\eta$ will be the same as that between τ_0 and τ_1, so the snapshot at time $\tau_1 + \gamma\eta$ will be the same as that at time τ_1, which by induction, is the same as that at time $\tau_1 + \gamma\xi$. On the other hand, if η is a limit ordinal, we have to use the extra condition: Since no register content or active line index is ever below that at time τ_0 before time δ and the snapshot at time τ_0 appears cofinally often before time δ, the state at time δ must be the same as that at time τ_0. Thus, the program is looping from time τ_1 on with period length γ. □

Definition 2.2.11. In the situation described in Lemma 2.2.10, i. e., if there are two ordinals $\tau_0 < \tau_1$ such that the snapshot at time τ_1 and τ_1 are the same and neither any register content nor the active program line drops below the value at these times between τ_0 and τ_1, we say that (τ_0, τ_1) "witnesses the looping of P."

Lemma 2.2.12 (Koepke, [108]). *If the α-wITRM-program P does not halt, then there are ordinals τ_0 and τ_1 such that (τ_0, τ_1) witnesses the looping of P.*

Proof. Assume that P does not halt. Thus, the computation $(c_\iota : \iota \in \mathrm{On})$ of P has proper class length. On the other hand, the possible configurations of P belong to $\omega \times \alpha^k$ when P uses k many registers, which is a set. It follows that some configuration must appear unboundedly often in the computation of P. Let C be the set of these configurations, so $C \neq \emptyset$; let us write $C = \{d_\iota : \iota < \beta\}$ for some $\beta \in \mathrm{On}$.

For every configuration c that appears only boundedly often in the computation of P, there is a supremum σ_c of the times at which it appears; and since the configurations appearing in the computation of P are a set, the supremum of all σ_c is again an ordinal δ. Thus $c_\iota \in C$ when $\iota > \delta$.

We consider a strictly increasing sequence $(\tau_\iota : \iota < \beta\omega)$ of ordinals such that $\tau_0 > \delta$ and $c_{\tau_{\beta i+\iota}} = d_\iota$ for all $i \in \omega$, $\iota < \beta$. That such a sequence exists is clear as all d_ι appear unboundedly often by assumption. Now let $\tau := \sup\{\tau_\iota : \iota < \beta\omega\}$. Then τ is a common limit of times at which d_ι appeared, for all $\iota < \beta$. By the liminf-rule, c_τ is componentwise $\leq d_\iota$ for all $\iota < \beta$. As $\tau > \delta$, we have $c_\tau \in C$, so there is $\tau' > \tau$ such that $c_{\tau'} = c_\tau$. Thus (τ, τ') witnesses the looping of P. □

2.3 Infinite time α-register machines

Though wITRMs are stronger than mere URMs (and in fact much stronger, as we will see in Chapter 3), they still have their shortcomings: The possibility that a computation may simply break off because, due to a register overflow, no new state of the computation is defined has no analogue in URM-computability and is somewhat unsatisfying. For this reason, a new version of wITRMs was introduced in [110], one that can continue at limit steps even in spite of a register overflow by simply resetting the content of the overflowing registers back to 0. This machine model is called an Infinite Time Register Machine (ITRM) and turned out to be considerably stronger than wITRMs.[5] Originally, ITRMs were restricted to registers that contain natural numbers. We work with the more general concept of an α-ITRM that was first mentioned by Koepke in [109].

Definition 2.3.1. Let $\alpha \in \mathrm{On}$ be a limit ordinal. An Infinite Time α-Register Machine (α-ITRM) is defined like an α-wITRM, with the following modification: If P is an α-ITRM-program (which is the same as an α-wITRM-program or a URM-program), λ is a limit ordinal, $(c_\iota : \iota < \lambda)$ is the partial computation of P on some input $(a_0, \ldots, a_k)$ in some parameter $\gamma < \alpha$ where $c_\iota = (l^\iota, r_0^\iota, \ldots, r_k^\iota)$ and $i \in \{0, 1, \ldots, k\}$ is such that $\liminf(r_i^\iota)$ is $\geq \alpha$, then $r_i^\lambda = 0$. If $x \subseteq \alpha$ is α-ITRM-computable from $y \subseteq \alpha$, we also call x α-ITRM-reducible to y and write $x \leq_{\mathrm{ITRM}} y$.

For β an ordinal, (α, β)-ITRMs are defined like α-ITRMs with the modification that a computation that takes $\geq \beta$ many steps counts as nonhalting; thus, (α, β)-ITRM-computability of a function $f : \alpha \to \alpha$ now means that, for some program P and some $\gamma < \alpha$, $P(\iota, \gamma)$ terminates in $< \beta$ many steps with output $f(\iota)$, for every $\iota < \alpha$.

The first thing to note about α-ITRMs is that they are at least as strong as α-wITRMs.

Lemma 2.3.2. *Let α be a limit ordinal, and let $f : \alpha \to \alpha$ be α-wITRM-computable. Then f is α-ITRM-computable.*

5 A historical remark may be in order here to avoid confusion: When wITRMs were originally introduced in [106], they were called ITRMs. Only later on, when the stronger type we just described was defined in [110], they were renamed.

Proof. First, note that a halting α-wITRM-computation with the program P in some parameter $\gamma < \alpha$ is also a halting α-ITRM-computation. It thus only remains to modify P in such a way to P' that, when the α-wITRM-computation of $P(\iota, \gamma)$ is undefined due to a register overflow for some $\iota \in \alpha$, then the α-ITRM-program $P'(\iota, \gamma)$ will not halt. To do this, we will along with $P(\iota, \gamma)$, run a routine on some additional registers that detects a register overflow in the run of $P(\iota, \gamma)$. We explain how this works for a particular register R used by P. First, we use flag registers as explained in the proof of Lemma 2.2.6 to determine whether P has arrived at a limit time. When that happens and R contains 0, we need to know whether this is due to R having overflown or having contained 0 cofinally often, and thus due to the liminf-rule. For this, we use an extra register R' that will be set to 0 if R contains 0 and otherwise will contain 1. It is easy to see that this behavior can be obtained (use a conditional jump and compare R to a register that always contains 0). Now, if the flags indicate a limit step and R contains 0, then R' containing 0 implies that this was due to cofinally many 0s, while R' containing 1 implies that it was due to an overflow. □

Proposition 2.3.3. *Let $\alpha < \beta$ be limit ordinals.*
(a) *Then β-wITRMs can simulate α-wITRMs and β-ITRMs can simulate α-ITRMs.*
(b) *Let $f : \alpha \to \alpha$ be α-ITRM-computable. Then f is also β-wITRM-computable in the parameter α.*

Proof. (a) Let P be an α-wITRM-program. Run P on a β-wITRM that contains α in some reserved register R_s. After each step of the computation of P, check whether some of the registers of P has now the same content as R_s. When that happens, start a subroutine that keeps adding 1 to R_0, which will eventually cause the computation to break off.

The proof for α-ITRMs is similar, with the difference that when the content of a register R_i is equal to that of R_s, that content is reset to 0 and then P is carried out further.

(b) In the parameter α, it is easy to simulate an α-ITRM-program P using n registers $R_1, \dots, R_n$ on a β-wITRM-program P' that uses the registers $R_1, \dots, R_n$, R_0 for storing the active program line of the simulated program P and some additional registers for scratch work: The first portion of P' compares the content of R_i with α and resets it to 0 when it is found to be equal for all $1 \leq i \leq n$. After that, one step of P is carried out and the contents of $R_0, R_1, \dots, R_n$ are updated accordingly. It is easy to see that this will correctly reflect the operations of P. □

Exercise 2.3.4. (a) Let us extend the definitions of α-(w)ITRMs to the case where $\alpha = \beta + 1$ is a successor ordinal: We take the usual definition, with the difference that, for α-wITRMs, attempting to increment the content of a register R containing β leads to an undefined state so that the computation breaks off while for α-ITRMs, it will cause the content of R to be reset to 0. Show that α-wITRMs and α-ITRMs can simulate each other.

(b) More generally, show that, when α is not additively indecomposable, i. e., there are $\beta, \gamma < \alpha$ such that $\beta + \gamma = \alpha$, then α-wITRMs and α-ITRMs can simulate each other. (Hint: Represent one register of an ITRM by two registers on an α-wITRM, one counting up to β, the other up to γ, to represent ordinals $\xi < \alpha$ as $(\xi, 0)$ when $\xi < \beta$ or as (β, ι) when $\xi \geq \beta$, where ι is such that $\beta + \iota = \xi$.)

Another simple, but important observation is the following: If $y \subseteq \alpha$ is α-ITRM-computable and $x \subseteq \alpha$ is α-ITRM-computable in the oracle y, then x is also α-ITRM-computable. Note that the obvious argument that one would use for Turing machines does not work: We cannot first compute y, store it somewhere and then use it as an oracle: α-ITRMs have no place to "store" an infinite set. Instead, we must run the program for computing y whenever the computation of x requests a bit from y. In particular, this means that we cannot expect the computation time of x to be bounded by $\beta + \gamma$ when y can be computed in β many steps and it takes γ many steps to compute x from y. However, we have the following.

Lemma 2.3.5. *Let α be multiplicatively closed, $y \subseteq \alpha$ be α-ITRM-computable by the program P in the parameter $\iota < \alpha$ in β many steps and let $x \subseteq \alpha$ be α-ITRM-computable by the program Q in the parameter $\iota' < \alpha$ and the oracle y in γ many steps. Then x is α-ITRM-computable in $\beta\gamma$ many steps.*

Proof. Run Q. Whenever Q makes an oracle call to determine whether $\iota \in y$, run $P(\iota)$ and return the result. This will clearly compute x in time bounded by $\beta\gamma$. □

However, the abilities of α-ITRMs reach much further. In Lemma 2.2.10, we proved a looping criterion for α-wITRMs, which also applies to α-ITRMs with the same proof. Since repetitions as described in the statement of Lemma 2.2.10 are not hard to detect, this leads to a crucial feature of ITRMs discovered by P. Koepke and R. Miller in [110], namely the solvability of the "bounded halting problem."

Definition 2.3.6. For each $k \in \omega$, fix some natural enumeration $(P_{i,k} : i \in \omega)$ of the ITRM-programs that mention at most k registers. Let $x \subseteq \omega$, $k \in \omega$. The k-bounded halting problem H_k^x for ITRMs in the oracle x is the set $\{i \in \omega : P_{i,k}^x\downarrow\}$.

Theorem 2.3.7 (Koepke, Miller [110]). *Let $n \in \mathbb{N}$. There is an ITRM-program H_n that solves the halting problem for ITRMs-programs using at most n registers. In fact, H_n works uniformly in the oracle: That is, if $(Q_i^n : i \in \omega)$ is a natural enumeration of the ITRM-programs using at most n registers, then for every $x \subseteq \omega$ and every $i \in \omega$, $H_n^x(i)\downarrow = 1$ if and only if $Q_i^n(0)$ stop when run in the oracle x and otherwise $H_n^x(i)\downarrow = 0$.*

Theorem 2.3.7 was proved by Koepke and Miller in [110] by a beautiful combinatorial argument that showed how an ITRM-program with a sufficient number of registers can "observe" the computation of an ITRM-program using n many registers and detect a strong loop. For the sake of brevity, we will not give this argument in detail here; it will follow in Chapter 3 from a much stronger analysis of ITRMs (Corollary 3.4.33).

For the interested reader, we offer here the idea from Koepke and Miller as a guided exercise.

Exercise 2.3.8. (Koepke and Miller, see Theorem 4 of [110]) This exercise will guide you through the "combinatorial" proof of the solvability of the bounded halting problem for ITRMs, which works uniformly in the oracle $x \subseteq \omega$. Chapter 3 will prove this from more advanced results on ITRMs.

Let $n \in \omega$ and $x \subseteq \omega$ be given. We want to decide, for an ITRM-program P_i using n registers, whether P_i^x halts, and we want this decision procedure to work on an ITRM in the oracle x. Let $(c_\iota : \iota \in \mathrm{On})$ be the computation of P_i^x.

For every $\tau \in \mathrm{On}$, denote by $C(\tau)$ the set of ITRM-configurations c_ι with $\iota < \tau$ such that, for all $\gamma \in [\iota, \tau)$, c_γ is componentwise $\geq c_\iota$.

(a) Using the looping criterion for register machines, show that, if $c_\tau \in C_\tau$ for some $\tau \in \mathrm{On}$, then P_i^x is looping, and thus $P_i^x \uparrow$.

(b) Suppose that P_i^x does not halt. Show that there is an ordinal δ such that every c_ι with $\iota > \delta$ occurs unboundedly often in the computation of P_i^x, i. e., there are arbitrarily large $\iota' \in \mathrm{On}$ such that $c_{\iota'} = c_\iota$.

(c) Show that, if $\iota, \iota' > \delta$, then there is $\gamma > \delta$ such that $c_\gamma \leq_{\mathrm{lex}} c_\iota$ and $c_\gamma \leq_{\mathrm{lex}} c_{\iota'}$. (Hint: Consider a common limit of the times at which c_ι and $c_{\iota'}$ appear and use the liminf-rule.)

(d) Show that there is $\gamma > \delta$ such that c_γ is lexically minimal among $\{c_\iota : \iota > \delta\}$.

(e) Let $\gamma' > \gamma$ be such that $c_{\gamma'} = c_\gamma$. Show that $c_{\gamma'} \in C_{\gamma'}$. Conclude that the necessary condition for $P_i^x \uparrow$ formulated in (a) is also sufficient.

The plan is now to check this condition—the existence of some τ such that $c_\tau \in C_\tau$—on an ITRM. To this end, we let P_i^x run and compute (codes for) C_τ along the way. The next parts of the exercise explain how C_τ is computed on an ITRM.

We code a configuration $c := (q, r_0, \ldots, r_{n-1})$ by $\mathrm{d}(c) := p_{n+1}(q, r_0, \ldots, r_{n-1})$; moreover, we code a finite set $\{c_i : 0 \leq i \leq k\}$ of configurations by $d(\{c_i : 0 \leq i \leq k\}) := \sum_{i=0}^{k} 2^{-\mathrm{d}(c_i)}$.

(f) Show that, if c and c' are configurations such that $c \leq_{\mathrm{lex}} c'$, then $\mathrm{d}(c) \leq \mathrm{d}(c')$. Moreover, show that, if $C \subseteq D$ are finite sets of configurations, then $\mathrm{d}(C) \leq \mathrm{d}(D)$.

The naive approach to check the criterion from part (e) would be run through the configurations in the computation of P_i^x one after the other, generating and storing C_τ along the way and checking for every configuration c_τ whether $c_\tau \in C_\tau$. However, C_τ will in general be an infinite set.

(g) Show that, if P_i is a program that just counts upwards in all of its registers, then C_ω is infinite.

Since we cannot encode infinite sets, we use the following trick: For $1 \leq i \leq n$, define $C_i(\tau)$ to be the subset of $C(\tau)$ for which every element of the form r_j is $\leq r_i$ for $i \in \{0, \ldots, (n-1)\}$, i. e., for which R_i contains the maximal register content.

(h) Show that, for any $1 \leq i \leq n$, $\tau \in \mathrm{On}$, $C_i(\tau)$ is finite.

We will encode a finite set $s := \{a_1, \dots, a_n\}$ of natural numbers by $c(s) := \sum_{i=1}^{n} 2^{a_i}$. The plan is now to compute (codes for) $C_i(\tau)$ for each $i \in \{1, 2, \dots, n\}$ separately along with the computation of P_i^x and to store them in n registers $R'_1, \dots, R'_n$. Let us fix $i \in \{1, 2, \dots, n\}$. Initially, we have $C_i(0) = \emptyset$, and hence $c(C_i(0)) = 0$.

(i) Let $C_i(\tau)$ and c_τ be given. Show how to obtain $C_i(\tau + 1)$ recursively from $c(C_i(\tau))$ and c_τ. Also show that, if $\tau > 0$ and $c_\tau = (0, \dots, 0)$, then P_i^x is looping.

(j) Let δ be a limit ordinal such that, at time δ, the ith register does not contain 0. Show that there is $\iota < \delta$ such that, for all $\gamma \in (\iota, \delta)$, $C_i(\gamma)$ contains a snapshot s that is componentwise $\leq c_\delta$ if and only if $s \in C_i(\delta)$.

(k) Using (h), show that $c(C_i(\delta))$ is recursive in $\liminf_{\iota<\delta} c(C_i(\iota))$ and c_δ.

(l) Show that $c(C(\tau))$ can be obtained recursively from $c(C_1(\tau)), c(C_2(\tau)), \dots, c(C_n(\tau))$.

(m) Use (g)–(j) to show how to compute $c(C_\tau)$ along with the computation of P_i^x. Noting that, while doing this, one will eventually either notice that P_i halted or that the looping criterion was satisfied, and conclude that the bounded halting problem for ITRMs with n registers is ITRM-computable.

Remark 2.3.9 (Question). As $C_\gamma(\tau)$ could be infinite when γ is and there is no (obvious) way to encode an infinite set of ordinals by a single ordinal that is compatible with inferior limits, the above argument does not work for α-ITRMs when $\alpha > \omega$. We do not know whether a similar approach also works to show the solvability of the bounded halting problem for α-ITRMs with much larger α nor whether it is even true. However, we can at least extend the Koepke–Miller argument above a bit, namely to the case $\alpha \in \{\omega \cdot k : k \in \omega\}$. This is the point of the next exercise.

Exercise 2.3.10. Let $n \in \omega$. We consider $\omega \cdot 2$-ITRMs with n registers. Let $(t_i : i < 2^n)$ be the lexicographical enumeration of the set $\{0, \omega\}^n$. Let us say that the $\omega \cdot 2$-ITRM-configuration $(q, r_0, \dots, r_{n-1})$ is of type i if and only if $t_i(j) \leq r_j < t_i(j)+\omega$ for all $j < n$ (e. g., $(3, \omega + 2, 7, \omega + 9)$ is of type 5, since $(\omega, 0, \omega)$ is the 5th element in the lexicographical enumeration of $\{0, \omega\}^3$).

For an ordinal $\beta < \omega \cdot 2$, we write $\beta = \omega a + b$ with $a, b \in \omega$ and call a the "ω-part" and b the "finite part" of β.

Now consider the following $\omega \cdot 2$-ITRM-program H: Given a $\omega \cdot 2$-ITRM-program P that uses n registers, use n registers to simulate P. Moreover, in each of another $2^n \cdot n$ many registers $R_0^0, \dots, R_0^{(n-1)}, R_1^0, \dots, R_1^{(n-1)}, \dots, R_{2^n-1}^0, \dots, R_{2^n-1}^{(n-1)}$, one for each type, do the following: In R_k^i, store the configurations arising in the course of the computation of P that are of type k and for which the finite parts of all register contents are $\leq$ the finite part of the ith register. The update rules, etc. work as in the ITRM-case.

(a) Finish the description of the algorithm H by adapting in detail the ideas from the ITRM-case to the new situation.

(b) Show that H solves the halting problem for $\omega \cdot 2$-ITRMs with n registers. Show that H works uniformly in the oracle.

(c) Generalize the preceding argument to $\omega \cdot k$-ITRMs for any $k \in \omega$.

We now draw some important consequences of Theorem 2.3.7.

Corollary 2.3.11 (Koepke and Miller, [110]). *There is no universal ITRM. In fact, the computational strength of ITRM-programs strictly increases with the number of registers.*

Proof. Suppose that U was a universal ITRM, and suppose that n is the number of registers that U uses. By Theorem 2.3.7, the halting problem for ITRMs using at most n registers is solvable by some ITRM-program H_n. Since U can by definition simulate H_n, this implies that there is an ITRM using n registers that solve the halting problem for ITRMs using at most n registers. But by the usual diagonalization argument, this cannot be. □

Another corollary of the above that we will frequently use is the following.

Definition 2.3.12. An ITRM-program P is "safe" if and only if $P^x\downarrow$ for every $x \subseteq \omega$. If $x, y \subseteq \omega$, then we say that x is "safely ITRM-computable from y" if and only if there is a safe ITRM-program P such that $P^x\downarrow = y$.

Corollary 2.3.13. *Let $x, y \subseteq \omega$, and suppose that x is ITRM-computable from y. Then x is safely ITRM-computable from y.*

Proof. Let P be an ITRM-program such that $P^x\downarrow = y$, and let n be the number of registers used by P. Then a safe ITRM-program that computes x from y works as follows: Given $z \subseteq \omega$, first use H_n^z to check whether P^z will halt. If not, halt with output 0. Otherwise, let P^z run and return whatever its output is.

This is clearly a safe ITRM-program that works like P on every oracle on which P halts. □

2.3.1 ITRMs with one register

Before moving on, let us pause for a moment to get a grasp on the "combinatorics" of ITRMs. In general, arguing about models of transfinite computability has several aspects; one are combinatorial arguments, in which one uses tools like the pigeonhole principle and combinatorial properties of ordinal arithmetic. The above looping criterion is a typical example. Another are algorithmic arguments that describe (usually on a rather high level) procedures for solving certain problems; the well-foundedness test in the next subsection is an example of such an argument. Finally, there are the arguments of a more set-theoretical or model-theoretical nature that usually start dominating after the computational strength of a model has been determined—many arguments in the chapter about randomness are of this kind. Quite often, these aspects interact.

In this subsection, we give an argument that, in our view, gives a good impression of the combinatorial reasoning in infinitary computability.

Consider an ITRM-program using only one register, $R = R_1$. That is, the commands of incrementing, copying and resetting to 0 are only applied to R, i. e., one may copy **to** R, but not **from** R. It is still allowed to use other registers in jump conditions of the form "IF $R_i = R_j$ THEN…"; however, as all registers will initially contain 0, this means that the condition is trivially satisfied if $i, j \neq 1$ and if one of them is 1, but the other is not, the condition is satisfied if and only if the current content of R is 0. We can thus think also of P as a program that uses unconditional jumps along with jumps that are carried out under the condition that $R = 0$. We call such a program a 1-ITRM-program.

1-ITRM-programs are nice as their behavior is simple enough to be describable with a minimum of set-theoretical apparatus, but still nontrivial. A careful reading of the following argument is an excellent preparation for the determination of the computational strength of ITRMs in the next chapter, where many of the ideas used here will reoccur in a more involved manner.

Theorem 2.3.14. *The possible halting times of a* 1*-ITRM-program are exactly the ordinals* $< \omega^\omega$.

Proof. We first show that, if P is a 1-ITRM-program that halts at time α, then $\alpha < \omega^\omega$. To do this, we will prove inductively that, if P uses n program lines, then $\alpha < \omega^{2n}$ and if P runs for more than ω^{2n} many steps, then it starts strongly looping before time ω^{2n}.

Observation. Let $\delta < \alpha$ be a limit ordinal. Then $r_1^\delta = 0$.

To see this, let us assume for a contradiction that $r := r_1^\delta > 0$. As $r_1^\delta = \liminf_{\iota<\delta} r_1^\iota$, this means that, for some $\beta < \delta$, we had $r_1^\iota \geq r$ for all $\iota \in [\beta, \delta)$. In particular, the content or R was never reset to 0 during this time, and thus was either incremented by 1 or left unchanged in every step between β and δ. If infinitely many incrementations would have taken place during that time interval, R would have overflown at time δ, so we would have $r = 0$, contradicting our assumption. Thus, the content of R reached a maximum value M in the time interval $[\beta, \delta)$, from which time on it remained unchanged.

Now let l be the active program line at time δ. As $l = \liminf_{\iota<\delta} l_\iota$, this implies that l must have occurred cofinally often before time δ, and that no smaller program line was active from some time $\gamma < \delta$ on. But this means that there were two distinct points after time γ in which the register content was M, the active program line was l, and between these two points, the register content was not changed and the program line index was always $\geq l$. In other words, the looping criterion is fulfilled, and P is in fact looping, and thus does not halt—a contradiction. The observation is proved.

Using the observation, the theorem is obvious for $n = 1$: If there is only one program line, it is the same at all configurations. At times ω and $\omega 2$, the register content is 0. Thus, for these two times, the looping criterion is satisfied. Thus P cannot run for more than ω^2 (in fact more than $\omega 2$) many steps without looping.

Now suppose that $\alpha \geq \omega^{2n}$, and let $l = l_{\omega^{2n}}$. Let $\gamma < \omega^{2n}$ be such that $l_\iota \geq l$ for $\iota \in [\gamma, \omega^{2n})$. We consider the time intervals of the form $I_\beta = [\omega^{2n-2}\beta, \omega^{2n-2}(\beta + 1))$ with $\beta < \omega^2$. We distinguish two cases:

Case I: There is $\beta < \omega^2$ such that l never was the active program line during I_β.

Consider the program P' that is like P, but with line l deleted (and the other lines relabeled and the jump instructions accommodated accordingly). During I_β, P runs exactly like P'. But, as P' has only $(n-1)$ program lines, P' must be strongly looping before reaching the end of I_β by the inductive assumption. However, if P' is strongly looping, then so is P.

Case II: l occurred as the active program line during I_β for every $\beta < \omega^2$.

In this case, consider the smallest $k \in \omega$ with $\tau := \omega^{2n-1}k > \gamma$. Then l was the active program line at time τ. But then the configurations at times τ and ω^{2n} satisfy the looping criterion, so again, P is looping.

Both cases thus lead to a contradiction. It follows that indeed P starts strongly looping before time ω^{2n} (and thus never halts) if it does not halt before this time.

On the other hand, we show that, for every $\alpha < \omega^\omega$, there is a 1-ITRM-program P with halting time α. Clearly, there is a 1-ITRM-program with halting time i for every $i \in \omega$: Just work through i many lines that, e. g., reset R to 0, after which there is no more line. There is also a 1-ITRM-program P that works for ω many steps:

1 $R_1 = R_1 + 1$
2 IF $R_1 = 0$ GOTO 5
3 $R_1 = R_1 + 1$
4 IF $R_1 = R_1$ GOTO 2

P simply increases the content of R by 1 until it overflows at time ω, at which point it halts.

Our claim will be proved if we additionally show the following:

(1) If α and β are 1-ITRM-halting times, then so is $\alpha + \beta$.

This is easily achieved: If P halts in α many steps, using n many lines and Q halts in β many steps, increase each line index in Q by n and append the result to the end of P. This new program will first run P and then Q, i. e., take $\alpha + \beta$ many steps in total.

(2) If α is a 1-ITRM-halting time, then so is $\alpha\omega$.

As $(\alpha + 1)\omega = \alpha\omega$ for all $\alpha > 0$, we may assume without loss of generality that α is a limit ordinal. Suppose that P halts in α many steps. Suppose that P uses n many program lines. Let P' arise from P by adding 3 to every program line index and accommodating the jump instructions accordingly. Now consider the following program Q:

1 $R_1 = R_1 + 1$
2 IF $R_1 = R_2$ THEN GOTO $(n + 7)$
3 $R_1 = 0$
P'
$(n + 4)$ $R_1 = 0$
$(n + 5)$ $R_1 = R_1 + 1$
$(n + 6)$ IF $R_1 = R_1$ GOTO 2

Thus Q works as follows: First, R_1 is increased by 1, so it will contain 1. The jump condition in the second line will then not be fulfilled, so R_1 is reset to 0 in line 3 and P' is started for the first time. We will then loop through the lines 2 until $(n+6)$, and the jump condition in line 2 will not be satisfied for any finite iteration of this loop due to line $(n+5)$. However, after ω many iterations, we are in line 2 and the register content will be 0: After all, it was set to 0 in line $(n+4)$ at least once in each run through the loop. Thus, after $\alpha\omega$ many steps, the jump condition in line 2 is satisfied and the program halts. □

Exercise 2.3.15. Let P be a 1-ITRM-program that has n program lines. Suppose that P is started on input 0 and, during the first ω many computation steps, the register R contains a number $> n$. Show that R overflows at time ω.

Exercise 2.3.16. Let P be a 1-ITRM-program and $0 < k \in \omega$ such that $P(k)$ halts with output 0 or 1. Show that the output does not depend on k. Use this to determine all subsets of ω that are 1-ITRM-decidable.

The last exercise shows that, in spite of having a rather rich spectrum of halting times, 1-ITRMs are computationally trivial. In fact, we have the following.

Theorem 2.3.17. *Fix a natural enumeration* $(Q_i : i \in \omega)$ *of the* 1-*ITRM-programs. Let* $H = \{i \in \omega : Q_i(0)\downarrow\}$. *Then* H *is recursive.*

Proof. Let P be a 1-ITRM-program with n lines $\{1, \ldots, n\}$. We show how to decide whether P halts. We start by running the n computations of P on input 0 when starting in line $i \in \{1, 2, \ldots, n\}$, each for $(n+1)^2$ many steps. For each such computation, one of the following cases must happen (assuming that we start in line i):

Case I: The register content is $> n$ at least once.

Let us see how this can come about. (If you still want to do the first exercise above, do not read further!) The register content is initially 0. It can only increase by the incrementation command, and only by 1 in each step. The only possibility for it to decrease is to be reset to 0 (either by the zero command or by copying the content of another register containing 0 to R). Thus, if the register content is $> n$ at some point, then this was preceded by a sequence of configurations in which the register content only increased or remained invariant in each step, from 0 up to n. Also, as the content increases by at most 1 in each step, this took $> n$ many steps. Consequently, some program line l must have repeated during this procedure while the content was already > 0. But now note that, after l appears for the second time, the same sequence of commands will be carried out as it was between the first and the second appearance of l: This is because all jump conditions are evaluated exactly as before, as the content of R is never 0. Now, if the content of R was the same at both appearances of l, the looping criterion applies, so P does not halt and we can output 0. Otherwise, the content was larger the second time l appeared than the first; as the sequence of commands is now repeated over and again, R will overflow at the next limit time. Moreover, the active line index at that limit time will be the minimal line index j that occurred during the repeating sequence of commands. Also, let k be the minimal line index that occurred from the start of the computation up to the second occurrence of l (i. e., the minimum of the repeating sequence together with the—possibly empty—initial segment before the period started). We store this information as $i \overset{\omega,l}{\mapsto} j$.

Case II: If Case I fails, the computation can assume only finitely many configurations (l, r) with $0 \leq l, r \leq n$ and $l > 0$. Thus, after $(n+1)^2$ many steps, some configuration must have appeared twice and so the computation will continue by repeating the sequence of configurations between these to occurrences until the next limit time. Let j be the smallest program line index occurring in this repeating sequence and let l be the smallest program line index occurring from the start of the computation up to the second occurrence of the repeating configuration. At the next limit time, the active program line index will be j. We again store this information as $i \overset{\omega,l}{\mapsto} j$.

Case III: If Cases I and II both fail, this means that the computation has halted in the simulated $(n+1)^2$ steps (and the arguments above show that if it has not halted within this number of steps, it will go on to the next limit time). We store this information as $i \overset{\omega}{\mapsto} S$.

Recall from Theorem 2.3.14 that the register content of a halting 1-ITRM-program at a limit time is always 0. Thus, after this step, we can determine from the stored information for each limit ordinal α the configuration at time $\alpha + \omega$ from the configuration at time α. We represent this information as a directed graph G_1, where we introduce a node for each possible limit configuration $(l, 0)$ (dropping the second component) and link i with j if and only if $i \overset{\omega,l}{\mapsto} j$ was stored for some $l \in \{1, 2, \ldots, n\}$. (For the halting "line," we introduce a special node labeled H.)

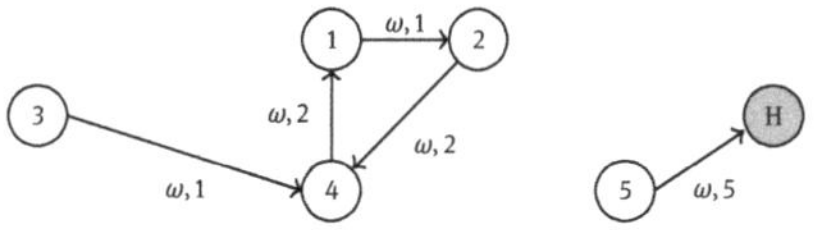

In G_1, each node, except S, has exactly one successor node. Thus each node either belongs to a path ending in S or to a cycle or to a path leading into a cycle. For nodes of the first kind, their labels clearly describe configurations that lead to a halting computation. If 1 is among them, we know that the computation will halt and output 1.

Otherwise, the next goal is to determine, for a limit configuration $(i, 0)$ occurring at time $\alpha \in \text{Lim}$, the configuration at time $\alpha + \omega^2$. The necessary information can be read of from G_1: If i belongs to a path leading to S, this configuration will be S. On the other hand, if i belongs to a cycle $(k_0, \ldots, k_m)$ (or to a path leading into a cycle), the computation starting in $(i, 0)$ will first pass through this path and then through the cycle ω many times. Afterwards, the new configuration will have the register containing 0 as always, and the new program line index will be the minimal edge j label occurring in the cycle $(k_0, \ldots, k_m)$. Thus, in the example graph above, if we start in 3, then, after ω^2 more steps, the new configuration will be $(1, 0)$. We also note the minimal edge label l

occurring both in the cycle and the path from i to the cycle (which may be of length 0) and store this information as $i \overset{\omega^2,l}{\mapsto} j$. These edges create a new graph G_2 on the same vertice set as G_1:

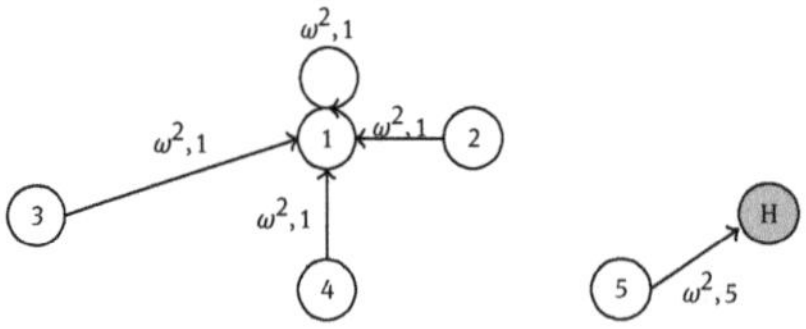

Iterating the last step, we obtain $G_3, G_4, \dots, G_{2n}$, showing us the transitions between the configurations in $\omega^3, \omega^4, \dots, \omega^{2n}$ steps.

The answer to the halting question can now be read off from G_{2n}: If $1 \overset{\omega^{2n}}{\mapsto} S$ for some $k \in \omega$ in G_{2n}, then the computation of P starting in $(1, 0)$ will reach S in the next ω^{2n} many steps, and thus halt. Otherwise, the computation goes on for at least ω^{2n} many steps and we know from Theorem 2.3.14 above that P will not halt. □

2.3.2 Coding transitive ∈-structures

The machines discussed in this book accept as inputs binary strings of ordinal length, which can be naturally interpreted as characteristic functions of classes (proper classes or sets) of ordinals. When, as we often will, we want our machines to work on other sets, we need to encode arbitrary sets as sets of ordinals. We will now describe such an encoding, which works if the set in question is transitive. This suffices for most applications in this book.

Definition 2.3.18. Let X be a transitive set.

If α is an ordinal and $f : \alpha \to X$ is a bijection, we say that the set $\{p(\beta, \gamma) : \beta, \gamma < \alpha \wedge f(\beta) \in f(\gamma)\}$ encodes X via f.

In general, we say that $S \subseteq \alpha$ encodes X if and only if there are $\alpha \in \mathrm{On}$ and a bijection $f : \alpha \to X$ such that S encodes X via f.

It is rather easy to extend this coding to arbitrary sets: If X is a set, we use the above technique to encode $\mathrm{tc}(X)$, the transitive closure of X, in such a way that X is represented by 0 in the code. However, we will rarely need to make use of this trick.

In many cases where we will apply this coding, X will be countable and α will be ω. In such a case, we are thus encoding transitive ∈-structures as subsets of ω, i. e., real numbers.

2.3.3 Testing well-foundedness

An important property of ITRMs (and hence all stronger models of computation) is their ability to test a (code for a) given relation for well-foundedness. This test uses a stack algorithm, a technique which will be frequently used in the later chapters as well. Along with it, some other techniques of programming α-ITRMs (and infinitary machines in general), such as the use of flags, will be explained in this section. We thus advise the reader to consider it carefully. For the rest of the section, unless stated otherwise, α is a multiplicatively closed ordinal.

The oracle of an α-ITRM is a subset of α. Thus, we first explain how linear orderings can be coded as subsets of α: Let $(X, \leq)$ be a linear ordering, and let $f : \alpha \to X$ be a bijection. Then $(X, \leq)$ is coded by the set $\{p(\iota, \xi) : f(\iota) < f(\xi)\} \subseteq \alpha$.

We will often need to compute on finite sequences of ordinals rather than single ordinals. Thus, we need a device for encoding finite sequences of ordinals $< \alpha$ in a unique way as ordinals $< \alpha$. The coding introduced in Definition 2.1.2 is not sufficient, as sequences of different length may still receive the same code.

Thus, we represent the sequence $(\gamma_0, \ldots, \gamma_n)$ by the pair $p_{n+1}(\gamma_0, \ldots, \gamma_n, (n+1))$, which is stored by putting the first component in one and the second in another register, called the "length register." However, to simplify the discussion, we will in the following usually ignore the length registers and identify sequences with their codes; thus, we, e. g., speak of $(\gamma_0, \ldots, \gamma_n)$ being contained in some register, where of course we really mean that this register contains $p_{n+1}(\gamma_0, \ldots, \gamma_n)$, while the corresponding length register contains $(n+1)$.

We will regard a sequence $(\gamma_0, \ldots \gamma_n)$ as a "stack," encoded by an ordinal. Basic operations like reading out, deleting or replacing the top element of the stack can be carried out on any machine that is able to compute the pairing function. A simple, but important observation is that codes for sequences increase with length.

Proposition 2.3.19. *Let $(\gamma_0, \ldots, \gamma_n)$ and $(\gamma_0, \ldots, \gamma_n, \gamma_{n+1}, \ldots, \gamma_{n+k})$ be sequences of natural numbers, where $k > 0$. Then $p_{n+k+1}(\gamma_0, \ldots, \gamma_{n+k}) > p_{n+1}(\gamma_0, \ldots, \gamma_n)$. Moreover, when $(\delta_0, \ldots, \delta_n)$ satisfies $\delta_i \geq \gamma_i$ for all $0 \leq i \leq n$, then $p_{n+1}(\delta_0, \ldots, \delta_n) \geq p_{n+1}(\gamma_0, \ldots, \gamma_n)$.*

Proof. An easy induction. □

Below, we will frequently exploit the fact that the pairing function has a rather humble continuity property.

Proposition 2.3.20. *Let $\alpha \in \mathrm{On}$, and let $\gamma < \alpha$ be a limit ordinal. Then $\lim_{\iota<\gamma} p(\alpha, \iota) = p(\alpha, \gamma)$.*

Proof. Assume otherwise. As $p(\alpha, \gamma) > p(\alpha, \iota)$ for $\iota < \gamma$, this implies that $p(\alpha, \gamma) > \lim_{\iota<\gamma} p(\alpha, \iota) =: \lambda$. Let $\beta_0, \beta_1 \in \mathrm{On}$ be the such that $p(\beta_0, \beta_1) = \lambda$. If $\max\{\beta_0, \beta_1\} < \alpha$, then (β_0, β_1) precedes all the pairs (α, ι) in the Cantor enumeration, a contradiction; similarly if $\max\{\beta_0, \beta_1\} > \alpha$. Thus $\max\{\beta_0, \beta_1\} = \alpha$. If $\beta_0 < \alpha$, then (β_0, α) again precedes all the pairs (α, ι) in the Cantor enumeration, a contradiction. Thus $\beta_0 = \alpha$. It follows that $\beta_1 > \iota$ for all $\iota < \gamma$ and at the same time $\beta_1 < \gamma$, a contradiction. □

This last rather trivial observation is necessary for controlling stack contents during searches through bounded name spaces. We note a particularly useful form of it.

Proposition 2.3.21 (See Koepke [107], Proposition 8). *Let γ be a limit ordinal, and let $(s_\iota : \iota < \gamma)$ be a sequence of finite sequences of ordinals, where $s_\iota = (\alpha_{0,\iota}, \dots, \alpha_{n_\iota,\iota})$. Also, let c_ι be the Cantor code for s_ι, for $\iota < \gamma$.*

(a) *Suppose that there is $\beta < \gamma$ such that, for $\iota > \beta$, s_ι starts with $(\alpha_0, \dots, \alpha_k)$; suppose moreover that $s_\iota = (\alpha_0, \dots, \alpha_k)$ for cofinally many values of ι below γ. Then $\lim_{\iota<\gamma} c_\iota = p(\alpha_0, \dots, \alpha_k)$.*
(b) *Suppose that there is $\beta < \gamma$ such that, for every $\iota > \beta$, s_ι starts with $(\alpha_0, \dots, \alpha_k, \beta)$ for some β Suppose moreover that when s_ι starts with $(\alpha_0, \dots, \alpha_k, \beta)$, then there is $\iota' \in [\iota, \gamma)$ such that $s_{\iota'} = (\alpha_0, \dots, \alpha_k, \beta)$ and let $\bar{\beta} = \liminf_{\iota<\gamma} \beta_\iota$. Then if $\bar{\beta} < p(\alpha_0, \dots, \alpha_k)$, we have $\liminf_{\iota<\gamma} c_\iota = p(\alpha_0, \dots, \alpha_k, \bar{\beta})$.*

Proof. (a) follows from the liminf-rule and Proposition 2.3.19. For (b), use Proposition 2.3.20. □

It will frequently be important that a program "knows" whether it is currently at a limit time, or more generally, at a limit of times at which a certain effect has taken place (e. g., the content of register 4 having been equal to the content of register 7). This is done by using "flags": For a flag, we reserve two separate registers R_0 and R_1, such that initially, R_0 contains 0 and R_1 contains 1. When both are 0, we increase the content of R_1 by 1 and continue. Then, whenever the effect in question shows up, we exchange the contents of R_0 and R_1. Now, when the order type of the set of times at which the effect has happened is a limit ordinal, both R_0 and R_1 have contained 0 cofinally often, and thus their contents will both be equal to 0, while otherwise, their contents will be different. So whether or not the effect has happened, a limited number of times can be read off by comparing the contents of R_0 and R_1.

Quite often when dealing with a model of computation with register bound or tape length α, we will want to "search" through α, i. e., test all elements of α one after the other for a certain decidable property P and halt and output the first example encountered in this way or output "no" if no example exists. For α-ITRMs and stronger models, this is possible by the following lemma. However, for α-wITRMs, we assume an extra condition. The reason is the following: We need some way to store the element ι of α currently being tested. If we do this by writing ι into a certain register and keep increasing that register when there is no example for the property in question, then the register will overflow and the computation will be undefined.

Lemma 2.3.22. *Let α be a limit ordinal.*

(i) *For every α-ITRM-program P, one can recursively in P obtain a program P_{search} such that the following holds for every $x \subseteq \alpha$ and every finite $\vec{p} \subseteq \alpha$: If $P^x(\iota, \vec{p})\downarrow \in \{0, 1\}$ for* *every $\iota \in \alpha$, then $P^x_{\text{search}}(\vec{p})\downarrow = (\iota, 1)$, where ι is minimal such that $P^x(\iota, \vec{p})\downarrow = 1$ if such a ι exists and otherwise, $P^x_{\text{search}}(\vec{p})\downarrow = (0, 0)$.*

(ii) *Suppose that either $\alpha = \omega$ or there are $\beta < \alpha$ and an α-wITRM-computable function $f : \beta \to \alpha$ such that $f[\beta]$ is unbounded in α. For every α-wITRM-program P, one can recursively in P obtain a program P_{wsearch} such that the following holds for every $x \subseteq \alpha$ and every finite $\vec{p} \subseteq \alpha$: If $P^x(\iota, \vec{p})\downarrow \in \{0, 1\}$ for every $\iota \in \alpha$, then $P^x_{\text{wsearch}}(\vec{p}, \beta)\downarrow = (\iota, 1)$, where ι is minimal such that $P^x(\iota, \vec{p})\downarrow = 1$ if such a ι exists and otherwise, $P^x_{\text{wsearch}}(\vec{p}, \beta)\downarrow = (0, 0)$.*

(iii) *For every α-wITRM-program P, one can recursively in P obtain a program P_{bsearch} ("bounded search") such that the following holds for every $\beta < \alpha$, every $x \subseteq \beta$ and every finite $\vec{p} \subseteq \alpha$: If $P^x(\iota, \vec{p})\downarrow \in \{0, 1\}$ for every $\iota \in \beta$, then $P^x_{\text{bsearch}}(\vec{p}, \beta)\downarrow = (\iota, 1)$, where $\iota < \beta$ is minimal such that $P^x(\iota, \vec{p})\downarrow = 1$ if such a ι exists and otherwise, $P^x_{\text{bsearch}}(\vec{p}, \beta)\downarrow = (0, 0)$.*

Proof. (i) Reserve a counting register R. Initially, R_2 contains 0. Run $P^x(0, \vec{p})$. If the output is 1, output $(0, 1)$ and halt. Otherwise, write 1 to a reserved register R', write 1 to R and start the following search routine: In each step, run $P^x(\iota, \vec{p})$, where ι is the current content of R_2. If the output is 1, write ι to R_0 and 1 to R_1 and halt. Otherwise, replace ι with $\iota + 1$ in R_2. Note that this algorithm repeatedly runs through the "testing" loop. After a limit number of values have been tested, the active program line will thus again be the start of the testing routine, which will thus be obtained to the limit ordinal that will then be contained in R_2. Thus, all elements of α are tried out in this way. When no ι exists for which $P^x(\iota, \vec{p})\downarrow = 1$, then R_2 will eventually overflow, and thus contain 0. This is the only way in which it can happen that R_2 contains 0 and R' contains 1, which we can detect and halt with output $(0, 0)$.

(ii) Here, the search routine works as follows: Start with 0 in R_2. Then proceed in phases. At the start of each phase, R_2 has a certain content $\iota \leq \beta$. Using the parameter β, test whether $\iota = \beta$. If that is the case, write $(0, 0)$ to the output and halt. Otherwise, compute $f(\iota)$ and write it to a new register R_3. Then, using an extra register R_4 to store the values currently being tested, perform the search described in (i) up to $f(\iota)$, i. e., start with R_4 containing 0 and then successively increase the content ι of R_4 and run $P^x(\xi, \vec{p})$ on some extra registers until the content of R_4 is equal to the content of R_3, i. e., $f(\iota)$. After each run of $P^x(\xi, \vec{p})$, erase all registers used in the run so that no "accidental" overflows can occur. When $P^x(\xi, \vec{p})\downarrow = 1$ for some $\xi < f(\iota)$, write ξ to R_0 and 1 to R_1 and halt. Otherwise, R_3 will eventually contain $f(\iota)$, in which case we increase the content of R_2 by 1 and start a new phase. It is easy to see that this algorithm works as desired.

When $\alpha = \omega$, we need to consider all $i \in \omega$. To this end, we start with two registers R_2 and R_3, initially both 0 and 0. At the beginning of a phase, let us say we have the natural number i in R_2. We then run $P^x(i, \vec{p})$. If the output is 1, we output $(1, 1)$ and halt. Otherwise, we increase the content of R_2 by 1, copy the content of R_2 to R_3, set the content of R_2 to 0, copy the content from R_3 back to R_2 and set the content of R_3

to 0. This way, after all elements of ω have been considered, both registers will contain 0 instead of overflowing.

(iii) This works as α-wITRMs can simulate β-ITRMs by Proposition 2.3.3. □

Being able to search through α is quite important. We thus define for later use the following.

Definition 2.3.23. An ordinal α is (w)ITRM-singular if and only if $\alpha = \omega$ or there are $\beta < \alpha$ and a cofinal function $f : \beta \to \alpha$ such that f is α-(w)ITRM-computable.

Remark 2.3.24. We will usually just say "search through α" to indicate applications of Lemma 2.3.22.

Theorem 2.3.25 (Koepke, Miller, [110]).[6]

(i) *Suppose that α is multiplicatively closed and ITRM-singular. Then there is an α-ITRM-program $P_{\alpha\text{-wf}}$ such that, for every $x \subseteq \alpha$ that codes a partial ordering $(X, \leq)$, $P^x_{\text{wf}}\downarrow = 1$ if and only if $(X, \leq)$ is well-founded and otherwise $P^x_{\text{wf}}\downarrow = 0$.*

(ii) *If α is merely multiplicatively closed, then there is an α-wITRM-program P_{bwf} such that, for every $\beta < \alpha$ and every $x \subseteq \beta$, $P^x_{\text{bwf}}(\beta)\downarrow = 1$ if and only if x codes a well-ordering and otherwise $P^x_{\text{bwf}}(\beta)\downarrow = 0$.*

Proof. (i) The idea of the algorithm is to exhaustively search through x for an infinite descending sequence by looking for ever smaller elements via a depth-first search. This will be done using a stack algorithm as described above.

We first deal with the case $\alpha = \omega$.

We reserve a register R_0 in which we will store the intermediate sequence built up so far, with corresponding length register R'_0. When R'_0 contains 0, this means that R'_0 has overflown and an infinite descending sequence has been found, in which case we terminate with the output 0. Below, we will find it convenient to mark the "bottom" of the stack stored in R_0; thus, initially, R_0 will contain (a code for) the sequence (0). Furthermore, we have two flag registers to determine limit steps of our computation. We begin by putting the sequence $(0, 0)$ on R_0.

The algorithm now works as follows:

Suppose we currently have the sequence $(0, y_0, \ldots, y_n)$ stored in R_0. In a separate register R_1, we have stored the top element y_n. Now, using some extra registers, we search through ω for some $\delta < \omega$ such that $p(\delta, y_n) \in x$ (which means that δ codes an element which is smaller than y_n). This search uses a register R_2 for storing the current candidate ι and then some auxiliary registers for computing $p(\iota, y_n)$ and testing for elementship in x.

6 See also the earlier paper by Hamkins and Lewis [81], where a similar idea was used for a well-foundedness test on ITTMs.

The search proceeds as follows: If the computation is in a limit state (as indicated by the flag registers) and R_0 contains $(0, y_0, \ldots, y_n)$, we first change the content of R_0 to $(0, y_0, \ldots, y_{n-1})$ and then change it again to $(0, y_0, \ldots, y_{n-1}, y_n + 1)$. (The seemingly pointless intermediate step has the function to guarantee the right behavior at limit times.) Then replace the content of R_1 by y_{n+1} and that of R_2 by 0.

If, at a limit time, R_0 contains (0), we output 1 and halt. If R_0 contains 0 at a limit time, we output 0 and halt.

At successor times, we proceed as follows: If R_0 contains $(0, y_0, \ldots, y_n)$ and R_2 contains δ, we check whether $p(\delta, y_n) \in x$. If not, we replace the content of R_0 first by $(0, y_0, \ldots, y_{n-1})$ and then again by $(0, y_0, \ldots, y_n)$, then we replace the content of R_2 by $\delta + 1$.

On the other hand, if some $\delta \in \omega$ with $p(\delta, y_n) \in x$ is found, replace the content of R_0 with $(y_0, \ldots, y_n, \delta)$, replace the content of R_1 with δ, replace the content of R_2 with 0 and continue at the start of the instructions for successor times.

We now show that this algorithm works as intended. Suppose first that $(X, \leq)$ has no infinite descending sequence, i. e., $(X, \leq)$ is a well-founded partial ordering. This means that every linear subordering of $(X, \leq)$ is a well-ordering, i. e., isomorphic to an ordinal. Let $\sigma := \sigma_{(X,\leq)}$ be the supremum of these ordinals. We will prove by induction on $\sigma_{(X,\leq)}$ that the algorithm terminates with output 1 in this case. This is obvious if $\sigma = 1$, for in this case, all elements of α will unsuccessfully be tried out as first elements of a descending sequence, none of which can be continued, so R_0 will contain (0) infinitely often, and thus eventually reach a limit time at which it contains (0), in which case we halt with output 1. If $\sigma > 0$ note that the first step is to pick (a code for) an element $x \in X$ and then only consider the ordering $(X_x, \leq_x)$, i. e., the ordering $(X, \leq)$ strictly below x. It is easy to see that $\sigma_{(X_x,\leq_x)} < \sigma_{(X,\leq)}$, so that, by induction, the algorithm searching $(X_x, \leq_x)$ for an infinite descending sequence terminates without success. But this means that (the code for) x is then removed from the top of R_0. As this holds for every x, R_0 will again contain (0) infinitely often, and thus also at some limit time, as desired.

Now assume that $(X, \leq)$ has an infinite descending sequence, coded by the sequence $x = (y_i : i \in \omega)$ of elements of ω. Let us furthermore suppose that x is lexically minimal, i. e., there is no code for an infinite descending sequence starting with an element smaller than y_0, or starting with y_0 but continuing with an element smaller than y_1, etc. The argument above shows that all elements below y_0 will eventually have been tried out as first elements and deleted from the stack; thus, at some point, R_0 contains $(0, y_0)$ (after which time y_0 will only be deleted finitely often, always to be replaced by itself). By the same argument, all elements below y_1 will eventually have been tried out as second elements and deleted from the stack; thus, at some point, R_0 contains $(0, y_0, y_1)$, and from now on, y_0 will never again be deleted from the stack (while y_1 may still be deleted and put on the stack again finitely often). Proceeding inductively, R_0 will at some point contain $(0, y_0, y_1, \ldots, y_n)$, from which point on $(0, y_0, \ldots, y_{n-1})$ will never again be deleted. But this means that the stack in R_0 will grow to infinite length, so that R_0' will eventually overflow, in which case the algorithm returns 0, as it should.

We now turn to ITRM-singular ordinals $\alpha > \omega$.

The idea for testing, given an initial stack $(\gamma_0, \ldots, \gamma_n)$ of elements of α, whether the descending sequence coded by its elements can be continued would be to run through all $\iota \in \alpha$, testing for each one whether or not $p(\iota, \gamma_n) \in x$. However, the naive approach used above does not work well with the stack representation: If we run through all elements of α on top of a stack coded, as it has to be, by some ordinal $< \alpha$, then in the limit, the coding will be distorted and all information will be lost.[7]

Thus, we have to be more careful and this is where ITRM-singularity is used. We describe how to adapt the above algorithm. Let $\beta < \alpha, f : \beta \to \alpha$ be of unbounded range in α and α-ITRM-computable. The idea is that f splits α into "managable portions": By putting $\beta + 1$ to the bottom of a stack, we can use the same idea as in the case $\alpha = \omega$ for a depth-first search through β on some stack. When this stack has top element ι, then on a different stack we first put $f(\iota)$ on the stack and then only search through elements $< f(\iota)$ "on top" of that. In this way, Proposition 2.3.21 always guarantees that limits work as desired. We proceed with the details.

Suppose that so far, a stack containing the sequence $(\delta_0, \gamma_0, \ldots, \delta_n, \gamma_n)$ is given in the stack register R_0. Along with R_0, we have built up another stack register R_1 which will be used for a search through β. In order to be able to use Proposition 2.3.21, we write $\beta + 1$ to the bottom of the stack. Thus, the ordinal coding the stack will always be $> \beta$ and we can use Proposition 2.3.21 to search through β on top of the stack. (The δ_i inserted in the stack in R_1 serve the same purpose.) R_1 will contain a stack $(\beta + 1, \iota_0, \ldots, \iota_n)$ of elements of β such that $f(\iota_i) = \delta_i$ for all $0 \leq i \leq n$.

We proceed by putting 0 on top of the stack in R_1 and $f(0)$ on top of the stack in R_0, thus changing them into $(\beta+1, \iota_0, \ldots, \iota_n, 0)$ and $(\delta_0, \gamma_0, \ldots, \delta_n, \gamma_n, f(0))$, respectively. Then we test for each $\iota < f(0)$ whether $p(\iota, \gamma_n) \in x$. When such a ι is found, we put it on top of the stack in R_0, thus changing it into $(\delta_0, \gamma_0, \ldots, \delta_n, \gamma_n, f(0), \iota)$. Otherwise, i. e., if no such ι is found below $f(0)$, we replace 0 with 1 on the top of R_1 and continue similarly. In general, when we have ι as the top element of the stack in R_0, we test whether $\iota < \beta$. If not, then the search for a next element was unsuccessful, so we delete the top element from the stack in R_1, increase the element below by 1 and also delete the two top elements from the stack in R_0. Otherwise, if $\iota < \beta$, we put $f(\iota)$ to the top of the stack in R_1, changing it to $(\delta_0, \gamma_0, \ldots, \delta_n, \gamma_n, f(\iota))$ and then successively put the elements ξ of $f(\iota)$ to the top of that stack, changing it to $(\delta_0, \gamma_0, \ldots, \delta_n, \gamma_n, f(\iota), \xi)$ and testing whether $p(\gamma, \xi) \in x$. If not, we replace ξ by $\xi + 1$. Since only elements $\xi < f(\iota)$

7 For example, if a natural number k is on the stack and one runs through all elements of ω in the next component, generating $[k, 0], [k, 1], [k, 2], \ldots$ without setting the content back to 0 after each unsuccessful attempt, then in the limit, the register will contain ω, independent of what k was.

are considered, the top element will be correct at limit times due to Proposition 2.3.21. When such a ξ is found, we write 0 to the top of the stack in R_1 and continue.

The argument that this procedure works is as in the case $\alpha = \omega$.

(ii) This is now an easy variant of the case $\alpha = \omega$. To ensure that the stack works correctly at limits, we start by writing β^2 to the bottom of the stack. Then the search through β works the same as the search through ω, using the parameter β to indicate when to stop. In particular, when β appears as the top element of the stack, we increase the element below by 1 if there is one and if not, i. e., if β is at the bottom of the stack, we output 1. □

Exercise 2.3.26. Show that, if $\mathrm{cf}(\alpha) > \omega$, then there is also a well-foundedness test for α-ITRMs. (Hint: Given an α-code $x \subseteq \alpha$ of a structure S via $f : \alpha \to S$, any infinite descending sequence $(a_i : i \in \omega)$ of elements of S will satisfy that $\{f^{-1}(a_i) : i \in \omega\}$ is bounded in α. One can thus use one register to count upwards through α, and for every $\beta < \alpha$ only search through the sequences of ordinals $< \beta$.)

An important observation on $P_{\alpha\text{-wf}}$ is that, when $x \subseteq \alpha$ coding an ordinal β, then $P^x_{\alpha-\mathrm{wf}}$ halts after at least β many steps. We can thus use the well-foundedness test as a kind of rough "stopwatch" on α-ITRMs. (We will see below that, for α-ITTMs, a much more precise stopwatch is available.)

Corollary 2.3.27 (Cf. Koepke, [32]). *Let α be multiplicatively closed and ITRM-singular, and let $x \subseteq \alpha$ code a well-ordering of order type $\beta \in \mathrm{On}$. Then $P^x_{\alpha\text{-wf}}$ halts after at least β many steps.*

Proof. Suppose that the claim fails, and let β be minimal such that, for some code x of β, $P^x_{\alpha\text{-wf}}$ halts after $< \beta$ many steps.

For every $\iota \in \alpha$, let x_ι denote the subset arising from x by deleting all elements $p(\gamma, \gamma')$ from x such that $\max\{\gamma, \gamma'\} \geq \iota$. It is easy to see that the configurations occurring in the computation of $P^{x_\iota}_{\alpha\text{-wf}}$ also occur in the computation of $P^x_{\alpha\text{-wf}}$ in the same order, and thus that the former computation takes at most as many steps as the latter. It follows that the computation time of $P^x_{\alpha\text{-wf}}$ is at least the supremum of the computation times of $P^{x_\iota}_{\alpha-\mathrm{wf}}$, $\iota \in \alpha$.

Now each initial segment γ of β is coded by some $x_{\iota(\gamma)}$. By minimality of β, $P^{x_{\iota(\gamma)}}_{\alpha-\mathrm{wf}}$ takes at least γ many steps. Thus, $P^x_{\alpha\text{-wf}}$ takes at least $\sup\{\gamma : \gamma < \beta\}$ many steps, which is $\geq \beta$ when β is a limit ordinal, contradicting the choice of β.

If $\beta = \gamma + 1$ is a successor ordinal, the inductive assumption implies that $P^x_{\alpha\text{-wf}}$ will run for at least γ many steps before putting γ at the bottom of the stack, which takes a further step, and thus runs for $\geq \gamma + 1 = \beta$ many steps, contradicting the choice of β. □

2.3.4 Evaluating truth predicates and other properties of structures

We have already seen above that, when α is multiplicatively closed, E is a binary relation on the set X and $f : \alpha \to X$ is bijective, then $\{p(\iota, \xi) : \iota, \xi \in \alpha \wedge f(\iota)Ef(\xi)\}$ codes the structure (X, E). An important feature of α-wITRMs is that, when α is wITRM-singular, then α-wITRMs can evaluate the truth predicate for (X, E) when a code $x \subseteq \alpha$ is given for it.

Moreover, α-ITRMs can determine the isomorphy of such structures for ITRM-singular α and also determine, when E is the $\in$ relation and X is transitive, which element of α represents a certain ordinal $\xi \in X$ given as the input. We will only deal explicitly with the case $\alpha = \omega$; the generalization to ITRM-singular α works as in the proof of Theorem 2.3.25 and we leave the details to the interested reader.

For simplicity, we assume that formulas to be tested only contain universal quantifiers, $\wedge$ and $\neg$ as logical symbols. By the simulation results in this chapter, the results in this section also hold for many other models of computation, such as ITTMs and ORMs. The results and proofs in this section can be found in Koepke and Miller [110] and Koepke [109].

In the following, if $x \subseteq \alpha \in \mathrm{On}$, then M_x denotes the structure (α, E_x) with domain α and the binary relation $E_x := \{(\iota, \xi) \in \alpha \times \alpha : p(\iota, \xi) \in x\}$.

Below, we will be concerned with evaluating formulas on infinitary machines. We therefore fix some relevant notation. Let $\phi(v_{i_0}, \ldots, v_{i_n})$ be an $\in$-formula with free variables $v_{i_0}, \ldots, v_{i_n}$. An assignment for ϕ is a function f with domain $\{v_{i_0}, \ldots, v_{i_n}\}$ that maps v_{i_j} to a set for each $j \in \{0, \ldots, n\}$. Usually, we will assume that $v_{i_j} = v_j$ for notational convenience; also, we will confuse the variable v_i with its index i and, e. g., write $f(1)$ for $f(v_1)$. When S is an $\in$-structure such that $f(v_{i_j}) \in S$ for all $j \in \{0, \ldots, n\}$, then we say that ϕ holds in S under the assignment f if and only if $S \models \phi(f(v_{i_0}), \ldots, f(v_{i_n}))$. If f is an assignment, $i \in \omega$ and x is a set, then $f[\frac{i}{x}]$ denotes the assignment g defined by $g(v_i) = x$ and $g(v_j) = f(j)$ for all $j \in \mathrm{dom}(f) \setminus \{i\}$. Since round brackets will be used for formulas and triples in stacks, we will from now on write $[a_0, \ldots, a_n]$ for a stack containing the sequence $(a_0, \ldots, a_n)$.

Theorem 2.3.28 (See Koepke and Siders [115], Koepke and Seyfferth [112]).

(i) *Let α be multiplicatively closed and wITRM-singular. Then there is an α-wITRM-program $P_{\alpha\text{-truth}}$ such that, for every $c \subseteq \alpha$ and every pair (ϕ, f) with ϕ an $\in$-formula and f a finite function mapping the free variables of ϕ to elements of α (both ϕ and f appropriately coded by elements of α), we have that $P^c_{\alpha\text{-truth}}(\phi, f)\downarrow = 1$ if and only if $M_c \models \phi$ under the assignment f and otherwise $P^c_{\alpha\text{-truth}}(\phi, f)\downarrow = 0$.*
Moreover, there is an α-ITRM-program with the same property whenever α is ITRM-singular, which we also call $P_{\alpha\text{-truth}}$.

(ii) *If α is merely multiplicatively closed, then there is an α-wITRM-program $P_{\alpha\text{-btruth}}$ (“bounded truth”) such that, for every $\beta < \alpha$, every $c \subseteq \beta$ and every pair (ϕ, f) with ϕ an $\in$-formula and f a finite function mapping the free variables of ϕ to elements* *of β, we have $P^c_{\alpha\text{-btruth}}(\phi,f,\beta)\downarrow = 1$ if and only if $M_c \models \phi$ under the assignment f and otherwise $P^c_{\alpha\text{-btruth}}(\phi,f,\beta)\downarrow = 0$.*

(iii) *For $n \in \omega$, let us call a formula ϕ "purely Σ_n" if and only if it starts with n quantifier alternations, starting with existential quantifiers, followed by a quantifier-free formula. Now, whenever α is multiplicatively closed and $n \in \omega$, there is an α-ITRM-program $P_{\alpha\text{-ntruth}}$ with the following property: For all $c \subseteq \alpha$ and every pair (ϕ,f) where c and f are as in (i) and ϕ is purely Σ_n, $P^c_{\alpha\text{-ntruth}}(\phi,f)\downarrow = 1$ if and only if $M_c \models \phi$ under the assignment f and otherwise $P^c_{\alpha\text{-ntruth}}(\phi,f)\downarrow = 0$.*

Proof. (i) As announced, we only deal with the case $\alpha = \omega$.

We describe a stack algorithm for the decision procedure. We have a main register $\mathcal{R}_{\text{main}}$ containing the stack. Initially, $\mathcal{R}_{\text{main}}$ contains (ϕ,f). We show how to manipulate the contents (using the notation $s \to t$ to denote that the (code of the) stack s is replaced with the (code of the) stack t) and leave it to the reader that these manipulations can actually be carried out by an ITRM (in fact a URM). In the following, R ("remainder") denotes a part of the stack content that is left untouched by the respective manipulation.

$[\mathrm{R},(\phi \wedge \psi,f)] \to [\mathrm{R},(\phi \wedge \psi,f),(\wedge,1),(\phi,f)]$

$[\mathrm{R},(\phi \wedge \psi,f),(\wedge,1),1] \to [\mathrm{R},(\phi \wedge \psi,f),(\wedge,2),(\psi,f)]$

$[\mathrm{R},(\phi \wedge \psi,f),(\wedge,2),1] \to [\mathrm{R},1]$

$[\mathrm{R},(x \in y,f)] \to [\mathrm{R},1]$ if $p(f(x),f(y)) \in c$ (which can be tested with an oracle call)

$[\mathrm{R},(x \subset y,f)] \to [\mathrm{R},0]$ if $p(f(x),f(y)) \notin c$ (which can be tested with an oracle call)

$[\mathrm{R},(\neg\phi,f)] \to [\mathrm{R},\neg,(\phi,f)]$

$[\mathrm{R},\neg,1] \to [\mathrm{R},0]$

$[\mathrm{R},\neg,0] \to [\mathrm{R},1]$

$[\mathrm{R},(\forall x\phi,f)] \to [\mathrm{R},(\forall x\phi,f),(\forall,0),(\phi,f[\frac{0}{x}])]$

$[\mathrm{R},(\forall x\phi,f),(\forall,i),0] \to [\mathrm{R},0]$

$[\mathrm{R},(\forall x\phi,f),(\forall,i),1] \to [\mathrm{R},1] \to [\mathrm{R},(\forall x\phi,f),(\forall,i+1),(\phi,f[\frac{i+1}{x}])]$

$[\mathrm{R},(x = y,f)] \to [\mathrm{R},(\forall z(z \in x \leftrightarrow z \in y),f)]$, where $\phi \leftrightarrow \psi$ abbreviates $\neg(\phi \wedge \neg\psi) \wedge \neg(\neg\phi \wedge \psi)$.

We prove by induction on the complexity of ϕ that, for each input (ϕ,f), $P_{\text{truth}}(\phi,f)$ will halt with $\mathcal{R}_{\text{main}}$ containing 1 if $(M_c,f) \models \phi$ and 0, otherwise.

If ϕ is of the form $x \in y$, this is immediate from the definition.

If $\phi = \neg\psi$, then the new top element of the stack after the manipulation will be ϕ. The algorithm will now run on ϕ, leaving the part of the stack below it unchanged until it is decided. By the inductive assumption, it will be decided correctly, so that we eventually have $[\dots,\neg,1]$ or $[\dots,\neg,0]$ on top of the stack, depending on whether $(M_c,f) \models \phi$ or not. Again by the above manipulation rules, this will be turned into $[\dots,0]$ or $[\dots,1]$, respectively, which is the truth value of $\neg\phi$ in (M_c,f).

If $\phi = \psi_1 \wedge \psi_2$, the algorithm first tests ψ_1 and then ψ_2, keeping track in the $(\wedge,\cdot)$-entry which part of the conjunction is being considered. Both tests give the correct result by induction. The final result is then 1 or 0, depending on whether or not both answers were 1.

If $\phi = \forall x\psi$, the algorithm will be run on $\psi(i)$ for all $i \in \omega$; that is, for every $i \in M_c$, it will be tested whether $\psi(i)$ holds. When the answer is 0 for some instance, 0 is returned with respect to ϕ, as it should be. On the other hand, when the answer is 1, the upper part of the stack including ϕ is deleted and replaced by 1 and only then inserted again. Thus, if all answers are 1, $\mathcal{R}_{\text{main}}$ will contain 1 in the limit state occurring after all these checks have been done, which is again the correct answer.

Finally, the case that ϕ is $x = y$ follows from the $\in$-case and the $\forall$-case, together with the axiom of extensionality.

(ii) This follows by an adaptation of the argument for (i) similar to that used in the proof of Theorem 2.3.25 and Lemma 2.3.22.

(iii) We prove this by induction on n. The algorithm given in (i) still works for evaluating quantifier-free formulas, i. e., purely Σ_0-formulas. For the quantifiers, we reserve n special registers $R_1, \dots, R_n$. Suppose that ϕ is of the form $\exists x_1, \dots, x_k \neg\psi$, where ψ is purely Σ_{n-1}; by induction, we assume that we have an α-ITRM-program $P_{\alpha\text{-}(n-1)\text{-truth}}$ for evaluating purely Σ_{n-1}-formulas and thus also their negations. Now consider an extra register R. In R, we run through α; each $\iota < \alpha$ is decoded into a k-tuple $(\iota_0, \dots, \iota_{k-1})$ and then $P_{\alpha\text{-}(n-1)\text{-truth}}$ is used to evaluate $\phi(f[\frac{x_0}{\iota_0} \cdots \frac{x_{k-1}}{\iota_{k-1}}])$ in M_c. □

We will sometimes need to compare elements between different structures. This is done by another stack algorithm, which we only sketch.

Lemma 2.3.29. *Let α be multiplicatively closed and ITRM-singular. There is an α-ITRM-program $P_{\alpha\text{-compare}}$ such that, if $x, y \subseteq \alpha$ are codes for transitive $\in$-structures M_x and M_y via the bijections $f : \alpha \to M_x$, $g : \alpha \to M_y$ and $\iota, \xi \in \alpha$, then $P^{x \oplus y}(\iota, \xi)$ stops with output 1 if and only $f(\iota) = g(\xi)$ (i. e., if ι codes in M_x the same set that ξ codes in M_y) and otherwise stops with output 0.*

Moreover, for multiplicatively closed α there is an α-wITRM-program with the same function on codes $x, y \subseteq \beta$ for $\beta < \alpha$, using β as a parameter.

Proof (Sketch). Again, let $\alpha = \omega$ and drop the subscript.

P_{compare} works as follows: On input $i, j \in \omega$, P_{compare} first checks whether $f(i) = \emptyset$ and whether $f(j) = \emptyset$. This can be done by testing, for each $k \in \omega$, whether $p(k, i) \in x$ and $p(k, j) \in y$, respectively. If both answers are positive, the algorithm returns 1. If the answers are not the same, it returns 0.

Otherwise, we use a recursive call of P_{compare} using a stack algorithm as in P_{truth} above as follows: We use two stacks, one, S_x, for storing (codes of) elements of M_x, the other, S_y, for (codes of) elements of M_y. Initially, S_x contains i and S_y contains j. For each $k \in \omega$, test whether $p(k, i) \in x$. When such a k is found, put in on the S_x on top of i. Now test, for each $l \in \omega$, whether $p(l, j) \in y$. Whenever the answer is positive, put l on S_y on top of j; now run $P^{x \oplus y}_{\text{compare}}(k, l)$, again using S_x and S_y for storing intermediate requests. If the answer is positive, delete k and l from the stacks and continue by testing whether $p(k+1,i) \in x$. If the answer is negative, delete l from S_y, then put $l+1$ on top of S_y and run $P^{x\oplus y}_{\text{compare}}(k, l+1)$. If all $l \in \omega$ have been unsuccessfully tried, return 0. On the other hand, if an appropriate l is found for each k, return 1.

Note that, by the well-foundedness of M_x and M_y, the algorithm will halt on every input (i,j) and that, for the same reason, the stack will only be of finite length at each time, so that the stack registers never overflow.

The second claim again follows from an adaptation to α-wITRMs using a β-bounded version of the stack algorithm. □

Corollary 2.3.30. *Let α be multiplicatively closed and ITRM-singular. Then there is an α-ITRM-program P such that, when $x, y \subseteq \alpha$ code transitive $\in$-structures M_x and M_y via bijections $f : \alpha \to M_x$, $g : \alpha \to M_y$ and $\iota \in \alpha$, then $P^{x\oplus y}(i)$ halts with output $\xi+1$ if and only if $f(\iota) = g(\xi)$ and with output 0 if and only if no such ξ exists.*

This can also be done by an α-wITRM for codes $x, y \subseteq \beta < \alpha$ in the parameter β when α is merely multiplicatively closed.

Proof. Search through α and use $P_{\alpha\text{-compare}}$ from Lemma 2.3.29. The second claim follows again by an adaptation similar to the one in the proof of Theorem 2.3.25. □

Corollary 2.3.31. *Let α be multiplicatively closed and ITRM-singular.*
(i) *There is an α-ITRM-program $P_{\alpha\text{-count}}$ such that, if $x \subseteq \alpha$ codes a transitive $\in$-structure M_x via the bijection $f : \alpha \to M_x$, then $P^x_{\text{count}}(\iota)$ halts with output $(\xi+1)$ if and only if $f(\xi) = \iota$, and with output 0 if there is no such ξ (i. e., if $\iota \notin M_x$).*
(ii) *There is an α-ITRM-program $P_{\alpha\text{-find}}$ such that, if $x \subseteq \alpha$ codes a transitive $\in$-structure M_x via the bijection $f : \alpha \to M_x$ and $y \subseteq \alpha$, then $P^{x\oplus y}(0)$ halts with output $(\iota+1)$ if $f(\iota) = y$ and with output 0 if no such ι exists.*

Again, (i) *and* (ii) *work on α-wITRMs for multiplicatively closed α for $x \subseteq \beta$ and $\iota < \beta$ when $\beta < \alpha$. Moreover, if α is wITRM-singular, there is an α-wITRM-program $P_{\alpha\text{-fcount}}$ such that, for any $x \subseteq \alpha$ coding a transitive $\in$-structure $M_x \supseteq \omega$ via $f : \alpha \to M_x$ and any $i \in \omega$, $P^x(i)$ halts with output $\xi+1$ if and only if $f(\xi) = i$.*

Proof. (i) Let $c := \{p(\gamma,\delta) : \gamma \in \delta \in \alpha\} \subseteq \alpha$. Thus c is a code for α in which every element of α is coded by itself and which is clearly easily computed an α-ITRM. Now run $P^{x\oplus c}_{\alpha\text{-compare}}(\iota)$.

(ii) Let $c_y := \{p(\iota+1, 0) : i \in y\} \cup \{p(\iota+1, \xi+2) : \iota < \xi < \alpha\}$. Thus c_y codes the transitive $\in$-structure $(\alpha \cup \{y\}, \in)$, in which y is coded by 0. Clearly, c_y can be computed from y on an α-ITRM. Now run $P^{x\oplus c_y}_{\alpha\text{-compare}}(\iota, 0)$ for all $\iota \in \alpha$. If this halts with output 1 for some $\iota \in \alpha$, return $(\iota+1)$, otherwise, return 0. This clearly works as desired.

For the second part of the last claim, one can, by wITRM-singularity of α, run through α to identify some $\xi_0 < \alpha$ such that $p(\iota, \xi_0) \notin x$ for any $\iota < \alpha$, thus identifying $f^{-1}(0)$. Similarly, one can again run through α to identify some $\xi_1 < \alpha$ such that $p(\iota, \xi_1) \leftrightarrow \iota = \xi_0$, i. e., $\xi_1 = f^{-1}(1)$. Iterating, one can obtain $f^{-1}(i)$ for any $i \in \omega$. □

2.3.5 Ordinal register machines

At the beginning of this section, we already mentioned a way to further generalize register machines: namely, to allow the registers to contain arbitrary ordinals. These models can be defined in much the same way as for (w)ITRMs. When arbitrary ordinals are allowed, no overflow can take place, as the inferior limit of a (set-sized) sequence of ordinals is always again an ordinal. In this case, one obtains Ordinal Register Machines (ORMs). All of these models were defined by P. Koepke, the ORMs in [115].

Definition 2.3.32. Setting α = On in the definition of an α-wITRM, we obtain Ordinal Register Machines (ORMs).

We will now show how to compute some of the basic operations of ordinal arithmetic with ORMs. Apart from giving an idea of how these machines work, this will be particularly important in the next chapter, where we determine their computability strength. These results and proofs come from ([107], Section 4). As it should be rather obvious that the algorithms we give work as desired, we will say nothing more on the proofs.

Lemma 2.3.33 (Cf. Koepke, [107], p. 13). *There is an ORM-program P_{add} for adding ordinals; i. e., for all $\beta, \gamma \in$ On, $P_{\text{add}}(\beta, \gamma)$ halts with output $\beta + \gamma$. Moreover, when α is additively closed, then P_{add} also works on an α-wITRM for $\beta, \gamma < \alpha$.*

Proof.

1 IF $R_2 = R_3$ GOTO 5
2 $R_1 + 1$
3 $R_3 + 1$
4 GOTO 1 □

In the following, we will freely use the operations for which we already gave algorithms in further programs. Thus, we will, e. g., write $R_1 = R_2 + R_3$ to indicate a subroutine that computes the sum of the contents of R_2 and R_3 and writes the result in R_1.

Lemma 2.3.34 (Cf. Koepke, [107], p. 14). *There is an ORM-program P_{mult} for multiplying ordinals; i. e., for all $\beta, \gamma \in$ On, $P_{\text{mult}}(\beta, \gamma)$ halts with output $\beta \cdot \gamma$. Moreover, when α is multiplicatively closed, then P_{mult} also works on an α-wITRM for $\beta, \gamma < \alpha$.*

Proof.

1 COPY(1,4)
2 IF $R_2 = R_3$ GOTO 6

3 $R_1 = R_1 + R_4$
4 $R_3 + 1$
5 GOTO 2 □

Lemma 2.3.35 (Cf. Koepke, [107], p. 13). *There is an ORM-program P_{exp} for the exponentiation of ordinals; i. e., for all $\beta, \gamma \in \text{On}$, $P_{\text{exp}}(\beta, \gamma)$ halts with output β^γ. Moreover, when α is exponentially closed, then P_{exp} also works on an α-wITRM for $\beta, \gamma < \alpha$.*

Proof. As for Lemma 2.3.34, just replace $+$ with $\cdot$ in line 3. □

Lemma 2.3.36.
(a) *There is an ORM-program $P_{\max}$ for determining the maximum of two ordinals; i. e., for all $\beta, \gamma \in \text{On}$, $P_{\max}(\beta, \gamma)$ halts with output $\max(\beta, \gamma)$.*
(b) *There is an ORM-program $P_<$ for determining whether $\beta < \gamma$; i. e., for all $\beta, \gamma \in \text{On}$, $P_<(\beta, \gamma)$ halts with output 1 if $\beta < \gamma$ and otherwise with output 0.*

Both $P_{\max}$ and $P_<$ also work on α-wITRMs, for $\beta, \gamma < \alpha$ for any limit ordinal α.

Proof.
(a) 1 COPY(1,3)
2 $R_1 = 0$
3 IF $R_1 = R_2$ GOTO 7
4 IF $R_1 = R_3$ GOTO 9
5 $R_1 + 1$
6 GOTO 2
7 COPY(3,1)
8 GOTO 10
9 COPY(2,1)
(b) Clear from (a). □

Lemma 2.3.37 (Cf. Koepke, [107], p. 15). *There is an ORM-program P_{pair} for computing Cantor's ordinal pairing function $p : \text{On} \times \text{On} \to \text{On}$. That is, for all $\beta, \gamma \in \text{On}$, $P_{\text{pair}}(\beta, \gamma)$ halts with output $p(\beta, \gamma)$. Moreover, when α is multiplicatively closed, then P_{pair} also works on an α-wITRM, for $\beta, \gamma < \alpha$.*

Proof. Consider the following program, which basically counts through the orderin $<^*$ from Definition 2.1.2 up to (β, γ). The idea is the following: R_7 contains the maximum of β and γ. In R_4, we store the current maximum of the pairs which we count through while R_5 and R_6 contain the first and the second element of that pair, respectively. Finally, R_3 contains the image of this pair under Cantor's pairing function.

The algorithm works in phases. At the beginning of each phase, it is tested whether R_4 has the same content as R_7, i. e., whether the maximum of β, γ has already been reached.

If not, we use R_5 and R_6 to count through all pairs with maximum r_4 by first letting $R_6 = R_4$ and counting from 0 up to r_4 in R_5 and then letting $R_5 = R_4$ and counting from 0 to r_4 in R_6. Whenever r_5 or r_6 is incremented during this procedure, r_3 is also incremented. After this is completed, we increment r_4 by 1 and continue with the next phase.

If yes, we distinguish two cases, according to whether $r_5 = r_4$ or $r_6 = r_4$ and then use a similar routine as just described to count through the remaining pairs.

We leave it to the reader to turn this sketch into an actual ORM-algorithm, should she desire so. □

Lemma 2.3.38. *There is an ORM-program P_{unpair} for computing the inverse of Cantor's ordinal pairing function; i. e., for $\beta, \gamma, \delta \in \mathrm{On}$ with $\beta = p(\gamma, \delta)$, $P_{\text{unpair}}(\beta)$ halts with output (γ, δ). Moreover, when α is multiplicatively closed, then P_{unpair} also works on an α-wITRM for $\beta < \alpha$.*

Proof. We leave the proof as an exercise to the reader. □

2.4 Infinitary analogues of Turing machines

The most prominent classical model of computation are **Turing machines.** We thus assume that the reader is familiar with this model, as well as its most important features such as the existence of universal Turing machines or the Church–Turing thesis. We only give a brief definition to recall notation and otherwise refer the reader to the literature [89, 167].

Definition 2.4.1. A Turing machine T is a quintuple $(\mathcal{A}, Q, P, k, q_0)$, where $\mathcal{A}$, the alphabet, is an arbitrary nonempty set (for us, it will usually be $\{0, 1\}$), Q is a nonempty finite set the elements of which are called "inner states," k is a natural number, the number of tapes of T, $q_0 \in Q$ is called the starting state and P is a Turing program, which we define next.

A Turing command is of the form $(q, i_0, \dots, i_k) \to (q', i'_0, \dots, i'_k, \mathrm{d}_0, \dots, \mathrm{d}_k)$, where $q, q' \in Q$, $i_0, \dots, i_k, i'_0, \dots, i'_k \in \mathcal{A}$ and $\mathrm{d}_0, \dots, \mathrm{d}_k \in \{-1, 0, +1\}$.

A non-deterministic Turing program is an arbitrary set of Turing commands. A non-deterministic Turing program P is a deterministic program, or simply a Turing program, if and only if, for every $(q, i_0, \dots, i_k) \in Q \times \mathcal{A}^{(k+1)}$, there is at most one element of the form $(q, i_0, \dots, i_k) \to (q', i'_0, \dots, i'_k, \mathrm{d}_0, \dots, \mathrm{d}_k)$ in P. A (deterministic or indeterministic) Turing program P is "complete" if and only if it contains a Turing command of the form $(q, i_0, \dots, i_k) \to (q', i'_0, \dots, i'_k, \mathrm{d}_0, \dots, \mathrm{d}_k)$ for every $(q, i_0, \dots, i_k) \in Q \times \mathcal{A}^{(k+1)}$.

Now imagine a device consisting of $(k + 1)$ "tapes," each subdivided into ω many cells, numbered $0, 1, 2, \dots$, from left to right, each of which contains one element of $\mathcal{A}$. Furthermore, each tape has a read/write-head, which at any time is located at some position of the tape.

A "snapshot" or "configuration" of T is given by $(k+1)$ assignment functions $t_0, \dots, t_k$ of the tape cells to elements of $\mathcal{A}$, i. e., functions $t_i : \omega \to \mathcal{A}$, the head positions on each of the $(k+1)$ tapes, i. e., a $(k+1)$-tuple $(h_0, \dots, h_k) \in \omega^k$ and an inner state $q \in Q$.

When $c = (q, t_0, \dots, t_k, h_0, \dots, h_k)$ is a snapshot and P is a deterministic Turing program, then the P-successor $c'_P(c)$ of c is determined as follows: For $a, b \in \mathbb{Z}$, let $a \dot{+} b = (a+b)$ if $a + b \geq 0$ and $a \dot{+} b = 0$, otherwise. Now, if P contains a Turing command of the form $(q, t_0(h_0), \dots, t_k(h_k)) \to (q', i'_0, \dots, i'_k, \mathrm{d}_0, \dots, \mathrm{d}_k)$, then $c' = (q', t'_0, \dots, t'_k, h_0 \dot{+} \mathrm{d}_0, \dots, h_k \dot{+} \mathrm{d}_k)$, where $t'_j(n) = \begin{cases} i'_j \text{ if } n = h_i \\ t_j(n), \text{ otherwise} \end{cases}$ for all $n \in \omega$. On the other hand, if P does not contain such a Turing command, then the successor configuration is undefined. See the following picture for an illustration.

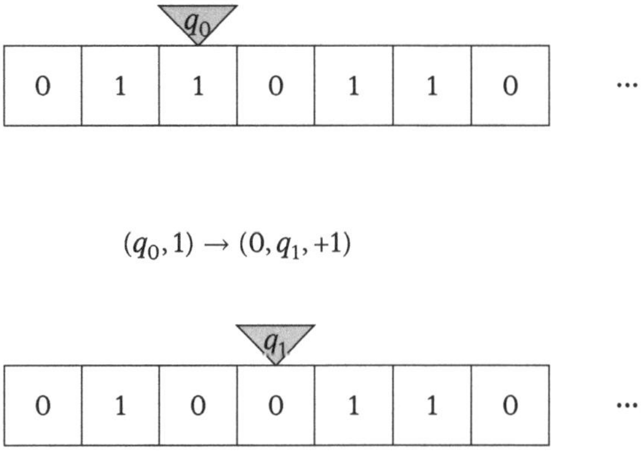

The computation of a Turing program P of length $n \in \omega$ on input $(t_0, \dots, t_k)$ is the sequence $(c_0, \dots, c_n)$ of configurations for P such that $c_0 = (q_0, t_0, \dots, t_k, h_0, \dots, h_k)$ and c_{j+1} is the P-successor of c_j for all $j \leq (n-1)$. If one of these P-successor configurations is undefined, then so is the computation of length n.

Finally, the full computation of P on input $(t_0, \dots, t_k)$ is the computation $(c_0, \dots, c_n)$ of P of length n on that input such that $c_P(c_n)$ is undefined if such an $n \in \omega$ exists or the sequence $(c_i : i \in \omega)$ of configurations such that, for any $j \in \omega$, $(c_0, \dots, c_j)$ is the computation of P of length j, if such a n does not exist.

2.5 Infinite time Turing machines

The historically first machine model of infinitary computability, invented by Hamkins and Kidder, are infinite time Turing machines. They were first described in 2000 by Hamkins and Lewis in [81], which is the paper that started the area of ordinal computability in the sense of the study of machine models of infinitary computability.

These machines have a tape of length ω, which is the special case that is best understood. However, as for register machines, there is no need to insist on tapes of length ω when considering infinitary analogues of Turing machines. We can consider tapes of length α for any ordinal α, and we also can restrict the working time to any ordinal β. Since it takes α many steps to even reach all cells of a tape of length α, this only makes sense when $\beta \geq \alpha$. Restricting the tape length to α and the working time to β leads to (α, β)-Turing machines, which up to now have not received much attention in this generality. If $\alpha = \beta$, we obtain the α-Turing machines (α-TMs) introduced and studied in [112]; if $\alpha \in \mathrm{On}$ and β is unbounded, we get the infinite time α-Turing machines (α-TMs) introduced and studied in Rin [149]. Finally, leaving both α and β unbounded yields Koepke's Ordinal Turing Machines (OTMs) (see [104]), which are one of the most extensively studied models of infinitary computability.

In the interest of generality, we usually work with a tape of length $\alpha \in \mathrm{On}$ and only specialize to ω where this is necessary or considerably simplifies the analysis. It will be convenient to assume that α is closed under ordinal multiplication, which we will do throughout.

Definition 2.5.1 (Hamkins, Lewis [81], Koepke [104, 112, 109]). An α-ITTM T is a quintuple $(Q, P, k, q_0, q_{\text{lim}})$, where the first four components are as for Turing machines except that $k \geq 3$ and q_{lim}, the limit state, is an additional element of Q (which may be equal to q_0); as α-ITTMs always work on the alphabet $\{0, 1\}$, the component $\mathcal{A}$ is dropped.[8]

Thus an α-ITTM has at least three tapes, namely an input tape T_{in}, at least one scratch tape T_{s} and an output tape T_{out}, the cells of which are indexed by ordinals $< \alpha$. Each cell contains either a 0 or a 1; when all cells of a tape contain 0, we say that the tape is "empty." We will see below that the models with one and any finite number of scratch tapes are equivalent and, therefore, usually assume that T uses only one scratch tape. On each tape, there is a read/write head, which we denote H_{in}, H_{s} and H_{out}.

Programs for α-ITTMs are just regular Turing programs (see above). A snapshot or a configuration for an α-ITTM is defined as for Turing machines, with the difference that the assignment functions representing the tape contents are now functions from α to $\{0, 1\}$; and by our convention above, we will usually have three such functions, denoted by t_{in}, t_{s} and t_{out}. Also, the successor configurations $c_P(c)$ for a program P are defined as for Turing machines, with the extra convention $\delta\dot{+}(-1) = 0$ when δ is a limit ordinal, i. e., when the head is moved to the left while currently occupying a limit position, it is reset to position 0.

A new point in the definition of α-ITTMs compared with classical Turing machines are the rules for limit configurations. If T is a Turing program, δ is a limit ordinal and

8 We will see in an exercise below that allowing finite alphabets yields no extra computational strength.

$s := (c_\iota : \iota < \delta)$ is a sequence of T-configurations with $c_\iota := (q_\iota, t_{\text{in}}^\iota, t_{\text{s}}^\iota, t_{\text{out}}^\iota, h_{\text{in}}^\iota, h_{\text{s}}^\iota, h_{\text{out}}^\iota)$, then the limit configuration $c_P(s)$ of P after s is given by $c_P(s) := (q_{\text{lim}}, t_{\text{in}}^\prime, t_{\text{s}}^\prime, t_{\text{out}}^\prime, h_{\text{in}}^\prime, h_{\text{s}}^\prime, h_{\text{out}}^\prime)$, where, for all $\xi < \alpha$, $t_{\text{in}}(\xi) = \liminf_{\iota<\delta} t_{\text{in}}^\iota(\xi)$, $t_{\text{s}}(\xi) = \liminf_{\iota<\delta} t_{\text{s}}^\iota(\xi)$, $t_{\text{out}}(\xi) = \liminf_{\iota<\delta} t_{\text{out}}^\iota(\xi)$. Concerning the head positions, we let $h_{\text{in}} = \liminf_{\iota<\delta} h_{\text{in}}^\iota$, $h_{\text{s}} = \liminf_{\iota<\delta} h_{\text{s}}^\iota$ and $h_{\text{out}} = \liminf_{\iota<\delta} h_{\text{out}}^\iota$ when these liminfs are $< \alpha$; otherwise, we set the respective head(s) back to the cell with index 0.

In the case $\alpha = \omega$, we simply speak of an ITTM and the rules are slightly modified by saying that the read/write-heads are always reset to position 0 at limit times.

Let $\gamma \in \text{On}$. The computation of length γ of an α-ITTM-program P on the input $t : \alpha \to \{0,1\}$ is the sequence $(c_\iota : \iota < \gamma)$ with $c_\iota := (q_\iota, t_{\text{in}}^\iota, t_{\text{s}}^\iota, t_{\text{out}}^\iota, h_{\text{in}}^\iota, h_{\text{s}}^\iota, h_{\text{out}}^\iota)$ of P-snapshots given by $c_0 := (q_0, t, \text{const}_0(\alpha), \text{const}_0(\alpha), 0, 0, 0)$ (where $\text{const}_0(\alpha)$ denotes the constant zero function on α), $c_{\xi+1} = c_P(c_\xi)$ for $\xi + 1 < \gamma$ and $c_\delta = c_P(c_\iota : \iota < \delta)$ when $\delta < \gamma$ is a limit ordinal.

Finally, if there is $\gamma \in \text{On}$ such that the computation of length $\gamma + 1$ of P on the input t is undefined, then the computation of P on input t is the computation of P on input t of length γ, while if no such γ exists, then the computation is the class-sized sequence $(c_\iota : \iota \in \text{On})$ such that, for any $\xi \in \text{On}$, $(c_\iota : \iota < \xi)$ is the computation of P of length ξ on input t.

We will consider α-ITTM-computations relative to ordinal parameters. An ordinal parameter for an α-ITTM is a finite set $\vec{p}$ of elements of α. The computation $P(\vec{p})$ of P in the input t and the parameter $\vec{p}$ is defined like the computation of P in the input t, with the difference that in the initial configuration c_0, $t_{\text{s}}(\iota) = \begin{cases} 1 \text{ if } \iota \in \vec{p} \\ 0, \text{ otherwise} \end{cases}$.

An (α, β)-ITTM works like an α-ITTM with the extra stipulation that computations that take $\geq \beta$ many steps count as nonhalting. (α, α)-ITTMs are also called α-Turing machines or α-TMs.

Remark 2.5.2. The "special rules" for the head positions of ITTMs are historically motivated: It is in this form that ITTMs were originally defined in [81]. We therefore decided to stick to this definition; in an exercise below, we will see that the two definitions yield equivalent models of computability anyway. Also note that, for ITTMs, there is no relevant difference between computations with and without parameters, as any finite set of natural numbers can easily be written to the scratch tape before a computation is started.

Instead of resetting the head back to cell 0 when the liminf of the sequence of earlier head positions is no valid cell index, we could also let the computation fail in this case, mimicking α-wITRMs. We have so far not considered these "unresetting α-ITTMs" and encourage the interested reader to do so.

To illustrate the behavior of ITTMs, we give the following table listing the first ω^2 many steps of the following simple ITTM-program, which, for the sake of simplicity, only uses one tape:

- $(q_0, 0) \rightarrow (q_0, 1, +1)$
- $(q_0, 1) \rightarrow (q_0, 0, +1)$
- $(q_{\text{lim}}, 0) \rightarrow (q_0, 0, 0)$
- $(q_{\text{lim}}, 1) \rightarrow (q_0, 1, 0)$

Time	State	Cell 0	Cell 1	Cell 2	Cell 3	Cell 4	Cell 5	...
0	q_0	**0**	0	0	0	0	0	...
1	q_0	1	**0**	0	0	0	0	...
2	q_0	1	1	**0**	0	0	0	...
...	...	...	...	...	...	...	...	...
ω	q_0	**1**	1	1	1	1	1	...
$\omega + 1$	q_0	0	**1**	1	1	1	1	...
$\omega + 2$	q_0	0	0	**1**	1	1	1	...
...	...	...	...	...	...	...	...	...
$\omega \cdot 2$	q_0	**0**	0	0	0	0	0	...
...	...	...	...	...	...	...	...	...
...	...	...	...	...	...	...	...	...
...	...	...	...	...	...	...	...	...
ω^2	q_0	**0**	0	0	0	0	0	...
...	...	...	...	...	...	...	...	...

Proposition 2.5.3. *Let $\alpha < \beta$. Then β-ITTMs can simulate α-ITTMs.*

Proof. Let P be an α-ITTM-program. Consider a β-ITTM with three extra tapes T'_{in}, T'_{s}, T'_{out}, with α marked on each of these tapes by writing a 1 in the αth cell. Carry out P on the three tapes T_{in}, T_{s}, T_{out}, moving along the head on T'_{in} with that on T_{in}, the head on T'_{s} with that on T_{s} and the head on T'_{out} with that on T_{out} (but without writing anything to these tapes). Whenever one of the heads on the extra tapes reads a 1, the head of the corresponding standard tape is reset to the start of the tape. □

Remark 2.5.4. Note that, as in Proposition 2.3.3, it is crucial that parameters are allowed. Without parameters, the proposition is false; see, e. g., Rin [149] or Carl, Schlicht and Rin [35].

2.5.1 Stack representation for tape models

Recall that, for register models, the depth-first search required, e. g., for evaluating truth predicates or a well-foundedness test used a stack algorithm, which lead us into somewhat inconvenient technical problems with limits and the introduction of (w)ITRMs-singularity as an extra condition.

For the tape models, things are much more convenient. Say α is multiplicatively closed. It is rather easy to split an α-tape into ω many disjoint portions of length α; e. g., we can take the ith portion to consist of cells with indices of the form $p(i, \iota)$, where $\iota < \alpha$.

Since α-ITTMs can evaluate the pairing function, we can also command the head to position $p(i, \iota)$ when i and ι are given.

Thus, a finite sequence $(\gamma_0, \ldots, \gamma_n)$ of ordinals $< \alpha$ can be represented on a tape by letting the value of the $p(i, \iota)$th cell be 1 if and only if $i \leq n$ and $\iota < \gamma_i$ and 0 otherwise. This stack representation has the marvelous property that, at limit times, the ith tape portion will represent the inferior limit of the sequence of ordinals represented by that portion so far. As a consequence, the analogues of the results from Theorem 2.3.25 up to Corollary 2.3.31 for α-ITTMs hold for all multiplicatively closed α.

2.5.2 Computability notions for tape models

For α-ITTMs, we have several notions of computability or writability, which we will now consider. They were originally introduced in [81] for ITTMs.

Definition 2.5.5. A subset $x \subseteq \alpha$ is α-ITTM-writable (α-writable) if and only if there are an α-ITTM-program P and a parameter $\vec{p} \subseteq \alpha$ such that $P(\vec{p})$ halts with x written on the output tape.

A subset $x \subseteq \alpha$ is α-ITTM-eventually writable (α-ew, eventually α-writable) if and only if there are an α-ITTM-program P and a parameter $\vec{p} \subseteq \alpha$ such that, from a certain point of time on, $P(\vec{p})$ has x written on the output tape and never modifies its output later on.

A subset $x \subseteq \alpha$ is α-ITTM-accidentally writable (α-aw, accidentally α-writable) if and only if there are an α-ITTM-program P and a parameter $\vec{p} \subseteq \alpha$ such that, at a certain point of time, $P(\vec{p})$ has x on the output tape.

When $x \subseteq \alpha$ is written by an α-ITTM in $< \beta$ many steps, we say that x is (α, β)-ITTM-writable, and an ordinal β is (α, β)-ITTM-writable if and only if it has an (α, β)-ITTM-writable α-code. Finally, we say that $x \subseteq \alpha$ is α-TM-computable if and only if there are an α-ITTM-program P and some $\xi < \alpha$ such that, for all $\iota \in \alpha$, $P(\iota, \xi)$ halts in $< \alpha$ many steps with output 1 on the first cell of the output tape when $\iota \in x$ and otherwise it halts in $< \alpha$ many steps with output 0 on the first cell of the output tape.

All of the above definitions relativize to arbitrary inputs $y \subseteq \alpha$ in the obvious way. When $x \subseteq \alpha$ is writable by an α-ITTM starting with input $y \subseteq \alpha$, we call x α-ITTM-reducible to y and write $x \leq_{\alpha\text{-ITTM}} y$.

Note the difference between writability and eventual writability: In the case of eventual writability, it is not required that the program halts. This makes eventual writability seem like a weaker notion than writability. We will see later that this is in fact the case. Furthermore, it is not ruled out that x is repeatedly written and overwritten later on, as long as it reappears at some point and is not changed afterwards. In contrast, in the case of accidental writability, the machine is allowed to overwrite x again after writing it, possibly never writing it again.

Proposition 2.5.6. *Every α-ITTM-writable set is α-ew. Every α-ew subset of α is α-aw.*

Proof. Trivial. □

As a machine model, α-ITTMs with a single scratch tape are easier to imagine and describe. Furthermore, when we argue that α-ITTMs can carry out certain operation, it is often convenient to work with several scratch tapes (and several read/write-heads). The crucial observation here is that for a α-ITTM-program with any finite number of scratch tapes, there is an equivalent α-ITTM-program with only one scratch tape.

Lemma 2.5.7. *Let α be multiplicatively closed.*

Let $1 \leq n \in \omega$, and let P be an α-ITTM-program using n scratch tapes. Then there is an α-ITTM-program P' using only one scratch tape that simulates P.

Proof (Cf. Winter [195], *Proposition 3.1).* We explain this in detail for the case $\alpha = \omega$.

Let Q be a n-ITTM-program. The rough idea is to split the scratch tape of an ITTM into n disjoint portions, writing the content of the ith cell of the jth scratch tape of Q to the cell $ni + j$. This does not quite work, as we need some extra space for organizing the simulation and directing the read/write-heads.

We thus refine the idea as follows: Imagine the scratch tape of the ITTM split into $(2n+1)$ many portions, where the ith portion consists of the cells with indices in $\{(2n+1)k + i : k \in \omega\}$. The 0th portion will be used for simulation scratch work; i. e., it will store the current inner state of Q as well as the symbols currently read out. Then a subprogram P' will determine which new state to assume and which new symbols to write. Such a program is easy to write; to make it work on the modified tape, each movement to the left or right is simply replaced with $(2n + 1)$ such movements.

Suppose that Q uses m inner states. Then the first m cells of the 0th portion will be used to store the inner state k in the form $\underbrace{1\ldots1}_{k\times}\underbrace{0\ldots0}_{(m-k)\times}$. After this, come the $(n + 2)$ symbols currently read out. For $1 \leq i \leq n$, the cells with indices in $\{(2n+1)k+2i : k \in \omega\}$ will store the content of the ith scratch tape of Q. Moreover, the cells with indices in $\{(2n + 1)k + (2i - 1) : k \in \omega\}$ will be used to store the position of the read/write-head on the ith scratch tape of Q as follows: If that head is on the kth position, then the $(2i - 1)$th portion will contain k many 1s, followed by 0s.

It should now be clear how to set up the simulation in such a way that these representations of snapshots of Q are updated correctly.

In the general case, the scratch tapes will have α many cells each, and the single scratch tape of P' will be split into $(2n + 1)$ many portions, where the ith portion consists of the cells of the form $\omega\gamma + (2n + 1)k + i$, $\omega\gamma\alpha$, $k \in \omega$. Apart from that, the simulation works exactly as in the case $\alpha = \omega$ (note that head positions at limit times are automatically correctly represented). □

Remark 2.5.8. In spite of Lemma 2.5.7, there are some subtleties with ITTMs when one wants to get rid of the input and output tape as well to strictly work with one tape. This model is not computationally equivalent to ITTMs, though the difference is not trifling; see, e. g., Welch [184] and Hamkins and Seabold [80] for details.

Exercise 2.5.9. Let $\alpha \in \mathrm{On}$ be multiplicatively closed and $1 \leq n \in \omega$. A n-symbol-α-ITTM works like an α-ITTM, but with the tape alphabet $\{0, 1, \ldots, n\}$. In particular, the content of cell ι at a limit time δ is again obtained as the inferior limit of the sequence of earlier contents of that cell. Show that $x \subseteq \alpha$ is n-α-ITTM-writable if and only if x is α-ITTM-writable. (Hint: Split the tape of an α-ITTM into intervals of length $(n+1)$ and use $\underbrace{1\ldots1}_{k\times}\ \underbrace{0\ldots0}_{(n+1-k)\times}$ to represent k.)

Theorem 2.5.10. *Let α be multiplicatively closed.*
- *If $x \subseteq \alpha$ is α-writable relative to a writable $y \subseteq \alpha$, then x is α-writable.*
- *If $x \subseteq \alpha$ is eventually α-writable relative to an eventually writable $y \subseteq \alpha$, then x is eventually α-writable. (Cf. Winter [195], Proposition 6.3)*
- *If $y \subseteq \alpha$ is accidentally α-writable, Q is an α-ITTM-program such that Q^z halts for all $z \subseteq \alpha$ and $Q^y\downarrow = x$, then x is accidentally α-writable.*

Proof. (i) If P is an α-ITTM-program that writes y in the parameter y and Q is an α-ITTM-program such that Q^y writes x in the parameter y', then x can be written by first running $P(y)$ and then running $Q(y')$ on the output of P.

For the sake of simplicity, we suppress the mentioning of parameters in (ii) and (iii).

(ii) Let P be an α-ITTM-program that eventually writes y and let Q be an α-ITTM-program such that Q^y eventually writes x. Then an α-ITTM-program R that eventually writes x works as follows: We reserve a special scratch tape which serves as the output tape of P. Then we run P and simultaneously, on another scratch tape, we run Q on the output of P (where "simultaneously" means that we alternately carry out one step in the computations of P and Q). Whenever the output of P changes, the second scratch tape is erased and Q is run on the second scratch tape, using the output of P as its input and writing its output to the output tape. By the choice of P, the output of P will eventually become y without changing anymore. Thus, from this point on, Q works with input y, so by the choice of Q, it will eventually write x to the output tape and not change its output afterwards. Thus x is eventually α-writable.

(iii) Suppose that P is an α-ITTM-program that accidentally writes y. Now modify P as follows: Whenever the output of P changes, run Q on the new output and write the output of Q to the output tape. At some point, P will write y to the output tape, and then Q will produce x. Thus x is accidentally α-writable. □

Remark 2.5.11. The condition on Q in part (iii) is necessary, and the statement fails without this condition. In fact, already the accidentally writable real numbers are not even closed under writability. See the chapter on degree theory, Theorem 6.2.4.

Definition 2.5.12. Let α, β be ordinals. Then $c \subseteq \alpha$ is a code for β if and only if there is an injection $f : \beta \to \alpha$ such that $c = \{p(f(\iota), f(\iota')) \mid \iota \in \iota' \in \beta\}$.

We say that β is α-writable/eventually α-writable/accidentally writable if and only if there is a code c for β which is α-writable/eventually α-writable/accidentally writable.

Proposition 2.5.13. *Let α be multiplicatively closed. Then the sets of α-writable, eventually α-writable and accidentally α-writable ordinals are downwards closed: That is, if γ is α-writable/eventually α-writable/accidentally α-writable and $\beta < \gamma$, then β is also α-writable/eventually α-writable/accidentally α-writable.*

Proof. Suppose that $\gamma \in \mathrm{On}$ is writable, and let $x \subseteq \alpha$ be an α-writable code for γ. That is, there is a bijection $f : \alpha \to \gamma$ such that $p(\iota, \xi) \in x$ if and only if $f(\iota) < f(\xi)$ for all $\iota, \xi \in \alpha$. Let $<^x$ denote the ordering on α given by $\iota <^* \xi$ if and only if $f(\iota) < f(\xi)$. Suppose that $\beta < \gamma$. This means that there is some $\xi \in \alpha$ such that the restriction of the ordering $<^*$ to the set $M_\xi := \{\iota \in \alpha : f(\iota) < f(\xi)\}$ has order type β. But it is easy to α-ITTM-compute a code for this restricted ordering from x when ξ is given as a parameter.

For eventual and accidental α-writability, we apply the above whenever the content of the output tape has changed. □

Remark 2.5.14. Note that the above proof crucially uses that we allow ordinal parameters in α-ITTM-computations. The parameter-free version even of the downwards closure of α-writable ordinals fails; see Carl, Schlicht and Rin [35].

An important fact in the theory of α-ITTM-computability is the existence of a 'universal' α-ITTM-program $\mathcal{U}_\alpha$ for multiplicatively closed α, which simultaneously simulates all α-ITTM-programs with all parameters $\gamma < \alpha$ on ω many disjoint parts of the scratch tape. The following proof is a slight generalization of the proof by Hamkins and Lewis for the existence of a universal ITTM.

Theorem 2.5.15 (Hamkins and Lewis, [81]). *Let α be closed under ordinal multiplication. Then there is an α-ITTM-program $\mathcal{U}_\alpha$ that simultaneously simulates each α-ITTM-program P_i in each parameter $\gamma < \alpha$.*

In particular, the following holds: For every $x \subseteq \alpha$, x appears on the output tape at some time in the computation of $\mathcal{U}_\alpha$ if and only if there are $i \in \omega$ and $\gamma < \alpha$ such that x appears on the output tape of the computation of $P_i(\gamma)$.

Proof. We will use several scratch tapes, namely $T_c, T_{c'}, T_{c''}, T_h, T_{h'}, T_{h''}$ and T, T', T''. (It then follows from Lemma 2.5.7 that the same can be done with one scratch tape.)

$T_c, T_{c'}, T_{c''}$ will store tape contents, $T_h, T_{h'}, T_{h''}$ will store head positions, T_s inner states, while T and T' are used for further scratch work.

We fix the injection $p_3 : \omega \times \alpha \times \alpha \to \alpha$ via iterated Cantor pairing. Then the tapes mentioned above will store information on a snapshot of a computation $P_i(\gamma)$ ($i \in \omega$, $\gamma < \alpha$) as follows:

The $p_3(i,\iota,\gamma)$th cell of T_c will contain the content of the ιth cell of the input tape of the computation $P_i(\gamma)$, similarly for $T_{c'}$, $T_{c''}$ and the scratch tape and the output tape.

The $p_3(i,\iota,\gamma)$th cell of T_h will contain 1 if and only if ι is $\leq$ the head position on the input tape in the computation $P_i(\gamma)$ and 0, otherwise; similarly for $T_{h'}$ and the head on the scratch tape and $T_{h''}$ and the head position on the output tape.

The $p_3(i,j,\gamma)$th cell of T_s will contain 1 if and only if j is $\leq$ the index of the inner state in the computation of $P_i(\gamma)$.

At the start of $\mathcal{U}_\alpha$, the tapes will be initialized as follows, starting with empty tapes: On T_c, a 1 is written in the $p_3(i,\gamma,\gamma)$th cell for every $i \in \omega$, $\gamma < \alpha$. This ensures that the portion representing the content of the input tape of $P_i(\gamma)$ starts with the parameter γ on the simulated input tape. $T_{c'}$ and $T_{c''}$ remain empty.

On T_h, $T_{h'}$ and $T_{h''}$, a 1 is placed in the $p_3(i,0,\gamma)$th cell for every $i \in \omega$, $\gamma < \alpha$. This ensures that the heads are initially on position 0 in the simulation. On T_s, a 1 is placed in the $p_3(i,0,\gamma)$th cell for every $i \in \omega$, $\gamma < \alpha$. This ensures that $P_i(\gamma)$ starts in the 0th state.

$\mathcal{U}_\alpha$ will now work by repeating the following routine, which consists of α many phases. At the start of each new run of the routine, T will be empty; whenever T is filled with 1s, it is erased and a new phase starts.

At the beginning of a new phase, a 1 is written in the cell of T with minimal index containing 0. Let us say that this is the βtch cell.

Now, on T', we compute $(i,\gamma) \in \omega \times \alpha$ such that $p_2(i,\gamma) = \beta$. After these have been found, use T'' to find the minimal $j \in \omega$ such that the $p_3(i,j,\gamma)$th cell of T_s contains 0. (This can be done by exhaustively searching through ω.) Store $(j-1)$ on T'.

Then, similarly, use T'' to search for the minimal $\iota,\iota',\iota'' < \alpha$ such that the $p_3(i,\iota,\gamma)$th cell of T_h, $T_{h'}$, $T_{h''}$ contains 0. By definition, these will be successor ordinals, and we also store $\iota-1$, $\iota'-1$, $\iota''-1$ on T''.

Then read out the $p_3(i,\iota-1,\gamma)$th bit b of T_c as well as the $p_3(i,\iota'-1,\gamma)$th bit b' of $T_{c'}$ and the $p_3(i,\iota''-1,\gamma)$th bit b'' of $T_{c''}$.

Now all information necessary for simulating the next step of $P_i(\gamma)$ has been determined. Given $(j-1)$, b, b' and b'', we can use i to determine which command of P_i (if any) is applicable. If there is no applicable command, the phase is finished and a new phase is started by writing 1 in the $(\beta+1)$th cell of T and erasing the other scratch tapes. Otherwise, let the applicable command be $(q_{j-1},b,b',b'') \to (q_k,c,c',c'',d,d',d'')$. Then modify the contents of the scratch tapes as follows:

On T_c, move the head to position $p_3(i,\iota-1,\gamma)$ and replace the content of this cell by b; similarly for $T_{c'}$, $\iota'-1$, b' and $T_{c''}$, $\iota''-1$ and b''.

On T_h, leave the content unchanged when $d=0$; if $d=1$, write a 1 in the $p_3(i,\iota+1,\gamma)$th cell. If $d=-1$ and $\iota-1$ is a successor ordinal, write a 0 in the $p_3(i,\iota,\gamma)$th cell; if $\iota-1$ is a limit ordinal, write a 0 in the $p_3(i,\xi,\gamma)$th cell for every $0<\xi<\alpha$; similarly for $T_{h'}$, ι', d' and $T_{h''}$, ι'', d''.

On T_s, write a 1 in every cell with index $p_3(i, j', \gamma)$ with $j' \leq k$ and 0 in every cell with index $p_3(i, j', \gamma)$, $j' > k$. If the content of $T_{c''}$ is changed, write the $p_3(i, \xi, \gamma)$th bit of $T_{c''}$ to the ξth cell of the output tape, for all $\xi < \alpha$.

This finishes the phase. After this, $\mathcal{U}_\alpha$ continues by writing 1 in the $(\beta + 1)$th cell of T. This concludes the description of $\mathcal{U}_\alpha$. Clearly, this program works as desired. □

By the obvious simplification of the proof, one obtains the following.

Corollary 2.5.16. *For every $\alpha \in$ On that is multiplicatively closed, there is an α-ITTM-program $\mathcal{U}'_\alpha$ such that the subsets $x \subseteq \alpha$ that appear at some time on the output tape of some parameter-free α-ITTM-computation are exactly those that appear at some time on the output tape of $\mathcal{U}'_\alpha$.*

Remark 2.5.17. For $\alpha = \omega$, all parameters are writable, and thus the outputs of the computations with and without parameters coincide. We will thus write $\mathcal{U}$ for $\mathcal{U}'_\omega$ and ignore $\mathcal{U}_\omega$.

In the analysis of the computational strength of α-ITTMs, the suprema of writable, eventually writable and accidentally writable ordinals will play important roles. Here, we make some basic observations.

Definition 2.5.18 (Cf. Hamkins and Lewis, [81]). Let $\alpha \in$ On, $x \subseteq \alpha$.
- $\lambda(\alpha)$ denotes the supremum of the α-writable ordinals in the oracle x.
- $\zeta(\alpha)$ denotes the supremum of the eventually α-writable ordinals in the oracle x.
- $\Sigma(\alpha)$ denotes the supremum of the accidentally α-writable ordinals in the oracle x.

When $x = \emptyset$, the superscript x is dropped.

Another important concept will be the idea of a "**grand sums**," originally defined in Hamkins and Lewis, [81]. For α multiplicatively closed, consider an α-ITTM-program $P_{\alpha\text{-gs}}$ that works as follows: The universal α-ITTM $\mathcal{U}_\alpha$ is run. After each step of $\mathcal{U}_\alpha$, the $\alpha = \omega\alpha$ many output tape contents of the ω many simulated programs, each with the α many parameter choices, are considered; say these are $(x_\iota : \iota \in \alpha)$, where $x_{\omega\beta+i}$ is the current content of the output tape of $P_i(\beta)$ for $i \in \omega$, $\beta < \alpha$. Now, for each x_ι, it is decided whether x_ι codes a well-ordering, using Theorem 2.3.25 above. Let

$$\rho(x_\iota) = \begin{cases} \beta, \text{ if } x_i \text{ codes a well-ordering of order type } \beta \\ 0, \text{ otherwise.} \end{cases}$$

Then $P_{\alpha\text{-gs}}$ computes a code for $\rho := \sum_{\iota\in\alpha} \rho_\iota$ and writes it to the output tape. This ordinal is the "grand sum" at the current running time τ of $\mathcal{U}_\alpha$. As every accidentally α-writable ordinal will appear on one of the simulated tapes during the run of $\mathcal{U}_\alpha$, there will be a grand sum $\gamma > \beta$ for every accidentally writable ordinal β, so that the grand sums occurring in this procedure will be unbounded in $\Sigma(\alpha)$.

In order to make this work, we need to see that α-ITTMs can compute infinite sums of ordinals. But this is not hard to see. First, we fix an encoding for sequence of subsets of α as subsets of α.

Definition 2.5.19. Let $s := (x_\iota : \iota \in \alpha)$ be a sequence of subsets of α. Then $x_S := \{p(\xi, \iota) : \iota, \xi \in \alpha \wedge \xi \in x_\iota\}$.

Proposition 2.5.20 (Cf. [81]). *Let α be multiplicatively closed. Then there is an α-ITTM-program $P_{\alpha\text{-sum}}$ with the following property: Let $s = (x_\iota : \iota \in \alpha)$ be a sequence of subsets of α, each of which codes a well-ordering (possibly including $\emptyset$), and let $c = x_S$. Suppose that γ_i is the order type of the well-ordering coded by x_ι, and let $\gamma = \sum_{\iota \in \alpha} \gamma_i$. Then P^c halts with output d, where d is a code for γ.*

Proof. For $\xi, \xi' \in \alpha$, $\xi = p(\gamma, \iota)$, $\xi' = p(\gamma', \iota')$, the desired ordering $\leq'$ is given by $\xi <' \xi'$ if and only if $\gamma < \gamma'$ or $\gamma = \gamma'$ and $p(\iota, \iota') \in x_\gamma$. This clearly codes the sum of the coded ordinals and is also clearly α-writable in s. □

We apply the notion of grand sums to determine some elementary information about $\lambda(\alpha)$, $\zeta(\alpha)$ and $\Sigma(\alpha)$ that will be used in the following without further mentioning.

Proposition 2.5.21.
(i) *$\lambda(\alpha)$, $\zeta(\alpha)$ and $\Sigma(\alpha)$ are limit ordinals. In particular, $\lambda(\alpha)$ is not α-writable, $\zeta(\alpha)$ is not eventually α-writable and $\Sigma(\alpha)$ is not accidentally α-writable.*
(ii) *We have $\lambda(\alpha) < \zeta(\alpha) < \Sigma(\alpha)$.*

Proof. We show this for the case $\alpha = \omega$, leaving the straightforward generalization to the reader.

(i) To see that Σ is not a successor ordinal, suppose for a contradiction that $\Sigma = \alpha + 1$. It follows that α is accidentally writable, say by the ITTM-program Q. Now modify Q to run P and, whenever P outputs a code for an ordinal (which can be determined using the well-foundedness test from Chapter 2) β, output a code for $\beta + 2$, which can be obtained recursively from a code for β. Then Q accidentally writes $\Sigma + 1$, which contradicts the definition of Σ. The arguments for λ and ζ are analogous. The second claim follows trivially.

(ii) To see that $\zeta < \Sigma$, consider the universal ITTM $\mathcal{U}$, let it run and write the grand sums to the output tape at each time. Now, as there are only countably many programs, there is a time τ at which every eventually writable ordinal is written on some portion of the output tape and will not be changed further. Hence, at this point, the grand sum is at least the sum of all eventually writable ordinals. As there is no maximal eventually writable ordinal by (i), this sum cannot be equal to any eventually writable ordinal. Thus, an ordinal $\geq \zeta$ is accidentally writable, so $\Sigma > \zeta$.

For $\lambda < \zeta$, we proceed similarly: As there are only countable many programs, there is an ordinal τ such that, at time τ, all programs that will halt have halted. We run the universal ITTM $\mathcal{U}$ and, at each time, pause to compute the sum of all ordinals written by simulated programs that have halted so far and write this sum to the output tape. At some time τ, this sum will be the sum of all writable ordinals and it will not change any more, so it is eventually writable. That this sum is $> \lambda$ follows again from (i). □

Exercise 2.5.22. Show that $\lambda(\alpha)$, $\zeta(\alpha)$ and $\Sigma(\alpha)$ are exponentially closed. (Hint: It may help to use that $\beta^\gamma \leq \max\{\beta,\gamma\}^{\max\{\beta,\gamma\}}$.)

We briefly mention the analogue of the looping criterion given above for register machines in the case of tape models.

Lemma 2.5.23 (Cf. [81]). *Suppose that P is an α-ITTM-program, $\rho < \alpha$, and let $\beta < \gamma$ be ordinals such that the configuration of $P(\rho)$ at times β and γ is the same and moreover, for every $\iota \in [\beta,\gamma)$:*
- *For every tape cell C of P, the content of C at time ι is at least as large as at time β.*
- *The inner state of P at time ι is at least as large as at time β.*
- *The head position at time ι is not to the left of the head position at time β.*

Then $P(\rho)$ never halts.

Proof. This is similar to Lemma 2.2.10: After time γ, the computation between times β and γ will be repeated, and due to the conditions on inner states, head positions and tape contents, the configuration after a limit of such repetitions is again the configuration at time β. Thus, $P(\rho)$ is looping, and hence does not halt. □

Definition 2.5.24. When β and γ are as in Lemma 2.5.23, we say that (β,γ) witnesses the looping of $P(\rho)$.

2.5.3 Clockable ordinals

An important topic in the study of infinitary models of computability is the possible halting times of programs. In the case of α-ITTMs, these are particularly important, as they play a significant role in the determination of their computability strength (see Chapter 3).

Definition 2.5.25 (Hamkins, Lewis, [81]). Let α be multiplicatively closed and let $x \subseteq \alpha$. An ordinal β is α-ITTM-clockable or α-clockable relative to x if and only if there is an α-ITTM-program P and a parameter $\gamma < \alpha$ such that $P^x(\gamma)$ halts in $\beta + 1$ steps on the empty input. If β is α-ITTM-clockable relative to $\emptyset$, β is called α-ITTM-clockable.

Why is $\beta + 1$ the halting time required for the clockability of β rather than β itself? Well, if we used β instead, then all limit ordinals would be ruled out from being clockable (since halting times are always successor ordinals), which would lead to an altogether less convenient theory.

Definition 2.5.26. As there are only countably many ITTM-programs, there are only countably many ITTM-clockable ordinals relative to x for each $x \subseteq \omega$. Consequently, there must be a supremum of these ordinals, which we denote by γ^x from now on.

More generally, for any infinite $\alpha \in \mathrm{On}$, there are only $\omega \cdot \alpha$ many pairs of an α-ITTM-program P and a parameter $\gamma < \alpha$; thus, there must be a supremum of the α-clockable ordinals in the oracle $x \subseteq \alpha$, which we denote by $\gamma^x(\alpha)$.

(Later on in Chapter 3, we shall see that $\gamma^x(\alpha) = \lambda^x(\alpha)$ for all $x \subseteq \alpha$ when α is sufficiently closed.)

Remark 2.5.27. As we will rarely talk about γ in the sense above and want to avoid losing γ as a variable, we make the convention that γ only denotes the supremum of the clockable ordinals when it is not clearly used as a variable.

To get a feel for the issue of clockability, the reader may try her hand at the following exercises.

Exercise 2.5.28. (a) Show that the following ordinals are ITTM-clockable: ω, $\omega \cdot 2$, ω^2, ω^ω:

(b) Show that, if α, β are ITTM-clockable, then so is $\alpha + \beta$.

(c) Show that, if α, β are ITTM-clockable, then so is $\alpha \cdot \beta$.

(d) Show that there is no largest ITTM-clockable ordinal. Deduce that γ^x is a limit ordinal for each $x \subseteq \omega$.

It is also not hard to show that ITTM-clockable ordinals must be countable. The following proof, however, makes use of some set-theoretic concepts and results that will only be introduced later in this book; the reader who is missing the background should hence skip it and return to it later on.

Lemma 2.5.29 (Hamkins and Lewis, [81]). *If α is ITTM-clockable, then α is countable.*

Proof. Let α be ITTM-clockable, and let P be an ITTM-program that halts in $\alpha+1$ many steps. Let $\beta \in \mathrm{On}$ be large enough so that the computation C of P is contained in V_β. Form the elementary hull $\mathcal{H}$ of C in V_β, and let M be the transitive collapse of $\mathcal{H}$. Clearly, $\mathcal{H}$ is countable, hence so is M. If $\bar{C}$ is the image of C under the collapsing map, then by elementarity, $\bar{C}$ is the computation of P in M. By absoluteness of computations (see Theorem 2.5.48 below), $\bar{C}$ is the computation of P in V as well. But M is transitive and countable, so all elements of M, including $\bar{C}$, are countable as well; thus the computation of P is countable after all. □

Exercise 2.5.30. Let α be multiplicatively closed. Show that, if β is α-ITTM-clockable, then the cardinality of β is at most the cardinality of α.

The first crucial thing to observe about clockable ordinals is the following.

Theorem 2.5.31 (The 'speedup theorem" for α-ITTMs (Hamkins, Lewis, [81])). *Let α be multiplicatively closed, δ a limit ordinal, $n \in \omega$, and suppose that $\delta + n$ is α-clockable. Then δ is also clockable.*

Proof. We first prove this for ITTMs, i. e., in the case $\alpha = \omega$.

Let us start by considering in some detail what it means that an ITTM-program P halts after $\delta + n$ many steps. At time δ, P is in the limit state and all three heads are at position 0. Now, if P halts after n additional steps, this can only be due to contents of the first n cells of T_{in}, T_{s} and T_{out} as no further cells can be reached, and thus read during n further steps. Let us denote the sequence of the first n bits on these tapes by s_{in}, s_{s} and s_{out}, respectively. In the following, we will usually consider them together and denote by s the combined information of these three finite sequences, which one may, e. g., think of as a $3 \times n$-matrix containing 0s and 1s. Similarly, we denote by s_τ these respective cell contents at time τ in the computation of P.

The idea is to build a program P' that halts after $\delta + 1$ many steps as follows: P' will (1) simulate P and (2) along the way (with only finitely many extra steps between two steps in the simulation of P) run a routine that "predicts" the occurrence of s in the following sense: Let c_0^{in} be the first cell of T_{in} and c_0^{s} the first cell of T_{s}. Then, at a limit time γ, c_0^{in} will contain 1 and c_0^{s} will contain 0 if and only if $s_\gamma = s$, i. e., when the tape contents of P at time γ are such that P will halt in another n steps. When P' observes that c_0^{in} and c_0^{s} have these contents while it is in the limit state, it will stop. Clearly, then P' will halt at time $\delta + 1$, and thus be as desired.

The crucial point is then to set up a "prediction routine" that works as described. To this end, let us call s_τ "0-correct" if s_τ has a 0 in all places where s does, and "1-correct" if it has a 1 in all places where s does. Clearly, if s_τ is both 0-correct and 1-correct, then $s_\tau = s$. The plan is to use c_0^{in} as an indicator for 1-correctness and c_0^{s} for 0-correctness. Thus, at a limit time γ, c_0^{in} should contain 1 if and only if s_γ is 1-correct and c_0^{s} should contain 0 if and only if s_γ is 0-correct.

We consider 1-correctness first. Initially, c_0^{in} contains 1. After each step of P, we consider the first n cells of each of its tapes. If the new configuration of P is 1-correct, the content of c_0^i is set to 1, whatever its current value is; otherwise, it is set to 0. Now, if we arrive at a limit time γ and c_0^i contains 1, all configurations of P were eventually 1-correct, hence so is s_γ. On the other hand, if c_0^{in} contains 0, it happened at cofinally many times $\beta < \gamma$ that s_β was not 1-correct, i. e., one of the $3n$ cells contained a 0 where s contains a 1. As $3n$ is finite, there must be one specific of these cells, say c, for which this happened cofinally often before time γ. But this means that c contains 0 at time γ, so s_γ is not 1-correct.

0-correctness is slightly more delicate (the reader may want to pause a minute to consider why). Here is the idea: Let $c_1, \ldots, c_k$ be the cells that contain 0 in s. At each moment in running P', we have one of these cells "under observation". After each step in the simulation of P, we do the following: If c_i is currently under observation and contains 0, then c_0^{s} is set to 0 and the new cell under observation is c_{i+1} for $i < k$

and c_1 if $i = k$. However, if c_i contains 1, then c_0^s is set to 1 regardless of its current content and c_i remains under observation.

We show that this leads to the desired behavior. Suppose first that, at a limit time γ, c_0^s contains 0. This means that c_0^s contained 0 cofinally often before. If one or several of $c_1, \ldots, c_k$ would contain 1 at time γ, it would have eventually contained 1 from some time $< \gamma$ on; so there is some time $\beta < \gamma$ from which on all of the cells among $c_1, \ldots, c_k$ that contain 1 at time γ contained 1. As we change the cell under observation whenever the currently observed cell contains 0, one of these cells would have come under and remained under observation after time β, which means that the content of c_0^s would have been 1 from this time on. But then c_0^s would contain 1 at time γ, contradicting our assumption. Thus, if c_0^s contains 0 at a limit time γ, then s_γ is 0-correct.

Now assume that c_0^s contains 1 at time γ. This means that c_0^s constantly contained 1 from some time $\beta < \gamma$ on. But this means that one of the cells $c_1, \ldots, c_k$, say c_i, came and remained under observation (as the content of c_0^s is set to 0 whenever the observed cell changes), and thus c_i always contained 1 from this time on, and hence also at time γ. Thus, if c_0^s contains a 1 at time γ, then s_γ is not 0-correct.

This completes the proof for ITTMs. In the general case, the head positions at limit times are not as easily dealt with. We modify the above construction as follows: Suppose that the α-ITTM-program P halts after $\delta + n$ many steps, and let $\xi_{\text{in}}, \xi_{\text{s}}, \xi_{\text{out}}$ be the head positions at time δ on the three tapes, respectively (the proof easily adapts to the case of several scratch tapes).

The new program that halts at time δ work with seven additional tapes, T'_{in}, T'_{s}, T'_{out}, T, T', T'' and T'''. On T'_{in}, T'_{s} and T'_{out}, ξ_{in}, ξ_{s} and ξ_{out} will be marked as parameters, respectively. On these tapes, the heads will be moved along with the movements of P on T_{in}, T_{s} and T_{out}, so that it will be detected when the heads are at the "right" positions. Moreover, the heads on T, T' and T'' will remain on the initial position and these cells will represent the 0-correctness and the 1-correctness of s and t, where s represents the first n bits of the tape (which is necessary in the case the head is set back to position 0 by moving left from a limit position) and t now represents the n bits to the right and all bits to the left up to the preceding limit ordinal of ξ, ξ', ξ'' on the tapes of P and are updated as in the ITTM-case. The proof now works analogously. □

Remark 2.5.32. One might object that this proof depends a bit too strongly on details of the machine architecture, such as having several tapes available, so that the speedup is more a property of the precise specification than of a robust class of computations. However, a very similar theorem can be proved for 1-tape machines, and only a further simplifying convention in the spirit of Definition 2.5.25 is necessary to fix things; see Hamkins and Seabold [80] and Welch [188]. Also, the speedup-theorem works as well, e. g., for wITRMs and ITRMs; it is thus much more robust than it might seem.

Having defined the notions of clockable and writable ordinals, it is natural to ask for the connection of the two concepts. Above we saw that set of α-writable ordinals is downwards closed. This, however, is not the case for clockable ordinals:

Definition 2.5.33 (Hamkins, Lewis [81]). A **gap** in the α-clockable ordinals is an ordinal interval $I := [\beta, \gamma)$ $(\beta < \gamma)$ such that no element of I is α-clockable, γ is α-clockable and β is either the successor of an α-clockable ordinal or a limit ordinal with cofinally many α-clockable ordinals below it. If $[\beta, \gamma)$ is a gap in the α-clockable ordinals, then we say that β starts a gap and γ ends a gap; moreover, if $\beta + \delta = \gamma$, then δ is called the length of the gap I.

Corollary 2.5.34 (Hamkins, Lewis [81]). *Let α be multiplicatively closed. If I is gap in the α-clockable ordinals, then the length of I is a limit ordinal. In particular, every gap has an infinite length.*

Proof. Suppose that $[\alpha, \alpha+\gamma+n)$ is a gap where γ is a limit ordinal and $n \in \omega$. Hence, by definition, $\gamma + n$ is clockable. By the speedup-theorem, γ is α-clockable. But γ belongs to the gap, which is a contradiction unless $n = 0$, in which case the gap has length γ, i. e., a limit ordinal. □

Theorem 2.5.35 (Hamkins, Lewis, [81] Theorem 3.4). *There are gaps in the ITTM-clockable ordinals. That is, there are ordinals $\alpha < \beta < \gamma$ such that α and γ are clockable, but β is not.*

Proof. We use the universal program $\mathcal{U}$ constructed above to simulate all ITTM-programs simultaneously in such a way that ω many steps of each program are simulated in ω many simulation steps. This has the consequence that, at a limit time δ, all programs have been simulated for δ many steps. We will call an interval of ω many simulation steps an ω-phase.

During the simulation, we keep track of whether any of the simulated program halt during such an ω-phase. We also reserve a special indicator cell c at the start of the scratch tape, which we set to 1 after the start of each ω-phase and in which we write 0 when some simulated program halts, while it otherwise remains at 1. Thus, at the beginning of the next ω-phase, we will know whether some program halted during the last phase by looking at c.

Now consider an ITTM-program Q that proceeds as described above, until c contains 1 at the start of a new ω-phase, in which case it halts. Clearly, as there are ordinal intervals of length ω that do not contain a clockable ordinal (there are only countable many clockable ordinals after all!), Q will halt.

Let us now assume that there are no gaps in the ITTM-clockable ordinals. What does that tell us about the halting time of Q? Well, under this assumption, every ordinal up to γ is clockable, so c will always contain a 0 at the start of a ω-phase until Q has simulated each ITTM-program for γ many steps, thus having itself run for γ many steps. But now, no ordinal $\geq \gamma$ is clockable, so no program will halt during the time interval $[\gamma, \gamma + \omega)$. This will be noticed by Q, prompting it to halt **after more than** γ **many steps.** Hence our assumption has led us to the existence of an ITTM-program, Q, that halts after more than γ many steps. But this contradicts the definition of γ! □

Exercise 2.5.36. Generalize the last proof to show that, if α is multiplicatively closed, then there are gaps in the α-clockable ordinals.

There is much more to say about the study of ITTM-clockable ordinals, such as the possible sizes of gaps, but we will not consider this topic in any detail here. The interested reader can find further information in [81, 184, 145]. We will return to the topic in Chapter 3, after determining the computational strength of α-ITTMs.

The following lemma will occasionally come in handy.

Lemma 2.5.37 (Hamkins, Lewis [81], Theorem 2.2). *Let α be multiplicatively closed. Then there is an α-ITTM-program $P_{\alpha\text{-stopwatch}}$ such that the following holds: If j is the index of an α-ITTM-program, $\beta < \alpha$ and $c \subseteq \alpha$ is the code of an ordinal γ, then $P^{c}_{\alpha\text{-stopwatch}}(j, \beta)$ carries out exactly γ many steps in the computation of $P_j(\beta)$ on some extra scratch tapes (alternatively, on some portions of the scratch tape) and then halts.*

Proof. Say c codes γ via the bijection $f : \alpha \to \gamma$, and let $\iota <^* \iota'$ if and only if $f(\iota) < f(\iota')$ for $\iota, \iota' \in \alpha$.

At each time of the procedure, we have a list $\mathcal{L}$ of elements of α on some portion of the scratch tape. Initially, $\mathcal{L}$ is empty. Now $P_{\alpha\text{-stopwatch}}$ runs through $\alpha \setminus \mathcal{L}$ and looks for the element of α that codes the $<^*$-minimal ι such that $\iota >^* \iota'$ for all $\iota' \in \mathcal{L}$. When such a ι is found, a step of $P_j(\gamma)$ is carried out and ι is added to $\mathcal{L}$. When none is found, then $\mathcal{L} = \alpha$, in which case we halt. □

2.5.4 α-ITTMs and α-ITRMs

In this section, we will show that, for multiplicatively closed α, α-ITTMs can "simulate" α-ITRMs in a rather strong sense. This is an important result, as it allows us to "export" all computational abilities obtained for α-ITRMs to α-ITTMs, just as Lemma 2.3.2 above allows us to conclude the α-ITRM-computability of $f : \alpha \to \alpha$ from its α-wITRM-computability.

Lemma 2.5.38. *Let α be a limit ordinal. Then each α-ITRM-program P can be simulated by an α-ITTM-program Q. More precisely, suppose that P uses n registers $R_1, \dots, R_n$; then Q has $(n+2)$ scratch tapes $T_1, \dots, T_{n+2}$, where T_i stores the content of R_i for $i \in \{1, 2, \dots, n\}$, T_{n+1} stores the active program line and T_{n+2} is used for scratch work.*

In particular, if $f : \alpha \to \alpha$ is α-ITRM-computable, then it is α-ITTM-computable. The same holds relative to any oracle $x \subseteq \alpha$.

Proof (Sketch). We use the extra scratch tape T_{n+2} to determine the necessary steps from the ITRM-program and then carry them out in the obvious manner: If a register content is incremented, the first 0 on the corresponding tape is replaced with 1. When a register content is decreased, either due to the zero command or due to a copying, the necessary number of 1s is deleted from right to left; similarly for the active program line. The liminf-rule for α-ITTMs guarantees that both the active program line and the register contents will be represented accurately by the tape contents. The reason for the "simulation delay" is that it, e. g., requires more than one step in general to delete all 1s from a tape when simulating a reset order in the α-ITRM-program. □

The same argument will be used below to show that OTMs can simulate ORMs. However, it will turn out in Chapter 3 that α-ITTMs are way stronger than α-ITRMs. A first glimpse of this vast difference can already be proved by the methods developed above:.

Exercise 2.5.39. Show that ITTMs can solve the halting problem for ITRMs. (Hint: Observe that the program H_n from Theorem 2.3.7 can be obtained uniformly from n in a recursive way, and thus by an ITTM. Now apply Lemma 2.5.38.) Show that, moreover, ITTMs can solve the halting problem for ITRMs with the halting problem for ITRMs as an oracle and in fact any finite iteration of the ITRM-halting problem.

Exercise 2.5.40. Show that α-TMs can simulate α-RMs.

2.5.5 Weak ITTMs

Again, it is worthwhile to pause for a minute to consider a weak variant of ITTMs, which gives a good grasp of the "combinatorics" of ITTMs. The model we will now examine are weak ITTMs (wITTMs). They were first defined by J. Hamkins and J. Kidder and much later discussed jointly by S. Friedman, J. Hamkins and P. Welch, who however never published their results safe for a post on MathOverflow by Hamkins responding to a question of the author,[9] sketching very briefly the two results we are about to give, along with their proofs.[10] The machines were later rediscovered by Bianchetti, who studied them in his Master's thesis [14] and in [15] and who also argued that they provide a "philosophically more satisfying model" of transfinite computability than ITTMs.

We give here full versions of the results and proofs sketched by Hamkins.

The idea of a weak ITTM is to take the definition of ITTMs and replace the liminf-rule with something much simpler, namely an "eventually constant" rule: Thus, at a limit time δ, a weak ITTM determines the content of the ith tape cell by considering the sequence $(C_i^\iota : \iota < \delta)$ of earlier contents of that cell and letting $C_i^\delta = 1$ if $C_i^\iota = 1$ for all large enough $\iota < \delta$ and $C_i^\delta = 0$ if $C_i^\iota = 0$ for all large enough $\iota < \delta$. If none of

9 See https://mathoverflow.net/questions/111902/transfinitely-iterated-limit-computability

10 We thank J. Hamkins for explaining the history of these machines to us in a private e-mail.

these two cases holds, the δth configuration is undefined and the computation fails, similar to the behavior weak ITRMs in the case of a register overflow.

We want to determine the computational strength of weak ITTMs. Weak ITTMs are sufficiently weak so that the arguments remain entirely combinatorial and algorithmic, without relying on model theory or higher set theory.

Theorem 2.5.41. *Let P be a wITTM-program. Then P, when run on the empty input, either halts in* $< \omega^2$ *many steps or does not stop at all.*

Proof. We first note the following fact: If τ is a limit ordinal and $n \in \omega$, then there is $\tau' < \tau$ such that, for all $\beta \in (\tau', \tau)$, the contents of the first n tape cells were the same as at time τ.

This follows from the limit rule of wITTMs: All n cells contents must become constant before time τ, and one of these finitely many cells will be the last to do so.

We will now show that P cannot halt at time $\omega^2 + n$ for any $n \in \omega$. Suppose otherwise. If P halts at time $\omega^2 + n$, the read/write head will occupy one of the first n tape cells, and until then, it will only have moved around on these first n cells after being at position 0 at time ω^2, (at which the inner state was the limit state). Thus, whenever, P starts in a configuration with the head on position 0 and the first n tape cells with the same contents as at time ω^2, it will halt after n more steps. However, by the observation above, the content of the first n cells was the same as at time ω^2 from some time $\tau' < \omega^2$ on; in particular, this tape content occurred at a limit time before ω^2. It follows that P already halted before time ω^2, a contradiction.

Now we extend the argument to ordinals of the form $\omega^2 + \omega k + l$, $k, l \in \omega$: As above, the halting at $\omega^2 + \omega k + l$ occurs only depending on the contents of the first l cells at time $\omega^2 + \omega k$. These l cell contents must have become constant at some time before $\omega^2 + \omega k$, say at time $\omega^2 + \omega(k-1) + l_1$, $l_1 \in \omega$. Moreover, by the same argument, there is $l_2 \in \omega$ such that the fact that these l cells had the contents they had at time $\omega_2 + \omega(k-1) + l_1$ only depends on the first l_2 many cells at time $\omega^2 + \omega(k-1)$. Iterating this argument, we find $l_k \in \omega$ such that, whenever the first l_k cell contents at some limit time τ are the same as at time ω^2, the machine will halt at time $\tau + \omega k + l$. But as above, this must have happened at some time $\tau < \omega^2$, so P halted at time $\tau + \omega k + l < \omega^2$, a contradiction.

It is now easy to see that the same argument works just as well for $\omega^2\alpha + \omega k + l$ with arbitrary $\alpha \in \mathrm{On}$, which shows that none of these ordinals can be the halting time of P. □

Let $x, y \subseteq \omega$. Following the usual notation, we will write $x \leq_T y$ to say that x is Turing-computable in the oracle y or that x is Turing-reducible to y.

Corollary 2.5.42. *If* $x \subseteq \omega$ *is wITTM-writable, then* $x \leq_T 0^{(n)}$ *for some* $n \in \omega$ *(where* $0^{(n)}$ *denotes the nth Turing jump).*

Proof. We prove by induction that the tape content of a wITTM-program P before time ωk is Turing reducible to $0^{(k-1)}$.

This is clear for $k = 1$, as a wITTM-computation of length $< \omega$ is just a regular Turing computation.

But the tape content $\hat{c}$ at time $\omega(k+1)$ is computable from the tape content c at time ωk with the help of c' as follows: For all $i, j \in \omega$, consider the Turing program Q that simulates P on input c and, after j simulation steps, starts noting whether the content of the ith tape cell changes. If such a change takes place, Q halts, otherwise, the simulation continues. Thus $Q^c(i,j)$ will halt if and only if the ith cell content in the computation of P^c is not constant after the jth step. Given c', we can determine, for each $i, j \in \omega$, whether $Q^c(i,j)$ halts. If it does not, we continue with the next j. $Q^c(i,j)$ must halt for some $j \in \omega$, for otherwise, the computation of P would not be defined. So such a j will eventually be found. But then, we only need to take the content of the ith cell after j many steps in the computation of P—this will be the content of the ith cell in the tape content of P at time $\omega(k+1)$. So $\hat{c} \leq_T c'$.

But, by the inductive assumption, we have $c \leq 0^{(k)}$, so $c' \leq_T (0^{(k-1)})' = 0^{(k)}$. □

We now show the other direction: If $x \leq_T 0^{(n)}$ for some $n \in \omega$, then x is wITTM-writable.

The first thing to note is that the wITTM-setup works quite well with an arbitrary finite alphabet $\mathcal{A}$ in the place of $\{0,1\}$: If $n \in \omega$ is such that $2^n \leq |\mathcal{A}| < 2^{n+1}$, just assign a $(n+1)$-bitstring to each element of $\mathcal{A}$ and consider the tape content as consisting of disjoint blocks of length $(n+1)$ each, whose contents represent elements of $\mathcal{A}$. This represents a tape with symbols from $\mathcal{A}$ in a straightforward manner, and the limit rule works just the same: The tape content is only defined at limit times when each cell content changed only finitely often before. Thus, in the following, we will freely use symbols besides 0 and 1.

Secondly, once $0^{(n)}$ is written to some sensible portion of the tape (such as the cells with even index), we can carry out a regular (halting, thus finite) Turing computation with this as an oracle with only finitely many changes of each cell content. It thus suffices to show that one can compute $0^{(n)}$ with a wITTM, for each $n \in \omega$.

The construction we give below shows how a wITTM can, for each $x \subseteq \omega$ given on a portion of the tape given by an arithmetical sequence of cell indices, write x' to another such portion. It is rather easy to iterate this program for a fixed finite number of steps, using some extra tape portions. We leave the details of this iteration to the reader.

Theorem 2.5.43. *There is a wITTM-program P such that the following holds: When $x \subseteq \omega$ and P is run on a tape whose $4n$th cell initially contains the nth bit of x, then P halts with the nth bit of x' on the $(4n+1)$th bit of the tape, for each $n \in \omega$.*

Proof. We regard the tape as split into 4 disjoint portions, according to the residues of the cell indices mod 4. The oracle x will be stored on the cells of the 0th portion, with cell indices $\equiv 0 \bmod 4$, the output will be written on the cells of the 1th portion, with cell indices $\equiv 1 \bmod 4$ and the 3rd portion will be used for the actual computation.

The second portion is an auxiliary tape that helps directing the read/write-head on the output portion 1.

The computation works in phases. At the beginning of the nth phase, the cells in the second portion will all contain 0 and the read/write-head will be at cell $(4n + 3)$, which contains a special marking symbol $\mathfrak{c}$. Further, the contents of portion 3 to the right of this cell (including it) will be as follows: Letting $n = p(i, j)$ (where p is Cantor's pairing function), it will contain $\mathfrak{c}1^n z_0 1^i z_1 1^j z_2 x_n z_3$. Here, z_0, z_1, z_2, z_3 are special distinct separation symbols, while x_n is the string containing of the first n bits of x.

Now the space of the 3rd portion to the right of z_3 is used to simulate P_i^x for j many steps. This can only make oracle calls for the first $j < n$ many bits of x, for which we use the x_n stored between z_2 and z_3. Thus, there is no need to move the read/write-head to the left of $\mathfrak{c}$ in the course of the simulation.

For the further behavior, we distinguish two cases:

Case 1: P_i^x does not halt in exactly j many steps.

In this case, we do not need to change the output, but only set up the next, i. e., the $(n + 1)$th phase. To do this, we first move the read/write-head to $\mathfrak{c}$ and then one place to the right, i. e., the cell $4(n+1)$ containing the $(n+1)$th bit of the oracle, which we append to the right of x_n to obtain x_{n+1}. Then the space to the right of z_3 is used to compute $(n + 1)$ as well as i', j' with $(n + 1) = p(i', j')$. Finally, $\mathfrak{c}$ is (deleted and) moved one place to the right, starting the following sequence: $\mathfrak{c}1^{(n+1)} z_0 1^{i'} z_1 1^{j'} z_2 x_{n+1} z_3$. The read/write-head is moved to $\mathfrak{c}$, and the $(n + 1)$th phase starts.

Case 2: P_i^x halts in *exactly* j many steps.

In this case, we want to change the content of the $(4i+1)$th cell (i. e., the ith output cell) to 1. For this, we need to move the read/write head to position $(4i + 1)$. We have $\mathfrak{c}$ at position $(4n + 3)$, followed by $1^n z_0 1^i$. We use the place on portion 3 to the right of z_3 to compute $1^{(n-i)}$. Then we move the read/write-head $4(n - i) + 2$ places to the left from $\mathfrak{c}$ as follows: Successively delete a 1 from the $1^{(n-i)}$-block and put a 1 in the first cell of the “auxiliary” tape portion, counting from $\mathfrak{c}$ to the left, that contains a 0. If all $(n - i)$ 1s are used up, write a 1 in the $(4i + 1)$th cell (which is two places to the left of cell $(4i + 3)$ where we just ended up). Then run to the right, replacing all 1s on the 2nd portion by 0 again, up to $\mathfrak{c}$. Now continue as in Case 1, adding one bit to the oracle carried along to obtain x_{n+1}, shifting $\mathfrak{c}$ to the right and setting up the next phase.

Note that only the cells to the right of $(4i + 1)$ are changed in this case.

It should be clear that, at time ω, we have x' on the “output” portion of the tape, provided the ωth configuration exists.

We now show that every cell content is only changed finitely often in this computation, which guarantees the existence of the ωth configuration.

This is clear for the 0th portion, i. e., the oracle part, which is not changed at all.

Also, on the output part, i. e., portion 1, each cell content changes at most once, namely from 0 to 1.

On the computation part, i. e., portion 3, after phase n, only cells with index $> n$ are used, and each phase, taking place in finite time, makes only finitely many changes. Thus cell $(4n + 3)$ is changed only finitely often in the first n phases, and never again after that.

Finally, for the "auxiliary" second portion, the $(4i + 2)$th cell content is only changed when some cell of the output portion with smaller index is modified; as each cell of the output changed is changed at most once, this happens at most i many times, i. e., again only finitely often.

Now all cells have been considered and we are done. □

2.5.6 Ordinal Turing machines

Ordinal Turing Machines (OTMs) arise from ITTMs in a similar way as ORMs arise from ITRMs, namely by removing any restriction on the working space. Thus, OTMs work like α-ITTMs when α is the whole class of ordinals. Clearly, an OTM can simulate an α-ITTM when the parameter α is given. Since it is easy to write an OTM-program that, when started on the empty tape, halts with a tape that has its only 1 on position ω, OTMs can in particular simulate ITTMs. When x and y are sets of ordinals such that x is OTM-computable (without parameters) relative to y, we write $x \leq_{\mathrm{OTM}} y$.

Corollary 2.5.44. *Let $n > 0$. An OTM-program P using n scratch tapes can be simulated by one using a single scratch tape.*

Proof. An easy adaptation of the proof of Lemma 2.5.7. □

When working with OTMs, we will usually ignore the input and output tape and merely talk of one tape.

Similar to Lemma 2.5.38, one can show that the behavior of ORMs can be simulated quite closely by OTMs. A similar statement holds in the reverse direction, though we will only be able to prove this in Chapter 3.

Lemma 2.5.45. *Every ORM-program P can be simulated by an OTM-program P'.*

Proof. The proof is an easy adaptation of the proof of Lemma 2.5.38. □

Exercise 2.5.46. A "one-way-OTM" works like an OTM, with the following modification: When the head is directed to the left while occupying a limit position, it remains where it is rather than being reset to the start of the tape.

(a) Prove that, if τ is the halting time of some one-way-OTM-program (run on empty tapes), then τ is the halting time of some ITTM-program.

(b) Prove that, if τ is the halting time of some ITTM-program (run on empty tapes), then τ is the halting time of some one-way-OTM-program.

Exercise 2.5.47. We consider the relation of ITTMs and OTMs. For simplicity, we consider ITTMs with only one tape; the arguments easily generalize to full ITTMs. The argument given here is a simplification of the analysis of ITTM-computability by Welch; see, e. g., [184].

(a) Show that there is an OTM-program Q with the following property: Whenever P_i is an ITTM-program and $\gamma \in \mathrm{On}$, then $Q(i, \gamma)$ will eventually halt and at the time of halting, the tape content between the $\omega\iota$th and the $\omega(\iota + 1)$th cell will be the tape content of P_i at time ι.

Consider a nonhalting ITTM-program P, and let $t : \omega \to \{0, 1\}$ be the tape content of P at time ω_1.

(b) Show that there is $\gamma < \omega_1$ such that, for any $\tau \in [\gamma, \omega_1)$ and any $i \in \omega$, if $t(i) = 1$, then the ith tape cell at time τ in the computation of P also contains 1.

(c) Show that, for any $i \in \omega$ such that $t(i) = 0$, there are cofinally many $\tau < \omega_1$ such that the ith tape cell at time τ in the computation of P contains 0. Show that the set C_i of times $\tau \leq \omega_1$ at which this holds is closed under limits.

(d) Show that there is an ordinal $\delta < \omega_1$ such that $\delta > \gamma$ and δ is a limit point of all the C_i defined in (c).

(e) Show that the configuration of P at times δ and ω_1 is the same. Moreover, show that, for any $\tau \in [\delta, \omega_1)$ and any $i \in \omega$, the content of the ith tape cell in the computation of P at time τ is $\geq$ the corresponding content at times δ and ω_1.

(f) Conclude that, when P is a nonhalting ITTM-program, then there are ordinals $\beta < \gamma$ such that (β, γ) witnesses the looping of P.

(g) Show that an OTM can solve the halting problem for ITTMs. (Hint: Adapt the algorithm from (a) to write more and more P-configurations to the tape; after each step, check whether a pair witnessing the looping of P has been found or whether P has stopped. By (f), one of the two possibilities must eventually occur.)

(h) Generalize the preceding argument to show that, for every limit ordinal $\alpha \in \mathrm{On}$, the halting problem for α-ITTM-programs (with parameters) can be solved by an OTM in the parameter α.

2.5.7 Absoluteness of computations

We close by mentioning a further feature of infinitary computations that will play an important role in all following chapters, namely their "absoluteness."

Theorem 2.5.48. *If M is a transitive class and it holds in M that C is the partial OTM-computation of P of length $\tau \in \mathrm{On} \cap M$, then the same holds in the set-theoretical universe V, i. e., C is in fact the initial OTM-computation of P of length τ. The same holds for α-wITRMs, α-ITRMs, α-ITTMs, ORMs and in fact all machine types considered here. In particular, when it holds in M that P holds with output x, then the same holds in the real* *world. Moreover, if $x \subseteq \alpha$ belongs to M, the same holds for computations relative to the oracle x.*

Proof. The condition is easily seen to be Δ_0, and thus absolute between transitive classes. (For a proof of the absoluteness of Δ_0-formulas between transitive classes see, e. g., Jech [92]. The reader who is unfamiliar with formula complexity altogether is referred to the next chapter.) □

2.6 Beyond computability: the jump operator

Roughly, the jump of an object x for a machine model M is the set of M-programs that halt when run in the input x. Jumps play an important role as they intuitively play the role of the "next difficult problem after x."

Definition 2.6.1. Let M be one of the tape models of computability, and suppose that M works on inputs of length α, where α is either an ordinal or the class of all ordinals. Then, for $x \subseteq \alpha$, the M-jump of x, denoted by x'_M, is the set of M-programs P such that P^x, when run on otherwise empty tapes, halts.

Similarly, if M is one of the register models with register contents bounded by α, where α is either an ordinal or the class of all ordinals, and let $x \subseteq \alpha$. Then x'_M is the set of M-programs P such that P^x, when run on otherwise empty tapes, halts.

Jumps can be iterated in the obvious way, letting $x_M^{(0)} = x$ and $x_M^{(k+1)} = (x_M^{(k)})'_M$ for $k \in \omega$.

Theorem 2.6.2. *If M is any of the machine models introduced above and $x \subseteq \alpha$ is a possible input for M, then x'_M is not M-computable from x.*

Proof. This is proved by the usual diagonal argument[11]: Suppose otherwise, and let P be an M-program that computes x'_M from x. We modify P slightly to obtain a program $\hat{P}$ that, when given the index $i \in \omega$ of an M-program P_i and the input x, first computes x'_M to check whether P_i^x will halt; after that, $\hat{P}$ will halt if the answer was "no" and otherwise will run into an endless loop. $\hat{P}$ will itself have some index j. But now, $\hat{P}^x(j)$ will by definition halt if and only if it does not, a contradiction. □

For the tape models, the jump operator defined above is associated with writability: A program index i belongs to the halting set if and only if the ith program writes a real number. However, the concept of eventual writability turned out to be quite relevant for the analysis of ITTMs above. This motivates the definition of a jump associated with eventual writability [180].

Definition 2.6.3. If M is either ITTM or OTM and $x \subseteq \omega$, we define ew-$x'_M := \{i : P_i^x$ eventually writes some real number$\}$.

11 which has probably never been written up in a more amusing way than in [146].

Theorem 2.6.4 (Welch, see [180]). *For $x \subseteq \omega$ and M either ITTM or OTM,* ew-x'_M *is not M-ew in the oracle x.*

Proof. We prove this in the case that $x = 0$; the argument relativizes straightforwardly. Suppose that Q is an M-program that eventually writes ew-$0'_M$. Now consider the M-program P that does the following: P runs Q. Whenever Q changes its output, say to $x \subseteq \omega$, Q starts simulating simultaneously all M-programs P_i with $i \in x$, while continuing to run P in parallel. The output of P is obtained by taking, in each simulation step, the ith bit of the current content of the output tape of P_i, switching it and writing it to the ith cell of the output tape if $i \in x$ and writing 0 in that position if $i \notin x$. If Q changes its output, the output is erased and the simulation is started anew with the new output.

Now, as Q eventually writes ew-$0'_M$, the output of Q will eventually stabilize at ew-$0'_M$. When this happens, the simulation will consider exactly those programs that eventually write a real number. Eventually, the outputs of all these programs will become stable. If $i \in$ ew-$0'_M$, and P_i stabilizes with output x, then after the stabilization time of P_i, the ith bit of the output of P will be different from the ith bit of x. Thus, after all P_i have stabilized, the output of P will be stable as well. On the other hand, this output will differ from every eventually writable real number by definition. Thus P eventually writes a real number that is not eventually writable, a contradiction. □

2.7 Further models

The above are not the only models of transfinite computations that have been considered so far. We briefly mention some more interesting examples.

2.7.1 Deterministic ordinal automata

Deterministic ordinal automata (DOAs) are a transfinite analogue of deterministic finite automata (DFAs). For the reader not familiar with the latter, we give a brief description and refer to [197].

A DFA $\mathcal{A}$ on the alphabet $\{0, 1\}$ is a finite directed graph $G = (V, E)$ whose edges are labeled with 0 or 1 such that every vertice has at most one outgoing vertice with each label. The vertices of G are called the "states" of the DFA. Also, a subset F of the vertices of G is specified, the elements of which are called "accepting states." Finally, one element $s_0 \in V$ is specified, called the "initial state" of $\mathcal{A}$.

The following is an example of a DFA, where q_1 and q_2 (gray) are accepting states and q_0 (white) is the initial state (note that the definition of a DFA does not exclude $s_0 \in F$).

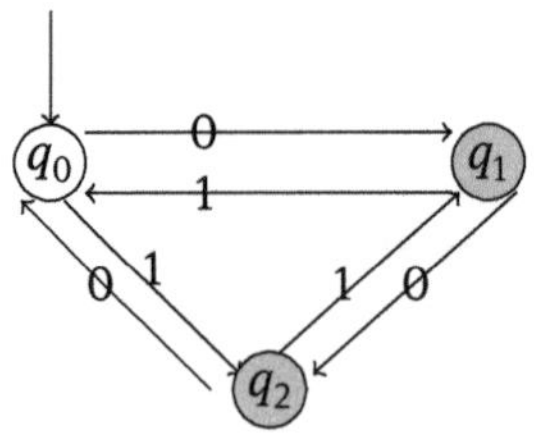

A DFA can be used to generate finite 0-1-sequences as follows: Starting in s_0, move along the vertices of $\mathcal{A}$, collecting the labels of the directed edges thereby traversed. The path may only end in an element of F. As an example, one can generate the word 01000 with the above DFA, but not the word 01001. The set of all words that one can generate with a DFA $\mathcal{A}$ is called the language accepted by $\mathcal{A}$. If a language is generated by some DFA, it is called "regular."

Regular languages play a quite important role in the theory of formal grammars and theoretical computer science. It is therefore tempting to consider a transfinite analogue. However, it is not straightforward how to deal with words of limit length. One way would be to order the set of states of the automaton and come up with some kind of a limit rule. This approach is taken in [157]. Here, we follow the idea that the crucial property of a DFA is its "ahistorical" nature or its "'forgetfulness": When one arrives in a certain state, it does not matter at all how one got there. Thus, if a word w is obtained as the concatenation of the words w_0 and w_1, then traversing the graph according to w, starting in state q, should lead to the same state as first traversing it according to w_0 and then continuing according to w_1. This is the only restriction we have on the accessibility relation of a DOA.

Definition 2.7.1. Let $\{0,1\}^{**}$ denote the proper class of ordinal-length binary sequences, i. e., functions $f : \alpha \to \{0,1\}$ with $\alpha \in \mathrm{On}$.

A Deterministic Ordinal Automaton (DOA) is a tuple (Q, s_0, F, D), where Q is an arbitrary (possibly infinite) nonempty set, $s_0 \in Q$, $F \subseteq Q$ and $D : Q \times \{0,1\}^{**} \to Q$ is a class function satisfying the following condition: If $w = w_0w_1 \in \{0,1\}^{**}$ and $q \in Q$, then $D(q, w_0w_1) = D(D(q, w_0), w_1)$.

The language $\mathcal{L}(\mathcal{A})$ accepted by a DOA $\mathcal{L}$ is the class $\{w \in \{0,1\}^{**} : D(s_0, w) \in F\}$. If $\mathcal{L}$ is the language accepted by some DOA, then we say that $\mathcal{L}$ is REG^∞.

Let us note that this is at least a formal analogue of DFAs in the sense that restricting DOAs to the finite realm leads to the notion of regularity in the usual sense.

Exercise 2.7.2. (a) Let $\mathcal{L} \subseteq \{0,1\}^{<\omega}$ be regular. Show that there is a DOA $\mathcal{A} = (Q, q_0, F, D)$ with Q finite such that $\mathcal{L} = \mathcal{L}(\mathcal{A}) \cap \{0,1\}^{<\omega}$.

(b) Conversely, let $\mathcal{A} = (Q, q_0, F, D)$ be a DOA such that Q is finite. Show that $\mathcal{L}(\mathcal{A}) \cap \{0,1\}^{<\omega}$ is regular.

Example 1. The language $\mathcal{L} := \{0^\alpha 1^\beta : \alpha, \beta \in \mathrm{On}\}$ is REG^∞.

Proof. Let $\mathcal{A} = (\{q_0, q_1\}, q_0, \{q_0, q_1\}, D)$, where $D(q_0, 0^\alpha) = q_0$ for all $\alpha \in \mathrm{On}$, $D(q_0, 0^\alpha 1^\beta) = q_1$ for all $\alpha, \beta \in \mathrm{On}$ with $\beta \neq 0$ and $D(q_1, q^\alpha) = q_1$ for all $\alpha \in \mathrm{On}$. It is easy to see that $\mathcal{L}(\mathcal{A}) = \mathcal{L}$. □

Example 2. The language $\mathcal{L} := \{0^\alpha 1^\alpha : \alpha \in \mathrm{On}\}$ is not REG^∞.

Proof. Suppose for a contradiction that $\mathcal{A} = (Q, s_0, F, D)$ is a DOA that generates $\mathcal{L}$. For $\alpha \in \mathrm{On}$, let $q^\alpha := D(s_0, 0^\alpha)$. As Q is a set, but On is a proper class, there must be $\alpha < \beta \in \mathrm{On}$ such that $q^\alpha = q^\beta$. But now, as $0^\alpha 1^\alpha \in \mathcal{L}$ but $0^\beta 1^\alpha \notin \mathcal{L}$, we have $F \ni D(s_0, 0^\alpha 1^\alpha) = D(D(s_0, 0^\alpha), 1^\alpha) = D(D(s_0, 0^\beta), 1^\alpha) = D(s_0, 0^\beta 1^\alpha) \notin F$, a contradiction. □

The next few exercises show how to adapt results from the theory of DFAs to DOAs.

Exercise 2.7.3. A DOA $\mathcal{A} = (Q, s_0, F, D)$ is "complete" if and only if $D : Q \times \{0,1\}^{**} \to Q$ is total, i. e., defined for every $(q, w) \in Q \times \{0,1\}^{**}$. Show that, for every REG^∞-language $\mathcal{L}$, there is a complete DOA $\mathcal{A}$ such that $\mathcal{L} = \mathcal{L}(\mathcal{A})$.

Exercise 2.7.4. A nondeterministic ordinal automaton (NOA) $\mathcal{A} = (Q, s_0, F, D)$ is like a DOA, but with the difference that $D : Q \times \{0,1\}^{**} \to \mathcal{P}(Q)$, i. e., a word may lead from one state to several other states. The coherence or "forgetfulness" condition is now that, for $w = w_0 w_1 \in \{0,1\}^{**}$ and $q \in Q$, we have that $D(q, w_0 w_1) = \bigcup_{q' \in D(q, w_0)} D(q', w_1)$. The language accepted by $\mathcal{A}$ consists of all $w \in \{0,1\}^{**}$ with $D(s_0, w) \cap F \neq 0$.

Prove that, if $\mathcal{L}$ is the language accepted by some NOA, then $\mathcal{L}$ is REG^∞. (Hint: The states of the desired DOA are the subsets of Q.)

Exercise 2.7.5. Show that, if $\mathcal{L}$ and $\mathcal{L}'$ are REG^∞, then so are $\mathcal{L} \cap \mathcal{L}'$ and $\mathcal{L} \cup \mathcal{L}'$.

Exercise 2.7.6. Show that the following languages are not REG^∞:

(a) $\{ww : w \in \{0,1\}^{**}\}$ (Hint: Consider words of the form $w = 0^\alpha 1$.)

(b) $\{0^{\alpha^2} : \alpha \in \mathrm{On}\}$.

(c) The class of $w \in \{0,1\}^{**}$ for which the $\mathrm{otp}(\{\iota < |w| : w(\iota) = 0\}) = \mathrm{otp}(\{\iota < |w| : w(\iota) = 1\})$.

2.7.2 Infinite time Blum–Shub–Smale machines

In [113], Koepke and Seyfferth introduced Infinite Time Blum–Shub–Smale Machines (ITBSSMs). These can be viewed as a further variant of the register machine model, but with the difference that the register contents are not ordinals, but real numbers. The finite-time versions of this model are known as Blum–Shub–Smale machines, which were introduced as a formal model for computations with real numbers. Taking limits of infinite sequences is a rather common operation in real analysis, which ITBSSMs can conveniently accommodate. The computational strength of IBSSMs was determined in [111] and [191].

The machine architecture of ITBSSMs resembles wITRMs; they have a finite number of registers $R_0, \dots, R_n$, and the programs consist of an enumerated finite list of program lines, which are of one of the forms

- $R_i = \phi(R_0, \dots, R_n)$, where ϕ is an arbitrary rational function in n variables (i. e., a quotient of polynomials in one variable with rational coefficients)
- IF $\phi(R_i) > 0$, THEN GOTO k ELSE GOTO l, where ϕ is an arbitrary rational function

At successor stages, an ITBSSM-program is carried out in the obvious way; if division by 0 occurs in the application of a rational function, the next configuration is undefined and the computation fails. At limit stages, the active program line is the inferior limit of the sequence of earlier active program lines, as for the register models discussed above. However, the content of the register R_i is now the Cauchy limit of the sequence of earlier contents of R_i, provided that this sequence converges; if it diverges, the next configuration is undefined and the computation fails.

Exercise 2.7.7 (Koepke, Seyfferth, [113]). (a) Show that there is an ITBSSM-program P such that P, starting on the empty input, halts with $\sqrt{2}$ in R_0. (Hint: Use an interval nesting.)

(b) A real number a is “algebraic” if and only if there is a nonzero polynomial p with rational coefficients such that $p(a) = 0$. By generalizing the idea from (a), show that, for every algebraic real number a, there is an ITBSSM-program P_a such that P_a, starting on the empty input, halts with a in R_0.

(c) By using $e = \sum_{l=0}^{\infty} \frac{1}{l!}$ and $\sum_{l=1}^{\infty} \frac{1}{l^2} = \frac{\pi^2}{6}$ or otherwise, show that there are such “multidimensional” ITBSSM-programs P_e and P_π that, when started on the empty input, P_e halts with e in R_0 and P_π halts with π in R_0.

An important step in determining the computational strength of ITBSSMs is showing that their halting times are $< \omega^\omega$. This was proved in [113]; the argument is quite prototypical for the analyses of computational strength in the next chapter, so we encourage the reader to try her hand at it.

Exercise 2.7.8 (Koepke, Seyfferth, [113]). Let P be an ITBSSM-program using $(n + 1)$ many registers.

(a) Show that, if P has the same configuration at times β and $\beta + 1$ for any $\beta \in \mathrm{On}$, then P will not halt.

(b) Show that an ITBSSM-program in which all program lines contain jump instructions will halt in finitely many steps or will never halt.

(c) Let α be an arbitrary ordinal. Show that, if P halts, then for some $i \in \omega$, some register content at time $\alpha + i$ in the computation of P must be different from the content at time α.

(d) Let α be a limit ordinal, and suppose that the program line $R_i = \phi(R_0, \dots, R_n)$ was carried out cofinally often before α. Let r_j be the content of R_j at time α for $j \in \{0, 1, \dots, n\}$. Show that $\phi(r_0, \dots, r_n) = r_i$. (Hint: Use the continuity of ϕ.)

(e) Let α be a limit ordinal, and suppose that the content of R_i in the computation of P is the first to change after time α. Suppose that this happens due to an application of the rational function ϕ to obtain the content of R_i. Show that ϕ was not applied to obtain the content of R_i cofinally often before time α.

(f) Let $0 < m \in \omega$, and suppose that the first register content to change after time ω^m in the computation of P is R_i, and that this is due to the application of the rational function ϕ. Show that there is $\beta < \omega^m$ such that ϕ was not applied to obtain the content of R_i between times β and ω^m. Observe that $\beta + \omega^{m-1} < \omega^m$. Thus, before time ω^m, the computation of P looked for more than ω^{m-1} many steps like a computation by a program with at least one program line less than P.

(g) Use induction to show that, if P has m program lines that do not contain jump instructions, then P will halt in less than ω^{m+1} many steps or not at all. Deduce that P will halt in $< \omega^\omega$ many steps or not at all.

Exercise 2.7.9 (Koepke, Seyfferth, [113]). (a) Show that, for each URM-program P, there is an ITBSSM-program P' such that, for all $i \in \omega$, if $P(i)$ halts with output j, then $P'(i)$ halts in finitely many steps with j in R_0, and if $P(i)$ does not halt, then the computation of $P'(i)$ is defined, but does also not halt. (Hint: To ensure that the computation of P' is defined at limit times, apply an appropriate rational function to the register contents in the computation of P.)

(b) Prove that ω is ITBSSM-clockable, i. e., that there is an ITBSSM-program P_ω that halts at time $\omega + 1$.

(c) Let $(P_i : i \in \omega)$ be a natural enumeration of the URM-programs, and let $h := \sum_{i\in\omega, P_i\downarrow} 2^{-i}$. Show that there is an ITBSSM-program P_{halt} which, when started on the empty input, halts with h in R_0.

We conclude with the following result on the computational strength of ITBSSMs; see Koepke and Morozov [111] and Welch [191].

Theorem 2.7.10 (See [111, 191]). *Say that a real number $x \subseteq \omega$ is ITBSSM-computable if and only if there is an ITBSSM-program P that, when started on the empty input, halts with x in register R_0. Then the set of ITBSSM-computable real numbers coincides with $\mathfrak{P}(\omega) \cap L_{\omega^\omega}$, the set of real numbers contained in the ω^ωth level of Gödel's constructible hierarchy.*[12]

2.7.3 Further models

We conclude with a list of some further models that for various reasons lead outside of the scope of this book. We nevertheless encourage the reader to explore these models, all of which open possibilities for further research.

12 See Chapter 3 for some basic background and references on the constructible hierarchy.

- In [63], S. Friedman and P. Welch introduced strengthened versions of ITTMs that use more complicated limit rules. The liminf-rule is easily seen to be Σ_2-definable, and it can in fact be shown that when one replaces the liminf-rule with and other Σ_2-definable rule for obtaining the machine configurations at limit stages, the resulting machine model is at most as strong as the ITTMs. In this sense, the liminf-rule can be said to be Σ_2-universal. Friedman and Welch then formulate universal Σ_n-rules for every $n \geq 2$ and study the corresponding machine models. The thus arising models are far more powerful than ITTMs, but it turns out that the theory of ITTMs that we will see in the next chapter "lifts" quite nicely to these stronger machines.
- Further important classical models of computations include λ-calculus, μ-recursive functions and Kripke's equational calculus. These were generalized to the transfinite by B. Seyfferth and T. Fischbach in [54, 55] and [53]; in each case, they turned out to be equivalent to OTMs in the sense of computability strength.
- In [133], R. Lubarsky pursued the question what happens when an ITTM is equipped with an oracle that codes the halting problem for ITTMs with that same oracle. Of course, in this form, the question is trivial: One gets the well-known contradiction showing that the halting problem is unsolvable. However, one can still try to see how far one can go without running into contradictions: Clearly, x may encode the halting behavior of programs that do not call the oracle. It may also encode the halting behavior of programs that call the oracle only to learn about the halting behavior of programs that do not call the oracle. Iterating this idea, one arrives at the notion of feedback ITTM-computability, which is not only much stronger, but also structurally quite different from ITTM-computability. For example, writability, eventual writability and accidental writability coincide for this new notion!
- In [65], L. Galeotti considered a generalization of ITBSSMs that operates on Conway's surreal numbers rather than on real numbers. As even bounded sequences of surreal numbers do in general not have limits, the limit rule needs to be replaced with a more subtle variant. The main result of [65] is that these "surreal Blum–Shub–Smale machines" are equivalent to OTMs when (appropriate representations of) the (surreal) coefficients allowed in the program for the surreal Blum–Shub–Smale machine are allowed as parameters for the OTM.

2.8 Exercises

Exercise 2.8.1 (Matzner, cf. [138], Section 4). For each $n \in \omega$, consider an "n-ITTM" that works like an ITTM, but has only one tape of length n. For each $n \in \omega$, determine the supremum of the n-ITTM-halting times, depending on n.

Exercise 2.8.2 (Cf. [32]). Define clockability for ITRMs as it was done above for ITTMs.

(a) Show that, for any $n \in \omega$, ω^n is ITRM-clockable.
(b) Show that, if α, β are ITRM-clockable, then so is $\alpha + \beta$.
(c) Deduce that every ordinal $< \omega^\omega$ is ITRM-clockable.

Exercise 2.8.3. Define clockability for OTMs as it was done above for ITTMs.

(a) Show that there are gaps in the OTM-clockable ordinals.

(b) Prove the analogue of the speedup-theorem for OTMs: If there is a (parameter-free) OTM-program P that halts in exactly $\delta + n$ many steps with δ a limit ordinal and $1 < n \in \omega$, then there is an OTM-program P' (possibly using more tapes) that halts in exactly $\delta + 1$ many steps.

Exercise 2.8.4 (Cf. [32]). (a) Formulate and prove an analogue of the speedup-theorem for wITRMs.

(b) Formulate and prove an analogue of the speedup-theorem for ITRMs.

Exercise 2.8.5 (Cf. [32]). (a) Show that, if an ITRM-computation repeats some configuration ω^ω many times, then it does not halt.[13]

(b) Suppose that β is ITRM-clockable with the program P and $\alpha < \beta$. Show that there is a P-configuration c and some $\gamma < \omega^\omega$ such that c occurs for the γth time at time β in the computation of P.

(c) Deduce that the ITRM-clockable ordinals are downwards closed, i. e., there are no gaps in the ITRM-clockable ordinals.

Exercise 2.8.6. (For the notions of admissibility and Σ_2-admissibility and their use in ordinal computability, see Chapter 3, in particular, Section 3.4.)

(a) Show that, if α is Σ_2-admissible, then α is not the halting time of any parameter-free OTM-computation.[14][15]

(b) A multitape-OTM works like an OTM, but with any finite number of tapes, each with their own read-write head. Show that there is a multitape-OTM that halts at time ω_1^{CK}.[16]

13 This is quite hard for an exercise. The reader who is lacking the patience to try it long enough is advised to take it for granted and continue with the much easier parts (b) and (c); the solution to this part can be found in [32].

14 Hint: Proceed by contradiction, assuming that P is a parameter-free OTM-program that halts at time α. Let ρ denote the head position at time α and distinguish three cases, depending on whether (i) $\rho < \alpha$ and the head position ρ has been assumed cofinally often before time α, (ii) $\rho < \alpha$ and the head position ρ has been assumed boundedly often before time α and (iii) $\rho = \alpha$. The assumption of Σ_2-admissibility is only fully used in case (ii).

15 After finishing this book, we learned from Jonas Reitz through personal communication that he and Steve Warner had basically worked this out during the Bonn International Workshop on Ordinal Computability (BIWOC) in 2007. However, the proof was never published.

16 Hint: Use (the result of) Exercise 2.8.5, combined with the results of Chapter 3, Section 3.4 on ITRMs to show that there is an ITRM-program that halts at time ω_1^{CK}. Then use a multitape-OTM to simulate this program without delay. Why does this strategy not work for ITTMs?

(c) (**Research** exercise): So far, the theory of OTM-clockability seems to have received no attention whatsoever.[17] Study the theory of OTM-clockability. A good starting point might be to read the main results on ITTM-computability in [81, 184, 31] and see which of them also work for OTMs. In particular, can parameter-free OTMs have admissible halting times?

Exercise 2.8.7. For $n \in \omega$, 0^n denotes the nth iteration of the Turing jump. Show (e. g., by induction) that, for each $n \in \omega$, there is a wITRM-program P that computes 0^n.

Exercise 2.8.8 (Cf. Winter, [195]). We consider the stability of ITTM-computability under changes of the underlying model.

(a) A "continuous ITTM" determines its head positions at limit positions not by resetting it to position 0, but as the inferior limit of the earlier head positions, if that happens to be finite; if it is infinite, the head is reset to 0. Show that continuous ITTMs can be simulated by ITTMs and vice versa.

(b) In addition to the rule for the head positions introduced in (a), a "very continuous ITTM" has its inner states numerated with natural numbers and determines its inner state at a limit time as the inferior limit of the series of earlier inner states. Thus, a "very continuous ITTM" is the special case of an α-ITTM with $\alpha = \omega$. Show that ITTMs can simulate very continuous ITTMs and vice versa.

17 The only exception seems to be the report of the above-mentioned BIWOC conference, which contains a questions concerning OTM-clockability on p. 42, namely question 9. See www.math.uni-bonn.de/ag/logik/publications/TR-2007-01.pdf

3 Computability strength

3.1 Preliminaries

In this chapter, we will determine the range of computable objects for the models of infinitary computability introduced in Chapter 2. To this end, we need some preliminaries from set theory, in particular descriptive and constructible set theory, which we recall here briefly. For the syntax of formal languages, we refer to Sacks [153]; we regard equality $=$ as a logical symbol that is by default part of all languages we will consider. The free variables $x_1, \dots, x_n$ of a formula ϕ relevant for the respective context will occasionally be pointed out by writing $\phi(x_1, \dots, x_n)$. We write $\vec{x}$ for finite tuples of variables. As usual, $\exists!\phi$ will abbreviate $\exists x\phi \wedge \forall x, y((\phi(x) \wedge \phi(y)) \rightarrow (x = y))$, i. e., "there is exactly one x such that $\phi(x)$ holds."

3.1.1 Hierarchies of sets and formulas

The language of first-order arithmetic $\mathcal{L}^1$ has two constant letters 0 and 1, one binary relation letter $<$ and two binary function letters $+$, $\cdot$, together with countably many variables $(x_i : i \in \omega)$ whose intended range is the set of natural numbers. Roughly, $\mathcal{L}^1$ is the appropriate language for areas like number theory and finite combinatorics. (This is not trivial: It requires a considerable bit of ingenuity to show that exponentiation of natural numbers is expressible in $\mathcal{L}^1$.) The formulas of $\mathcal{L}^1$ are called arithmetical formulas.

The language $\mathcal{L}^2$ of second-order arithmetic, on the other hand, extends $\mathcal{L}^1$ by another binary relation letter $\in$ and countably many extra variables $(X_i : i \in \omega)$ which range over sets of natural numbers, i. e., real numbers.

Consider an arithmetical sentence, i. e., an arithmetical formula ϕ that contains no free variables. If ϕ has no quantifiers, it is a Boolean combination of inequalities and equalities between terms that are formed by adding and multiplying natural numbers. Such a statement is easily decidable by simply calculating all occurring terms and comparing the results. On the other hand, no such procedure suggests itself for a statement of the form $\exists x_1\phi$ or $\forall x_1\phi$; and indeed, the set of true $\mathcal{L}^1$-formulas of this form is not recursive as, e. g., the halting problem is Turing-reducible to it. An important intermediate class of formulas is the class of **bounded** arithmetical formulas. In a bounded arithmetical formula, all existential quantifiers occur in the form $\exists x(x < t \wedge \phi)$ and $\forall x(x < t \rightarrow \phi)$, where t is a $\mathcal{L}^1$-term; in the following, these will be written as $\exists x < t\phi$ and $\forall x < t\phi$, respectively. A bounded statement is still decidable by trying out all of the finitely many possible values for x whenever a quantifier occurs and evaluating the quantifier-free parts as above.

https://doi.org/10.1515/9783110496154-003

Definition 3.1.1. An arithmetical formula ϕ belongs to Δ^0_0 if and only if it is bounded. Furthermore, we let $\Sigma^0_0 = \Pi^0_0 = \Delta^0_0$.

For $n \in \omega$, ϕ belongs to Σ^0_{n+1} if and only if it is of the form $\exists x\phi'$, where ϕ' is Π^0_n; ϕ belongs to Π^0_{n+1} if and only if it is of the form $\forall x\phi'$, where ϕ' is Σ^0_n.

If M is an $\mathcal{L}^1$-structure and ϕ is an arithmetical formula such that there are a Σ^0_n-formula ψ and a Π^0_n-formula ψ' such that $M \vDash \forall x(\phi(x) \leftrightarrow \psi(x) \leftrightarrow \psi'(x))$, then we say that ϕ is Δ^0_n over M. Likewise, if T is an $\mathcal{L}_1$-theory that implies $\forall x(\phi(x) \leftrightarrow \psi(x) \leftrightarrow \psi'(x))$, then ϕ is Δ^0_n over T.

The superscript 0 will frequently be dropped.

The formulas of second-order arithmetic can be stratified in a similar way:

A $\mathcal{L}^2$-formula ϕ belongs to $\Delta^1_0 = \Sigma^1_0 = \Pi^1_0$ if and only if it is arithmetical. For $n \in \omega$, ϕ is Σ^1_{n+1} if and only if it is of the form $\exists X\psi$ with $\psi \in \Pi^1_n$ and ϕ is Π^1_{n+1} if and only if it is of the form $\forall X\psi$ with $\psi \in \Sigma^1_n$.

Finally, if M is a $\mathcal{L}^2$-structure and ϕ is such that there are $\psi \in \Sigma^1_n$ and $\psi' \in \Pi^1_n$ with $M \vDash \forall\vec{x}\forall\vec{X}(\phi(\vec{x},\vec{X}) \leftrightarrow \psi(\vec{x},\vec{X}) \leftrightarrow \psi'(\vec{x},\vec{X}))$, we say that ϕ is Δ^i_n over M; similarly, when T is a $\mathcal{L}^2$-theory that implies this equivalence, we say that ϕ is Δ^1_n over T.

Remark 3.1.2. Note that, while the Σ- and Π-classes are entirely syntactical, the Δ-classes have a semantic aspect; it may, e. g., depend on the axiomatic background whether a formula is Δ^0_n, but not whether it is Σ^0_n. Nevertheless, when M or T are clear from the context, they will not always be made explicit.

Definition 3.1.3. Let $i \in \{0,1\}$, $n \in \omega$. A set $x \subseteq \omega$ is $\Sigma^i_n/\Pi^i_n/\Delta^i_n$ if and only if there are a $\Sigma^i_n/\Pi^i_n/\Delta^i_n$-formula ϕ such that $x = \{k \in \omega : \phi(k)\}$.

Similarly, for $n \in \omega$, a set $X \subseteq \mathfrak{P}(\omega)$ is $\Sigma^1_n/\Pi^1_n/\Delta^1_n$ if and only if there is a $\Sigma^1_n/\Pi^1_n/\Delta^1_n$-formula ϕ and a parameter $\vec{p}$ such that $X = \{Y \subseteq \omega : \phi(Y,\vec{p})\}$. If $\vec{p} = \emptyset$, X is parameter-free $\Sigma^1_n/\Pi^1_n/\Delta^1_n$.

Besides the languages of first- and second-order arithmetic, the language $\mathcal{L}_\in$ of set theory, which has countably many variables $(v_i : i \in \omega)$ and a single binary predicate letter $\in$ as its only nonlogical symbol, will play a crucial role in the development of infinitary computability. Again, we have a hierarchy of formulas based on quantifier alternations. In the language of set theory, $\exists x(x \in y \wedge \phi)$ and $\forall x(x \in y \to \phi)$ are abbreviated as $\exists x \in y\phi$ and $\forall x \in y\phi$, respectively. Again, a formula in which all quantifiers are of this form is called **bounded**.

Definition 3.1.4. Δ_0, Σ_0 and Π_0 denote the set of bounded $\mathcal{L}_\in$-formulas.

For $n \in \omega$, a formula ϕ belongs to Σ_{n+1} if and only if it is of the form $\exists x\phi'$, where ϕ' is Π_n; it belongs to Π_{n+1} if and only if it is of the form $\forall x\phi'$, where ϕ' is Σ_n. For M a $\mathcal{L}_\in$-structure and T a $\mathcal{L}_\in$-theory, ϕ is Δ_n over M / over T if and only if there are a Σ_n-formula ψ and a Π_n-formula ψ' such that $\phi \leftrightarrow \psi \leftrightarrow \psi'$ holds in M / is implied by T.

For a set or a class Y, $X \subseteq Y$ is Σ_n/Π_n over Y if and only if there are a Σ_n/Π_n-formula ϕ and a parameter $\vec{p} \in Y$ such that $X = \{x \in Y : Y \vDash \phi(x,\vec{p})\}$. If X is both Σ_n and Π_n over

Y, then it is Δ_n over Y. If $\vec{p} = \emptyset$, X is parameter-free $\Sigma_n/\Pi_n/\Delta_n$; we occasionally write $\boldsymbol{\Sigma}_n/\boldsymbol{\Pi}_n/\boldsymbol{\Delta}_n$ to emphasize that the use of parameters is allowed.

For an $\in$-structure M and $X \subseteq M$, we say that X is $\Sigma_n/\Pi_n/\Delta_n$ over M if and only if there are a Σ_n-formula ϕ / a Π_n-formula ψ / both a Σ_n-formula ϕ and a Π_n-formula ψ such that $X = \{x \in M : M \models \phi(x)\}$ / $X = \{x \in M : M \models \psi(x)\}$ / $X = \{x \in M : M \models \phi(x)\} = \{x \in M : M \models \psi(x)\}$.

Again, M and T will be dropped when they are clear from the context. Often, T will be ZFC, ZF or KP (see below).

For $n \in \omega$, X a class and $f : X \to X$ a (possibly class-sized, partial) function, we say that f is Σ_n over X if and only there are $\vec{p} \subseteq X$ and a Σ_n-formula $\phi(x,y,z)$ such that, for all $a, b \in X$, we have $f(a) = b$ if and only if $X \models \phi(a, b, \vec{p})$.

We will frequently use the following facts without further mentioning.

Proposition 3.1.5. *Let X be a class, f and g (possibly class-sized, partial) functions mapping X to X.*
(i) *Suppose that f and g are Σ_1 over X. Then so is $f \circ g$.*
(ii) *Suppose that f and g are Σ_2 over X. Then so is $f \circ g$.*
(iii) *Let $p, q \in X$, $\phi, \phi' \in \Sigma_2$ be such that p is the unique element a of X such that $X \models \phi'(a, q)$. Then there is a Σ_2 formula ϕ'' that can be recursively obtained from ϕ and ϕ' such that $X \models \phi''(q) \leftrightarrow \phi(p)$.*

Proof. We only prove (ii); (i) is an easier variant of the same proof. Let $\psi_f, \psi_g \in \Delta_0$, $\vec{p}, \vec{q} \subseteq X$ such that, for all $a, b \in X$, we have $f(a) = b$ if and only if $\exists x \forall y \psi_f(a, b, x, y, \vec{p})$ and $g(a) = b$ if and only if $\exists u \forall v \psi_g(a, b, u, v, \vec{q})$.

Then $(f \circ g)(a) = b$ if and only if $\exists c, x, u \forall y, v (\psi_g(a, c, u, v, \vec{q}) \wedge \psi_f(c, b, x, y, \vec{p}))$, which is Σ_2.

(iii) If ϕ is $\exists x \forall y \psi(x, y, z)$ and ϕ' is $\exists u \forall v \psi'(u, v, w, w')$, then ϕ'' is $\exists x, u, z \forall y, v$ $[\phi'(u, v, z, q) \wedge \phi(x, y, z)]$. □

Exercise 3.1.6. Show more generally that, for any $n \in \omega$, if the functions f and g are Σ_n over X, then so is $f \circ g$.

3.1.2 Basics on the constructible hierarchy

We recall some basic properties of Gödel's constructible universe L that will be of continuing importance for the treatment of infinitary computability. The reader who is not familiar with L at all is referred to Devlin [47], Schindler and Zeman [156] or the paper of Jensen [93].

Definition 3.1.7 (Gödel). For X a set, $\vec{p}$ a sequence of elements of X and ϕ an $\in$-formula, denote by $I(X, \phi, \vec{p})$ the set $\{x \in X : (X, \in) \models \phi(x, \vec{p})\}$. Moreover, by Def($X$), we denote the set $\{I(X, \phi, \vec{p}) : \phi \text{ an } \in\text{-formula}, \vec{p} \subseteq L_\alpha \text{ finite}\}$

$$\begin{aligned} L_0 &= \emptyset \\ L_{\alpha+1} &= \mathrm{Def}(L_\alpha) \\ L_\lambda &= \bigcup_{\iota<\lambda} L_\iota, \quad \text{for } \lambda \text{ a limit ordinal.} \\ L &= \bigcup_{\iota\in\mathrm{On}} L_\iota. \end{aligned}$$

If we allow a predicate P as a new primitive symbol in $\in$-formulas, and interpret P as a class C in the above definition, we obtain $L[C]$.

Remark 3.1.8. For most applications of the constructible universe, it has become customary since the seminal work of Jensen [93] to use the J-hierarchy instead of the L-hierarchy. This indeed offers various technical advantages; however, the definition is more involved and it requires quite a few technicalities to get things going. We hence decided to stick to the L-hierarchy. We refer to [156] and [47] for proofs of the theorems cited here; although there the proofs are given for the J-hierarchy, they can easily be adapted to the L-hierarchy.

What does L have to do with computability? Well, very roughly, in (transfinite) computability, we specify some basic operations, like register commands or head movements, and then consider the closure of a (possibly empty) set of initial objects under (transfinite) iterations of these basic operations. On the other hand, L is obtained by transfinite iteration of a rather basic set operation, namely bounded comprehension, which ultimately comes down to unions, intersections, complements, pairings and projections. There is thus a reason to expect connections between constructibility and transfinite computability, and indeed, this turns out to be the case.

Lemma 3.1.9 (Gödel). *L has a definable well-ordering. In fact, there is a Σ_1-formula $\phi(x,y)$ such that the class $<_L := \{x,y \in L : L \models \phi(x,y)\}$ is a well-ordering of L. Moreover, $<_L$ respects the L-hierarchy: That is, if $\alpha < \beta$, $x \in L_\alpha$ and $y \in L_\beta \setminus L_\alpha$, then $x <_L y$.*

Proof. (Sketch) One first shows by induction that every L_α has such a well-ordering. For each $x \in L$, there are $\alpha \in \mathrm{On}$, $i \in \omega$ and a finite sequence $\vec{v}$ of elements of L_α such that $x = \{y \in L_\alpha : L_\alpha \models \phi_i(y, \vec{v})\}$. Assuming that $<_L$ is defined on L_α, let $(\alpha, i, \vec{v})$ be the lexically minimal such triple. Then elements of L are ordered by the lexical ordering of their minimal names. □

We observe that it actually suffices to use ordinals as parameters in building up the levels of L. This will be helpful in encoding elements of L as tuples of ordinals later on.

Lemma 3.1.10. *For every $x \in L_{\alpha+1}$, there are $k \in \omega$, $\beta_0 < \cdots < \beta_k < \alpha \in \mathrm{On}$ and an $\in$-formula ϕ such that $x = \{y \in L_\alpha : L_\alpha \models \phi(y, \beta_0, \ldots, \beta_k)\}$.*

Proof. Assume otherwise, and let $x \in L$ be a $<_L$-minimal counterexample. As $x \in L$, there is a minimal $\alpha \in \mathrm{On}$, a finite sequence $(b_0, \dots, b_n) = \vec{b} \subseteq L_\alpha$ and an $\in$-formula ψ such that $x = \{y \in L_\alpha : L_\alpha \models \psi(y, \vec{b})\}$. By minimality of α, we have $x \in L_{\alpha+1} \setminus L_\alpha$ while all elements of $\vec{b}$ are elements of L_α, so $b <_L x$ for every $b \in \vec{b}$. By minimality of x, there are ordinals $\alpha_0, \dots, \alpha_n < \alpha$, formulas $\phi_0, \dots, \phi_n$ and finite sequences of ordinals $\vec{\beta_i} \subseteq \alpha_i$ for $i \in \{0, \dots, n\}$ such that $b_i = \{y \in L_{\alpha_i} : L_{\alpha_i} \models \phi_i(y, \vec{b_i})\}$. Let $\phi(y, \vec{b_0}, \dots, \vec{b_n})$ be the formula $\exists b'_0, \dots, b'_n((\bigwedge_{i \in \{0,\dots,n\}} b'_i = \{y \in L_{\alpha_i} : L_{\alpha_i} \models \phi_i(y, \vec{b_i})\}) \wedge \psi(y, b'_0, \dots, b'_n))$. Then $x = \{y \in L_\alpha : L_\alpha \models \phi(y, \vec{b_0}, \dots, \vec{b_n})\}$, a contradiction to the assumption that x is a counterexample to our lemma. □

Lemma 3.1.11 (See, e. g., Schindler and Zeman, [156]).
(i) *There is a Σ_1-formula $\phi_{\Delta_0\text{-truth}}(x, y)$ such that, whenever δ is a limit ordinal, ϕ_i is a Δ_0-formula and $\vec{p} \subseteq L_\delta$, then $L_\delta \models \phi_{\text{truth}}(i, \vec{p})$ if and only if $L_\delta \models \phi_i(\vec{p})$.*
(ii) *There is a Π_1-formula $\phi'_{\Delta_0\text{-truth}}$ such that, whenever δ is a limit ordinal, ϕ_i is a Δ_0-formula and $\vec{p} \subseteq L_\delta$, then $L_\delta \models \phi'_{\text{truth}}(i, \vec{p})$ if and only if $L_\delta \models \phi_i(\vec{p})$.*
Thus, the set $\{(i, \vec{p}) \in \omega \times L_\delta^{<\omega} : \phi_i \in \Delta_0 \wedge L_\delta \models \phi_i(\vec{p})\}$ is uniformly Δ_1 over L_δ for limit ordinals δ.
(iii) *There is a Σ_1-formula $\phi_{\Sigma_1\text{-truth}}(x, y)$ such that, whenever δ is a limit ordinal, ϕ_i is a Σ_1-formula and $\vec{p} \subseteq L_\delta$, then $L_\delta \models \phi_{\Sigma_1\text{-truth}}(i, \vec{p})$ if and only if $L_\delta \models \phi_i(\vec{p})$.*

Proof. See, for example, Proposition 1.12 and 1.13 of [156]; there the lemma is stated and proved for Jensen's J-hierarchy. The same proof works in our context, as we restrict ourselves to limit levels of the L-hierarchy. □

Lemma 3.1.12 (Gödel). *The relation $x = L_\alpha$ is Σ_1. Moreover, there is a formula ϕ such that, for all transitive sets X, $(X, \in) \models \phi$ if and only if $x = L_\alpha$ for some $\alpha \in \mathrm{On}$. We will write ϕ as $V = L$. In fact, ϕ is Π_2.*

Proof. For the first claim see, e. g., [156] or [47]. Very roughly, the idea is that Lemma 3.1.11 suffices to express the bounded satisfaction relation $x \models \phi_k(\vec{p})$ by a Δ_1-formula $\mathrm{Sat}(y, k, \vec{p})$. The relation between a set X and $\mathrm{Def}(X)$ is definable in terms of the bounded satisfaction relation. Now $x = L_\alpha$ is expressible as

"There is a sequence $(z_\iota : \iota \leq \alpha)$, $z_0 = \emptyset$, $z_\alpha = x$, $z_{\iota+1} = \mathrm{Def}(z_\iota)$ for $\iota < \alpha$, $z_\delta = \bigcup_{\iota<\delta} z_\iota$ for $\delta \leq \alpha$ a limit ordinal."

For the second claim, let $V = L$ be $\forall x \exists \alpha \in \mathrm{On} \exists y \exists k \exists \vec{p}((y = L_\alpha \wedge \vec{p} \subseteq y \wedge \forall z \in x\, \mathrm{Sat}(y, k, \vec{p})) \wedge \forall z \in y(z \in x \leftrightarrow \mathrm{Sat}(y, k, \vec{p})))$. As $y = L_\alpha$ is Σ_1 by the first claim, this is Π_2. □

An important tool when dealing with the constructible hierarchy is Gödel's condensation lemma (see, e. g., Section 5 of [47]).

Lemma 3.1.13 (Gödel). *For classes X and Y, let us say that $X \prec_{\Sigma_1} Y$ (X is a Σ_1-submodel of Y) if and only if, for every Σ_1-statement ϕ and all finite $\vec{p} \subseteq X$, we have $X \models \phi(\vec{p})$ if and*

only if $Y \models \phi(\vec{p})$. *If* $X \prec_{\Sigma_1} L_\alpha$, *then there is some* $\beta \leq \alpha$ *such that* $(L_\beta, \in)$ *is isomorphic to* $(X, \in)$.

The following property, known as the "acceptability" (see, e. g., Section 1 of [156]) of the constructible hierarchy, will also be frequently used.

Lemma 3.1.14. *Let* $\beta \in \mathrm{On}$ *be such that* $(L_{\alpha+1} \setminus L_\alpha) \cap \mathfrak{P}(\beta) \neq \emptyset$, *i. e.,* $L_{\alpha+1}$ *contains a 'new' subset of* β. *Then* $L_{\alpha+1}$ *contains a bijection* $f : \beta \to L_\alpha$ *and moreover* $L_{\alpha+1}$ *contains a set* $X \subseteq \beta \times \beta$ *such that* (β, X) *is isomorphic to* $(L_\alpha, \in)$.

A case of the last lemma particularly relevant for our purposes is $\beta = \omega$.

Definition 3.1.15. $\alpha \in \mathrm{On}$ is an "index" if and only if $(L_{\alpha+1} \setminus L_\alpha) \cap \mathfrak{P}(\omega) \neq \emptyset$.

Corollary 3.1.16. *If* α *is an index, then* $L_{\alpha+1}$ *contains a real number coding* L_α.

Definition 3.1.17 (Jensen, see [93]). For $\beta \in \mathrm{On}$, $h_1^{L_\beta} : \omega \times L_\beta^{<\omega} \to L_\beta$ denotes the canonical Σ_1-Skolem function for L_β. That is, when $\phi_i(y) \equiv \exists x \psi(y, x, \vec{z})$, $\psi \in \Delta_0$, is a Σ_1-formula and $\vec{p}$ is a finite sequence of elements of L_β, then $h_1^{L_\beta}(i, \vec{p})$ is the $<_L$-smallest $b \in L_\beta$ such that $L_\beta \models \exists x(b, x, \vec{p})$, if it exists; otherwise, $h_1^{L_\beta}(i, \vec{p})$ is undefined.

An important feature of the L-hierarchy is that $h_1^{L_\delta}$ is uniformly definable without parameters over L_δ when δ is a limit ordinal.

Lemma 3.1.18.
(i) *There is a* Σ_1*-formula* ψ_{sf} *("Skolem function") such that, whenever* δ *is a limit ordinal,* ϕ_i *is a* Σ_1*-formula, and* $x, \vec{p} \in L_\delta$, *then* $L_\delta \models \phi_{\mathrm{sf}}(i, \vec{p}, x)$ *if and only if* $h_1^{L_\delta}(i, \vec{p}) = x$.
(ii) *Let* $\beta < \delta$ *be limit ordinals. Then* $h_1^{L_\beta} = h_1^{L_\delta} \upharpoonright \omega \times L_\beta^{<\omega}$.

Proof. (i) See, e. g., Theorem 1.15 of [156]. (ii) is an easy consequence of (i). □

Lemma 3.1.19 (Jensen). *Let* β *be a limit ordinal. Then* L_β *has a* Σ_2*-Skolem function that is* Σ_2 *with parameters over* L_β. *That is, there are a* Σ_2*-formula* ϕ *and* $\vec{p} \subseteq L_\beta$ *such that, for all* $\phi_i \in \Sigma_2$ *and all* $x, q \in L_\beta$, *we have* $L_\beta \models \phi(i, q, x, \vec{p})$ *if and only if* $L_\beta \models \phi_i(x, q)$.

Proof. This works by an adaptation of the proof given in [93] from the J-hierarchy to the L-hierarchy, which we skip here. □

Definition 3.1.20. Let $n \in \omega$, $\alpha \in \mathrm{On}$, $X \subseteq L_\alpha$. Then the set $H \subseteq L_\alpha$ that, for each Σ_n-formula $\phi(x, y)$ and each finite sequence $\vec{p} \subseteq X$, contains the $<_L$-minimal element a of L_α such that $L_\alpha \models \phi(a, \vec{p})$ (if there is one) is called the Σ_n-hull of X in L_α, written $\Sigma_n^{L_\alpha}\{X\}$. The union $\bigcup_{n \in \omega} \Sigma_n^{L_\alpha}\{X\}$ is called the 'elementary hull' of X in L_α and denoted $\Sigma_\omega^{L_\alpha}\{X\}$.

Lemma 3.1.21. *For every* $\beta \in \mathrm{On}$, *every* $X \subseteq L_\beta$ *and every* $n \in \omega$, $\Sigma_n^{L_\beta}(X) \prec_{\Sigma_n} L_\beta$.

Proof. This is not hard to see and can, e. g., be found in [47]. □

3.1.3 Admissibility

A set is called "admissible" when a certain axiom system, called KP, holds in it. The concept of admissibility will frequently occur when characterizing computability strength of machine models of infinitary computations. Very roughly, it is a "constructive" version of ZFC.[1] Imagine a being equipped with ordinal time that constructs sets, and consider the realm R of sets that it can produce. Should one expect separation to hold in R, i. e., why is $\{y \in x : \phi(y)\}$ in R whenever x is in R and ϕ is an $\in$-formula? If yes, why so? Well, one could go through the elements of x, checking whether they satisfy ϕ, and collect together those which do. Now, if ϕ contains unbounded quantification over all sets, checking whether ϕ holds of an element of x would require going through all of R, i. e., one would need to have finished constructing all sets in R. However, when we arrive at constructing $\{y \in x : \phi(y)\}$, R is not yet finished. There is thus something "nonconstructive" or "impredicative" about ZFC, and one might try to get obtain a constructive version by only allowing bounded quantifiers in separation and collection. For similar reasons, the power set axiom needs to go, and one arrives at the axiomatic system of Kripke–Platek set theory. We state some basic facts about KP, but only give proofs when these are short and instructive in the sense that they give the reader unfamiliar with KP a sense of how it "works." For detailed treatments, we refer to Chong and Yu [40], Barwise [10] and the excellent exposition by Mansfield and Weitkamp [135]; all our material on admissibility can be found in these sources.

Definition 3.1.22. Kripke–Platek set theory, abbreviated KP, is the following theory (here cited from [40], with the addition of infinity)

- Axiom of Extensionality: For all x, y, $x = y$ if and only if $\forall z(z \in x \leftrightarrow z \in y)$
- Axiom of Induction: For all $\in$-formulas ϕ, the following holds: If $\phi(x)$ is true whenever $\phi(y)$ is true for all $y \in x$, then $\phi(x)$ holds for every set x.
- Axiom of Pairing: For all x, y, there is a set z such that, for all u, u belongs to z if and only if $u = x$ or $u = y$ (i. e., $z = \{x, y\}$).
- Axiom of Union: Every set x has a union $\bigcup x$, i. e., for all x, there is y such that $\forall z(z \in y \leftrightarrow \exists u(z \in u \wedge u \in x))$.
- Axiom of Δ_0-Separation/Comprehension: For every set x, every Δ_0-formula ϕ and every finite sequence of parameters $\vec{p}$, there is a set y such that $\forall z(z \in y \leftrightarrow (z \in x \wedge \phi(z, \vec{p})))$.
- Axiom of Δ_0-Collection: For every Δ_0-formula ϕ, every set x and every finite sequence of parameters $\vec{p}$, the following holds: If there is, for every $y \in x$, some z with $\phi(y, z, \vec{p})$, then there is a set a that collects such candidates together, i. e., such that, for every $y \in x$, there is some $z \in a$ such that $\phi(y, z, \vec{p})$.

1 For the following idea of an "infinite mind" and the criticism of unbounded comprehension based on it, see Takeuti [171] page 159; see also Chapter 8.

- Axiom of Infinity: There is a set x containing $\emptyset$ and such that, for all $z \in x$, we also have $z \cup \{z\} \in x$.

Definition 3.1.23. A transitive set X is called **admissible** if and only if $(X, \in) \models \mathrm{KP}$. An ordinal α is admissible if and only if $L_\alpha \models \mathrm{KP}$. Similarly, for $x \subseteq \mathrm{On}$, α is x-**admissible** if and only if $L_\alpha[x] \models \mathrm{KP}$.

Remark 3.1.24. Infinity is often omitted from KP. We include it, as this is the relevant version for us.

A first observation is that, though only Δ_0-collection and Δ_0-separation are explicitly demanded, we get Σ_1-collection and Δ_1-separation for free.

Lemma 3.1.25 (See, e. g., Chong and Yu, [40], Chapter 3, Theorems 3.1.6 and 3.1.7). *Let $M \models \mathrm{KP}$, $Y, \vec{p} \in M$.*

(i) *Suppose that $\phi(y, z, \vec{p}) \equiv \exists x \psi(x, y, z, \vec{p})$ (ψ a Δ_0-formula) is a Σ_1-formula such that $M \models \forall y \in Y \exists z \phi(y, z, \vec{p})$. Then there is $Z \in M$ such that $M \models \forall y \in Y \exists z \in Z \phi(y, z, \vec{p})$.*

(ii) *Suppose that $\phi(y, \vec{p})$ is Δ_1 over M and that $Y \in M$. Then there is $Z \in M$ such that $M \models \forall x (x \in Z \leftrightarrow (x \in Y \wedge \phi(x, \vec{p})))$.*

Proof. (i) By assumption, we have that, for all $y \in Y$, there are $z, x \in M$ such that $\psi(x, y, z, \vec{p})$ holds in M; as M is closed under (unordered, and thus under ordered) pairs, this implies that, for all $y \in Y$, there is a pair $(z, x) \in M$ such that $\psi(x, y, z, \vec{p})$. By Δ_0-collection in M, there is a set X such that, for all $y \in Y$, there is a pair $(z, x) \in X$ such that $\psi(x, y, z, \vec{p})$. By union and Δ_0-separation, M contains the set $Z := \{v : \exists u (u, v) \in X\}$. But now Z is clearly as desired.

(ii) As ϕ is Δ_1 over M, there are a Σ_1-formula ψ_0 and a Π_1-formula ψ_1 such that, for all $x \in M$, $M \models \phi(x, \vec{p})$ if and only if $M \models \psi_0(x, \vec{p})$ if and only if $M \models \psi_1(x, \vec{p})$. Now from the tautology that $\forall x (\psi_0(x, \vec{p}) \vee \neg \psi_0(x, \vec{p}))$, it follows that $M \models \forall x (\psi_0(x, \vec{p}) \vee \neg \psi_1(x, \vec{p}))$. Let $\psi_0(x, \vec{p}) \equiv \exists y \psi_0'(x, \vec{p}, y)$ and $\psi_1(x, \vec{p}) \equiv \forall y \psi_1'(x, \vec{p}, y)$, respectively, where ψ_0' and ψ_1' are Δ_0. Then $M \models \forall x (\exists y \psi_0'(x, \vec{p}, y) \vee \exists y \neg \psi_1'(x, \vec{p}, y))$, or, equivalently, $M \models \forall x \exists y (\psi_0'(x, \vec{p}, y) \vee \neg \psi_1'(x, \vec{p}, y))$. In particular, $M \models \forall x \in Y \exists y (\psi_0'(x, \vec{p}, y) \vee \neg \psi_1'(x, \vec{p}, y))$. As $\psi_0'(x, \vec{p}, y) \vee \neg \psi_1'(x, \vec{p}, y)$ is Δ_0, it follows from (a) that there is $Z \in M$ such that $M \models \forall x \in Y \exists y \in Z (\psi_0'(x, \vec{p}, y) \vee \neg \psi_1'(x, \vec{p}, y))$. Pick such a set $Z \in M$.

Now, for $x \in Y$, we have $\phi(x, \vec{p})$ if and only if $\exists y \psi_0'(x, \vec{p}, y)$ if and only if $\neg \exists y \neg \psi_1'(x, \vec{p}, y)$. It follows that, for $x \in Y$, we have $\phi(x, \vec{p})$ if and only if $\exists y \in Z \psi_0'(x, \vec{p}, y)$, which is Δ_0. Thus, by Δ_0-separation in M, we have $\{x \in Y : \phi(x, \vec{p})\} \in M$, as desired. □

Exercise 3.1.26 (Fitting, [56], Kap. 9). Let M, Y, $\vec{p}$ as in Lemma 3.1.25, and let ϕ be a Σ_1-formula such that, for all $y \in Y$, there is a **unique** $z \in M$ such that $M \models \phi(y, z, \vec{p})$. Show that $\{z \in M : \exists y \in M \phi(y, z, \vec{p})\} \in M$. Moreover, show that there is a function $f : Y \to M$ in M such that, for all $y \in Y$, we have $M \models \phi(y, f(y), \vec{p})$.

Admissibility comes up in many considerations in transfinite computability, often for the following reason, which we call the uniform bounding of halting times: Suppose that γ is admissible, and let P be a program for any of our machine types. Suppose that $P(\iota)$ halts in $< \gamma$ many steps for each $\iota \in \alpha$, where $\alpha < \gamma$. Now, the function sending ι to the halting time of $P(\iota)$ is Σ_1-definable over L_γ and thus, by the last lemma, cannot be cofinal in γ. Hence, there must be $\beta < \gamma$ such that all of these computations halt before time β. But then, $\{\iota : P(\iota)\downarrow = 1\}$ is definable over L_β and thus contained in $L_{\beta+1} \subseteq L_\gamma$. Admissibility is thus an important tool for analyzing the computational strength of our models.

Although it will not be used in the forthcoming development, we highly encourage the reader to consider the following occurrence of admissibility in the theory of ITTMs, which gives a good impression of how it can be applied.

Theorem 3.1.27 (Hamkins and Lewis, see [81], Theorem. 8.8). *If α is admissible, then α is not ITTM-clockable.*

Proof. Suppose for a contradiction that P is an ITTM-program that halts at time $\alpha + 1$, where α is admissible. Now, by the limit rule for ITTMs, the read/write-heads were at position 0 at time α and the inner state was the special limit state q. Thus, if the there was a transition to the halting state from time α to time $\alpha+1$, this is due to the contents of the 0th cells of the three tapes. Let $(x_0, x_1, x_2) \in \{0,1\}^3$ be the respective contents. In particular, these contents cannot have occurred before at a limit time $\tau < \alpha$, for otherwise, the transition to the halting state would have taken place from time τ to time $\tau + 1$ and the computation would have already halted at time $\tau + 1$.

Note that, as an admissible ordinal, α is a limit of limit ordinals. If $x_0 = x_1 = x_2 = 1$, then, by the liminf-rule for limit times, there is $\beta < \alpha$ such that the contents of the 0th tape cells were equal to 1 at all times $\tau \in [\beta, \alpha)$. In particular, this must then have occurred at a limit time $< \alpha$, so again, the computation would have halted before time α. Thus, at least one of the x_i is equal to 0. Moreover, there is $\beta < \alpha$ such that, for all $i \in \{0,1,2\}$, if $x_i = 1$, then the 0th cell of the ith tape contained 1 at all times in the interval $[\beta, \alpha)$. So let us assume without loss of generality that $x_0 = x_1 = x_2 = 0$, the other cases being easily handled in an analogous way (by additionally letting $\tau_0 > \beta$). We then know that, for $i \in \{0,1,2\}$, there are cofinally in α many times at which the 0th tape of the ith cell contained 0.

For $i \in \{0,1,2\}$ and $j \in \omega$, let τ_{3j+i} be the first time after $\max\{\tau_k : k < 3j + i\}$ that the 0th cell of the ith tape contained 0. Then the map $f : \omega \to \alpha$ that sends k to τ is Σ_1-definable over L_α as 'There is a P-computation and a strictly increasing sequence $(\tau_i : i \leq k)$ of limit ordinals such that $\tau_k = \tau$, and if $j \equiv r \bmod 3$, then the 0th cell of the rth tape contains 0 at time τ_j and if $j > 0$, then τ_j is the minimal ordinal with this property $> \tau_{j-1}$'.

As α is admissible, f is bounded in α, let $\sup f[\omega] = \gamma < \alpha$. Then γ is a limit ordinal and, by the liminf-rule, the 0th cells of all three tapes contain 0 at time γ. But then the computation stops at time $\gamma + 1$, a contradiction. □

Another reason for the crucial importance that admissibility has in infinitary computability is the fact that admissible sets satisfy a weak form of the set-theoretical recursion theorem (see, e. g., [47]), namely recursion for Σ_1-definable class functions (see, e. g., [40]). We will make use of a slightly more general variant, namely recursion along well-founded partial orderings instead of well-founded linear orderings (i. e., well-orderings): If $(X, \leq)$ is well-founded and the value of $F(x)$ is given by a Σ_1-formula depending on the values of $F(y)$ for all $y < x$, then there is exactly one function F satisfying all these conditions.

Definition 3.1.28. Let $(X, \leq)$ be a well-founded partial ordering. Then $\mathrm{ht}(X, \leq)$ is the supremum of those ordinals α such that there is an order-preserving embedding of α into $(X, \leq)$. When $\leq$ is clear from the context, we will also write $\mathrm{ht}(X)$ instead of $\mathrm{ht}(X, \leq)$.

Theorem 3.1.29 (See, e. g., Barwise [10], Chapter V, Section 3). *Suppose that M is a transitive model of KP. Let $(X, \leq) \in M$ be a well-founded partial ordering, and $X_y := \{x \in X : x < y\}$ for $y \in X$. Finally, let $\vec{p} \in M$ and $\phi(x, y, z, \vec{p})$ be a Σ_1-formula that is functional in M, i. e., such that $M \models \forall x \forall y \exists! z \phi(x, y, z, \vec{p})$. Then there is exactly one function $F : X \to M$ in M such that we have $M \models \phi(x, F \restriction X_x, F(x), \vec{p})$ for all $x \in X$.*

Proof. We will prove the theorem by induction on $\mathrm{ht}(X)$. By considering the minimal point of divergence between two functions F, F' satisfying the condition and working toward a contradiction, it is easy to see that there is at most one such F.

Note that, for all $x \in X$, $\mathrm{ht}(X_x) < \mathrm{ht}(X)$. Suppose for a contradiction that $(X, \leq) \in M$ is a counterexample of minimal height to our claim, and let $\vec{p} \in M$ and $\phi(x, y, z, \vec{p}) = \exists u \psi(x, y, z, \vec{p}, u)$ (with $\psi \in \Delta_0$) be a Σ_1-formula for which the claim fails.

Let $x \in X$. As $\mathrm{ht}(X_x) < \mathrm{ht}(X)$, we know by induction that there is a unique function $F_y : X_y \to M$ satisfying the conditions for all $y < x$. Note that the uniqueness of F_y implies that $F_u = F_y \restriction X_u$ whenever $u < y$.

We now aim at characterizing the relation between x and F_x. By the definition of F_x, we have that, for all $u \in M$:

$u = F_x$ if and only if $\forall x' \in X \exists y (x' < x \to \psi(x', u \restriction x', u(x'), \vec{p}, y))$.

In particular, by Δ_0-collection, there is a set $Y^x \in M$ such that

$$\forall x' \in X \exists y \in Y^x (x' < x \to \psi(x', F_x \restriction x', F_x(x'), \vec{p}, y)).$$

As F_x exists for all x by assumption, we have

$$\forall x \in X \exists u \exists Y' [\forall x' \in X \exists y \in Y' (x' < x \to \psi(x', u \restriction x', u(x'), \vec{p}, y))].$$

Thus, again by Δ_0-collection, there is $Z \in M$ such that

$$\forall x \in X \exists u, Y' \in Z [\forall x' \in X \exists y \in Y' (x' < x \to \psi(x', u \restriction x', u(x'), \vec{p}, y))].$$

In particular, Z contains F_x for all $x \in X$.

By the assumption on ϕ, we have $\forall x \in X \exists a, b \exists u, Y' \in Z[\forall x' \in X \exists y \in Y'(x' < x \rightarrow \psi(x', u \upharpoonright x', u(x'), \vec{p}, y)) \wedge \psi(x, u, a, \vec{p}, b)]$. Once again by Δ_0-collection, there is $A \in M$ such that, for all $x \in X$, there are $a, b \in A$ with this property. In particular, A contains all candidates for elements of $F[X]$.

Now consider the set $X \times A$, and form the subset F of elements (x, a) satisfying $\exists u, Y' \in Z \exists b \in A[\forall x' \in X \exists y \in Y'(x' < x \rightarrow \psi(x', u \upharpoonright x', u(x'), \vec{p}, y)) \wedge \psi(x, u, a, \vec{p}, b)]$, which exists in M by Δ_0-separation. It is now easy to verify that F is as desired. □

We note the following particularly useful special case.

Corollary 3.1.30. *Let $M \models$ KP be transitive, $\alpha \in M \cap \mathrm{On}$, $\vec{p} \in M$ and $\phi(x, y, z, \vec{p})$ a Σ_1-formula such that $M \models \forall \iota \in \alpha \forall y \exists! z \phi(\iota, y, z, \vec{p})$. Then there is exactly one function $F : \alpha \rightarrow M$ in M such that we have $M \models \phi(\iota, F \upharpoonright \iota, F(\iota), \vec{p})$ for all $\iota < \alpha$.*

Exercise 3.1.31. For a set x and $i \in \omega$, define $\bigcup^0 x = x$, $\bigcup^{i+1} x = \bigcup \bigcup^i x$, and $\mathrm{tc}(x) := \bigcup_{i \in \omega} \bigcup^i x$. Thus, $\mathrm{tc}(x)$ contains all elements of x along with all elements of elements of x, all elements of elements of x, etc. $\mathrm{tc}(x)$ is therefore called the "transitive closure" of x. Show that, if $M \models$ KP is transitive and $x \in M$, then $\mathrm{tc}(x) \in M$.

An important aspect of KP-models is that they can "decode" the codes for structures that we defined in Chapter 2.

Lemma 3.1.32. *If $M \models$ KP is transitive, $\alpha \in \mathrm{On} \cap M$ and M contains a subset $c \subseteq \alpha$ that codes a transitive $\in$-structure $(S, \in)$, then $S \in M$.*

Proof. Let $\leq_c := \{(\iota, \xi) \in \alpha \times \alpha : p(\iota, \xi) \in c\}$. Then $(\alpha, \leq_c)$ is a well-founded partial ordering in M. We now define our decoding by recursion on $(\alpha, \leq_c)$ by letting $F(\iota) := \{F(\xi) : \xi <_c \iota\}$ for all $\iota \in \alpha$. This can be done by a Σ_1-recursion using the formula ϕ given by $\phi(x, y, z) \equiv [\iota \in \alpha \wedge$ "y is a function" $\wedge \mathrm{dom}(y) = \{\xi \in \alpha : \xi <_c \iota\} \wedge x = \mathrm{rng}(y)] \vee [\iota \notin \alpha \wedge x = \emptyset]$, which is clearly Σ_1.

Thus, by Theorem 3.1.29, there is a unique function $F \in M$ satisfying the recursion, which means that, for all $\iota \in \alpha$, $F(\iota)$ is the set coded by ι in c. Now $S = \bigcup_{\iota \in \alpha} F(\iota)$, which belongs to M by the union axiom. □

Most of the time, all KP-models that we consider will be transitive, i. e., their $\in$-relation will be well-founded. However, we will occasionally need to consider ill-founded KP-models as well. We now work out some basic observations on such models.

Definition 3.1.33. Let M be a set and E a binary relation on M. The well-founded part $\mathrm{wf}(M, E)$ is the set of all $x \in M$ such that there is no infinite chain $(y_i : i \in \omega)$ of elements of M with $y_0 = x$ and $y_{i+1} \mathrm{E} y_i$ for all $i \in \omega$.

Proposition 3.1.34. *Well-founded parts are transitive. More precisely, let M be a set, E a binary relation on M, $x, y \in M$. If $xEy \in \mathrm{wf}(M, E)$, then $x \in \mathrm{wf}(M, E)$.*

Proof. Trivial. □

By renaming elements if necessary, we may assume without loss of generality that E coincides with $\in$ on the well-founded part of (M, E). This will be done tacitly in the following.

Proposition 3.1.35. *Let $(M, E) \models \mathrm{KP}$, and let $\alpha = \mathrm{wf}(M, E) \cap \mathrm{On}$. Then $\alpha \notin M$.*

Proof. Otherwise, we would have $\alpha \in \mathrm{wf}(M, E) \cap \mathrm{On} = \alpha$, a contradiction. □

3.1.4 Admissibility of constructible levels

We will frequently need to check whether levels of the (relativized) L-hierarchy are admissible. We note that most axioms of KP hold under rather mild conditions; usually, it is only collection and separation that need to be checked.

Lemma 3.1.36. *Let $z \subseteq \alpha$, and let $\beta > \alpha$ be a limit ordinal. Then $L_\beta[z]$ satisfies the axioms of extensionality, induction, pairing, union, infinity and Δ_0-separation.*

Proof. Extensionality is clear, as $L_\beta[z]$ is transitive for every $z \subseteq \alpha$, $\beta \in \mathrm{On}$. Also, as $\beta > \omega$, we have $\omega \in L_\beta[z]$, so the axiom of infinity holds in $L_\beta[z]$. Suppose that induction fails, let ϕ be such that the assumption of the induction axiom holds and (by well-foundedness of $L_\beta[z]$) let x be $\in$-minimal such that $\neg\phi(x)$. By $\in$-minimality, $\phi(y)$ holds for all $y \in x$. By the assumption, it thus holds for x, a contradiction.

If $a, b \in L_\beta[z]$, then $a, b \in L_\gamma[z]$ for some $\gamma < \beta$, as β is a limit ordinal. Now $\{a, b\} = \{x \in L_\gamma : x = b \lor x = a\}$, so $\{a, b\} \in L_{\gamma+1}[z] \subseteq L_\beta[z]$. So $L_\beta[z]$ satisfies pairing. Similarly, $\bigcup a = \{x \in L_\gamma : \exists y \in a(x \in y)\} \in L_{\gamma+1}[z] \subseteq L_\beta[z]$, so $L_\beta[z]$ also satisfies union. Δ_0-separation follows by the same idea. □

The following characterization of admissible ordinals will be frequently used. The proof can be found in the given references.

Lemma 3.1.37 (See, e. g., [10, 40, 135]). *An ordinal $\alpha > \omega$ is admissible if and only if it is a limit ordinal and there are no $\beta < \alpha$ and a total map $f : \beta \to \alpha$ which is Σ_1-definable over L_α such that the range of f is cofinal in α. In other words, if α is admissible, $\beta < \alpha$ and $f : \beta \to \alpha$ is Σ_1 over L_α and total, then there is $\gamma < \alpha$ such that $f \subseteq \beta \times \gamma$. This also holds relativized: For $x \subseteq \alpha$, α is x-admissible if and only if there is no $\beta < \alpha$ such that some total map f which is Σ_1 over $L_\alpha[x]$ maps β cofinally into α.*

Lemma 3.1.37 motivates the following strengthening of admissibility, of which Σ_2-admissibility will play an important role.

Definition 3.1.38. Let $n \geq 1$. An admissible ordinal α is Σ_n-admissible if and only if there is no total function $f : \beta \to \alpha$ with $\beta < \alpha$ and $f[\beta]$ unbounded in α that is Σ_n-definable over L_α.

Lemma 3.1.39 (See, e. g., [40], Lemma 3.5.4). *Let* $(M, E) \models \mathrm{KP}$, $x \subseteq \mathrm{On}$, $x \in M$ *and* $\alpha = \mathrm{wf}(M) \cap \mathrm{On}$. *Then* α *is* x*-admissible.*

Proof. Suppose otherwise, and let $\beta < \alpha, f : \beta \to \alpha$ be total and Σ_1 over $L_\alpha[x]$ with range cofinal in α. Let $\gamma \in \mathrm{On}^M$ be such that $M \models \gamma > \alpha$, and let $\phi(x, y, \vec{p}) \equiv \exists z \psi(x, y, \vec{p}, z)$ (with ψ a Δ_0-formula) be a Σ_1-formula that defines f, using some parameter $\vec{p} \in L_\alpha[x]$. Note that $L_\alpha[x] \subseteq \mathrm{wf}(M)$. What we would like to do now is to use ϕ in $(L_\gamma[x])^M$ to define α by just taking the union of $\mathrm{rng}(f)$, which would imply $\alpha \in (L_{\gamma+1})^M[x] \subseteq M$, contradicting Proposition 3.1.35. However, it is not clear that ϕ still defines a function in $(L_\gamma[x])^M$: For $(L_\gamma[x])^M$ contains more potential witnesses for statements of the form $\exists z \psi(x, y, \vec{p}, z)$, so that there could be some $\iota < \beta$ and some $\delta \in \mathrm{On}^M \setminus \mathrm{wf}(M, E)$ such that $(L_\gamma[x])^M \models \phi(\iota, \delta, \vec{p})$. Fortunately, there is a way out: For the witnesses contained in $L_\alpha[x]$ will always be smaller in the sense of $<_{(L_\gamma[x])^M}$ than these undesirable new witnesses in $(L_\gamma[x])^M$, and hence we can still define f over $(L_\gamma[x])^M$ by restricting ourselves to minimal witnesses.

Hence, we can define f over $(L_\gamma[x])^M$ by letting $(\iota, \zeta) \in f$ if and only if $(L_\gamma[x])^M \models$ 'There is an ordinal δ and $z \in L_\delta[x]$ such that $\psi(\iota, \zeta, \vec{p}, z)$ and there are no $z' <_{L[x]} z$ in $L_\delta[x]$ and no ordinal ζ' such that $\psi(\iota, \zeta', \vec{p}, z')$'. With this definition, it follows as just explained that $f \in M$, hence $\mathrm{rng}(f) \in M$ by Δ_0-separation, and hence $\alpha = \bigcup \mathrm{rng}(f) \in M$ by the axiom of union, which now leads to the desired conflict with Proposition 3.1.35. □

Admissibility suffices for obtaining the constructible hierarchy. We summarize the relevant aspects of this general claim in the following lemma:

Lemma 3.1.40 (See [135], Theorems 5.13, 5.14 and the remark before). *Let* M *be transitive and admissible,* $\alpha \in M \cap \mathrm{On}$, *then* $L_\alpha \in M$. *Moreover, the function* $\alpha \mapsto L_\alpha$, *restricted to* $M \cap \mathrm{On}$, *is* Σ_1 *over* M. *If* $x \in M$, *then the same holds for the function* $\alpha \mapsto L_\alpha[x]$. *If* $x \in M$ *and* $M \models x \in L$, *then* $x \in L$. *Finally, if* $M \models V = L$, *then* $M = L_\alpha$ *for some* $\alpha \in \mathrm{On}$.

Definition 3.1.41. When α is an ordinal and x is a set, then $\alpha^{+,x}$ denotes the first x-admissible ordinal $> \alpha$. Iteratively, we let $\alpha^{+(i+1),x} := (\alpha^{+i,x})^{+,x}$ for $i \in \omega$ and $\alpha^{+\omega,x} := \sup_{i \in \omega} \alpha^{+i,x}$. When $x = \emptyset$, the superscript x is dropped.

We will frequently make use of a particularly beautiful characterization of the minimal x-admissible ordinal for $x \subseteq \omega$.

Definition 3.1.42. An ordinal α is "recursive" if and only if there is a recursive real number c that codes α. Similarly, for $x \subseteq \omega$, α is x-recursive if there is such a code which is recursive in x.

Exercise 3.1.43. Let $x \subseteq \omega$. Prove that the set of x-recursive ordinals is downwards closed: That is, if $\alpha < \beta$ and β is x-recursive, then α is also x-recursive.

By definition, there are, for every $x \subseteq \omega$, at most as many x-recursive ordinals as there are programs, i. e., countably many. It is also obvious that any x-recursive ordinal is countable. It thus follows that, for every $x \subseteq \omega$, the x-recursive ordinals must have a countable supremum. Intuitively, if one only believes in recursive bijections, then this supremum is the first uncountable ordinal under this belief. This supremum plays a crucial role in the following.

Definition 3.1.44. For $x \subseteq \omega$, $\omega_1^{\mathrm{CK},x}$ (known as "ω_1-Church–Kleene") denotes the supremum of the x-recursive ordinals. Equivalently, $\omega_1^{\mathrm{CK},x}$ is the minimal ordinal which is not x-recursive.

Theorem 3.1.45 (See [135], Corollary 5.18). *For each $x \subseteq \omega$, $\omega_1^{\mathrm{CK},x}$ is the smallest x-admissible ordinal. Moreover, when x codes $\omega_i^{\mathrm{CK},y}$ for some $i \in \omega$, $y \subseteq \omega$, then $\omega_1^{\mathrm{CK},x\oplus y} \geq \omega_{i+1}^{\mathrm{CK},y}$.*

Proof. For the first part, see [135], Corollary 5.18. For the second part, note that $\omega_1^{\mathrm{CK},x\oplus y}$ is by definition y-admissible and $> \omega_i^{\mathrm{CK},y}$. □

It can in fact be shown that, for every countable admissible ordinal α, there is a x such that $\alpha = \omega_1^{\mathrm{CK},x}$ (see, e. g., Theorem 4.1 of Simpson and Weitkamp [166]; see also Durand and Lafitte [50], where a new proof of this via ITTMs is given). But we will not use this result.

Definition 3.1.46. Let $x \subseteq \omega$. We define $\omega_\iota^{\mathrm{CK},x}$ for every $0 < \iota \in \mathrm{On}$ by the following recursion: $\omega_1^{\mathrm{CK},x}$ was defined above (and shown to be the minimal x-admissible ordinal). $\omega_{\iota+1}^{\mathrm{CK},x}$ is the minimal x-admissible ordinal strictly larger than $\omega_\iota^{\mathrm{CK},x}$. If δ is a limit ordinal, then $\omega_\delta^{\mathrm{CK},x} = \sup\{\omega_\iota^{\mathrm{CK},x} : \iota < \delta\}$.

Exercise 3.1.47. Show that $\omega_\omega^{\mathrm{CK}}$ is not admissible. Thus, limits of admissible ordinals are not necessarily admissible. (Hint: Show that the function mapping $i \in \omega$ to $L_{\omega_i^{\mathrm{CK}}}$ is Σ_1 over $L_{\omega_\omega^{\mathrm{CK}}}$.)

It will occasionally be convenient to use the notion of hyperarithmeticity, which is also a central concept of higher recursion theory (see, e. g., Sacks, [153]).

Definition 3.1.48. Let $x, y \subseteq \omega$. x is hyperarithmetic in y, written $x \leq_h y$ iff x is $\Delta_1^1(y)$.

Lemma 3.1.49 (Cf. Simpson, [163] or Sacks, [153]). *Let $x, y \subseteq \omega$. Then we have:*
(1) $x \in L_{\omega_1^{\mathrm{CK},y}}[y]$ *if and only if x is $\Delta_1^1(y)$, i. e., x is hyperarithmetic in y.*
(2) $x \leq_h y$ *if and only if there is $\alpha < \omega_1^{\mathrm{CK},y}$ such that x is recursive in $y^{(\alpha)}$, the αth iterated Turing jump of y.*[2]

2 Transfinite iterations of the Turing jump work by taking the jump at successors and joining the results achieved so far together in some reasonable manner at limits; for details, we refer to [153].

For lower bounds on the computational strength of the register models, we also need a very weak version of admissibility.

Definition 3.1.50. If $x \subseteq \alpha$, an ordinal $\beta \geq \alpha$ is called "x-suitable" when β is additively indecomposable and there is no total cofinal map $f : \omega \to \beta$ which is Σ_1 over $L_\beta[x]$.

For α and ordinal and $x \subseteq \alpha$, let $\alpha^{*,x}$ denote the first x-suitable ordinal $\geq \alpha$. Moreover, let us recursively define $\alpha^{*1,x} = \alpha^{*,x}$ and $\alpha^{*(i+1),x} = (\alpha^{*i,x})^{*,x}$ for $i \in \omega$. Also, let $\alpha^{*\omega,+} = \sup_{i\in\omega} \alpha^{*i,x}$.

When $x = \emptyset$, we merely say that β is "suitable" and write α^{*i} for $\alpha^{*i,\emptyset}$.

Proposition 3.1.51. *Let $x \subseteq \alpha$, and let $\beta > \omega$ be x-admissible. Then x is x-suitable.*

Proof. Clearly, an admissible ordinal β is additively indecomposable; for if $\beta = \gamma_0 + \gamma_1$ with $\gamma_0, \gamma_1 < \beta$, then $f : \gamma_1 \mapsto \beta$, defined by $f(\iota) = \gamma_0 + \iota$, is a total cofinal function which is Σ_1 over L_β.

By Lemma 3.1.25 above, x-admissibility implies the condition that there is no cofinal map $f : \gamma \to \beta$, i. e., Σ_1 over $L_\beta[x]$ with $\gamma < \beta$. x-suitability is the specialization of this condition to $\gamma = \omega$. □

Exercise 3.1.52. (a) Show that no suitable ordinal is ITTM-clockable.

(b) Show that no suitable ordinal is ORM-clockable (without parameters).

Lemma 3.1.53. *Let $x \subseteq \omega$. Then $\omega^{*i,x} = \omega_i^{\mathrm{CK},x}$ and thus $\omega^{*\omega,x} = \omega_\omega^{\mathrm{CK},x}$.*

Proof. As admissibility is a stronger condition than suitability, it is clear that $\omega^{*i,x} \leq \omega_i^{\mathrm{CK},x}$ for all $i \in \omega$.

We now work by induction on i (for the sake of the argument, we let $\omega^{*0,x} = \omega_0^{\mathrm{CK},x} = 0$). Suppose that $\omega^{*i,x} = \omega_i^{\mathrm{CK},x}$. We want to show that $\omega^{*(i+1),x} \geq \omega_{(i+1)}^{\mathrm{CK},x}$.

Claim. *If $\alpha < \omega^{*(i+1),x}$, then α is countable in $L_{\omega^{*(i+1),x}}[x]$.*

Proof. Suppose for a contradiction that α is the minimal counterexample. Now, $L_{\omega^{*i,x}+1}[x]$ is the minimal $L[x]$-level that believes in the existence of an ith x-suitable ordinal; it follows that $L_{\omega^{*i,x}+2}[x]$ contains a bijection $f : \omega \to \omega^{*i,x}$. Thus $\alpha > \omega^{*i,x}$. By definition of $\omega^{*(i+1),x}$, there must be a cofinal function $f : \omega \to \alpha$ that is Σ_1 over L_α. Moreover, by minimality of α, $f(j)$ is countable in $L_{\omega^{*(i+1),x}}[x]$ for any $j \in \omega$; let g_j be the $<_{L[x]}$-minimal bijection between ω and $f(j)$, for any $j \in \omega$. The function mapping $j \in \omega$ to the smallest ordinal β_j such that $g_j \in L_{\beta_j}[x]$ is Σ_1 over $L_{\omega^{*(i+1),x}}[x]$, and thus by x-suitability of $\omega^{*(i+1),x}$, we have $\beta := \sup\{\beta_j : j \in \omega\} < \omega^{*(i+1),x}$. Now, by mapping $p(j,k)$ to $g_j(k)$, we obtain a surjection $F : \omega \to \alpha$, which is definable over $L_\beta[x]$. It follows that $L_{\beta+1}[x] \subseteq L_{\omega^{*(i+1),x}}[x]$ contains a bijection $G : \omega \to \alpha$, so α is countable in $L_{\omega^{*(i+1),x}}[x]$, as claimed. □

Now, let us assume for a contradiction that $\omega^{*(i+1),x} < \omega_{(i+1)}^{\mathrm{CK},x}$, i. e., that $\omega^{*(i+1),x}$ is not x-admissible. Hence, there are $\alpha < \omega^{*(i+1),x}$ and a cofinal map $f : \alpha \to \omega^{*(i+1),x}$ which is Σ_1-definable over $L_{\omega^{*(i+1),x}}[x]$. By the claim, there is a bijection $g : \omega \to \alpha$

in $L_{\omega^{*(i+1),x}}[x]$, and so $(f \circ g) : \omega \rightarrow \omega^{*(i+1),x}$ is a total cofinal function that is still Σ_1-definable over $L_{\omega^{*(i+1),x}}[x]$, contradicting the definition of $\omega^{*(i+1),x}$. Hence $\omega^{*(i+1),x}$ is indeed x-admissible and by induction $> \omega_i^{\mathrm{CK},x}$, i. e., $\omega^{*(i+1),x} \geq \omega_{i+1}^{\mathrm{CK},x}$. □

3.1.5 Computations and the constructible hierarchy

In particular, for determining the computational strength of α-TMs, it will be important to know where computations appear in the L-hierarchy.

Lemma 3.1.54 (Cf. e.g. Winter, [195]). *Let P be an OTM-program, and let C be the computation of P. Suppose that C halts in α many steps. Then $C \in L_{\alpha+\omega}$. In general, the partial computation of P in the parameter $\beta \in$ On of length α is contained in $L_{\alpha+\omega}$. The same holds for wITRMs, ITRMs, ITTMs, α-TMs and ORMs instead of OTMs.*

Proof. We begin by showing the following: Let C_γ be the initial segment of C of length γ. Then $C_\gamma \in L_{\beta+\gamma+\omega}$. We proceed by induction. It is clear that the initial configuration with the head at the start of the tape and the tape contents being 0 everywhere expect at position β is contained in $L_{\beta+\omega}$. Now, if $\gamma = \delta + 1$ and C_δ is contained in $L_{\beta+\delta+\omega}$, then there is $i \in \omega$ such that $C_\delta \in L_{\beta+\delta+i}$. The next configuration c is then easily definable over $L_{\beta+\delta+i}$, and thus contained in $L_{\beta+\delta+i+j}$ for some $j \in \omega$ (note that we might have $j > 1$, as we form ordered tuples, which only appear in later L-levels). But C_γ is obtained from C_δ by adding $\{(\gamma, c)\}$, and thus $C_\gamma \in L_{\beta+\delta+i+j+3} \subseteq L_{\beta+\delta+\omega} = L_{\beta+\gamma+\omega}$. If γ is a limit ordinal, we have $\delta + \omega \leq \gamma$ for all $\delta < \gamma$, so inductively, we have $C_\delta \subset L_{\beta+\gamma}$ for all δ such that $\beta + \delta < \gamma$. For the same reason, the ordered pairs (ι, C_ι) will also be contained in $L_{\beta+\gamma}$. The relation "x is the P-computation in parameter β of length $\iota < \gamma$" is definable over $L_{\beta+\gamma}$, and hence so is the function sending $\iota < \gamma$ to C_ι. Let $\phi_\gamma(\iota, x, \beta)$ be an $\in$-formula defining this function over $L_{\beta+\gamma}$. Using ϕ_γ and the liminf-rule, one can define the tape content of P at time γ over $L_{\beta+\gamma}$. It follows that this tape content is contained in $L_{\beta+\gamma+i}$ for some $i \in \omega$, and thus, for some $j \in \omega$ the whole configuration is contained in $L_{\beta+\gamma+i+j} \subseteq L_{\beta+\gamma+\omega}$, as desired. □

Corollary 3.1.55. *Let P be an OTM-program that halts in α many steps. Then $L_{\alpha+\omega} \models P\downarrow$. The same holds for wITRMs, ITRMs, ITTMs, α-TMs and ORMs instead of OTMs.*

Proof. This is clear from Lemma 3.1.54, as the computation of P is contained in $L_{\alpha+\omega}$. □

3.2 The computability strength of ordinal register machines and α-register machines

In this section, we determine the computational strength of ordinal register machines with and without parameters. From the methods involved, a characterization of the computational strength of α-RMs for exponentially closed α will follow smoothly. The results about ORMs and their proofs come from Koepke [107] and Koepke and Siders [114], while those about OTMs can be found in Koepke [104]. The results about α-RMs and their proofs are variants of the results about α-Turing machines obtained by Koepke and Seyfferth in [112].

We start with upper bounds.

Definition 3.2.1 (See, e. g., Barwise, [10]). An ordinal α is Σ_1-fixed if and only if there is a Σ_1-statement ϕ (without parameters) such that α is minimal with the property that $L_\alpha \models \phi$. We denote the supremum of the Σ_1-fixed ordinals by σ.

Exercise 3.2.2. Show that σ is a limit ordinal.

Proposition 3.2.3.
(i) $L_\sigma \prec_{\Sigma_1} L$.
(ii) *σ is admissible.*

Proof. (i) Clearly, if $\vec{p}$ is a finite tuple from L_σ and ϕ is a Σ_1-formula such that $L_\sigma \models \phi(\vec{p})$, then $L \models \phi(\vec{p})$ by upwards absoluteness of Σ_1-formulas. We now show the converse.

Let $H = \Sigma_1^{L_\sigma}\{\emptyset\}$. We claim that $H = L_\sigma$. To see this, first note that, by definition of σ, H will contain cofinally in σ many ordinals α that are Σ_1-fixed. By an easy fine-structural argument, such a L_α is countable in L, and thus L believes that there is a bijection $f : \omega \to L_\alpha$; this is again a Σ_1-formula which is true in L, and thus in L_σ. Thus H also contains a bijection $g : \omega \to L_\alpha$. Moreover, H clearly contains all natural numbers, and thus $L_\alpha = g[\omega] \subseteq H$ for each such α. As these α are cofinal in σ, we have $L_\sigma \subseteq H$, as desired.

It follows that each $x \in L_\sigma$ is the image of some natural number under the canonical Σ_1-Skolem function h_1, and thus has a parameter-free Σ_1-definition in L_σ. Thus, for each Σ_1-formula $\phi(\vec{p})$ with parameters $\vec{p} \in L_\sigma$, we can find a parameter-free Σ_1-formula ϕ' such that $L_\sigma \models \phi(\vec{p}) \leftrightarrow \phi'$. By upwards absoluteness of h_1, we also have $L \models \phi(\vec{p}) \leftrightarrow \phi'$. Thus, if $L \models \phi(\vec{p})$, then $L \models \phi'$. By definition of σ, there is $\alpha < \sigma$ such that $L_\alpha \models \phi'$ and thus, by upwards absoluteness of Σ_1-formulas, $L_\sigma \models \phi'$. But then, we also have $L_\sigma \models \phi(\vec{p})$, as desired.

(ii) Otherwise, let $\alpha < \sigma$ and $f : \alpha \to \sigma$ be a total function which is Σ_1 over L_σ and has cofinal range in σ; let $\phi_f(x, y, \vec{p})$ be a Σ_1-formula that defines f, where $\vec{p}$ is a finite sequence of elements of σ. It follows that L believes that there is an L-level L_β such that $\phi_f(x, y, \vec{p})$ defines a total function with domain α over L_β; the existence of such a level is a Σ_1-formula with parameters in L_σ which is true in L. Thus, by (i), it is true in L_σ, and hence, $\beta < \sigma$. Consequently, we have $f[\alpha] \subseteq \beta < \sigma$, so σ is admissible. □

Theorem 3.2.4.
(i) *Let $x \subseteq \alpha$ be ORM-computable without parameters, where $\alpha < \sigma$. Then $x \in L_\sigma$*
(ii) *Let $x \subseteq$ On be ORM-computable with parameters. Then $x \in L$.*

Proof. (i) Let P be an ORM-program that decides x. For each $\iota < \alpha$, we have that $P(\iota)\downarrow$, which is a Σ_1-statement in the parameter $\iota < \alpha < \sigma$, and thus holds in L_σ by Proposition 3.2.3. Thus $P(\iota)$ halts in $\alpha_\iota < \sigma$ many steps for each $\iota < \alpha$. The map $f : \alpha \to \sigma$ that sends ι to α_ι is total and Σ_1 over L_σ. By admissibility of σ, there is $\beta < \sigma$ such that $f[\alpha] \subseteq \beta$. Let us assume without loss of generality that $\beta > \alpha$. Then x is definable over L_β in the parameter α by the formula saying that $\iota \in x$ if and only if $\iota \in \alpha$ and $P(\iota)\downarrow = 1$. Hence $x \in L_{\beta+1} \subseteq L_\sigma$.

(ii) Let P be an ORM-program, $\xi \in \mathrm{On}$ such that, for all $\iota \in \mathrm{On}$, $P(\iota, \xi)\downarrow = \chi_x(\iota)$. Let $\alpha = \sup(x) + 1$. For $\iota < \alpha$, let α_ι be the halting time of $P(\iota, \xi)$, $\bar{\alpha} := \sup\{\alpha_\iota : \iota < \alpha\}$. Then x is definable over $L_{\bar{\alpha}}$ in the parameters ξ and α by the formula saying that $\iota \in x$ if and only if $\iota \in \alpha$ and $P(\iota, \xi)\downarrow = 1$. Thus $x \in L_{\bar{\alpha}+1}$, so $x \in L$. □

3.2.1 Lower bounds for ORMs

We now want to show that the upper bound L for ORM-computability with parameters is actually attained. The idea of the proof is that elements of L can be "named" by tuples of ordinals and that such names can be processed by ORMs. The results and proofs are due to Koepke and Siders; see, e. g., [107, 114] and [115].

By Lemma 3.1.10 and given the natural enumeration $(\phi_i : i \in \omega)$ of the $\in$-formulas we fixed above, a constructible set can uniquely be coded by an element of $\mathrm{On} \times \omega \times \mathrm{On}^{<\omega}$, where the first component is the index of the L level over which it is defined, the second is the index of the defining formula and the third is a parameter sequence.

Definition 3.2.5 (Koepke, see [104]). A **parameter-formula** (p-formula) is defined like an $\in$-formula, but besides the variables $\{v_i : i \in \omega\}$, one also has the constant symbols $\{p_i : i \in \omega\}$, called "parameter variables" or simply "parameters."

A p-formula is **nice** if it has no free variable occurring as a bound variable or a bound variable being quantified over twice.

Definition 3.2.6. A **name** is a triple (α, ϕ, f) where $\alpha \in \mathrm{On}$, f is a finite function mapping parameters of ϕ to elements of α and ϕ is a (nice) parameter-formula.

When $\alpha \leq \beta$, the triple is called a β-name; when $\alpha < \beta$, the triple is a $<\beta$-name.

If $\eta = (\alpha, \phi, f)$ is a name, then we let $\delta(\eta) = \{x \in L_\alpha : (L_\alpha, f) \models \phi(x)\}$, i. e., the set of all $x \in L_\alpha$ such that $\phi(x)$ is true in L_α under the assignment f. $\delta(\eta)$ is the "interpretation" of η; if x is a set such that $\delta(\eta) = x$, then η is called a "name for x."

A **question** is a triple (α, ϕ, f), where $\alpha \in \mathrm{On}$, ϕ is a (nice) p-formula and f is a finite function mapping each parameter of ϕ either to a $<\alpha$-name or to an ordinal $<\alpha$.

The intended interpretation of a question $q = (\alpha, \phi, f)$ is whether or not $(L_\alpha, \hat{f}) \models \phi$, where $\hat{f}(p) = \delta(f(p))$. Questions are thus about truth in initial segments of the constructible hierarchy.

We are now going to show below (in Lemma 3.2.13) that ORMs can decide such questions, i. e., there is a program P_{truth} such that, if an ordinal α, an $\in$-formula ϕ with parameters and an assignment function f are given, then $P_{\text{truth}}(\alpha, \phi, f)\downarrow = 1$ if and only if $(L_\alpha, \hat{f}) \models \phi$, and otherwise $P_{\text{truth}}(\alpha, \phi, f)\downarrow = 0$.

This will be done recursively: Thus, deciding whether a formula ϕ holds in some L_α given some parameters $p_0, \ldots, p_n$ will be reduced to deciding (possibly infinitely many instances of) the truth of statements about an earlier L-level, or in the same L-level, but for a simpler formula.

The idea in evaluating questions with an ORM is thus to recursively reduce questions to easier questions until one arrives at primitive questions that can be answered directly, such as the question whether one ordinal is an element of another. To realize this idea, we define well-founded orderings on the sets of names and questions that reflect the intuition of "simpler."

Definition 3.2.7. We define a well-founded partial ordering $\leq_p$ on the set of $\in$-formulas with parameters: In this ordering, we demand that $\phi, \psi <_p \phi \wedge \psi$, $\phi <_p \neg\phi$, $\phi(q) <_p \forall x \phi$ where q is any parameter.

We now pick and fix a recursive well-ordering $\leq_f$ on the set of these formulas that respects $\leq_p$ (which will obviously not be or order type ω). (It is easy to see that such an ordering indeed exists.)

Definition 3.2.8. Let $\eta_0 = (\alpha_0, \phi_0, f_0)$ and $\eta_1 = (\alpha_1, \phi_1, f_1)$ be names. Then $\eta_0 <_n \eta_1$ if and only if one of the following holds:
(i) $\alpha_0 < \alpha_1$.
(ii) $\alpha_0 = \alpha_1$ and $\phi_0 <_f \phi_1$.
(iii) $\alpha_0 = \alpha_1$, $\phi_0 = \phi_1$ and $\text{rng}(f_0) <_{\text{lex}} \text{rng}(f_1)$.
(iv) $\alpha_0 = \alpha_1$, $\phi_0 = \phi_1$, $\text{rng}(f_0) = \text{rng}(f_1)$ and if i is minimal with $f_0(p_i) \neq f_1(p_i)$, then $f_0(p_i) < f_1(p_i)$.

If β is an ordinal and $\eta = (\alpha, \phi, f)$ is a name, then $\beta <_n \eta$ if and only if $\beta \leq \alpha$, and otherwise $\beta >_n \eta$.

Convention. Unless stated otherwise, η_ι will below denote the ιth name in the well-ordering defined above.

Proposition 3.2.9. *If $x \in y$, then the $<_n$-smallest name of x is $<_n$ the $<_n$-smallest name of y.*

Proof. If $x \in y$, then x appears for the first time at an earlier L-level than y, so the first component of the $<_n$-smallest name of x is smaller than the first component of any name for y, which is enough. □

Finite functions mapping indices of parameters to ordinals are finite sets of pairs of ordinals; since pairs of ordinals can be encoded as ordinals via Cantor pairing, such functions can also be regarded as finite sets of ordinals, which can in turn be encoded with ordinal codes for finite sequences. When f is such a function, we write $c(f)$ for the ordinal coding f in this sense. Now, a name $\eta := (\alpha, \phi_i, f)$ can be encoded as $p_3(\alpha, i, c(f))$, which we will denote by $c(\eta)$; so elements of L are now coded by ordinals. Likewise, we will write η_ι for the name η with $c(\eta) = \iota$.[3]

Lemma 3.2.10. *There is an ORM-program P_{name} such that, for every $\beta \in \text{On}$, if $\eta_\beta = (\alpha, \phi_i, f)$, then $P_{\text{name}}(\beta)$ halts with output $p_3(\alpha, i, c(f))$.*

Proof. Suppose that initially, R_0 contains β.

The idea is now to successively count through all names with first component γ, for all ordinals $\gamma < \beta$. To this end, we reserve three registers R_1, R_2, R_3 to count through the first, second and third components of names, respectively. The content of a separate register R_{count}, initially containing 0, is incremented by 1 whenever a new name is considered; when the content of R_{count} is equal to the content of R_0, we halt and output the triple given by the current contents of R_1, R_2 and R_3.

We leave the details to the reader. □

The ordering of questions now works similarly.

Definition 3.2.11. Let $q_0 = (\alpha_0, \phi_0, f_0)$ and $q_1 = (\alpha_1, \phi_1, f_1)$ be questions. Then $q_0 <_q q_1$ if and only if the following holds:
(i) $\alpha_0 < \alpha_1$.
(ii) $\alpha_0 = \alpha_1$ and $\phi_0 <_f \phi_1$.
(iii) $\alpha_0 = \alpha_1$, $\phi_0 = \phi_1$ and $\text{rng}(f_0) <_{\text{lex}} \text{rng}(f_1)$, where $<_n$ is used to determine the lexicographical ordering on names and ordinals.
(iv) $\alpha_0 = \alpha_1$, $\phi_0 = \phi_1$, $\text{rng}(f_0) = \text{rng}(f_1)$ and if i is minimal with $f_0(p_i) \neq f_1(p_i)$, then $f_0(p_i) < f_1(p_i)$.

The names are ordered lexicographically, first by their first, then by their second component. This ordering is clearly well-founded and the reductions lead to a decrease in the sense of this ordering.

Proposition 3.2.12. *If η is an α-name, then the code $c(\eta)$ of η is $< \alpha^\omega$.*

Proof. Let $\eta = (\alpha, \phi_i, f)$. By Proposition 2.1.5 from Chapter 2, $c(f) < \alpha^\omega$. Hence, by the same proposition, $p(\alpha, i, c(f)) < \alpha^\omega$. □

The significance of this last observation is that one can search through all names with first component bounded by α by searching through all ordinals $< \alpha^\omega$ and decoding them.

3 This encoding gives us another well-ordering of the set of names, and one might wonder why we do not take this one instead of the more cumbersome $<_n$ defined above. However, we will make use of some very specific properties of $<_n$ below. We thus need to work with two ways of coding names as ordinals. We thus need to work with two ways of coding names as ordinals.

We now describe an ORM-algorithm for deciding questions, i. e., evaluating the bounded truth predicate in L. We will freely use the techniques for stack manipulation developed in Chapter 2.

Theorem 3.2.13 (See Koepke and Siders [114]). *There is an ORM-program P_{truth} such that, for every question $q = (\alpha, \phi, f)$ with code $\gamma = c(q)$, $P(\gamma)$ halts with output 1 if and only if ϕ holds in L_α under the assignment f, and otherwise $P(\alpha)$ halts with output 0.*

Proof. We will prove this by giving a high-level description of the desired algorithm, together with an inductive proof on the well-ordering of questions that it works correctly. The algorithm uses a main stack register $\mathcal{R}$, which initially contains q and then subsequently build up a finite sequence of subquestions, to be answered successively. We will describe the algorithm by showing how the content of $\mathcal{R}$ need to be changed. Here, the notation $[\alpha_0, \dots, \alpha_k] \mapsto [\beta_0, \dots, \beta_l]$ indicates that the code for the stack on the left-hand side is replaced by the code for the stack on the right-hand side in one step (i. e., with one application of the copy instruction). The necessary subcalculations for determining $[\beta_0, \dots, \beta_l]$ are done in the background with the help of numerous extra registers; we will not go into any detail there.

The general situation will be that $\mathcal{R}$ contains some top elements $(\alpha_0, \dots, \alpha_k)$ relevant for the next step, preceded by a finite sequence $(\beta_0, \dots, \beta_m)$ that do not influence and will remain unchanged during the next step. We will abbreviate this "rest" sequence $(\beta_0, \dots, \beta_m)$ by R.

We also make some conventions for simplifying the notation and enhancing the readability. First, we will use the words "true" and "false" as stack elements; one may imagine these as replaced by, e. g., 0 and 1. Occasionally, we will plug names directly into formulas and write, e. g., $\phi(\eta)$ for the instantiation of $\forall x \phi$ with $\delta(\eta)$ instead of introducing a new parameter p, writing $\phi(p)$ and amending the assignment function f with $\{(p, \eta)\}$. This should be regarded as an abbreviation. Also, we will write η_ι for the ιth element of the well-ordering of the names, where ι is an element computed earlier. In this case, we assume that a subroutine was used to obtain η_ι from ι.

Finally, evaluating q will only require reference to questions $q' < q$ which will be α-questions and thus have codes $< \alpha^\omega$. Hence, by putting α^{ω^2} to the bottom of the stack, the stack techniques from Chapter 2 will work. To simplify the discussion, we will ignore this point and, e. g., write $[\alpha_0, \dots, \alpha_n]$ for a stack that really contains $[\alpha^\omega, \alpha_0, \dots, \alpha_n]$.

Case 0: If the question is answered, the question is deleted and the answer is returned:

$[(\alpha, \phi, f), \text{false}] \mapsto [\text{false}]$

$[(\alpha, \phi, f), \text{true}] \mapsto [\text{true}]$

Case 1: The formula to be evaluated is a conjunction of two formulas. We apply the following rules:

C0 $[R, (\alpha, \phi \wedge \phi, f)] \mapsto [R, (\alpha, \phi, f)]$

From now on, we assume that $\phi \neq \psi$.

C1 $[R, (\alpha, \phi \wedge \psi, f)] \mapsto [R, (\alpha, \phi \wedge \psi, f), (\alpha, \phi, f)]$

C2 $[R, (\alpha, \phi \wedge \psi, f), (\alpha, \phi, f), \text{false}] \mapsto [R, (\alpha, \phi \wedge \psi, f), \text{false}]$

C3 $[R, (\alpha, \phi \wedge \psi, f), (\alpha, \phi, f), \text{true}] \mapsto [R, (\alpha, \phi \wedge \psi, f), (\alpha, \psi, f)]$

C4 $[R, (\alpha, \phi \wedge \psi, f), (\alpha, \psi, f), \text{false}] \mapsto [R, (\alpha, \phi \wedge \psi, f), \text{false}]$

C5 $[R, (\alpha, \phi \wedge \psi, f), (\alpha, \psi, f), \text{true}] \mapsto [R, (\alpha, \phi \wedge \psi, f), \text{true}]$

In each case, a new question is put on the stack, this is done in accordance with the reduction rules explained above. We may thus assume inductively that these questions are handled correctly by the algorithm. Thus, when (α, ϕ, f) is put on top of the stack, the algorithm will eventually write true or false on top of it, depending on whether or not ϕ is true in L_α under the assignment f; and likewise for (α, ψ, f). The rules then makes sure that "true" is put on top of $(\alpha, \phi \wedge \psi, f)$ if and only if both questions are answered with "true," and otherwise the value "false" is put on top of $(\alpha, \phi \wedge \psi, f)$.

Case 2: The formula to be evaluated is a negated formula. We apply the following rules:

N1 $[R, (\alpha, \neg\phi, f)] \mapsto [R, (\alpha, \neg\phi, f), \neg, (\alpha, \phi, f)]$

N2 $[R, \neg, (\alpha, \phi, f), \text{false}] \mapsto [R, \text{true}]$

N3 $[R, \neg, (\alpha, \phi, f), \text{true}] \mapsto [R, \text{false}]$

Again, the new question put on top of the stack (α, ϕ, f) is a reduction of the original question $(\alpha, \neg\phi, f)$ and is thus answered correctly by induction. The other rules then deal with the negation operator by switching the answer.

Case 3: The formula to be evaluated is universally quantified. We apply the following rules:

Q1 $[R, (\alpha, \forall x\phi, f)] \rightarrow [R, (\alpha, \forall x\phi, f), \alpha^\omega + 1, 0]$

Q2 $[R, (\alpha, \forall x\phi, f), \alpha^\omega + 1, \iota, (\alpha, \phi(\eta_\iota), f), \text{false}] \mapsto [R, (\alpha, \forall x\phi, f), \text{false}]$

Q3 $[R, (\alpha, \forall x\phi, f), \alpha^\omega + 1, \iota, (\alpha, \phi(\eta_\iota), f), \text{true}] \mapsto [R, (\alpha, \forall x\phi, f), \alpha^\omega + 1, \iota + 1]$

Q4 $[R, (\alpha, \forall x\phi, f), \gamma] \mapsto [R, (\alpha, \forall x\phi, f), \gamma, (\gamma)_0 < \alpha]$

Q5 $[R, (\alpha, \forall x\phi, f), \alpha^\omega + 1, \gamma, (\gamma)_0 < \alpha, \text{true}] \mapsto [R, (\alpha, \forall x\phi, f), \alpha^\omega + 1, \gamma, (\alpha, \phi(\eta_\gamma), f)]$

Q6 $[R, (\alpha, \forall x\phi, f), \alpha^\omega + 1, \gamma, (\gamma)_0 < \alpha, \text{false}] \mapsto [R, (\alpha, \forall x\phi, f), \alpha^\omega + 1, \gamma + 1]$

Q7 $[R, (\alpha, \forall x\phi, f), \alpha^\omega + 1, \alpha^\omega] \mapsto [R, (\alpha, \forall x\phi, f), \text{true}]$

In Q4, $(\gamma)_0$ is the first component when γ is interpreted as a triple, i. e., $(\gamma)_0$ is such that there are δ, δ' with $p(p((\gamma)_0, \delta), \delta') = \gamma$. Of course, the question whether $(\gamma)_0 < \alpha$ is easily decidable by an ORM; therefore, we do not dedicate a special case to questions of this form.

A universally quantified statement $\forall x\phi$ is thus evaluated in L_α by running through all ordinals below α^ω, testing whether they code a name with first component $< \alpha$ and if yes, whether ϕ holds when applied to the object named by it. If this test ever fails, the answer "false" is returned. On the other hand, if this routine runs up to α^ω

without encountering a counterexample, "true" is returned. The new questions put on top have one quantifier less, are thus smaller than $(\alpha, \forall x\phi, f)$ in the sense of our reduction relation, and hence answered correctly by our algorithm by the inductive assumption. An important observation is that, if γ is a limit ordinal and $(\alpha, \phi(\eta_\iota, f))$ is not answered with "no" for any $\iota < \gamma$, then the content of $\mathcal{R}$ after all these statements have been checked is $[R, (\alpha, \forall x\phi, f), \gamma]$. This is a consequence of Lemma 2.3.21(b) in Chapter 2, as our stack is coded by iterating the pairing function; the assumption that the first component of the pair is larger than the limit in question is satisfied due to our insertion of α^{ω^2} into the stack. Note that Proposition 3.2.12 ensures that considering the first α^ω many ordinals will indeed lead to a complete search through the α-names.

Case 4: Formulas of the form $p_i \in p_j$. Suppose the question is of the form $(\alpha, p_i \in p_j, f)$, where f is a function with domain $\{p_i, p_j\}$ the range of which is the class of $< \alpha$-names together with the set of ordinals $< \alpha$. We distinguish several subcases depending on $f(p_i)$ and $f(p_j)$ (compare with the definition of the reduction relation above). We fix some notation in advance: If $f(p_i)$ is a name, we denote it by η_0 and its components by $\eta_0 = (\alpha_0, \phi_0, f_0)$; if $f(p_i)$ is an ordinal, we denote it by α_0. Likewise, if $f(p_j)$ is a name, it is denoted by $\eta_1 = (\alpha_1, \phi_1, f_1)$ and if $f(p_j)$ is an ordinal, it is denoted by α_1. Occasionally, we abuse notation by plugging the names into the formulas directly instead of using the assignment function writing, e. g., $\eta_0 \in \eta_1$.

Subcase 1: $f(p_i)$ and $f(p_j)$ are names, $\alpha_0 < \alpha_1$. We apply the following rules:

E1.1 $[R, (\alpha, p_i \in p_j, f)] \mapsto [R, (\alpha, p_i \in p_j, f), (\alpha_1, \phi_1(\eta_0), f_1)]$
E1.2 $[R, (\alpha, p_i \in p_j, f), (\alpha_1, \phi_1(\eta_0), f_1), \text{false}] \mapsto [R, (\alpha, p_i \in p_j, f), \text{false}]$
E1.3 $[R, (\alpha, p_i \in p_j, f), (\alpha_1, \phi_1(\eta_0), f_1), \text{true}] \mapsto [R, (\alpha, p_i \in p_j, f), \text{true}]$

By definition, we have $\delta(\eta_0) \in \delta(\eta_1)$ if and only if $L_{\alpha_1} \models \phi_1(\delta(\eta_0))$ under the assignment f_1. The new question put on top of the stack is smaller than the original question in the sense of our reduction relation, and thus answered correctly by induction; this correct answer is then passed on.

Subcase 2: $f(p_i)$ and $f(p_j)$ are names, where $\alpha_0 \geq \alpha_1$. This means that $a = \delta(\eta_0)$ is defined in η_0 over a higher or the same L-level than $b = \delta(\eta_1)$ via η_1. If a is still supposed to be an element of b, a must possess a smaller name η_0'. The crucial observation is that L_{α_0} must then be able both to see this smaller name and to know that it names the same set as η_0. Thus, if $a \in b$, then L_{α_0} contains some z such that (i) $z = \{u \in L_{\alpha_0} : (L_{\alpha_0}, f_0) \models \phi_0(u)\}$ and (ii) z satisfies the defining formula of b, i. e., $(L_{\alpha_1}, f_1) \models \phi_1(z)$. This, however, is expressible as a question about L_{α_0}, which is thus a reduction in the first component compared to $(\alpha, p_i \in p_j, f)$.

We apply the following rules:

E2.1 (i) $[R, (\alpha, p_i \in p_j, f)] \mapsto [R, (\alpha, p_i \in p_j, f), (\alpha_0, \exists u(\phi_1(u) \wedge \forall z(z \in u \leftrightarrow \phi_0(z))), f_0 \cup f_1)]$, if $\alpha_0 = \alpha_1$
(ii) $[R, (\alpha, p_i \in p_j, f)] \mapsto [R, (\alpha, p_i \in p_j, f), (\alpha_0, \exists u([\phi_1(u)]^{L_{\alpha_1}} \wedge \forall z(z \in u \leftrightarrow \phi_0(z))), f_0 \cup f_1)]$ if $\alpha_0 > \alpha_1$

E2.2 $[R, (\alpha, p_i \in p_j, f), \psi, \text{true}] \mapsto [R, (\alpha, p_i \in p_j, f), \text{true}]$
E2.3 $[R, (\alpha, p_i \in p_j, f), \psi, \text{false}] \mapsto [R, (\alpha, p_i \in p_j, f), \text{false}]$

Note that, strictly speaking, the expression $[\phi_1(u)]^{L_{\alpha_1}}$ in E2.1(ii) does not belong to our language. To be precise, we should introduce another variable v via the existential quantifier, write out the formula saying that v is the α_1th L-level and then bound all quantifiers of ϕ_1 by v. The notation above should be read as an abbreviation of this.

Clearly, as $\alpha_0 < \alpha$, the new questions put on top of the stack are smaller in the sense of our reduction relation than the original question and are thus answered correctly by the inductive assumption. By our explanation above, these new questions are equivalent to the original one under the current case assumption; E2.2 and E2.3 then pass on the correct answer.

Subcase 3: Both $f(p_i)$ and $f(p_j)$ are ordinals. In this case, we can answer the question by an easy subroutine right away, without needing to reduce to other questions.

Subcase 4: $f(p_i)$ is an ordinal and $f(p_j)$ is a name, where $\alpha_0 < \alpha_1$. Then we proceed as in Subcase 1.

Subcase 5: $f(p_i)$ is an ordinal and $f(p_j)$ is a name, where $\alpha_0 \geq \alpha_1$. In this case, we return "false": If $\alpha_0 \geq \alpha_1$, it cannot be an element of a subset of L_{α_1}.

Subcase 6: $f(p_i)$ is a name and $f(p_j)$ is an ordinal, where $\alpha_0 < \alpha_1$. Then we apply the following rules:
E6.1 $[R, (\alpha, p_i \in p_j, f)] \mapsto [R, (\alpha, p_i \in p_j, f), (\alpha_1, p_i \in \text{On}, \{(p_i, \eta_0)\})]$
E6.2 $[R, (\alpha, p_i \in p_j, f), (\alpha_1, x \in \text{On}, \{(p_i, \eta_0)\}), \text{true}] \mapsto [R, (\alpha, p_i \in p_j, f), \text{true}]$
E6.3 $[R, (\alpha, p_i \in p_j, f), (\alpha_1, p_i \in \text{On}, \{(p_i, \eta_0)\}), \text{false}] \mapsto [R, (\alpha, p_i \in p_j, f), \text{false}]$

Clearly, $\delta(\eta_0)$ can only be an element of an ordinal if it is an ordinal; moreover, as $\alpha_0 < \alpha_1$, it will have to be an ordinal $< \alpha_1$ in this case, and thus be an element of α_1. As $\alpha_1 < \alpha$, the new question is smaller in the sense of our ordering than the original one.

Subcase 7: $f(p_i)$ is a name and $f(p_j)$ is an ordinal, where $\alpha_0 \geq \alpha_1$. We apply the following rules:
E7.1 $[R, (\alpha, p_i \in p_j, f)] \mapsto [R, (\alpha, p_i \in p_j, f), (\alpha_0, \exists y(y \in \alpha_1 \wedge \forall x(\phi_0(x) \leftrightarrow x \in y)), f_0)]$
E7.2 $[R, (\alpha, p_i \in p_j, f), (\alpha_0, \exists y(y \in \alpha_1 \wedge \forall x(\phi_0(x) \leftrightarrow x \in y)), f_0), \text{true}] \mapsto [R, (\alpha, p_i \in p_j, f), \text{true}]$
E7.3 $[R, (\alpha, p_i \in p_j, f), (\alpha_0, \exists y(y \in \alpha_1 \wedge \forall x(\phi_0(x) \leftrightarrow x \in y)), f_0), \text{false}] \mapsto [R, (\alpha, p_i \in p_j, f), \text{false}]$

Clearly, the new question put on top of the stack is equivalent to the original one, and also clearly, it is smaller in the sense of our ordering. This hence works correctly by induction.

Case 5: Formulas of the form $p_i = p_j$. The idea is to eliminate equality by using extensionality. We only deal with the case that $f(x)$ and $f(y)$ are both names; the cases where one or both are ordinals are easily dealt with analogously, as in case 4.

Let $\bar{\alpha} = \max\{\alpha_0, \alpha_1\}$. We apply the following rules:

$$[R, (\alpha, p_i = p_j, f)] \mapsto [R, (\alpha, p_i = p_j, f), (\bar{\alpha}, \forall x(\phi_0(x) \leftrightarrow \phi_1(x)), f_0 \cup f_1)]$$
$$[R, (\alpha, p_i = p_j, f), (\bar{\alpha}, \forall x(\phi_0(x) \leftrightarrow \phi_1(x)), f_0 \cup f_1), \text{true}] \mapsto [R, (\alpha, p_i = p_j, f), \text{true}]$$
$$[R, (\alpha, p_i = p_j, f), (\bar{\alpha}, \forall x(\phi_0(x) \leftrightarrow \phi_1(x)), f_0 \cup f_1), \text{false}] \mapsto [R, (\alpha, p_i = p_j, f), \text{false}]$$

It should be clear that this works as desired.

All cases have now been dealt with (the other junctors and the existential quantifier can be assumed to have been eliminated according to the usual rules without loss of generality). The proof is completed. □

Lemma 3.2.14 (Koepke). *There is an ORM-program P_{decide} such that, for $\beta \in \text{On}$ and a name η, $P_{\text{decide}}(\beta, \eta)\downarrow = 1$ if and only if $\beta \in \eth(\eta)$, and otherwise $P_{\text{decide}}(\beta, \eta)\downarrow = 0$.*

Proof. Given $\beta \in \text{On}$ and $\eta = (\alpha, \phi, f)$, run $P_{\text{truth}}(\max\{\alpha, \beta\} + 1, p_0 \in p_1, \{(p_0, \beta), (p_1, \eta)\})$. This will return the desired result by Lemma 3.2.13. □

We can now characterize the ORM-computable sets of ordinals.

Theorem 3.2.15 (Koepke). *There is an ORM-program P such that the following holds: For every constructible set of ordinals X, there is $\alpha \in \text{On}$ such that, for all $\beta \in \text{On}$, $P(\beta, \alpha)\downarrow = 1$ if and only if $\beta \in X$, and otherwise $P(\beta, \alpha)\downarrow = 1$. In other words, every constructible set of ordinals is ORM-decidable relative to some ordinal parameter.*

Proof. $P(\beta, \alpha)$ works by first computing the αth name η_α and then running $P_{\text{decide}}(\beta, \eta_\alpha)$, using Lemma 3.2.14. As every name appears among the η_α and every constructible set has a name, this works as desired. □

3.2.2 ORMs without parameters

Without parameters, we have the following.

Theorem 3.2.16 (Schlicht and Seyfferth, [160]).
(i) *Let $X \in L_\sigma$ be a real number. Then there is an ORM-program P such that, for all $i \in \omega$, we have $P(i)\downarrow = 1$ if and only if $i \in X$, and otherwise $P(i)\downarrow = 0$. We emphasize that P does not use any parameter.*
(ii) *Let $X \in L_\sigma$ be a subset of an ordinal $\alpha < \sigma$. Then there is an ORM-program P such that, for all $\iota \in \alpha$, we have $P(\iota)\downarrow = 1$ if and only if $\iota \in X$ and otherwise $P(\iota)\downarrow = 0$. We emphasize that P does not use any parameter.*
(iii) *Let $X \subseteq \omega$ be ORM-decidable without parameters. Then $X \in L_\sigma$.*
Thus, a real number is ORM-decidable without parameters if and only if $X \in L_\sigma$.

Proof. (i) As $X \in L_\sigma$, there is a Σ_1-fixed ordinal α such that $X \in L_\alpha$; let ϕ be a Σ_1-sentence such that α is minimal such that $L_\alpha \models \phi$. By an easy finestructural argument, we have that $L_{\alpha+1}$ contains a bijection between ω and L_α, and thus $L_{\alpha+1}$ contains the $<_L$-minimal real number c coding L_α.

It is not hard to see that X is ORM-computable without parameters from c: For in c, X is coded by some natural number k, which, like all natural numbers, is ORM-computable. Thus, when c is given, to decide whether some $i \in \omega$ belongs to X, one identifies the natural number j that codes i in the sense of c (this can, e. g., be done by evaluating truth predicates in c, as we can easily compute from i an $\in$-formula $\psi_i(x)$ such that, for any transitive Y and any $a \in Y$, we have $Y \models \psi_i(a)$ if and only if $a = i$). When this is done, one simply checks whether $p(j,k) \in c$ and returns the result.

It thus remains to see that c is ORM-computable. Suppose that $i \in \omega$ is given and we want to decide whether $i \in c$. To this end, we search through the ordinals and, for each $\beta \in \mathrm{On}$, use P_{truth} to evaluate the statement 'There is a $<_L$-minimal code c for an L-level L_α such that $L_\alpha \models \phi$'. Eventually, a β will be found such that this statement will come out true. Once this has happened, one again uses P_{truth} to evaluate the statement 'There is a $<_L$-minimal code c for a L-level L_α such that $L_\alpha \models \phi$ and $i \in c$' in L_β. If it is true, then $i \in c$, otherwise $i \notin c$.

(ii) Pick $\alpha < \sigma$ such that $X \in L_\alpha$. By (i), there is a parameter-freely ORM-computable real number c that codes L_α. In c, X is coded by some natural number j. Now, given $\iota < \alpha$, we can use the ORM-variant of the program P_{find} from Corollary 2.3.31 to identify an $i \in \omega$ that codes ι in the sense of c. Then we simply check whether $p(i,j) \in c$ and return the result.

(iii) Let P be an ORM-program that decides X without parameters. Consider the ORM-program P' that successively runs $P(\iota)$ for all $\iota \in \alpha$ and halts afterwards. Now, the sentence $P'\downarrow$ is Σ_1 in the parameter $\alpha < \sigma$ and true in L, so, by definition of σ, it holds in L_σ, and thus its halting time τ is $< \sigma$. In particular, $P(\iota)$ halts in $< \tau < \sigma$ many steps for all $\iota \in \alpha$. Thus, X is definable over L_τ by $\iota \in X$ if and only if $P(\iota)\downarrow = 1$. Hence, $X \in L_{\tau+1} \subseteq L_\sigma$. □

Remark 3.2.17. It can be shown (see Koepke and Siders [116]) that there is a universal ORM using 4 registers and that fewer registers do not suffice.

3.3 The computational strength of α-register machines

We now turn to the first bounded version of ORMs, namely α-RMs, which, as we recall, are register machines with register contents and working time bounded by the ordinal α.

In [112], Koepke and Seyfferth observed that the time it takes an OTM to evaluate an instance I the bounded truth predicate is bounded by an exponential ordinal term t in the parameters occurring in I. Below, we show the same for ORMs, along with an explicit example for t. This will then allow us to analyze the computational strength of α-RMs (and more).

Lemma 3.3.1 (Koepke and Seyfferth, see [112]). *Let (α, ϕ, f) be a question. Then $P_{\text{truth}}(\alpha, \phi, f)$ halts in $< f(\alpha, \phi) := (\omega\alpha)^{\omega^2\alpha+\omega|\phi|}$ many steps, where $|\phi|$ denotes the length of ϕ, i. e., the number of symbols in ϕ.*

In particular, if ε is closed under ordinal exponentiation, then, for all questions (α, ϕ, f) with $\alpha < \varepsilon$, $P_{\text{truth}}(\alpha, \phi, f)$ halts in $< \varepsilon$ many steps.

Proof. We show this by induction on the ordering of questions; the case $\alpha < \omega$ is trivial, as these questions can be decided by a regular Turing machine, so we only consider questions $q := (\alpha, \phi, f)$ with $\alpha \geq \omega$ from now on. For the sake of simplicity, we ignore the time taken up by stack operations and consider these as trivial; the reader may verify that our time bound still holds when they are taken into account.

We distinguish the following cases:

Case 1: ϕ is atomic.

Thus, P_{truth} will reduce q to a question $q' := (\beta, \psi, g)$ with $\beta < \alpha$. By induction, P_{truth} will run for $< (\omega\beta)^{\omega^2\beta+\omega|\psi|}$ many steps on q', which is a limit ordinal. It thus suffices to show that $(\omega\beta)^{\omega^2\beta+\omega|\psi|} < (\omega\alpha)^{\omega^2\alpha+\omega|\phi|}$ when $\beta < \alpha$. But $|\phi|$ and $|\psi|$ are natural numbers and $\beta < \alpha$ implies $\omega^2\beta + \omega^2 \leq \omega^2\alpha$, so we have $\omega^2\beta + \omega|\psi| < \omega^2\alpha + \omega|\phi|$, and hence $(\omega\beta)^{\omega^2\beta+\omega|\psi|} < (\omega\alpha)^{\omega^2\alpha+\omega|\phi|}$.

(Clearly, it can be determined even in time $< \alpha$ whether or not $\beta < \gamma$ for given $\beta, \gamma < \alpha$.)

Case 2: $\phi \equiv \phi_0 \wedge \phi_1$.

In this case, P_{truth} will first consider $q' := (\alpha, \phi_0, f)$ and then $q'' := (\alpha, \phi_1, f)$. By induction, this will take $< (\omega\alpha)^{\omega^2\alpha+\omega|\phi_0|}$ and $< (\omega\alpha)^{\omega^2\alpha+\omega|\phi_1|}$ many steps, respectively. But now $(\omega\alpha)^{\omega^2\alpha+\omega|\phi_0|} + (\omega\alpha)^{\omega^2\alpha+\omega|\phi_1|} = (\omega\alpha)^{\omega^2\alpha}((\omega\alpha)^{\omega|\phi_0|} + (\omega\alpha)^{\omega|\phi_1|}) \leq (\omega\alpha)^{\omega^2\alpha}((\omega\alpha)^{\omega|\phi_0|} \cdot (\omega\alpha)^{\omega|\phi_1|}) = (\omega\alpha)^{\omega^2\alpha+\omega(|\phi_0|+|\phi_1|)} < (\omega\alpha)^{\omega^2\alpha+\omega(|\phi_0|+|\phi_1|+1)} = (\omega\alpha)^{\omega^2\alpha+\omega|\phi|}$.

Case 3: $\phi \equiv \neg\psi$.

Then P_{truth} will run on (α, ψ, f) and one step afterwards; and we have $(\omega\alpha)^{\omega^2\alpha+\omega|\psi|} + 1 < (\omega\alpha)^{\omega^2\alpha+\omega(|\psi|+1)} = (\omega\alpha)^{\omega^2\alpha+\omega|\neg\psi|}$.

Case 4: $\phi \equiv \forall x\psi$.

In this case, P_{truth} will consider α^ω many questions (β, ψ, g) with $\beta < \alpha$. By induction, P_{truth} will run for $< (\omega\alpha)^{\omega^2\alpha+\omega|\psi|}$ many steps on each of these questions. And now we have $(\omega\alpha)^{\omega^2\alpha+\omega|\psi|}\alpha^\omega \leq (\omega\alpha)^{\omega^2\alpha+\omega(|\psi|+1)} < (\omega\alpha)^{\omega^2\alpha+\omega(|\psi|+2)} = (\omega\alpha)^{\omega^2\alpha+\omega|\phi|}$.

Now all cases have been considered and the first claim follows.

Clearly, if ε is exponentially closed and $\omega \leq \alpha < \varepsilon$, then $(\omega\alpha)^{\omega^2\alpha+\omega|\phi|} < \varepsilon$, so the second claim follows; for $\alpha < \omega$, it is clear anyway. □

Corollary 3.3.2. *If ε is exponentially closed, then an ε-code $c \subseteq \varepsilon$ for L_ε is ORM-computable in $< \varepsilon$ many steps, and thus ε-RM-computable.*

Proof. For $\iota < \varepsilon$, we imagine that ι codes the ιth minimal name $\eta(\iota)$, where a name is "minimal" if and only if there is no name with a smaller index for the same object.

By exponential closure of ε and Lemma 3.3.1, $\eta(\iota)$ can be obtained by a ε-RM from ι by counting through the names and using P_{truth}. More specifically, given ι, we use two separate registers R and R', with initial values 0. In R, we count upwards; when R contains γ, we run through γ and test, using P_{truth}, whether there is $\xi < \gamma$ such that η_ξ and η_γ code the same set. If that is the case, we increment R. Otherwise, we increment R' and then increment R. When R' contains ι, then R contains an index for $\eta(\iota)$ in the well-ordering of names.

Now let $p(\iota, \xi) \in c$ if and only if $\delta(\eta(\iota)) \in \delta(\eta(\xi))$, which can be determined using P_{truth} once again. □

Theorem 3.3.3 (Cf. Koepke and Seyfferth [112]).[4] *Let ε be exponentially closed. Then $x \subseteq \varepsilon$ is ε-RM-semidecidable if and only if it is $\mathbf{\Sigma}_1(L_\varepsilon)$. Moreover, $x \subseteq \varepsilon$ is ε-RM-decidable if and only if it is $\mathbf{\Delta}_1(L_\varepsilon)$.*

Proof. Suppose that x is Σ_1 over L_ε with parameter $\rho < \varepsilon$ via the formula $\phi \equiv \exists \psi \psi(\iota, x)$, $\psi \in \Delta_0$. Then an ε-RM-program P that semidecides x in the parameter ρ works as follows: Let $\iota < \varepsilon$ be given. Search through all $< \varepsilon$-names η, which is possible on an ε-RM as ε is exponentially closed, and thus $\alpha^\omega < \varepsilon$ for all $\alpha < \varepsilon$. For any $< \varepsilon$-name η, use P_{truth} to decide whether or not $\psi(\iota, \delta(\eta))$ holds (note that ψ is Δ_0, and thus absolute between all L-levels containing the relevant parameters.) For any η, use P_{truth} to decide whether or not $\psi(\iota, \delta(\eta))$. If the answer is "yes," then $\iota \in x$ and we return "yes," Otherwise, the search continues. This will clearly halt in $< \varepsilon$ many steps if and only if $\iota \subset x$.

If both x and $\bar{x} = \varepsilon \setminus x$ are $\mathbf{\Sigma}_1$ over L_ε (i. e., if x is $\mathbf{\Delta}_1$ over L_ε), then we can run the programs $P_x, P_{\bar{x}}$ simultaneously and halt with output "yes" when the former halts and "no" when the latter does. Thus x is indeed ε-RM-decidable.

On the other hand, assume that $X \subseteq \alpha$ is α-RM-semidecidable, say by the program P in the parameter $\gamma < \alpha$. Then, for any $\iota < \alpha$, the computation of $P(\iota, \gamma)$ will halt in $\tau < \alpha$ many steps. As $\alpha > \omega$ is additively indecomposable, we have $\tau + \omega < \alpha$ and thus the whole computation of $P(\iota, \gamma)$ is contained in L_α.

But then, for $\iota < \alpha$, the statement $\iota \in X$ is expressible as 'there is a computation of P on input ι that halts with output 1," which is a Σ_1-formula over L_α, as desired.

Moreover, when P decides X in the parameter γ, then $P(\iota, \gamma)\downarrow = 1$ is both expressible as 'there is a computation of P on input ι that halts with output 1," which is Σ_1 over L_α and as "all computations of P in the parameter γ on input ι halt with output 1," which is Π_1 over L_α. Thus X is in fact Δ_1 over L_α. □

Exercise 3.3.4. Let ε be exponentially closed, $x \subseteq \varepsilon$. Show that x is ε-RM-semidecidable without parameters if and only if x is Σ_1 over L_ε (i. e., without parameters). Also, show

4 The analogous theorem for α-TMs is due to Koepke and Seyfferth (see [112]); the argument for α-RMs we give below is the same.

that x is ε-RM-decidable without parameters if and only if it is Δ_1 over L_ε (i. e., without parameters).

3.4 α-wITRMs and α-ITRMs

We now consider the computational strength of α-wITRMs and α-ITRMs for various values of α. A particularly important case is the case $\alpha = \omega$, in which case we obtain Koepke's unresetting and resetting "Infinite Time Register Machines" (see Chapter 2). The results in this section, along with their proofs, are mostly generalizations of the results of Koepke and Miller on this case; see [109, 110] and [106]. There, it is proved that a real number x is wITRM-computable if and only if $x \in L_{\omega_1^{\mathrm{CK}}}$ and ITRM-computable if and only if $x \in L_{\omega_\omega^{\mathrm{CK}}}$. The reason is that halting times of wITRMs are bounded by ω_1^{CK} and those of ITRMs by $\omega_\omega^{\mathrm{CK}}$. A tempting conjecture would then be that analogous time bounds hold in general and that the α-wITRM-computable subsets of α are those in L_{α^+} and that the α-ITRM-computable such sets coincide with those in $L_{\alpha^{+\omega}}$. This, however, fails badly: For example, we will see below that if α is Σ_2-admissible, then α-wITRM-computability is the same as α-RM-computability; and when κ is a regular cardinal, then all α-ITRM-computable subsets of κ belong to $L_{\kappa+1}$. Thus α^+ and $\alpha^{+\omega}$ are far too big in many cases. On the other hand, we will see that the halting times of ω_1^{CK}-wITRMs are unbounded in $\omega_\omega^{\mathrm{CK}}$.[5]

There are at important differences between α-(w)ITRMs in the cases $\alpha = \omega$ and $\alpha > \omega$ that the analysis needs to take into account. For $\alpha = \omega$, when $k \in \omega$ appears in some register R at limit time δ due to the liminf-rule, then k appeared cofinally often in R before and the content of R was $\geq k$ from a certain time $\tau < \delta$ on. These two features play an important role in Koepke's and Miller's analysis of the computational strength of (w)ITRMs. This fails for $\alpha > \omega$: When we, e. g., simply count upwards in some register R on an $\omega \cdot 2$-ITRM, starting with 0, then at time ω, R will contain ω, even though all earlier contents of R were finite. "Morally," much of Koepke's and Miller's analysis can still be used, but a somewhat more general framework is required.

Proposition 3.4.1. *Let P be a register program, α an ordinal, $x \subseteq \alpha$, $\xi \in \alpha$, and suppose that P decides x in the parameter ξ such that $P(\iota, \xi)$ runs for $< \tau$ many steps for every $\iota \in \alpha$. Then $x \in L_{\tau+1}$. If $\tau > \alpha$ is admissible, then $x \in L_\tau$.*

Proof. The set x is definable over L_τ via $\iota \in x \Leftrightarrow P(\iota, \xi)\downarrow = 1$, and thus contained in $L_{\tau+1}$.

5 One reason why the arguments from the case $\alpha = \omega$ do not generalize is that, if $\alpha \neq \omega$, then the supremum of the α-recursive ordinals can be different from α^+. Ordinals for which equality holds are known as Gandy ordinals, see e.g. [2], [11], [69].

If τ is also admissible, then consider the function f mapping ι to the halting time of $P(\iota, \xi)$. Then f is a Σ_1-function from α to τ, and thus, $f[\alpha]$ is bounded in τ, say by $\bar{\tau} < \tau$. Then, as above, x is definable over $L_{\bar{\tau}}$, and thus contained in $L_{\bar{\tau}+1} \subseteq L_{\tau}$. □

We can thus get upper bounds on the computational strength of a machine by proving upper bounds on its halting times. This is the point of the following analysis.

Definition 3.4.2. Let $x \subseteq \alpha$. An ordinal τ satisfies (weak) x-α-bounding if it is a limit ordinal and the following holds for every α-(w)ITRM-computation $P^x(\iota)$, $\iota < \alpha$: When the configuration of $P^x(\iota)$ at time τ is defined and equal to $(l, r_0, \dots, r_n) \in \omega \times \alpha^{(n+1)}$, then there is $\bar{\tau}$ such that every configuration of $P(\iota)$ arising between times $\bar{\tau}$ and τ is componentwise $\geq (l, r_0, \dots, r_n)$.

Proposition 3.4.3. *The ordinal $\tau \geq \alpha$ satisfies (weak) α-bounding if and only if, for any (weak) register program P, any $\iota < \alpha$ and any register R_i used by P, if R contains $\beta < \alpha$ at time τ in the computation of $P(\iota)$, then there is some $\bar{\tau}_i < \tau$ such that R had no content $< \beta$ between times $\bar{\tau}_i$ and τ.*

Proof. Clearly, (weak) α-bounding implies the condition in question.

In the other direction, suppose that P uses the registers $R_0, \dots, R_n$ and let $\bar{\tau}_i$ be as in the statement of the proposition for $0 \leq i \leq n$. Now consider the program P' that, in the parameter ι, runs $P(\iota)$ and keeps the active line indices of P in a separate register R. By the assumption of the statement applied to $P'(\iota)$, there is an ordinal $\hat{\tau} < \tau$ such that the content of R does not drop below the content of R at time τ between times $\hat{\tau}$ and τ. Then $\bar{\tau} := \max\{\hat{\tau}, \bar{\tau}_0, \dots, \bar{\tau}_n\}$ witnesses that τ satisfies α-bounding for $P(\iota)$. □

Proposition 3.4.4. *Let $x \subseteq \alpha$. If τ satisfies (weak) x-α-bounding and β is an ordinal, then $\beta + \tau$ satisfies (weak) x-α-bounding.*

Proof. We can regard a computation of $P(\iota)$ length $\beta + \tau$ as a computation of length τ which starts in the configuration that $P(\iota)$ had at time β. □

Suppose that τ satisfies (weak) α-bounding, P is a register program, $\iota < \alpha$ and $\bar{\tau} < \tau$ witnesses that τ is (weakly) α-bounding with respect to $P(\iota)$. Then, if the configuration of $P(\iota)$ at time τ appears at some time ξ in the interval $(\bar{\tau}, \tau)$, the pair (ξ, τ) witnesses the looping of $P(\iota)$. This is the idea of the following lemma.

Lemma 3.4.5 (Koepke, [106]). *Let $x \subseteq \alpha$.*

Assume that $\tau \geq \alpha$ satisfies weak x-α-bounding and is suitable. Then every halting x-α-wITRM-computation halts in $< \tau$ many steps. Moreover, if τ also satisfies α-bounding and an α-ITRM-computation $P(\xi)$ has no overflowing register at time τ, then $P(\xi)$ does not halt.

Proof. For simplicity, we let $x = \emptyset$; the relativization is straightforward.

Let P be an α-ITRM program, $\xi < \alpha$. We show that, if $P(\xi)$ reaches time τ and no register overflows at time τ, then $P(\xi)$ is looping. As every α-wITRM-computation is also an α-ITRM-computation without overflowing registers, the first claim follows.

By weak α-bounding, there is $\bar{\tau} < \tau$ such that every configuration of $P(\xi)$ arising in the time interval $(\bar{\tau}, \tau)$ is in every component at least as large as the configuration at time $\bar{\tau}$. Let $(l, y_0, \dots, y_n)$ be the configuration at time τ.

Now, due to the liminf-rule, there must be, for each $i \in \{0, 1, \dots, n\}$, cofinally in τ many ι such that R_i contains y_i at time ι in the computation of $P(\xi)$; and similarly, the active program line must have been l unboundedly often before τ.

Now define a function $f : \omega \to \tau$ recursively by letting $f(0) = \bar{\tau}$ and, for $0 < m \in \omega$, $m = k(n+2) + r$ with $0 \leq r < (n+2)$, let $f(m)$ be the smallest ordinal $\iota > f(m-1)$ such that at time ι in the computation of $P(\xi)$, R_r contains y_r when $0 \leq r \leq n$ and such that the active program line is equal to l when $r = (n+1)$.

It is not hard to see that f is Σ_1 over $L_\tau[x]$ in the parameter $\bar{\tau}$: For we have $f(k(n+2)+r) = \beta$ if and only if there are a function g with domain $k(n+2)+r+1$ and an ordinal $\eta < \tau$ such that g satisfies the recursive condition given above in L_η up to $k(n+2)+r+1$ and $g(k(n+2)+r) = \beta$.

Let $\beta = \sup\{f(i) : i \in \omega\}$. As f is a total function from ω to τ which is Σ_1 over L_τ, it follows from the suitability of τ that $\beta < \tau$.

Now β is a joint limit of times at which R_i had the content y_i and at which the active program line was l in the computation of $P(\xi)$. As $\beta > \bar{\tau}$, none of the R_i contained an ordinal $< y_i$ between times $\bar{\tau}$ and β, and neither was the index of the active program line $< l$. It follows that, at time β, R_i contains y_i and the active program line is l, so that the configuration at time β is the same as at time τ.

Moreover, as $\beta > \bar{\tau}$, no component of any configuration arising in the computation of $P(\xi)$ between times β and τ ever dropped below its corresponding value at time τ. Thus (β, τ) witnesses the looping of $P(\xi)$, so $P(\xi)\uparrow$. □

Combining Lemma 3.4.5 with Proposition 3.4.1, we get a first upper bound the computational strength of α-wITRMs.

Corollary 3.4.6. *Let $x, y \subseteq \alpha$, and suppose that x is α-wITRM-computable from y. If $\tau \geq \alpha$ is y-suitable and satisfies weak y-α-bounding, then $x \in L_{\tau+1}[y]$. If $\tau > \alpha$ is also admissible, then $x \in L_\tau$.*

Exercise 3.4.7. Suppose that τ is suitable and that P is an ITTM-program such that, at time τ, the tapes contain only finitely many 1s in the computation of P. Show that P does not halt. Why does this argument not work for infinitely many 1s?

Below, we will combine this with various criteria for weak α-bounding. But first, we generalize the preceding idea to α-ITRMs.

Lemma 3.4.8 (Koepke, [109]). *Let $x \subseteq \alpha$, and suppose that $(\tau_i : i \in \omega)$ is an increasing sequence of x-suitable ordinals $> \alpha$ that satisfy x-α-bounding. Also, let $y < \alpha$, P an α-ITRM-program whose computation in the oracle x reaches a time of the form $\tau = \beta + \tau_n$ and such that, at time τ, the number of registers containing 0 is smaller than n. Then $P^x(y)\uparrow$.*

Proof. As in the proof of Proposition 3.4.4, we may without loss of generality assume that $\beta = 0$. We now proceed by induction.

If $n = 1$, then the condition means that no register overflows at time τ. Consequently, $P^x(y)$ does not stop by Theorem 3.4.5.

Assume now that the lemma holds for n and that $P^x(y)$ reaches time τ_{n+1} with $l < n + 1$ registers containing 0 at that time. If none of these zeros was produced by a register overflow, Theorem 3.4.5 implies that $P^x(y)$ does not halt. So assume that some register has overflown at time τ, say R_0. This implies that there is $\delta < \tau_{n+1}$ such that R_0 did not contain the value 0 between times δ and τ. Also, by α-bounding, let $\bar{\tau} < \tau_{n+1}$ be such that, between times $\bar{\tau}$ and τ_{n+1}, all arising configurations were at least as large as the configuration at time τ_{n+1} in each component. We may pick δ large enough so that $\delta \geq \bar{\tau}$.

Now consider $\tau' = \delta + \tau_n$. As τ_{n+1} is additively indecomposable, we have $\tau' < \tau$. As $\tau' > \delta$, neither R_0 nor any of the registers not containing 0 at time τ can contain 0 at time τ'. Thus the number of registers containing 0 at time τ' is at most $l - 1$; as $l < n + 1$, we have $l - 1 < n$. Hence at time $\tau' = \delta + \tau_n$, less than n registers contain the value 0. Again, we can regard the computation of $P(y)$ of length $\delta + \tau_n$ as a computation of length τ_n starting in the configuration that $P(y)$ had at time δ. Hence, by induction, $P^x(y)$ does not stop. □

Corollary 3.4.9. *Let $x \subseteq \alpha$, and suppose that $(\tau_i : i \in \omega)$ is an increasing sequence of x-suitable ordinals $> \alpha$ that satisfy α-bounding; let $\tau = \sup_{i \in \omega} \tau_i$. If P is an α-ITRM-program using n registers and $\iota < \alpha$ is such that $P(\iota)$ halts, then $P(\iota)$ halts in $< \tau_{n+1}$ many steps. In particular, all halting α-ITRM-computations halt in $< \tau$ many steps.*

Thus, if $x, y \subseteq \alpha$, and x is α-ITRM-computable from y, then $x \in L_\tau[y]$.

Proof. If P uses only n many registers, then only n registers can contain 0 at time τ_{n+1}, which implies that $P(\iota)$ is looping by Lemma 3.4.8. Thus $P(\iota)$ either halts before τ_{n+1} or not at all. The second claim follows since every program uses some finite number of registers.

For the last claim, note that $x \in L_{\tau_{n+1}+1}[y]$ when x is α-ITRM-computable by a program using n many registers; but $L_{\tau_{n+1}+1}[y] \subseteq L_\tau$. □

These bounds are rather abstract unless we can say more about (weak) α-bounding. We now give some sufficient criteria.

Lemma 3.4.10.
(i) *Every limit ordinal satisfies x-ω-bounding for every $x \subseteq \omega$.*
(ii) *If $\tau \geq \alpha$ is Σ_2^x-admissible, then τ satisfies weak x-α-bounding, for $x \subseteq \alpha$.*
(iii) *If $\tau > \alpha$ is Σ_2^x-admissible, then τ satisfies x-α-bounding, for $x \subseteq \alpha$.*
(iv) *If κ is a regular cardinal, then κ^n satisfies κ-bounding for every $n \geq 2$.*

Proof. (i) Let τ be a limit ordinal and suppose that P is a ITRM-program such that the register R_i contains $b \in \omega$ at time τ in the computation of P^x. If R_i would contain a value $< b$ cofinally often before time τ, then as there are only finitely many ordinals $< b$, one of them, say b', would occur cofinally often. But then, by the liminf-rule, R_i would contain a value $\leq b' < b$ at time τ, a contradiction. By Proposition 3.4.3, τ has x-ω-bounding.

(ii) We work with $x = \emptyset$ for simplicity. Let P be an α-wITRM-program, $\gamma < \alpha$, R_i some register used by P and let $\beta < \alpha$ be the content of R_i at time τ in the computation of $P(\gamma)$. Suppose for a contradiction that, cofinally often before time τ, R_i contains an ordinal $< \beta$. Clearly, no particular $\iota < \beta$ can occur cofinally often as the content of R_i: For otherwise, by the inferior limit rule, the content of R_i would be $\leq \iota < \beta$ at time τ. Thus, for every $\iota < \beta$, the supremum $\tau(\iota)$ of the times $< \tau$ at which R_i contains ι is strictly smaller than τ. Now the function $f : \iota \mapsto \tau(\iota)$ is definable over L_τ as follows: $f(\iota) = \xi$ if and only if it holds in L_τ that L_ξ either believes that R_i contains ι unboundedly often or $\xi = \xi' + 1$ is a successor ordinal and R_i contains ι at time ξ' in the computation of $P(\nu)$, and moreover, R_l has a content different from ι in all configurations of $P(\nu)$ after time ξ. This is expressible as a Π_1-formula over L_τ. By Σ_2-admissibility of τ, $f[\beta]$ is bounded in τ; let $\bar{\tau} = \sup f[\beta]$. Then $\bar{\tau} < \tau$ and, between times $\bar{\tau}$ and τ, R_i cannot contain any $\iota < \beta$. Thus $\bar{\tau}$ is as desired.

(iii) If $\tau > \alpha$, then we can define the function $f : \alpha \to \tau$ as in (ii) and the proof goes through.

(iv) It is easy to see that $\mathrm{cf}(\kappa^n) = \mathrm{cf}(\kappa)$ for every $n \geq 1$. Now suppose that $\mathrm{cf}(\tau) \geq \kappa$ and that the register R contains β at time τ in the computation of $P(\gamma)$, $\gamma < \kappa$. For $\iota < \beta$, let S_ι denote the set of times $\xi < \tau$ such that R contains ι at time ξ, and let $S := \bigcup_{\iota<\beta} S_\iota$.

If S_ι is unbounded in τ, then the liminf-rule implies that R has a content $\leq \iota < \beta$ at time τ, a contradiction. Thus each S_ι has a supremum $\tau_\iota < \tau$. As $\beta < \tau$ and $\mathrm{cf}(\tau) \geq \kappa$, $\{\tau_\iota : \iota < \beta\}$ is bounded in S, say by $\bar{\tau}$. But then R will have no content $< \beta$ between times $\bar{\tau}$ and τ, a contradiction. By Proposition 3.4.3, τ satisfies κ-bounding. □

Exercise 3.4.11. Show that, if $\tau > \alpha$ is Σ_2-admissible, P is an α-ITRM-program and $\iota < \alpha$, then no register of $P(\iota)$ can overflow at time τ.

We can now deduce the upper estimates in the case $\alpha = \omega$ proved by Koepke and Miller; below, we will see that these are strict.

Corollary 3.4.12 (Koepke [106, 109], Koepke and Miller [110]). *Let $x, y \subseteq \omega$.*
(i) *The halting times of wITRM-computations in the oracle y are bounded above by $\omega_1^{\mathrm{CK},y}$. If x is wITRM-computable from y, then $x \in L_{\omega_1^{\mathrm{CK},y}}[y]$.*
(ii) *The halting times of ITRM-computations in the oracle y are bounded above by $\omega_\omega^{\mathrm{CK},y}$. If x is ITRM-computable from y, then $x \in L_{\omega_\omega^{\mathrm{CK},y}}[y]$.*

Proof. (i) Since y-admissible ordinals are y-suitable, $\omega_1^{\mathrm{CK},y}$ is y-suitable. Since it is a limit ordinal, it satisfies weak x-ω-bounding by Lemma 3.4.10(i). Thus halting times in the oracle y are bounded by ω_1^{CK} and $x \in L_{\omega_1^{\mathrm{CK},y}}[y]$ by Corollary 3.4.6.

(ii) As in (i), $\omega_i^{\mathrm{CK},y}$ is y-admissible, and thus y-suitable for every $i > 0$. Moreover, every such $\omega_i^{\mathrm{CK},y}$ is a limit ordinal $> \omega$, and thus satisfies x-ω-bounding by Lemma 3.4.10(i). Thus halting times in the oracle y are bounded by $\omega_\omega^{\mathrm{CK},y}$ and $x \in L_{\omega_\omega^{\mathrm{CK},y}}[y]$ by Corollary 3.4.9. □

Perhaps surprisingly, the rather mild closure condition of Σ_2-admissibility already implies that the unbounded working time granted to α-wITRMs does not help them in any way.

Corollary 3.4.13. *If α is Σ_2-admissible, then $x \subseteq \alpha$ is α-wITRM-computable if and only if it is α-RM-computable. Thus, $x \subseteq \alpha$ is α-wITRM-computable if and only if x is $\mathbf{\Delta}_1$ over L_α.*

Proof. By Lemma 3.4.10, α satisfies weak α-bounding; clearly, α is admissible. By Lemma 3.4.5, every halting α-wITRM-computation halts in $< \alpha$ many steps and is thus in fact an α-RM-computation.

On the other hand, if x is α-RM-computable by the program P, then P clearly also computes x on an α-wITRM. □

We note that, as a consequence of Corollary 3.4.13, we get that α-wITRMs cannot in general decide truth predicates over α-coded structures.

Corollary 3.4.14. *There are no ω_1-wITRM-program P, $\rho < \omega_1$ such that, for every $\in$-sentence ϕ_i, every $\iota < \omega_1$ and every ω_1-code c for a well-founded $\in$-structure S via $f : \omega_1 \to S$, we have $P^c(i, \iota, \rho)\downarrow = 1$ if and only if $(S, \in) \models \phi_i(f(\iota))$, and otherwise $P^c(i, \iota, \rho)\downarrow = 0$.*

Proof. Suppose for a contradiction that there was such a program P and such a parameter ρ. By Theorem 3.3.3, some ω_1-code c for L_{ω_1} is ω_1-wITRM-computable. Moreover, the halting set

$$H := \{(i, \iota) \in \omega \times \omega_1 : \text{The } i\text{th } \omega_1\text{-RM program halts in the input } \iota\}$$

for ω_1-RMs is clearly Σ_1-definable over L_{ω_1}, say by the formula $\phi_j(x, y)$.

It follows that H is ω_1-wITRM-decidable: To determine whether $(i, \iota) \in H$, just use $P(\rho)$ in the oracle c to test whether $L_{\omega_1} \models \phi_j(i, \iota)$. By Theorem 3.4.13, H is also ω_1-RM-decidable, a contradiction. □

In particular, Corollary 3.4.13 shows that, in contrast to the case $\alpha = \omega$, α-wITRMs can in general not compute up to L_{α^+}. For α-ITRMs, we have the following, which shows that they can in general not reach up to $L_{\alpha^{+\omega}}$, and in fact not even to L_{α^+}.

Corollary 3.4.15. *Let $\kappa > \omega$ be a regular cardinal, P a κ-ITRM-program, $\xi < \kappa$. If P uses n registers, then $P(\xi)$ halts in $< \kappa^{(n+1)}$ many steps or not at all. Thus, the halting times of κ-ITRM-computations are bounded above by κ^{ω} and all κ-ITRM-computable subsets of κ are contained in $L_{\kappa^{\omega}}$.*

Proof. By Lemma 3.4.10(iv), κ^n satisfies κ-bounding for all $n \in \omega$. Clearly, as $\mathrm{cf}(\kappa^n) = \mathrm{cf}(\kappa) = \kappa > \omega$, κ^n is suitable for all $n \in \omega$. By Corollary 3.4.9, the halting time of a program P using n registers is bounded by $\kappa^{(n+1)}$. The second claim follows since $\kappa^{\omega} = \sup\{\kappa^i : i \in \omega\}$ and the last claim follows again from Corollary 3.4.9. □

Exercise 3.4.16. Show that every $x \in L_{\omega_1+1} \cap \mathfrak{P}(\omega_1)$ is ω_1-ITRM-computable.[6]

With a bit of extra effort, we can now completely determine the computational strength of κ-ITRMs for uncountable L-regular cardinals κ.

Lemma 3.4.17. *Let $\kappa > \omega$ be a regular cardinal in L, and let $x \subseteq L_{\kappa}$ be constructible such that $L \models \mathrm{Card}(x) < \kappa$. Then $x \in L_{\kappa}$.*

Proof. Let $\alpha := \sup\{\iota < \kappa : (L_{\iota+1} \setminus L_{\iota}) \cap x \neq \emptyset\}$. By regularity of κ, we have $\alpha < \kappa$. Note that $x \subseteq L_{\alpha}$. Now let $\beta > \alpha$ be large enough so that $x \in L_{\beta}$ and consider $H := \Sigma_1^{L_{\beta}}\{L_{\alpha} \cup \{x\}\}$. By Lemma 3.1.13, H is isomorphic to some L-level $L_{\bar{\alpha}}$ via the collapsing map π. As $\mathrm{card}^L(L_{\alpha}) = \mathrm{card}(\alpha) < \kappa$, we have $\mathrm{card}^L(H) < \kappa$, and thus $\bar{\alpha} < \kappa$. As $x \subseteq L_{\alpha} \subseteq H$, we have $\pi(y) = y$ for all $y \in x$, so $\pi(x) = x$. Hence $x \in L_{\bar{\alpha}} \subseteq L_{\kappa}$, as desired. □

The next theorem was obtained in joint work with Philipp Schlicht.

Theorem 3.4.18. *Let $\kappa > \omega$ be a regular cardinal, $x \subseteq \kappa$. Then x is κ-ITRM-computable if and only if $x \in L_{\kappa+1}$.*

Proof. That all $x \in \mathfrak{P}(\kappa) \cap L_{\kappa+1}$ are κ-ITRM-decidable follows from the facts that a code for L_{κ} is κ-ITRM-computable and Theorem 2.3.28(iii) from Chapter 2; see Exercise 3.4.16 above.

We now turn to the other direction, so suppose that $x \subseteq \kappa$ is computable by the κ-ITRM-program P, where P uses $(n+1)$ registers $R_0, \ldots, R_n$ and $s \in \omega$ many program lines. We want to show that $x \in L_{\kappa+1}$. By Corollary 3.4.15, $P(\xi)$ halts in $< \kappa^{(n+2)}$ many steps for all inputs $\xi < \kappa$.

The central ingredient of our argument is the following claim:

6 If you want to do the exercise, do it before reading the proof of Theorem 3.4.18, which is a generalization of the statement here.

Claim. *For every $k \in \omega$, let F^P_k be the function that sends an $(n+2)$-tuple $c := (j, \alpha_0, \dots, \alpha_n) \in s \times \kappa^{(n+1)}$ coding a P-configuration c to the pair (c', c'') of tuples $c' := (l, \beta_0, \dots, \beta_n)$ and $c'' := (m, \mu_0, \dots, \mu_n)$ such that c' codes the P-configuration at time κ^k when the computation is started in the configuration c while m is the smallest program line index appearing in this computation and μ_i is the minimal content of R_i during this computation for $i \in \{0, \dots, n\}$. (Thus in particular, the first component of $F^P_0(c)$ is the successor configuration of c according to P.) Then F^P_k is definable over L_κ.*

Proof. We proceed by induction on k. Clearly, F^P_0 is definable over L_κ.

Now suppose that F^P_k is definable over L_κ. We are given a configuration $c := (j, \alpha_0, \dots, \alpha_n)$ and want to determine the configuration after $\kappa^{(k+1)}$ more steps of P, along with the minimal program line index and register contents occurring along the way.

We can imagine the tuples in $s \times \kappa^{(n+1)}$ as the vertices of a labeled directed graph G with the vertices c and c' joined by an edge labeled c'' if and only if $F^P_k(c) = (c', c'')$; so G codes what happens in κ^k many steps of P. (Note the similarity to our treatment of ITRMs with one register in Chapter 2.) Our goal is to "draw" the $\kappa^{(k+1)}$-edges in a way definable over L_κ and to determine their labels along the way.

Given c, the P-computation starting with c of length $\kappa^{(k+1)}$ is a sequence $(c_\iota : \iota < \kappa^{(k+1)})$ of P-configurations with $c_0 = c$. Let us define a sequence (basically a "thinned out" version consisting of every κth configuration in that computation) $S^P_{c,(k+1)} := ((d_\iota, d'_\iota) : \iota < \kappa)$; here, we let $d_\iota = c_{\kappa^k \iota}$, $d'_{\iota+1} := (l^{\iota+1}, \mu_0^{\iota+1}, \dots, \mu_n^{\iota+1})$, where $l^{\iota+1}$ is the minimal program line index and $\mu_i^{\iota+1}$ the minimal content of R_i occurring between times $\kappa^k \iota$ and $\kappa^k(\iota+1)$ for $\iota < \kappa$ and finally $d'_\delta = d_\delta$ when $\delta < \kappa$ is a limit ordinal.

By Lemma 3.4.17, every proper initial segment of $S^P_{c,(k+1)}$ is contained in L_κ. Moreover, the function G mapping $\gamma < \kappa$ to $S^P_{c,(k+1)} \restriction \gamma$ is uniformly in c, k and P definable over L_κ; namely, we have $G(\gamma) = ((a_\iota, b_\iota) : \iota < \gamma)$ if and only if:

- $a_\iota, b_\iota \in s \times \kappa^{(n+1)}$ for every $\iota < \gamma$.
- $a_0 = b_0 = c$.
- If $\iota + 1 < \gamma$, then $(a_{\iota+1}, b_{\iota+1}) = F^P_k(a_\iota)$.
- If $\delta < \gamma$ is a limit ordinal, then $a_\delta = b_\delta$ is the componentwise inferior limit of the sequence $(b_\iota : \iota < \delta)$.

As F^P_k is definable over L_κ by assumption, so is G; and clearly, the definition is uniform in c, k and P. For $\gamma < \kappa$, let $H(\gamma) := (G(\gamma+1))(\gamma)$; thus $H(\gamma) = S^P_{c,(k+1)}(\gamma)$ for all $\gamma < \kappa$. Clearly, H is definable over L_κ using G.

Finally, given the definability of H over L_κ, the definability of $F^P_{(k+1)}$ works as follows. We have $F^P_{(k+1)}(c) = (c', c'')$ if and only if c' is the componentwise inferior limit of the second components of the sequence $(H_\iota : \iota < \kappa)$ and c'' is the componentwise limit of the first components of the sequence $(H_\iota : \iota < \kappa)$, which is clearly definable over L_κ using the definability of H. □

Now, given the claim, x is definable over $L_{\kappa+1}$ by saying that, for $\xi < \kappa$, we have $\xi \in x$ if and only if, with $F^P_{(n+2)}(0, \xi, 0, \ldots, 0) = (c', c'')$, the second component of c' (which is the content of R_0 at time $\kappa^{(n+2)}$, which in turn is above the halting time of $P(\xi)$ as we remarked above) is equal to 1. By definability of $F^P_{(n+2)}$ over L_κ, x is also definable over L_κ, and thus contained in $L_{\kappa+1}$. □

Exercise 3.4.19. (i) Let α be a limit ordinal. Show that, for every $n \in \omega$, α^n is α-ITRM-clockable.

(ii) Show that, when κ is a regular uncountable cardinal, then every κ-ITRM-clockable ordinal is also κ-ITRM-computable. Thus, 'clockables are writable' also holds for κ-ITRMs.

Exercise 3.4.20. By adapting the proof of Theorem 3.4.18, show that, if $\alpha > \omega$ is regular in L_{α^ω} (i. e., there is no function $f : \beta \to \alpha$ in L_{α^ω} such that $\beta < \alpha$ and $f[\beta]$ is unbounded in α), then the α-ITRM-computable subsets of α are exactly those in $L_{\alpha+1}$.[7]

Remark 3.4.21. By Theorem 3.4.18 and the exercise above, κ-ITRMs only compute subsets of κ in $L_{\kappa+1}$, but have halting times up to κ^ω when κ is an uncountable regular cardinal. Thus, the rule of thumb that holds, e. g., for wITRMs, ITRMs, α-ITTMs, OTMs and ORMs that computable objects are those contained in L_α when α is the supremum of the halting times fails badly in this case.

Lemma 3.4.22. *Let α be a limit of exponentially closed ordinals, $\beta < \alpha$, and let $f : \beta \to \alpha$ be a total function that is Σ_1 over L_α in the parameter $\vec{p} \subseteq \alpha$. Then f is α-wITRM-computable in the parameter $\vec{p}$.*

Proof. Suppose that $\xi < \alpha$ is given; we want to compute $f(\xi)$. Let $\phi \equiv \exists z \psi(x, y, z, \vec{p})$, $\psi \in \Delta_0$, be a Σ_1-formula that defines $f(x) = y$ over L_α in the parameter $\vec{p}$. Now we proceed as follows:

In a certain register R_0 that initially contains $\max(\vec{p})$, we keep counting upwards. For every new exponentially closed value ι in R_0, we compute a ι-code c for L_ι, which is possible by Corollary 3.3.2 above. We then identify the elements of ι that code ξ and the elements of $\vec{p}$ in c (using Corollary 2.3.31 from Chapter 2) and check, using c to evaluate truth in L_ι, whether $L_\iota \models \exists z, y \psi(\xi, y, z, \vec{p})$. If not, we continue with $\iota + 1$. Otherwise, we search through ι to find the element that codes the respective y and return the ordinal coded by that element, again using the methods described in Chapter 2.

Now, as f is total, such a ι will eventually be found, at which point the search terminates. Thus, R_0 will never overflow, and the computation will thus terminate with output $f(\xi)$. □

We can now investigate the conditions of Corollary 3.4.13 a bit further.

---

7 The required analogue of Lemma 3.4.17 is the following: If α is regular in L_{α^ω}, $\beta < \alpha$ and $f : \beta \to L_\alpha$ is contained in L_{α^ω}, then $f \in L_\alpha$.

Definition 3.4.23. An ordinal α is r-weak if and only if the halting times of α-wITRMs are bounded above by α.

We saw above that Σ_2-admissible ordinals are r-weak. In the other direction, we have the following.

Lemma 3.4.24. *Suppose that α is a limit of exponentially closed ordinals. If α is r-weak, then α is admissible.*

Proof. Otherwise, let $f : \beta \to \alpha$ be total, cofinal and Σ_1 over L_α, where $\beta < \alpha$. By Lemma 3.4.22, f is α-wITRM-computable. Now we proceed as follows: In some register R, we run through β. For every $\iota < \beta$, we compute $f(\iota)$, then, in another register, we count upwards from 0 to ι once. Once R reaches β, we halt. This will use at least $\Sigma_{\iota<\beta} f(\iota) \geq \alpha$ many steps, so α is not r-weak, a contradiction. □

Lemma 3.4.25. *Let α be exponentially closed. Then the halting set H_α that consists of the pairs $(i, \iota) \in \omega \times \alpha$ for which the ith α-RM-program halts on the input ι is α-ITRM-decidable. In particular, there is a subset of α that is α-ITRM-computable, but not α-RM-computable.*

Proof. By Theorem 3.3.3, L_α has an ω_1-wITRM-computable code c. The statement $\phi_{i,\iota}$ that the α-RM-program P halts in the input ι is purely Σ_n for some $n \in \omega$, where n is independent of i and ι. By Theorem 2.3.28 from Chapter 2, there is an α-ITRM-program that uses c to decide, on input (i, ι), whether $\phi_{i,j}$ holds in L_α. Thus, we compute H_α. □

3.4.1 Lower bounds

We now turn to the lower bounds on the computational strength of α-(w)ITRMs. Most importantly, we will see that, when $\alpha = \omega$, then the upper bounds proved above are actually attained.

We start with some preliminary observations linking α-(w)ITRM-computability and constructibility. We formulate them for α-wITRMs under the condition of wITRM-singularity when possible; but the same proofs will show that they also hold for α-ITRMs under the assumption of ITRM-singularity, and we will use these variants below as well.

Lemma 3.4.26. *Let α be multiplicatively closed.*
(i) *If α is wITRM-singular, then there is an α-wITRM-program $P_{\alpha\text{-level}}$ such that the following holds: If $c \subseteq \alpha$ codes an ordinal β, then $P^c_{\alpha\text{-level}}$ computes an α-code for L_β.*
(ii) *Moreover, if α is ITRM-singular, then there is an α-ITRM program $P_{\alpha\text{-rlevel}}$ such that, if $a \subseteq \alpha$ and $c \subseteq \alpha$ codes an ordinal β, then $P^{c\oplus a}_{\alpha\text{-rlevel}}$ computes an α-code for $L_\beta[a]$. The same holds for wITRMs.*

Proof. (i) Let $g : \alpha \to \beta$ be bijective and $c = \{p(\iota, \xi) : f(\iota) \in f(\xi)\}$. Thus, a $< \beta$-name $(\bar{a}, \phi_i, f)$ can be encoded as $p_3(g^{-1}(\bar{a}), i, g^{-1}(c(f)))$, which is an element of α.

Similarly, $< \beta$-questions can be encoded as elements of α. But now, P_{truth} works on these coded names as well, and all questions arising in the process will again be coded by elements of α. By wITRM-singularity of α, the stack algorithm used in P_{truth} can also be implemented on an α-wITRM; we skip the details, which are similar to the proof of Lemma 2.3.25 in Chapter 2.

(ii) For the relativized version, we proceed similarly; but now, new questions of the form $g(\iota) \in a$ turn up, where $\iota < \alpha$. In this case, we do the following: In a separate register, run through α. For each $\xi < \alpha$, use the program $P_{\alpha\text{-count}}$ from Corollary 2.3.31 to find $g^{-1}(\xi)$; unless $g^{-1}(\xi) = \iota$, continue running through α; when the run finishes without encountering such a ξ, output 0. Otherwise, use the oracle a to check whether $\xi \in a$ and return the result. For the part about wITRMs, we use P_{fcount} instead of $P_{\alpha\text{-count}}$. □

Corollary 3.4.27. *Let α be multiplicatively closed, and suppose that $c \subseteq \alpha$ codes an ordinal β.*
(i) *If α is wITRM-singular and $x \in \mathfrak{P}(\omega) \cap L_\beta$, then x is α-wITRM-computable in the oracle c.*
(ii) *If α is ITRM-singular, and $x \in \mathfrak{P}(\alpha) \cap L_\beta$, then x is α-ITRM-computable in the oracle c.*

Proof. (i) Given c, use the program $P_{\alpha\text{-level}}$ from Lemma 3.4.26 to compute a code d for L_β. In d, the set x is coded by some $\iota \in \alpha$. Now, for any given $i \in \omega$, we can use P_{fcount} from Corollary 2.3.31 in Chapter 2 to identify the element ξ of α coding i in the sense of d and then check whether $p(\xi, \iota) \in d$.

(ii) This works like (i), but now using the full power of P_{count}. □

3.4.1.1 Lower Bounds for α-wITRMs

By Lemma 3.4.26, the question about the α-wITRM-computable subsets of ω thus leads us to inquire which ordinals have α-wITRM-computable codes. For wITRMs, this is easily achieved: Recalling that ω_1^{CK}, the first admissible ordinal, is also the supremum of the recursive ordinals; it is clear that the wITRM-computable ordinals are unbounded in ω_1^{CK}. On the other hand, by Corollary 3.4.12, every wITRM-computable real number belongs to $L_{\omega_1^{\text{CK}}}$. By Lemma 3.1.32 above, if $c \subseteq \omega$ codes an ordinal β and $c \in L_{\omega_1^{\text{CK}}}$, then $\beta \in L_{\omega_1^{\text{CK}}}$, so $\beta < \omega_1^{\text{CK}}$. Consequently, the wITRM-computable ordinals are exactly those below ω_1^{CK}, and by relativizing this argument to an oracle y and applying Corollary 3.4.27, we obtain the following characterization of wITRM-computability.

Theorem 3.4.28 (Koepke, [106]). *$x \subseteq \omega$ is wITRM-computable in the oracle $y \subseteq \omega$ if and only if $x \in L_{\omega_1^{\text{CK},y}}[y]$.*

As a consequence, admissibility is not sufficient for r-weakness.

Proposition 3.4.29. *The halting times of ω_1^{CK}-wITRMs are unbounded in $\omega_\omega^{\mathrm{CK}}$ and all elements of $\mathfrak{P}(\omega) \cap L_{\omega_\omega^{\mathrm{CK}}}$ are ω_1^{CK}-wITRM-computable. In particular, ω_1^{CK} is admissible, but not r-weak.*

Proof. By Proposition 2.3.3 from Chapter 2, ω_1^{CK}-wITRMs can simulate ITRMs. The result now follows from Theorem 3.4.31. □

In general, we get the following from Corollary 3.4.27:

Theorem 3.4.30. *Let α be wITRM-singular, suppose that β has an α-wITRM-computable code, and let $x \in \mathfrak{P}(\omega) \cap L_\beta$. Then x is α-wITRM-computable.*

3.4.1.2 Lower bounds for α-ITRMs

We now turn to the computational strength of α-ITRMs.

For $\alpha = \omega$, we can proceed as follows; see Koepke [109]: For each $x \subseteq \omega$, we can use P_{wf} from Chapter 2 to decide whether it codes a well-ordering. Hence, we can use an ITRM to decide the set S of all $i \in \omega$ for which the ith classical Turing program computes a well-ordering. For $j \in \omega$, if the jth Turing program computes a real number coding a well-ordering, denote this well-ordering by x_j; otherwise, let $x_j = 0$. Now, for $i, i', j, j' \in \omega$, let

$$c_1(p(i,j), p(i',j')) = \begin{cases} 0, & \text{when } j < j' \\ 1, & \text{when } j > j' \\ x_j(l), & \text{otherwise.} \end{cases}$$

Thus, c_1 is a code for the sum of all recursive ordinals, which is ω_1^{CK}. By Theorem 3.1.45 above, the supremum of the ordinals with real codes that are recursive in c_1 is at least ω_2^{CK}, so the same technique allows us to ITRM-compute a code c_2 for ω_2^{CK} as well. Iterating this argument, we obtain that, for every $i \in \omega$, ω_i^{CK} has an ITRM-computable real code c_i. As ITRMs are stronger than wITRMs, it follows from Lemma 3.4.26 that all $x \subseteq \omega$ in some $L_{\omega_i^{\mathrm{CK}}}$, $i \in \omega$, are ITRM-computable. The argument easily relativizes to arbitrary oracles $y \subseteq \omega$. Thus, we get that the upper bound proved above is actually attained.

Theorem 3.4.31 (Koepke, [109]). *Let $x, y \subseteq \omega$. Then x is ITRM-computable relative to y if and only if $x \in L_{\omega_\omega^{\mathrm{CK},y}}[y]$.*

Proof. The direction from left to right is shown in Corollary 3.4.12. On the other hand, if $x \in L_{\omega_\omega^{\mathrm{CK},y}}[y]$, then there is $i \in \omega$ such that $x \in L_{\omega_i^{\mathrm{CK},y}}[y]$, so the other direction follows from the argument above. □

In general, we have, by Corollary 3.4.27, the following.

Theorem 3.4.32. *Let α be ITRM-singular, suppose that β has an α-ITRM-computable code, and let $x \in \mathfrak{P}(\alpha) \cap L_\beta$. Then x is α-ITRM-computable.*

We can finally prove an important property of ITRMs, which was already mentioned in Chapter 2.

Corollary 3.4.33 (Koepke and Miller, see [110]). *Let $n \in \omega$, and let $(Q_i : i \in \omega)$ enumerate in some natural manner the ITRM-programs mentioning at most n registers. Then there is an ITRM-program P such that, for all $i \in \omega$, $P(i)\downarrow = 1$ if and only if $Q_i\downarrow$, and otherwise $P(i)\downarrow = 0$. Moreover, P can be chosen in such a way that, for any $x \subseteq \omega$ we have $P^x(i)\downarrow = 1$ if and only if $Q_i^x\downarrow$, and otherwise $P^x(i)\downarrow = 0$.*

Proof. The proof of Theorem 3.4.31 shows how to construct, for each $n \in \omega$, an ITRM-program $\bar{P}$ such that, whenever $x \subseteq \omega$, $\bar{P}^x$ computes a code c for $\omega_n^{\mathrm{CK},x}$.

By Theorem 3.4.12, Q_i^x will either halt in $< \omega_{n+1}^{\mathrm{CK},x}$ many steps or not at all. Thus, we can simply use c as a stopwatch: By Corollary 2.3.27, running P_{wf}^c will take at least $\omega_{n+1}^{\mathrm{CK},x}$ many steps.

Now, we run one step of P_{wf}^c after each step of Q_i^x. When Q_i^x halts before the test does, output 1, otherwise, i. e., when P_{wf}^c halts but Q_i^x has not halted so far, output 0. □

Exercise 3.4.34. (a) Show by induction on n that ω_1^n is ω_1-ITRM-clockable for any $n \in \omega$.

(b) Conclude that, for any $n \in \omega$, the halting problem for ω_1-ITRMs using at most n registers is ω_1-ITRM-computable.

Remark 3.4.35. Our knowledge of the computational strength of α-(w)ITRMs is rather sparse: We know the cases $\alpha = \omega$ and $\alpha = \mathrm{On}$ from the work of Koepke, and we saw above what happens when α is Σ_2-admissible. Otherwise, all we have are some upper and lower bounds. Here is a lot of space for further investigations, a space that is even expanded when also considering (α,β)-(w)ITRMs, i. e., α-(w)ITRMs that need to stop before time β; this question was first posed in [109].

For the case $\alpha = \omega$ and $\beta \in \{\omega_i^{\mathrm{CK}} : i \in \omega\}$, we have the following result, which is proved in the subsequent exercise.

Proposition 3.4.36. *Let $\beta \in \{\omega_i^{\mathrm{CK}} : i \in \omega\}$. Then a real number x is (ω,β)-ITRM-computable if and only if $x \in L_\beta$.*

Exercise 3.4.37. Since the strength of ITRMs increases with the number of registers as more and more admissible ordinals become computable it becomes natural to attempt to make this relation more precise. A natural conjecture would be that a code for ω_n^{CK} is computable by an ITRM with n registers. This is certainly not true, as we saw in Chapter 2 that ITRMs with one register are extremely weak. However, we can make the following modification: For $m, n \in \omega$, an ITRM of type (m, n) has m main registers $R_1, \dots, R_m$ and an additional n "auxiliary" registers $R_1', \dots, R_n'$. These registers are treated differently at limits: Main registers are treated like ITRM-registers (i. e., one forms the liminf of earlier contents and resets the content to 0 when this limit is not finite), while auxiliary registers work like wITRM-registers (i. e., the liminf is formed, and if this is not finite, the computation fails).

(a) Show that, for every $i \in \omega$, there is $n(i) \in \omega$ such that a code for ω_i^{CK} is computable by an ITRM of type $(i, n(i))$.

(b) Suppose that P is an ITRM-program of type (k, n), $\tau \in \mathrm{On}$ and that, in the computation of P, less than k registers contain 0 at time $\tau + \omega_{k+1}^{\mathrm{CK}}$. Show that P does not halt. Conclude that an ITRM-program of type (k, n) halts in $< \omega_{k+1}^{\mathrm{CK}}$ many steps or diverges.

(c) Suppose that $x \subseteq \omega$ is computable by an ITRM of type (k, n). Show that $x \in L_{\omega_{k+1}^{\mathrm{CK}}}$.

(d) Show that, if $x \in L_{\omega_k^{\mathrm{CK}}} \cap \mathfrak{P}(\omega)$, then x is computable by an ITRM of type (k, n) for some $n \in \omega$.

(e) Show that $x \subseteq \omega$ is computable by an ITRM of type (k, n) for some $n \in \omega$ if and only if $x \in L_{\omega_{k+1}^{\mathrm{CK}}}$.

(f) Show that x is ITRM-computable in $< \omega_i^{\mathrm{CK}}$ many steps if and only if $x \in L_{\omega_i^{\mathrm{CK}}}$.

3.5 α-TMs and OTMs

The computability strength of OTMs and α-TMs (which, as we recall here, are obtained from OTMs by imposing α as a bound both on working time and space) for exponentially closed α is rather easy to determine, since we already know about ORMs and α-RMs and due to the simulation results from Chapter 2.

Exercise 3.5.1. Show that during an OTM-computation that takes $< \alpha$ many steps, the heads cannot attain a position with index $\geq \alpha$.

Again, we start with upper bounds.

Lemma 3.5.2 (Koepke).
(i) *If X is an OTM-writable set of ordinals, then $X \in L$.*
(ii) *If X is a parameter-freely OTM-writable set of ordinal, then $X \in L_\sigma$.*
(iii) *If α is exponentially closed and $X \subseteq \alpha$ is α-TM-computable, then X is $\boldsymbol{\Delta}_1$ over L_α.*
(iv) *If α is exponentially closed and $X \subseteq \alpha$ is α-TM-computable without parameters, then X is Δ_1 over L_α.*

Proof. All of these proofs work by straightforwardly adapting the corresponding proofs about ORMs and α-RMs. More specifically:

(i) If P is an OTM-program and $\iota \in \mathrm{On}$ is such that $P(\iota)$ halts with output X, then the halting computation of $P(\iota)$ is definable by recursion, and thus contained in L, and hence so is X. (For more details, see the proof of Theorem 3.2.4 above.)

(ii) If P is an OTM-program that halts with output X on the empty input, then as in the proof of (i), the statement that P halts is true in L. Moreover, it is a Σ_1-statement without parameters, and thus holds in L_σ. Hence L_σ contains the halting computation of P, and thus also X.

(iii) If P is an α-TM-program and $\xi < \alpha$ is such that $P(\iota, \xi)$ halts with output $\chi_X(\iota)$ for every $\iota < \alpha$ in $< \alpha$ many steps, then all such computations $P(\iota, \xi)$ are contained in L_α by exponential closure of α (this actually only requires that α is a multiple of ω^2). As in the proof of Theorem 3.3.3, X is thus Δ_1 in the parameter ξ over L_α.

(iv) Works like (iii), but dropping the parameter. □

Fortunately, these are already the lower bounds as well.

Lemma 3.5.3 (Koepke). *There is a nonhalting OTM-program P_L such that, when run without parameters on the empty input, the following holds: For every $\iota \in \mathrm{On}$, there is $\tau(\iota) \in \mathrm{On}$ such that at time $\tau(\iota)$, the first cell of the output tape contains 1 and the remaining content of the output tape is a code for L_ι. Moreover, the $\tau(\iota)$ form a strictly increasing sequence of ordinals.*

Proof. By Theorem 3.2.13, there is an ORM-program P such that, for all $\iota \in \mathrm{On}$, $P(\iota)$ halts with the index of a name for a code of L_ι written to the output tape. P works by simply running through all names and using P_{truth} to test, for each given name, whether it names L_ι, which is possible in the parameter ι. Since such a name exists, it will eventually be found, in which case P halts.

P_L now simply works by successively for each $\xi \in \mathrm{On}$ first using Lemma 2.5.45 from Chapter 2 to compute a code for L_ι, then writing it to the output tape starting in the second cell, changing the content of the first cell of the output tape to 1 and then back to 0. □

Lemma 3.5.4.
(i) (Koepke [104]) *If X is a constructible set of ordinals, then X is OTM-writable.*
(ii) (Seyfferth and Schlicht, [160]) *If $X \in L_\sigma \cap \mathfrak{P}(\mathrm{On})$, then X is OTM-writable without parameters.*
(iii) (Koepke and Seyfferth, [112]) *If α is exponentially closed and $X \subseteq \alpha$ is $\boldsymbol{\Delta}_1$ over L_α, then X is α-TM-computable.*
(iv) (Koepke and Seyfferth, [112]) *If α is exponentially closed and $X \subseteq \alpha$ is Δ_1 over L_α, then X is α-TM-computable without parameters.*

Proof. (i) By Theorem 3.2.15, there are an ORM-program $\bar{P}$ and an ordinal γ such that, for all $\xi \in \mathrm{On}$, $\bar{P}(\xi, \gamma)\downarrow = \chi_X(\xi)$. By Lemma 2.5.45 in Chapter 2, there is an OTM-program Q with the same property. Now, P works as follows: Let α be large enough such that $X \subseteq \alpha$. Given the parameters ξ and α, P runs through α and for every $\iota < \alpha$ uses Q to determine whether or not $\iota \in X$. If the answer is yes, then a 1 is written in the ιth cell of the output tape, otherwise a 0. When the run through α is finished, the program stops. Clearly, P writes X.

(ii) This works by dropping ξ in the proof of (i), expect that now, the upper bound α for X cannot be given as a parameter, and thus also needs to be computed. But this is easy: As $X \in L_\sigma$, there is $\beta < \sigma$ such that $X \subseteq \beta$. By definition of σ, there is $\alpha \in [\beta, \sigma)$ such that, for some Σ_1-formula ϕ, α is minimal with the property that $L_\alpha \models \phi$. Now use P_L from Lemma 3.5.3 to successively write codes for all L-levels and for each coded L-level L_γ, use OTM-simulation of P_{truth} to determine whether $L_\gamma \models \phi$. Eventually, such a γ will be found, which is thus $\geq \alpha$ and can hence serve as an upper bound for X in the computation.

(iii) We saw in Chapter 2 that α-TMs can simulate α-RMs. It is not hard to see that, when α is exponentially closed and Q is an α-RM-program such that $P(\xi)$ halts in $< \alpha$ many steps for some $\xi < \alpha$, then the same holds for the α-TM-program simulating Q. The claim now follows from Theorem 3.3.3.

(iv) This works by dropping the parameters in (iii). □

Thus, we get a complete characterization of the computational strength of OTMs and α-TMs, both with and without parameters.

Theorem 3.5.5 (Koepke [104] and Koepke and Seyfferth, [112]).
(i) *A set X of ordinals is OTM-writable if and only if $X \in L$.*
(ii) *A set X of ordinals is OTM-writable without parameters if and only if $X \in L_\sigma$.*
(iii) *If α is exponentially closed, then $X \subseteq \alpha$ is α-TM-computable if and only if X is $\boldsymbol{\Delta}_1$ over L_α.*
(iv) *If α is exponentially closed, then $X \subseteq \alpha$ is α-TM-computable without parameters if and only if X is Δ_1 over L_α.*

Finally, we note that ORMs can simulate OTMs in a certain sense.

Exercise 3.5.6 (Cf. Miller and Hamkins [140], Lemma 9). Show that there is an ORM-program P_{sim} with the following property: If Q is an OTM-program and $\iota, \xi, \tau \in \text{On}$, Then $P_{\text{sim}}(\xi, \tau)$ halts with output $(i, j) \in \omega \times \{0, 1\}$, where i is the index of the inner state and j is the content of the ιth tape cell at time τ of the computation of $Q(\xi)$.

3.6 α-ITTMs and the Σ_2-machine[8]

We now turn to the computational strength of α-ITTMs. For the whole section, α is a multiplicatively closed ordinal.[9]

---

8 The material in this section is adapted by permission from Springer Nature from the article "Taming Koepke's Zoo" [34] by S. Ouazzani, P. Welch and the author that appeared in 2018 in the LNCS proceedings "Sailing Routes in the World of Computation" for the CiE 2018. We thank Springer Nature as well as our coauthors S. Ouazzani and P. Welch for the kind permission to reuse it here.

9 Multiplicative closure of the tape length is rather a technical convenience than a necessary assumption. For example, since finite tuples of ordinals $< \alpha$ can be encoded by elements of α when α is multiplicatively closed, one can assume without loss of generality that parameters only consist of one single

The construction we describe below, the Σ_2-theory machine, is due to Philip Welch and Sy-David Friedman; see [62] for the case $\alpha = \omega$. Generalizing their construction to multiplicatively closed α leads to a few minor extra subtleties, but is mostly straightforward.[10] The generalized version described here was given in Carl, Ouazzani and Welch [34], but, mainly due to the limited space, the proofs were only sketched. Here, we give full details of the generalized versions.

Recall that, in the analysis of α-(w)ITRMs the important role played by admissible ordinals was due to Π_2-reflection, which implied that all machine configurations at sufficiently high admissible times must already have occurred so often that the machine is in fact looping. This allowed us to obtain upper bounds on the computability strength of α-(w)ITRMs.

Can be do something similar to obtain at least upper bounds on the computability strength of α-ITTMs? Indeed, recall Exercise 2.5.47 from Chapter 2, where we showed by a similar reflection argument that an ITTM-program either halt in $< \omega_1$ many steps or starts looping before time ω_1. The crucial observation was the following: Any tape cell of an ITTM-program P that contains 1 at time ω_1 must have constantly had that value from some earlier time τ on, and cells that contain 0 at time ω_1 have already simultaneously contained 0 at some time τ' after τ, but before ω_1, and thus (τ', ω_1) witnesses the looping of P.

Since ITTM-halting times are countable and there are only countably many of them, there must be some countable upper bound to all ITTM-halting times; thus ω_1 is quite rough as an upper estimate. So let us look a bit closer what is really needed for the argument just sketched. To this end, we first introduce some terminology.

Definition 3.6.1 (Welch). Let $\iota \in \alpha$, P an α-ITTM-program, and C its computation in the parameter $\vec{p} \subseteq \alpha$. For any of the tapes of P, we say that the ιth cell C_ι "stabilizes" if and only if there is an ordinal δ such that the content of C_ι in C remains unchanged after time β. C_ι "stabilizes before δ" if there is $\beta < \delta$ such that the content of C_ι in C remains unchanged after time δ.

We say that C_ι 'stabilizes **at** δ' if and only if there is an ordinal $\beta < \delta$ such that the content of C_ι in C remains unchanged between times β and δ.

Finally, for $\delta < \beta$, C_ι "stabilizes **at** β **before** δ" if there is $\gamma < \delta$ such that the content of C_ι does not change between times γ and β.

We note that the stabilization of C_ι is a Σ_2-property: It says that there is some δ such that, for all $\eta > \delta$, the content of C_ι in the computation $P(\vec{p})$ at time η is the same

element. However, dropping the closure conditions on the tape length would not change much of the results.

10 We are indebted to Philip Welch for helpful discussions and ideas concerning these "extra subtleties." The approach presented below is similar to one described to us by Welch, but designed in order to minimize the need for fine-structural background knowledge.

as at time δ. Likewise, stabilization at δ is Σ_2 over L_δ and stabilization at β before δ is Σ_2 over L_β in the (extra) parameter δ.

What we need for the argument sketched above to work at some ordinal δ in place of ω_1 is the existence of some $\gamma < \delta$ such that every cell that stabilizes at δ in fact stabilizes at δ before γ. As stabilization is Σ_2, this would in particular hold when every Σ_2-statement that holds in L_δ also holds in L_γ; we are thus led to the notions of a Σ_2-submodel and of Σ_2-extendibility.

Definition 3.6.2. Let X and Y be $\in$-structures, $i \in \omega$. We say that X is a "Σ_i-submodel" of Y, written $X \prec_{\Sigma_i} Y$, if and only if $X \subseteq Y$ and, for every tuple $\vec{p} \subseteq X$ and every Σ_i-formula ϕ, $X \models \phi(\vec{p})$ is equivalent to $Y \models \phi(\vec{p})$. If this holds for every $i \in \omega$, then X is an "elementary submodel" of Y.

For $\alpha \in \mathrm{On}$, let $\Sigma'(\alpha)$ denote the smallest ordinal $\beta > \alpha$ such that L_β has a proper Σ_2-submodel that contains $\alpha + 1$ as a subset. By the condensation lemma, if H is that submodel, then H is isomorphic to an L-level $L_{\bar{\beta}}$. The smallest $\bar{\beta} > \alpha$ such that $L_{\bar{\beta}} \prec_{\Sigma_2} L_{\Sigma'(\alpha)}$ will be denoted by $\zeta'(\alpha)$. We also say that $\zeta'(\alpha)$ is the smallest Σ_2-extendable ordinal $> \alpha$. Finally, we denote by $\lambda'(\alpha)$ the smallest ordinal $\beta > \alpha$ such that $L_\beta \prec_{\Sigma_1} L_{\zeta'(\alpha)}$.

We can also relativize this notation to a subset x of α and write $\Sigma'^{,x}(\alpha)$ for the minimal $\beta > \alpha$ such that $L_\beta[x]$ has a proper Σ_2-submodel containing $\alpha + 1$ as a subset and similarly $\zeta'^{,x}(\alpha)$ for the minimal ordinal height of such a submodel and $\lambda'^{,x}(\alpha)$ for the minimal ordinal height $> \alpha$ of a Σ_1-submodel of $L_{\Sigma'^{,x}(\alpha)}[x]$.

The names of these ordinals are somewhat suggestive; and indeed, it will turn out below that $\lambda'(\alpha) = \lambda(\alpha)$, $\zeta\prime(\alpha) = \zeta(\alpha)$ and $\Sigma'(\alpha) = \Sigma(\alpha)$ whenever α is multiplicatively closed. The next section will show that $\lambda'(\alpha)$, $\zeta'(\alpha)$ and $\Sigma'(\alpha)$ provide upper bounds for the writability, eventual writability and accidental writability strength of α-ITTMs.

An observation that will come in handy below is that, when L_β is a Σ_1-submodel of L_γ, then Σ_2-truth will be preserved upwards from L_α to L_β.

Lemma 3.6.3. *Let β, γ be ordinals such that $L_\beta \prec_{\Sigma_1} L_\gamma$ and let ϕ be a Σ_2-formula. Suppose that $\vec{p} \in L_\beta$ is such that $L_\beta \models \phi(\vec{p})$. Then $L_\gamma \models \phi(\vec{p})$.*

Proof. Let $\phi \equiv \exists x \forall y \psi(x, y, \vec{z})$, where ψ is Δ_0. Since $L_\beta \models \phi(\vec{p})$, there is $a \in L_\beta$ such that $L_\beta \models \forall y \psi(a, y, \vec{p})$. Thus $L_\beta \models \neg\exists y \neg\psi(a, y, \vec{p})$. Hence, if $L_\gamma \models \exists y \psi(a, y, \vec{p})$, we have a contradiction to $L_\beta \prec_{\Sigma_1} L_\gamma$. Thus $L_\gamma \models \forall y \psi(a, y, \vec{p})$, i. e., $L_\gamma \models \exists x \forall y \psi(x, y, \vec{p})$, hence $L_\gamma \models \phi(\vec{p})$, as desired. □

Exercise 3.6.4. Show more generally that, if $L_\alpha \prec_{\Sigma_n} L_\beta$ and $L_\alpha \models \phi(\vec{p})$ where ϕ is Σ_{n+1} and $\vec{p} \in L_\alpha$, then $L_\beta \models \phi$.

Lemma 3.6.5.
(i) *If $\beta < \Sigma'(\alpha)$ and $\beta > \alpha$, then $L_{\beta+\omega}$ contains a surjection $f : \alpha \to L_\beta$.*
(ii) *If $\beta < \Sigma'(\alpha)$ is a limit ordinal greater than α and $X \subseteq L_\beta$, then $H := \Sigma_1^{L_\beta}(\alpha + 1 \cup X)$ is transitive. More generally, any Σ_1-submodel of L_β is transitive. In fact, it is equal to L_ξ for some $\xi \le \beta$.*

Proof. (i) Let $H := \Sigma_2^{L_\beta}(\alpha)$. By condensation, there is $\gamma \le \beta$ such that $H \simeq L_\gamma$. If $\gamma < \beta$, then $\alpha \subseteq L_\gamma \prec_{\Sigma_2} L_\beta$ by Lemma 3.1.21. But this contradicts the minimality of $\Sigma'(\alpha)$. Thus $H \simeq L_\beta$.

Now it is easy to see that the canonical Σ_2-Skolem $h_2^{L_\beta}$ function for L_β is definable over L_β from Lemma 3.1.19, and hence contained in $L_{\beta+\omega}$. Thus, we also have $\Sigma_2^{L_\beta}(\alpha) \in L_{\beta+\omega}$. By definition, $h_2^{L_\beta}$ will be a partial surjection from $\omega \times \alpha$ to H.

Moreover, one can see by induction on γ that the restriction π_γ of the collapsing map π that sends sets to their transitive collapse, is an element of $L_{\gamma+\omega}$ for every ordinal γ. By the above, we have $\pi(H) = \pi[H] = L_\beta$.

Now $\pi_\beta \circ h_2^{L_\beta}$ will map $\omega \times \alpha$ surjectively onto L_β and is contained in $L_{\beta+\omega}$. This suffices.

(ii) Let $x \in H$. Since β is a limit ordinal, we have $L_\beta \models \exists \gamma(x \in L_\gamma)$, which is Σ_1 over L_β by Lemma 3.1.12. Thus, there is $L_\gamma \in H$ such that $x \in L_\gamma$. By (i), there is $f \in H$ such that $f : \alpha \to L_\gamma$ is surjective. As $\alpha \subseteq H$, we have $f(\iota) \in H$ for every $\iota < \alpha$, i.e., $x \subseteq L_\gamma - f[\alpha] \subseteq H$. Thus $x \subseteq H$, and as x was an arbitrary element of H, H is transitive. The apparently more general claim is proved in the same way. By condensation, H is isomorphic to some L_ξ with $\xi \le \beta$. As two transitive sets are isomorphic if and only if they are equal, the last claim also follows. □

Theorem 3.6.6 (Hamkins and Lewis, see [81], Theorem 8.3). *Let $x \subseteq \alpha$.*
- *$\lambda'^{,x}(\alpha)$ is an x-admissible limit of x-admissible ordinals. Moreover, it is a limit of x-admissible limits of x-admissible limits of x-admissible ordinals and so on. This remains true for any finite iteration of "limit of."*[11]
- *$\zeta'^{,x}(\alpha)$ is an x-admissible limit of x-admissible ordinals. Like for $\lambda'^{,x}(\alpha)$, this can be iterated. Moreover, $L_{\zeta'^{,x}(\alpha)}[x]$ satisfies Σ_2-collection: If $\beta < \zeta'^{,x}(\alpha)$ and $f : \beta \to L_{\zeta'^{,x}(\alpha)}[x]$ is Σ_2 over $L_{\zeta'^{,x}(\alpha)}[x]$, then there is $\gamma < \zeta'^{,x}(\alpha)$ such that $f[\beta] \subseteq L_\gamma[x]$. That is, $\zeta'^{,x}(\alpha)$ is Σ_2-admissible.*
- *$\Sigma'^{,x}(\alpha)$ is a limit of x-admissible ordinals.*

Proof. We restrict ourselves to the unrelativized versions.

(a) We first verify the admissibility of $\lambda'(\alpha)$ by checking that the KP-axioms hold in $L_{\lambda'(\alpha)}$. So let $\xi < \lambda'(\alpha)$, and let $f : \xi \to L_{\lambda'(\alpha)}$ be Σ_1 over $L_{\lambda'(\alpha)}$; say $f(\iota) = x$ if and

11 In fact, it also holds for many infinite iterations; but we will not need this.

only if $L_{\lambda'(\alpha)} \models \exists y\phi(\iota, x, y, \vec{p})$ for some finite parameter sequence $\vec{p}$ from $L_{\lambda'(\alpha)}$ and some Δ_0-formula ϕ. As every $\iota < \xi$ has an f-image in $L_{\lambda'(\alpha)}$ and $\lambda'(\alpha) < \zeta'(\alpha)$, it follows that the Δ_0-formula $\forall\iota \in \xi\exists x \in L_{\lambda'(\alpha)}\exists y \in L_{\lambda'(\alpha)}\phi(\iota, x, y, \vec{p})$ holds in $L_{\zeta'(\alpha)}$. But this means that $L_{\zeta'(\alpha)} \models \exists\beta\forall\iota \in \xi\exists x \in L_\beta\exists y \in L_\beta\phi(\iota, x, y, \vec{p})$, which is a Σ_1-statement in the parameters $\xi, \vec{p} \in L_{\lambda'(\alpha)}$ and hence also holds in $L_{\lambda'(\alpha)}$ by Σ_1-elementarity of $L_{\lambda'(\alpha)}$ in $L_{\zeta'(\alpha)}$. Let $\beta \in L_{\lambda'(\alpha)}$ witness the truth of this statement. Then $\forall\iota \in \xi\exists x \in L_\beta\exists y \in L_\beta\phi(\iota, x, y, \vec{p})$, hence all f-images of elements of ξ are contained in L_β, where $\beta < \lambda'(\alpha)$, which is what we wanted to show. So $\lambda'(\alpha)$ is admissible.

Now, for $\xi < \lambda'(\alpha)$, $L_{\zeta'(\alpha)}$ believes that there exists an admissible ordinal $> \xi$ (certainly $\lambda'(\alpha)$ is a witness for this statement). This is again a Σ_1-statement and thus already holds in $L_{\lambda'(\alpha)}$. So $L_{\lambda'(\alpha)}$ contains an admissible ordinal above every ordinal and so $\lambda'(\alpha)$ is a limit of admissible ordinals. Iterating this idea gives the result.

(b) We show Σ_2-collection in $L_{\zeta'(\alpha)}$, which is stronger than admissibility. Say f is as in the assumptions and $f(\iota) = a$ for $\iota < \beta$ if and only if $L_{\zeta'(\alpha)} \models \exists x\forall y\phi(x, y, \iota, a)$, where ϕ is Δ_0. Now, for given $\iota \in \beta$ and $a \in L_{\zeta'(\alpha)}$, we have $L_{\zeta'(\alpha)} \models \exists x\forall y\phi(x, y, \iota, a)$ if and only if $L_{\Sigma'(\alpha)} \models \exists x\forall y\phi(x, y, \iota, a)$ by Σ_2-elementarity. So $\forall\iota \in \beta\exists a \in L_{\zeta'(\alpha)}[L_{\zeta'(\alpha)} \models \exists x\forall y\phi(x, y, \iota, a)]$, and hence $\forall\iota \in \beta\exists a \in L_{\zeta'(\alpha)}[L_{\Sigma'(\alpha)} \models \exists x\forall y\phi(x, y, \iota, a)]$, so that $L_{\Sigma'(\alpha)} \models \forall\iota \in \beta\exists a \in L_{\zeta'(\alpha)}\exists x\forall y\phi(x, y, \iota, a)$.

In particular, then

$$L_{\Sigma'(\alpha)} \models \exists\gamma\forall\iota \in \beta\exists a \in L_\gamma\exists x\forall y\phi(x, y, \iota, a).$$

Let g be the function mapping each $\iota < \beta$ to the $<_L$-minimal pair $(\hat{a}, \hat{x})$ such that $L_{\zeta'(\alpha)} \models \forall y\phi(\hat{x}, y, \iota, \hat{a})$. Then, by Σ_2-elementarity, we also have $L_{\Sigma'(\alpha)} \models \forall y\phi(x, y, \iota, a)$ for all $\iota < \beta$, $(\hat{a}, \hat{x}) = g(\iota)$. But g is clearly definable over $L_{\zeta'(\alpha)}$ and thus contained in $L_{\zeta'(\alpha)+1} \subseteq L_{\Sigma'(\alpha)}$. Thus

$$L_{\Sigma'(\alpha)} \models \exists\gamma, g\forall y\forall\iota \in \beta\phi((g(\iota))_0, y, \iota, (g(\iota))_1),$$

where $(p)_0$ and $(p)_1$ denote the first and second component of the ordered pair p, as usual.

This is a Σ_2-formula, so, as it holds in $L_{\Sigma'(\alpha)}$, it holds in $L_{\zeta'(\alpha)}$. Consequently, $L_{\zeta'(\alpha)} \models \exists\gamma\forall\iota \in \beta\exists a \in L_\gamma\exists x\forall y\phi(x, y, \iota, a)$. But this is just what we wanted to show.

The other statements now follow as in the proof of (a).

(c) Suppose that $\Sigma'(\alpha)$ was not a limit of admissible ordinals. Then there would be $\xi < \Sigma'(\alpha)$ such that the Π_1-statement 'For all $\beta > \xi$, β is not admissible' would be true in $L_{\Sigma'(\alpha)}$. Hence the statement 'There is β such that, for all $\beta > \xi$, β is not admissible', which is Σ_2, would hold in $L_{\Sigma'(\alpha)}$. By Σ_2-elementarity, it would then hold in $L_{\zeta'(\alpha)}$, so that $\zeta'(\alpha)$ would not be a limit of admissible ordinals either, contradicting (b). □

Remark 3.6.7. We will see below that $\Sigma'(\alpha)$ is not admissible.

Lemma 3.6.8.
(i) *$\lambda'(\alpha)$ is the supremum of ordinals β that are minimal with the property that L_β contains a witness x for some Σ_1-statement $\exists x\psi(x, y)$ such that $L_{\zeta'(\alpha)} \models \exists x\psi(x, y)$, where ψ is Δ_0 and $y < \alpha$.*
(ii) *Let us say that an ordinal ξ is Σ_2-fixed in $\Sigma'(\alpha)$ if and only if there is a Σ_2-formula $\phi(z) \equiv \exists x\forall y\psi(x, y, z)$, $\psi \in \Delta_0$-formula, with a free variable z and some $y < \alpha$ such that L_ξ is minimal with the property that there is $a \in L_\xi$ such that $L_{\Sigma'(\alpha)} \models \forall y\psi(a, y, y)$. Then $\zeta'(\alpha)$ is the supremum of the ordinals that are Σ_2-fixed in $\Sigma'(\alpha)$.*

Proof. (i) By definition of $\lambda'(\alpha)$, when $\exists x\psi(x, y)$ holds in $L_{\zeta'(\alpha)}$, then it holds in $L_{\lambda'(\alpha)}$. Thus, $\lambda'(\alpha)$ is larger than any such β. Now suppose that the supremum ρ of these β was strictly smaller than $\lambda'(\alpha)$. Considering $\exists x(x = y)$ for every $y < \alpha$, we see that $\rho \geq \alpha$. Clearly, any Σ_1-statement that holds in L_ρ also holds in $L_{\lambda'(\alpha)}$ by upwards absoluteness of Σ_1-formulas. On the other hand, we have $L_{\lambda'(\alpha)} = \Sigma_1^{L_{\lambda'(\alpha)}}\{\alpha\}$ by Lemma 3.6.5, so $\rho \geq \lambda'(\alpha)$ as well.

(ii) Since $L_{\zeta'(\alpha)} \prec_{\Sigma_2} L_{\Sigma'(\alpha)}$, every ordinal that is Σ_2-fixed in $\Sigma'(\alpha)$ is smaller than $\zeta'(\alpha)$. Now suppose that the supremum ρ of these ordinals is strictly smaller than $\zeta'(\alpha)$. Then the Σ_2-Skolem hull of α in $L_{\Sigma'(\alpha)}$ is a subset of L_ρ, and thus isomorphic to L_β for some $\beta \leq \rho$, which contradicts the minimality of $\zeta'(\alpha)$. □

3.6.1 Upper bounds

We are now ready to prove upper bounds on the computational strength of α-ITTMs.

Lemma 3.6.9 (Welch, [184, 62]). *Let P be an α-ITTM-program, $y, \iota < \alpha$.*
(i) *Suppose that C_ι of $P(y)$ stabilizes at $\Sigma'(\alpha)$. Then C_ι stabilizes at $\zeta'(\alpha)$.*
(ii) *Suppose that $\beta < \zeta'(\alpha)$ is such that the content of C_ι does not change between times β and $\zeta'(\alpha)$ in the computation of $P(y)$. Then the content of C_ι remains unchanged up to time $\Sigma'(\alpha)$. In other words, if C_ι stabilizes at $\zeta'(\alpha)$, then it stabilizes at $\Sigma'(\alpha)$ before $\zeta'(\alpha)$.*

Proof. (i) We observed above that the statement that C_ι stabilizes at $\Sigma'(\alpha)$ in the computation of $P(y)$ is Σ_2 in the parameters y and ι over $L_{\Sigma'(\alpha)}$. Since $L_{\zeta'(\alpha)} \prec_{\Sigma_2} L_{\Sigma'(\alpha)}$, the same statement holds in $L_{\zeta'(\alpha)}$. Thus C_ι stabilizes at $\zeta'(\alpha)$.

(ii) The statement "there is $\delta > \beta$ such that, at time δ, C_ι has a different content than at time β in the computation of $P(y)$" is Σ_1 in the parameters $\iota, \beta, y < \zeta'(\alpha)$. If the content of C_ι changes between times β and $\Sigma'(\alpha)$, then this statement holds in $L_{\Sigma'(\alpha)}$, and thus, by Σ_2-elementarity, it also holds in $L_{\zeta'(\alpha)}$. But this implies that the content of C_ι changes between times β and $\zeta'(\alpha)$, which contradicts the assumption. □

Corollary 3.6.10 (Welch, see [184]; also see [62]). *Let P be an α-ITTM-program, $y < \alpha$. If $P(y)$ runs for $\Sigma'(\alpha)$ many steps, then it does not halt. In fact, it will loop from time $\zeta'(\alpha)$ on, repeating the sequence of configurations between $\zeta'(\alpha)$ and $\Sigma'(\alpha)$.*

Proof. We will show that $(\zeta'(\alpha), \Sigma'(\alpha))$ witnesses the looping of $P(y)$. Consider a cell C_ι, $\iota < \alpha$.

Claim. *If C_ι contains 1 at time $\zeta'(\alpha)$, then it contains 1 at all times $\tau \in [\zeta'(\alpha), \Sigma'(\alpha)]$.*

To see this, note that, by the liminf-rule and since $\zeta'(\alpha)$ is a limit ordinal, C_ι will only contain 1 at time $\zeta'(\alpha)$ when C_ι has stabilized at $\zeta'(\alpha)$ with value 1. But then, by (ii) of Lemma 3.6.9, the content of C_ι remains unchanged up to $\Sigma'(\alpha)$, which establishes the claim.

Similarly, it follows from Lemma 3.6.9 that, if C_ι stabilizes at $\zeta'(\alpha)$ with value 0, then C_ι does not change its content up to time $\Sigma'(\alpha)$. Thus, no cell that has stabilized at $\zeta'(\alpha)$ changes its content up to time $\Sigma'(\alpha)$. Moreover, part (i) of Lemma 3.6.9 now implies that every cell the value of which has stabilized at $\Sigma'(\alpha)$ has also stabilized at $\zeta'(\alpha)$ and is thus constant from some time $\tau < \zeta'(\alpha)$ on up to time $\Sigma'(\alpha)$.

Let us finally consider the case that C_ι does not stabilize at $\zeta'(\alpha)$. Then Lemma 3.6.9 implies that it does not stabilize at $\Sigma'(\alpha)$ either, and hence the liminf-rule implies that C_ι contains 0 at both time $\zeta'(\alpha)$ and $\Sigma'(\alpha)$.

As the above cases for the behavior of C_ι at $\zeta'(\alpha)$ are exhaustive, the cell contents at times $\zeta'(\alpha)$ and $\Sigma'(\alpha)$ are the same. Moreover, it follows that, if C_ι contains 1 at time $\zeta'(\alpha)$ or $\Sigma'(\alpha)$, then it contains 1 during the whole interval $[\zeta'(\alpha), \Sigma'(\alpha)]$, and thus, for every time τ in that interval, the content of C_ι at time τ is at least as large at time $\zeta'(\alpha)$. We have checked that, as far as tape contents are concerned, the looping criterion is satisfied.

We still need to take care of inner states and head positions. These, however, can easily be dealt with using the following trick: We modify P a bit to a program P' that, on some extra tapes, stores the inner states and head positions during the computation of $P(y)$ by writing a sequence of n many 1s followed by 0s to represent the nth inner state and also writing ι many 1s followed by 0s to represent that the head is at position ι. It is easy to see that these tapes can be updated along with the computation of $P(y)$ and that the representation is automatically correct at limit levels by the liminf-rule. But now, applying the analysis above to $P'(y)$, we get that the inner states and head positions at times $\zeta'(\alpha)$ and $\Sigma'(\alpha)$ are also identical, and moreover, that neither does the index of the inner state drop below its value at these times at any time $\tau \in [\zeta'(\alpha), \Sigma'(\alpha)]$ nor does the head occupy a position to the left of its position at times $\zeta'(\alpha)$ and $\Sigma'(\alpha)$ at such a time τ.

Thus, all criteria for strong looping have been checked and the corollary is proved. □

Corollary 3.6.11 (Welch, cf. [184, 62]). *Let P be an α-ITTM-program, $y < \alpha$. Then $P(y)$ either halts in $< \lambda'(\alpha)$ many steps or not at all.*

Proof. From Corollary 3.6.10, it follows that, if β is the halting time of any α-ITTM-computation, then $\beta < \Sigma'(\alpha)$. But now, the statement that $P(y)$ halts is a Σ_1-statement in the parameter y, and if $P(y)$ halts, then this statement is true in $L_{\Sigma'(\alpha)}$ as $L_{\Sigma'(\alpha)}$ will contain the halting computation of $P(y)$. As a Σ_1-submodel of a Σ_2-submodel of $L_{\Sigma'(\alpha)}$, $L_{\lambda'(\alpha)}$ is also a Σ_1-submodel of $L_{\Sigma'(\alpha)}$, hence the same statement holds in $L_{\lambda'(\alpha)}$. It follows that the halting computation of $P(y)$ is contained in $L_{\lambda'(\alpha)}$. By admissibility of $\lambda'(\alpha)$, the length of that computation must be an element of $L_{\lambda'(\alpha)}$, and thus must be $< \lambda'(\alpha)$. Thus, if $P(y)$ halts, it takes less than $\lambda'(\alpha)$ many steps. □

Lemma 3.6.12. *Let $x \subseteq \alpha$.*
1. *If x is accidentally α-writable, then $x \in L_{\Sigma'(\alpha)}$.*
2. *If x is eventually α-writable, then $x \in L_{\zeta'(\alpha)}$.*
3. *If x is α-writable, then $x \in L_{\lambda'(\alpha)}$.*

Proof. (i) Suppose that P is an α-ITTM-program and $y < \alpha$ is such that $P(y)$ accidentally writes x. If $P(y)$ halts, then it does so before time $\lambda'(\alpha) < \Sigma'(\alpha)$. If $P(y)$ does not halt, then let τ be minimal such that $P(y)$ has x on the output tape at time τ. Then, as $P(y)$ is looping from time $\Sigma'(\alpha)$ on, no new configurations will show up after that time, and thus $\tau < \Sigma'(\alpha)$. Hence the partial computation of $P(y)$ up to time τ is contained in $\Sigma'(\alpha)$, and thus so is x.

(ii) Let P be an α-ITTM-program, $y < \alpha$ and τ an ordinal such that the computation of $P(y)$ has x on the output tape from time τ on. (If $P(y)$ halts, we can argue as in (i).) In particular, this means that every cell C_ι of the output tape of $P(y)$ stabilizes. Since $P(y)$ is looping from time $\Sigma'(\alpha)$ on, repeating the sequence of configurations between times $\zeta'(\alpha)$ and $\Sigma'(\alpha)$, the content of C_ι cannot change in this time interval, and thus, this stabilization must have taken place before time $\zeta'(\alpha)$. We now know that each cell of the output tape stabilizes before time $\zeta'(\alpha)$, but this is not good enough: If these stabilization times were unbounded in $\zeta'(\alpha)$, we could still not conclude that x appears on the output tape before that time, and thus that $x \in L_{\zeta'(\alpha)}$. We thus need to rule out this possibility.

Hence, we consider the function $f : \alpha \to \zeta'(\alpha)$ that maps each $\iota \in \alpha$ to the first time $\tau_\iota < \zeta'(\alpha)$ from which on C_ι is stable. In other words, we have $f(\iota) = \beta$ if and only if, for every $\delta > \beta$, the content of C_ι at time δ is the same as at time β, while there are unboundedly in β many ordinals $\delta < \beta$ such that the content of C_ι at time δ is different from that at time β. Thus $f(\iota) = \beta$ is Π_1-expressible in the parameter y by saying that, for every $\delta > \beta$, $L_{\delta+\omega}$ believes that the condition just formulated holds. So f is in particular Σ_2-definable over $L_{\zeta'(\alpha)}$. But now, the Σ_2-admissibility of $\zeta'(\alpha)$ proved above implies that there must be an upper bound $\rho < \zeta'(\alpha)$ for the range of f, which means that all cells of the output tape have indeed stabilized before time ρ, and hence that x is on the output tape at time ρ. As $L'_\zeta(\alpha)$ contains the partial computation of $P(y)$ of length $\rho + 1$, it also contains x, as desired.

(iii) This was already observed as a part of the proof of (i): When $P(\gamma)\downarrow = x$, then Corollary 3.6.11 implies that the whole computation of $P(\gamma)$ has length $< \lambda'(\alpha)$ and thus that it is contained in $L_{\lambda'(\alpha)}$, hence so is x. □

Corollary 3.6.13. *Let β be an ordinal.*
1. *If β is α-writable, then $\beta < \lambda'(\alpha)$.*
2. *If β is eventually α-writable, then $\beta < \zeta'(\alpha)$.*
3. *If β is accidentally α-writable, then $\beta < \Sigma'(\alpha)$.*

Proof. (i) If β is α-writable, then there is an α-writable $c \subseteq \alpha$ that codes β. By Lemma 3.6.12, we have $c \in L_{\lambda'(\alpha)}$. By admissibility of $L_{\lambda'(\alpha)}$ and decoding in KP-models, we have $\beta \in L_{\lambda'(\alpha)}$ and thus $\beta < \lambda'(\alpha)$, as desired.

(ii) now works exactly as (i). For (iii), note that, since $\Sigma'(\alpha)$ is a limit of admissible ordinals, there must be some admissible $\delta < \Sigma'(\alpha)$ such that L_δ contains an α-code for β; now, as in (i), we have $\beta < \delta$ and the claim follows from $\delta < \Sigma'(\alpha)$. □

3.6.2 The Σ_2-machine

The above reflection arguments led us to upper bounds on the computability strength of α-ITTMs. In the case of register models, we found that these bounds were indeed attained for certain values of α. We will now see that, for α-ITTMs, no further conditions on α besides multiplicative closure are needed.[12] The ideas, results and proofs in this subsection are easy generalizations of the analysis of the case $\alpha = \omega$ given by S. D. Friedman and P. Welch in [62].

One idea for obtaining lower bounds on α-ITTM-writability, eventual writability and accidental writability is to construct a machine that successively constructs α-codes for L-levels L_ξ, where $\xi < \Sigma'(\alpha)$. Since we can decide truth predicates in given structures, computing an α-code for $L_{\xi+1}$ from an α-code for L_ξ is not hard. However, after writing an infinite sequence of such codes to a certain tape T, there is no reason to expect the content of T at a limit time δ to encode L_δ; rather, it will be a jumbled mess, heavily depending on the details of the codings used below; in fact, it is easy to arrange things in such a way that all content of T at time δ will be 0, so all information is lost.

The crucial observation by Friedman and Welch in [62] that still makes this approach viable is that (i) levels below $\Sigma'(\alpha)$ can be encoded by their Σ_2-theory and (ii) that codes for theories are much more well behaved at limits than level codes. To illustrate the second point, suppose that, instead of writing codes for L-levels, the machine

12 Even multiplicative closure is only motivated by conveniences, such as being able to replace a finite sequence of parameters by a single parameter. We conjecture that dropping it will leave the following analysis intact, but make it messier. The interested reader is invited to try her hand at it.

would write their Σ_1-theories to the output tape T, writing 1 in cell $i \in \omega$ at stage ι if and only if $L_\iota \models \phi_i$ and 0 otherwise. Then, at a limit time δ of such ι, the content of T will indeed be the Σ_1-theory of L_δ, for a Σ_1-sentence holds in L_δ if and only if it holds in all sufficiently large L_ι with $\iota < \delta$. If in addition we knew some way to reconstruct a code for L_δ from its Σ_1-theory, the enumeration of L-levels would work just fine: Given a code for L_ι, we write its Σ_1-theory to T, compute a code for $L_{\iota+1}$, write its Σ_1-theory to the output tape, etc.; at limit levels δ, we obtain a code for L_δ from the Σ_1-theory of L_δ that will happen to be present on the output tape, and we can proceed.

Indeed, under the right circumstances, it is not hard to obtain a code for L_β from its Σ_1-theory: Namely, suppose that $L_\beta = \Sigma_1^{L_\beta}\{\emptyset\}$, i. e., every element of L_β is the image of some $n \in \omega$ under the canonical Σ_1-Skolem function h_1. Thus, we can take $n \in \omega$ to represent $h_1^{L_\beta}(n)$, and all elements of L_β will be represented. Moreover, to determine whether $h_1^{L_\beta}(n) \in h_1^{L_\beta}(m)$, one only needs to look up the truth value of the Σ_1-formula $\exists x, y(h_1^{L_\beta}(n) = x \wedge h_1^{L_\beta}(m) = y \wedge x \in y)$ in the Σ_1-theory of L_β; and we put $p(n, m)$ in the code for L_β if and only if this truth value is 1. The arising set is not yet quite what we want, as some elements of L_β will have multiple representations, while on the other hand, some elements of ω will fail to represent an element of L_β. But it is not hard to "clean up" the code to take care of this. In the general case of α-tapes, we would work with the Σ_1-theory with parameters in α and represent elements of L_β as Skolem values of the form $h_1^{L_\beta}(n, \iota)$ with $n \in \omega$ and $\iota < \alpha$.

However, this approach can only work up to the first γ for which there is a $\beta < \gamma$ with $L_\beta \prec_{\Sigma_1} L_\gamma$: For at this point, the Σ_1-theory for L_β and L_γ will be the same and thus the subroutine that computes structures from theories will again yield a code for L_β instead of one for L_γ, so that the machine will loop from that point on. In fact, it can be seen that one can work upwards exactly up to this point.

Since we want to write level codes up to the first γ for which there is $\beta > \alpha$ such that $L_\beta \prec_{\Sigma_2} L_\gamma$, we need to strengthen the above idea. This leads us to the idea of the Σ_2-theory machine, which will do for the Σ_2-theories what we sketched above for the Σ_1-theories, and this is exactly the idea of [62].

Definition 3.6.14 (Cf. [34]). Let $\beta, \gamma \in \mathrm{On}$. Then $\Sigma_2 - \mathrm{Th}(L_\gamma, \beta + 1)$ denotes the Σ_2-theory of L_γ with parameters $\leq \beta$, i. e., the set $\{(i, \iota) \in \omega \times \beta : L_\gamma \models \psi_i(\iota, \beta)\}$, where $(\psi_i : i \in \omega)$ enumerates the Σ_2-formulas in some natural way.

Theorem 3.6.15 (Friedman and Welch, see [62]). *Let α be multiplicatively closed. Then there is a program $P_{\alpha-\Sigma_2}$ with the following property: For any limit ordinal $\delta = \omega\tau < \Sigma'(\alpha)$, the output tape of $P_{\alpha-\Sigma_2}$ contains at time $\alpha^{\omega+2}\tau$ the Σ_2-theory of L_δ with parameters in $(\alpha + 1)$; more precisely, for all $i \in \omega$, $\iota < \alpha$, the $p(i, \iota)$th cell of the output tape contains 1 if and only if $(i, \iota) \in \Sigma_2 - \mathrm{Th}(L_\tau, \alpha)$ and otherwise, it contains 0.*

Moreover, $P_{\alpha-\Sigma_2}$ has a second tape T such that, for any limit ordinal $\delta = \omega\tau < \Sigma'(\alpha)$, the tape T contains an α-code for L_δ at time $\alpha^{\omega+2}\tau$.

We write $P_{\alpha\text{-level}}$ for the program that works like $P_{\alpha-\Sigma_2}$, but with the roles of T and the output tape exchanged.

The proof of this theorem will occupy the final part of this section.

But let us first see how Theorem 3.6.15 yields lower bounds on α-ITTM-computability.

3.6.3 α-ITTM-computability and the theory machine

Lemma 3.6.16. *Let $x \subseteq \alpha$.*

- *For each $x \subseteq \alpha$, if $x \in L_{\Sigma'(\alpha)}$, then x is α-aw.*
- *If $x \in L_{\lambda'(\alpha)}$, then x is α-writable.*
- *If $x \in L_{\zeta'(\alpha)}$, then x is α-ew.*

Proof. (i) Let $\tau < \Sigma'(\alpha)$ be such that $x \in L_\tau$. By Theorem 3.6.15, an α-code c for L_τ will appear on the output tape of $P_{\alpha-\text{level}}$. In c, x is coded by some $\iota < \alpha$. Thus, x is α-writable in the oracle c by a program that halts for every oracle. As the α-aw subsets of α are closed under α-writability by such programs (see Chapter 2, Theorem 2.5.10), it follows that x is α-aw.

(ii) Suppose that $a \in L_{\lambda'(\alpha)} \cap \mathfrak{P}(\alpha)$. By Lemma 3.6.8, there are $\beta < \lambda'(\alpha)$, $\gamma < \alpha$ and a Σ_1-statement $\phi \equiv \exists x\psi(x,\gamma)$ such that $a \in L_\beta$ and β is minimal with the property that $L_\beta \models \exists x\psi(x,\gamma)$. Now consider the following algorithm Q: We let $P_{\alpha\text{-level}}$ run. Whenever a (code for a new) L-level L_δ is written to the output tape, we test whether $L_\delta \models \phi$ using P_{truth}. If not, $P_{\alpha\text{-level}}$ continues. Otherwise, we must have $\delta \geq \beta$, and since $P_{\alpha\text{-level}}$ enumerates the L-levels one after the other and β is minimal with this property, we must have $\delta = \beta$ and we have written an α-code c for L_β. Now, since $a \in L_\beta$, a is coded by some ordinal $\iota < \alpha$ in c. Thus, in the parameter ι, a is α-writable from c and as we already saw that c is α-writable, it follows that a is α-writable.

(iii) This is a slightly more intricate reuse of the idea in (ii). Let $a \in L_{\zeta'(\alpha)}$. It follows from Lemma 3.6.8 that there are $\beta < \zeta'$, $\gamma < \alpha$ and a Σ_2-statement $\phi \equiv \exists x\forall y\psi(x,y,z)$ (where ψ is Δ_0) such that $a \in L_\beta$ and β is minimal with the property that it contains some b such that $L_{\Sigma'(\alpha)} \models \forall y\psi(b,y,\gamma)$. We will show that an α-code c for L_β is α-ew. It then follows that a is α-writable from c and thus, by closure of the α-ew sets under α-writability, that a is α-ew.

First note that, since $P_{\alpha\text{-level}}$ successively writes α-codes for all L-levels L_δ with $\delta < \Sigma'(\alpha)$, it will eventually write a code c for L_β. What we need to achieve, then, is that once this happens, c will remain on the output tape, so that c is eventually written.

So consider the following algorithm Q: We let $P_{\alpha\text{-level}}$ run; for reasons soon to become apparent, we will call this the "outer run" of $P_{\alpha\text{-level}}$. We arrange the whole program so that the inner states of Q are numerated in such a way that all states directing this run are smaller than all states for the subroutine to be described next. Whenever a new α-code d for an L-level L_δ (where the coding works via the bijection $f : \alpha \to L_\delta$) is written to the output tape, we interrupt the computation of $P_{\alpha\text{-level}}$ and, on a sufficient number of extra tapes, run the subroutine P_{ct} ("candidate testing") that works as follows:

P_{ct} has a separate 'counting tape' T_{count}, which is initially empty and which, at any time during the run of P_{ct}, will contain a sequence of 1s, followed by 0s. Say T_{count} contains a ι-string of 1s while the rest of the tape is filled with 0s. If $\iota = \alpha$, i. e., if there are no 0s left on T_{count}, the outer run of $P_{\alpha-\text{level}}$ is continued and all tapes used by P_{ct} are erased. Otherwise, we start a new run of $P_{\alpha\text{-level}}$ on some extra tapes, called the "inner run." (Note that this will not modify in any way the tape contents from the outer run.) When this writes an α-code e for an L-level L_η (where the coding works via the bijection $f' : \alpha \to L_\eta$) to the output tape, we first test whether $\eta \geq \delta$. If not, we continue the inner run of $P_{\alpha\text{-level}}$. If yes, we use the program P_{count} from Corollary 2.3.31 in Chapter 2 to compute $\iota' < \alpha$ such that $f(\iota) = f'(\iota')$; also, we use P_{fo} to identify $\xi < \alpha$ such that $f'(\xi) = y$. Then we use P_{truth} in the oracle e to check whether $L_\eta \models \forall y\psi(f'(\iota'), y, y)$. If yes, then the inner run of $P_{\alpha\text{-level}}$ is continued. Otherwise, all tapes used in the inner run are deleted and we again start P_{ct} (note that, due to our having added a 1 on T_{count}, this will now run with $\iota + 1$ in place of ι.)

We claim that Q eventually writes an α-code for L_β. To see this, first consider the case that a code d for an L_δ with $\delta < \beta$ is generated by the outer run of $P_{\alpha\text{-level}}$. In this case, P_{ct} will run through all elements of $\iota < \alpha$, and for each of these, the following will happen: As $L_{\Sigma'(\alpha)} \not\models \forall y\psi(f(\iota), y, y)$ by minimality of β, the inner run of $P_{\alpha\text{-level}}$ will eventually produce a code for an L-level L_η containing some $\hat{a}$ such that $L_{\Sigma'(\alpha)} \models \neg\psi(f(\iota), \hat{a}, y)$; as ψ is Δ_0, we then also have $L_\eta \models \neg\psi(f(\iota), \hat{a}, y)$, so $L_\eta \not\models \forall y\psi(f(\iota), \hat{a}, y)$. This will be detected by P_{truth} and cause P_{ct} to continue with $\iota + 1$. After a limit number of repetitions of P_{ct}, P_{count} will either contain a sequence of 1s of limit length ρ, followed by 0s, in which case what we just described for ι will happen for ρ, or it will be filled with 1s, in which case P_{ct} is stopped, its scratch work is erased and the outer run is continued to produce a new code for an L-level. After this has happened a limit number of times, say τ many times, the tapes used by P_{ct} will be blank by the liminf-rule (having cofinally often been erased) while the tapes used in the outer run of $P_{\alpha\text{-level}}$ will have the content that the tapes of $P_{\alpha\text{-level}}$ would usually have after τ many L-levels have been produced; and by our choice of the enumeration of the inner states, the inner state of Q will correspond to the inner state that $P_{\alpha\text{-level}}$ assumes after producing τ many L-levels, so that it can continue to produce a code for L_τ.

Proceeding in this way, any L_δ with $\delta < \beta$ produced in the outer run will eventually be discarded and replaced. Hence, eventually, the second case must occur, namely that the outer run produces a code d for L_β. In this case, P_{ct} will search through the elements of L_β, discarding any $\hat{a} \in L_\beta$ for which $L_{\Sigma'(\alpha)} \not\models \forall y\psi(\hat{a}, y, y)$. By definition of β, it will eventually arrive at some $\iota < \alpha$ such that $L_{\Sigma'(\alpha)} \models \forall y\psi(f(\iota), y, y)$. From this point on, d and ι will never be changed again, and the inner run will just continue producing codes for ever higher L-levels L_η in which $\forall y\psi(f(\iota), y, y)$ will hold, and eventually loop. Thus, d will remain on the output tape. So d is eventually written by Q (in the parameter y), as desired. □

Exercise 3.6.17. Welch's first analysis of the computational strength of ITTMs in [184] used the universal ITTM-program $\mathcal{U}$ rather than the theory machine. Let us for the sake of this exercise take (i) of Lemma 3.6.16 for granted. Thus, $\mathcal{U}_\alpha$ will accidentally write codes for all L_ξ with $\xi < \Sigma'(\alpha)$, though it will do so in a much less orderly manner than the theory machine; in particular, earlier levels might be written much later than later ones, levels will be written repeatedly, etc.

(a) Show (ii) of Lemma 3.6.16 using $\mathcal{U}_\alpha$ rather than the theory machine.

(b) Show (iii) of Lemma 3.6.16 using $\mathcal{U}_\alpha$ rather than the theory machine.

Combining Lemma 3.6.16 and Corollary 3.6.12, we finally get Welch's characterization of α-ITTM-computability, known as the λ-ζ-Σ-theorem.

Theorem 3.6.18 (Welch, [184]). *Let $x \subseteq \alpha$.*
- *x is α-aw if and only if $x \in L_{\Sigma'(\alpha)}$.*
- *x is α-ew if and only if $x \in L_{\zeta'(\alpha)}$.*
- *x is α-writable if and only if $x \in L_{\lambda'(\alpha)}$.*

Corollary 3.6.19 (Welch, [184]). *Let β be an ordinal.*
- *β is α-writable if and only if $\beta < \lambda'(\alpha)$.*
- *β is eventually α-writable if and only if $\beta < \zeta'(\alpha)$.*
- *β is accidentally α-writable if and only if $\beta < \Sigma'(\alpha)$.*

Proof. By Lemma 3.6.13, we only need to show the directions from right to left. For (i), this works as follows: By Theorem 3.6.15, $P_{\alpha\text{-level}}$ will produce a code c for L_β in $< \beta^+$ many steps. As $\beta < \lambda$ and λ is a limit of admissible ordinals by Lemma 3.6.6, this will happen in $< \lambda'(\alpha)$ many steps. Thus, the computation leading up to c, and hence c itself will be contained in $L_{\lambda'(\alpha)}$. It then easily follows that $L_{\lambda'(\alpha)}$ also contains an α-code for β, which is hence α-writable by Theorem 3.6.18. (ii) and (iii) now work analogously. □

Corollary 3.6.20 (Welch, [184]). *$\lambda(\alpha)$ is the supremum of the α-clockable ordinals. Thus $\lambda(\alpha) = \gamma(\alpha)$ (in the notation of Chapter 2). In particular, every α-ITTM-clockable ordinal is also α-ITTM-writable.*

Proof. First, suppose that $\beta < \lambda'(\alpha)$. Then, by Corollary 3.6.19, β is α-writable. Now, given an α-code b for β, we can run through b, deleting in each step the remaining element of α that is minimal in the sense of b, and thus run for β many steps and halt afterwards. This will take $\geq \beta$ many steps, and thus, there is an α-clockable ordinal $\geq \beta$, so $\beta < \gamma(\alpha)$. As $\beta < \lambda'(\alpha)$ was arbitrary, it follows that $\lambda'(\alpha) \leq \gamma(\alpha)$.

On the other hand, suppose that $\beta < \gamma(\alpha)$. Thus, there is an α-ITTM-clockable limit ordinal $\gamma > \beta$ (note that, if γ is clockable, then so is $\gamma + \omega$; we can thus assume without loss of generality that γ is a limit ordinal). Let P be an α-ITTM-program, $\iota < \alpha$ such that $P(\iota)$ clocks γ. Now run $P(\iota)$ along with $P_{\alpha\text{-level}}$ in the sense that each time $P_{\alpha\text{-level}}$ writes a new level code to its output tape, one step of $P(\iota)$ is carried out and halt when $P(\iota)$ halts. This will halt with an α-code c for L_γ on the output tape. Now, from c, it is easy to compute a code for γ. Thus γ is writable, and hence $< \lambda'(\alpha)$ by Corollary 3.6.19. It follows that $\gamma(\alpha) \leq \lambda'(\alpha)$.

Taken together, we obtain $\lambda'(\alpha) = \gamma(\alpha)$, as desired. □

We now turn to the proof of Theorem 3.6.15. We begin by showing how to (quickly) compute $\Sigma_2 - \text{Th}(L_{\beta+1}, \alpha)$ from $\Sigma_2 - \text{Th}(L_\beta, \alpha)$.

3.6.3.1 The theory machine at successor levels

At successor levels, the theory machine works as follows: Given $T_\beta := \Sigma_2 - \text{Th}(L_\beta, \alpha)$, we first compute an α-code c for L_β from T_β, then we compute an α-code d for $L_{\beta+1}$ from c and finally, we compute $\Sigma_2 - \text{Th}(L_{\beta+1}, \alpha)$ from d. The last two steps were already dealt with: Computing a code for $L_{\beta+1}$ from a code for L_β is comparably easy using P_{truth}, and computing the Σ_2-theory of $L_{\beta+1}$ with parameters in α from a code for $L_{\beta+1}$ is again possible by using P_{truth}. The main difficulty is thus the first step, to pass from the Σ_2-theory to a code for the structure. This problem will now be attacked.

We need some prerequisites.

Definition 3.6.21. For $\beta < \gamma$, we say that β is Σ_1-stable in L_γ if and only if $L_\beta \prec_{\Sigma_1} L_\gamma$. Let $S_\gamma := \{\beta < \gamma : L_\beta \prec_{\Sigma_1} L_\gamma\}$.

Proposition 3.6.22. *Let β be a limit ordinal. Then the condition that β' is Σ_1-stable in β is Π_1-expressible over L_β.*

Proof. The condition that $L_{\beta'} \prec_{\Sigma_1} L_\beta$ means that, for every Σ_1-formula $\phi_i \equiv \exists x \psi_j(x, y)$ with $\psi_j \in \Delta_0$ and all $x \in L_\beta$, $y \in L_{\beta'}$, if $L_\beta \models \psi_j(x, y)$, then $L_{\beta'} \models \exists z \psi_j(z, y)$. Using the Δ_1-truth predicate from Lemma 3.1.11 above to express $\psi_j(x, y)$ uniformly in j, x and y as Σ_1 and $L_{\beta'} \models \exists z \psi_j(z, y)$ uniformly in β', j and y as Π_1 (and replacing $A \rightarrow B$ with $\neg A \vee B$), we obtain the desired Π_1-formula. □

Lemma 3.6.23. *For every γ, S_γ is closed; i. e., if $S_\gamma \cap \delta$ is unbounded in δ for some $\delta < \gamma$, then $\delta \in S_\gamma$. Consequently, S_γ is either unbounded in γ or has a maximal element.*

Proof. Let $\delta < \gamma$ be such that $S_\gamma \cap \delta$ is unbounded in δ, let $\vec{p} \subseteq L_\delta$ and let ϕ be a Σ_1-statement such that $L_\gamma \models \phi(\vec{p})$. Since δ is a limit ordinal and $\vec{p}$ is finite, there is $\delta' < \delta$ such that $\vec{p} \subseteq L_{\delta'}$. By assumption, there is $\gamma' \in S_\gamma \cap \delta$ such that $\gamma' > \delta'$. Hence $\vec{p} \subseteq L_{\gamma'} \prec_{\Sigma_1} L_\gamma$, so $L_{\gamma'} \models \phi(\vec{p})$. By upwards absoluteness of Σ_1-formulas, we have $L_\delta \models \phi(\vec{p})$, as desired. □

Proposition 3.6.24. *There is a surjective map $s : \alpha \rightarrow \omega \times \alpha^{<\omega}$ which is Σ_1 over L_α.*

Proof. We can define a surjection $t : \alpha \to \omega \times \alpha$ by writing every $\beta < \alpha$ as $\beta = p(\gamma, \delta)$ and further writing γ in the form $\gamma = \omega\gamma' + k$, where $k \in \omega$; then, we can send β to (k, δ). As α is closed under Cantor pairing, t is surjective. Moreover, we have a surjection $t' : \alpha \to \alpha^{<\omega}$ that works as follows: Given $\beta < \alpha$, write it in the form $\beta = p(\omega\gamma + n, \delta)$ with $\gamma, \delta \in \alpha$, $n \in \omega$, then let $\delta = p(\delta_0, \dots, \delta_{n-1})$ and let $t(\beta) = (\delta_0, \dots, \delta_n)$. First, applying t and then applying t' to the second component of the resulting pair is the desired map s, which is clearly Σ_1 over L_α. □

Lemma 3.6.25. *Let $\delta < \Sigma'(\alpha)$ be a limit ordinal with $\delta > \alpha$.*
(i) *If S_δ has a maximal element β or if $S_\delta \cap (\alpha, \delta) = \emptyset$, then $\Sigma_1^{L_\delta}(\alpha + 1 \cup \{\beta\}) = L_\delta$.*
(ii) *In any case, we have $H := \Sigma_1^{L_\delta}(\alpha + 1 \cup S_\delta) = L_\delta$.*
(iii) *In fact, if $x \in L_\delta$, then there are $n \in \omega$, $\iota \in \alpha$ and $\beta \in S_\delta \cup \{0\}$ such that $x = h_1^{L_\delta}(n, (\iota, \alpha, \beta))$.*

Proof. (i) Suppose first that S_δ has a maximal element. By condensation, $\Sigma_1^{L_\delta}(\alpha+1\cup\{\beta\})$ is isomorphic to some L-level L_γ, and moreover, by Lemma 3.6.5(ii), it is transitive; thus $\Sigma_1^{L_\delta}(\alpha + 1 \cup \{\beta\}) = L_\gamma$, $\gamma \leq \delta$ and we only need to exclude $\gamma < \delta$. So suppose $\gamma < \delta$ for a contradiction. Now, by Lemma 3.6.5, L_δ, and hence $\Sigma_1^{L_\delta}(\alpha + 1 \cup \{\beta\})$, will contain a surjection $f : \alpha \to \beta$, and thus $\beta = f[\alpha] \subseteq L_\gamma$ as in the proof of Lemma 3.6.5(ii). As $\beta \in L_\gamma$ by definition of L_γ, we have $\gamma > \beta$. However, we have $L_\gamma \prec_{\Sigma_1} L_\delta$ by Lemma 3.1.21, so we have contradicted the maximality of β.

If $S_\delta \cap (\alpha, \delta) = \emptyset$, consider $H := \Sigma_1^{L_\delta}(\alpha+1)$. By Lemma 3.6.5(ii), H is transitive. By the condensation lemma, it is isomorphic to some L_γ with $\gamma \leq \delta$ and thus, by transitivity we in fact have $H = L_\gamma$. Also, $H \prec_{\Sigma_1} L_\delta$ and as $\alpha + 1 \subseteq H = L_\gamma$, we have $\gamma > \alpha$. As there are no ordinals in (α, δ) that are Σ_1-stable in L_δ, we must have $\gamma = \delta$.

(ii) If S_δ has a maximal element, this follows from (i). If S_δ does not have a maximal element, then it follows from Lemma 3.6.23 that S_δ is unbounded in δ. Hence, if $\gamma < \delta$, then there is $\gamma' \in S_\delta$ with $\gamma' > \gamma$. By Lemma 3.6.5, L_δ contains a surjection $f : \alpha \to L_{\gamma'}$, and thus $\gamma \in \gamma' = f[\alpha] \subseteq H$. Thus, every $\gamma < \delta$ is an element of H, so $\delta \subseteq H$, and hence H cannot be isomorphic to any L_γ with $\gamma < \delta$. It follows from condensation that $H \simeq L_\delta$ and from Lemma 3.6.5(ii) that $H = L_\delta$.

(iii) First, note that, since α is closed under the pairing function, quantification over finite sequences of elements of α can be replaced with quantification over elements of α. Now, if S_δ has a maximal element or $S_\delta \cap (\alpha, \delta) = \emptyset$, then the claim follows from (i). On the other hand, if S_δ is unbounded in δ, then there is $\beta \in S_\delta$ such that $x \in L_\beta$. As in the proof of (ii), we have $x \in L_\beta \subseteq \Sigma_1^{L_\delta}(\alpha + 1 \cup \{\beta\})$ and the claim follows again. □

This suggests that any element $a \in L_\delta$ can be "named" by the smallest triple $(n, \iota, \beta) \in \omega \times \alpha \times (S_\delta \cup \{0\})$ such that $a = h_1^{L_\delta}(n, (\iota, \alpha, \beta))$.

Lemma 3.6.26 (See [62] and [34]). *There is an α-ITTM-program $P_{\alpha-\text{lft}}$ ("level from theory") such that, for every limit ordinal $\gamma < \Sigma'(\alpha)$, $P_{\alpha-\text{lft}}$ computes an α-code of L_γ from $\Sigma_2 - \text{Th}(L_\gamma, \alpha + 1)$ in $< \alpha^3$ many steps.*

Proof. This proof is a generalized version of the proof of Lemma 2 of [62].

First, for any Π_1-formula ϕ_n and any $\iota < \alpha + 1$ such that $\phi_n(x, \iota)$ has a witness $a \in L_\gamma$, we pick such a witness as follows: By Lemma 3.6.25, there are $\xi < \alpha + 1$, $m \in \omega$, $\beta \in S_\gamma \cup \{0\}$ such that $a = h_1^{L_\gamma}(m, (\xi, \alpha, \beta))$. We now pick such a triple (m, ξ, β) lexically minimal (i. e., first minimize m, then ξ and finally β) and let $h_1^{L_\gamma}(m, (\xi, \alpha, \beta))$ be our witness. Let W be the set of these witnesses for all Π_1-formulas ϕ_n and all $\iota < \alpha + 1$. Note that $\alpha + 1 \subseteq W$, as each formula $x = \iota$, $\iota \in \alpha + 1$, will have a witness in W.

Now, for any $\xi \in \alpha + 1$, $m \in \omega$, the statement $A^{n,\iota}_{m,\xi}$:

"There is a $\beta \in S_\gamma \cup \{0\}$ such that $h_1^{L_\gamma}(m, (\xi, \alpha, \beta)) = x$ is defined and $\phi_n(x, \iota)$ holds" is Σ_2 in the parameters m, ξ, ι, α, namely it can be written as

$$\exists\beta\exists x[x = h_1(m, (\xi, \alpha, \beta)) \wedge (\beta = 0 \vee \beta \in S_\gamma) \wedge \phi_n(x, \iota)]$$

where $x = h_1(m, (\xi, \alpha, \beta))$ is Σ_1, $\beta = 0 \vee \beta \in S_\gamma$ is Π_1 by Proposition 3.6.22, and $\phi_n(x, \iota)$ is Π_1.

Thus, we can, for all $m \in \omega$, $\xi \in \alpha + 1$, determine from $\Sigma_2 - \text{Th}(L_\gamma, \alpha + 1)$ whether $A^{n,\iota}_{m,\xi}$ holds in L_γ. In this way, we decide can whether there is a pair (m, ξ) for which $A^{n,\iota}_{m,\xi}$ holds in L_γ and if there is one, we can determine the lexically minimal of these pairs. Moreover, we can store all the quadruples (n, ι, m, ξ) where (m, ξ) is lexically minimal with $A^{n,\iota}_{m,\xi}$ on a separate tape T; let us write F for the function sending (n, ι) to this (m, ξ); then we have now seen that F is α-ITTM-computable from $\Sigma_2 - \text{Th}(L_\gamma, \alpha + 1)$.

We now form the Σ_1-hull of $(\alpha + 1) \cup W$ in L_γ. By Lemma 3.6.5, this will be some L-level L_δ. We claim that $L_\delta \prec_{\Sigma_2} L_\gamma$, which implies $\delta = \gamma$ by the assumption that L_γ has no proper transitive Σ_2-submodels that contain $\alpha + 1$.

First, $L_\delta \prec_{\Sigma_1} L_\gamma$, as it is a Σ_1-hull of a subset of L_γ. It follows that Σ_2-formulas are preserved upwards between L_δ and L_γ by Lemma 3.6.3.

So let $\phi(\vec{p}) \equiv \exists x\forall y\psi(x, y, \vec{p}) \in \Sigma_2$, $\psi \in \Delta_0$ with parameters $\vec{p} \in L_\delta$ such that $L_\gamma \vDash \phi(\vec{p})$. We need to show that $L_\delta \vDash \phi(\vec{p})$. We do this in two steps:

STEP 1: Getting rid of the parameters.

We want to obtain an equivalent Σ_2-formula that only has parameters in $\alpha + 1$. To this end, we need to define $\vec{p}$ by a Σ_2-formula. For simplicity, let us say that $\vec{p}$ consists of a single element p. By definition of L_δ, there is a finite sequence $\vec{w}$ of elements of W and some $k \in \omega$ such that $p = h_1^{L_\gamma}(k, \vec{w})$. Let us for the sake of simplicity also assume that $\vec{w}$ only has a single element w. As $w \in W$, there are a Π_1-formula ϕ_n and some $\iota \in \alpha + 1$ such that w is the minimal (in the sense of having a lexically minimal preimage under $h_1^{L_\gamma}$ as explained above) with $L_\gamma \vDash \phi_n(w, \iota)$. Also, we know that there are lexically minimal $(m, \xi) \in \omega \times (\alpha + 1)$ such that $w = h_1(m, (\xi, \beta))$ for some $\beta \in S_\gamma \cup \{0\}$; and we can use m and ξ as parameters, as they are elements of $\alpha + 1$.

Moreover, w is also minimal in the same sense with the property that $L_\delta \models \phi_n(w,\iota)$: For if L_δ knew a smaller witness, then it would also be one in L_γ, as Σ_1-stables in L_δ are also Σ_1-stable in L_γ and Π_1-statements are absolute between L_δ and L_γ since $L_\delta \prec_{\Sigma_1} L_\gamma$ by Lemma 3.6.3.

Thus, w is characterized uniquely (both in L_γ and L_δ) as the only x satisfying the following formula ϕ_w:

$$\begin{aligned}&\exists\beta[x = h_1(m,(\xi,\beta)) \wedge \phi_n(x,\iota) \wedge \text{`}\beta > \alpha \text{ is } \Sigma_1\text{-stable} \vee \beta = 0\text{'}\\ &\wedge [\beta = 0 \vee L_\beta \models \forall\beta'(\text{`}\beta' \text{ is not } \Sigma_1\text{-stable'}\\ &\vee \forall y(y \neq h_1(m,(\xi,\beta'))) \vee (\neg\phi_n(h_1(m,(\xi,\beta')),\iota)))]].\end{aligned}$$

Here, $x = h_1(m,(\xi,\beta))$ is Σ_1, $\phi_n(x,\iota)$ is Π_1, "$\beta > \alpha$ is Σ_1-stable $\vee\ \beta = 0$" is Π_1 and the last conjunct is Δ_0 since all quantifiers are bounded by L_β. Thus, the formula is Σ_2.

Now, since h_1 is Σ_1-definable, $p = h_1^{L_\gamma}(k,w)$ is uniquely characterized via a Σ_2-formula ϕ_p by Proposition 3.1.5.

But then, we can express $\phi(\vec{p})$ as $\exists x,z\forall y(\phi_p(z) \wedge \psi(x,y,z))$, which is now Σ_2 in parameters from $(\alpha+1)$.

STEP 2: Preservation for Σ_2-formulas with parameters in $\alpha+1$.

Suppose that $L_\gamma \models \exists x\forall y\psi(x,y,p)$ for some $p \in \alpha+1$. Deleting the existential quantifier, we obtain $\forall y\psi(x,y,p)$, which is Π_1 with parameters from $(\alpha+1)$. Now, by assumption, L_γ has a witness for this, and hence such a witness w was included in $W \subseteq L_\delta$. But then, $L_\delta \models \forall y\psi(w,y,p)$, i. e., $L_\delta \models \exists x\forall y\psi(x,y,p)$, as desired. It follows that a Σ_2-statement with parameters in $\alpha+1$ holds in L_γ if and only if it holds in L_δ.

Combining step 1 and step 2, it now follows that $L_\delta \prec_{\Sigma_2} L_\gamma$.

Now we show how to compute a surjection from α to $L_\delta = L_\gamma$ on an α-ITTM from $\Sigma_2 - \text{Th}(L_\gamma,\alpha+1)$. Recall that F is α-ITTM-computable from $\Sigma_2 - \text{Th}(L_\gamma,\alpha+1)$.

Elements of L_δ are obtained as $h_1(k,\vec{w})$, where $k \in \omega$ and $\vec{w} = (w_0,\dots,w_l) \subseteq W$. Elements of W are in turn given by quadruples (n,ι,m,ξ) with $F(n,\iota) = (m,\xi)$. Thus, elements of L_δ can be represented as pairs $(k,(q_0,\dots,q_l))$, where $k \in \omega$ and the q_i are such quadruples. Clearly, we can compute a bijection between α and these pairs.

Thus, all that remains to be done to compute a code for L_δ is to determine when, for two such pairs $p := (k,(q_0,\dots,q_l))$ and $p' := (k',(q'_0,\dots,q'_l))$, the first represents an element of the set represented by the second.

Let us say that $a(p)$ is the element represented by p and $a(p')$ the one represented by p'. Then we already saw above (in STEP 1) how $a(p)$ and $a(p')$ can be uniquely characterized by Σ_2-formulas ϕ, ϕ' with parameters in $\alpha+1$; moreover, ϕ and ϕ' can be effectively obtained on an α-ITTM from p and p' as described in STEP 1 above.

Thus, $a(p) \in a(p')$ is equivalent to $\exists x,y(\phi(x) \wedge \phi'(y) \wedge x \in y)$, which is Σ_2 with parameters in $(\alpha+1)$ and thus can be read off from $\Sigma_2 - \text{Th}(L_\gamma,\alpha+1)$.

Hence, we obtain our α-code c for L_γ by putting $p(\iota,\iota')$ in c if and only if ι and ι' code pairs p, p' such that the Σ_2-statement expressing $a(p) \in a(p')$ belongs to $\Sigma_2 - \text{Th}(L_\gamma,\alpha+1)$.

Concerning the time bound, let us first note that, in order to compute F, the worst case is that one has to run through $\Sigma_2 - \text{Th}(L_\gamma, \alpha + 1)$ once for every $(n, \iota) \in \omega \times \alpha$, which takes $\leq \omega\alpha^2$, and hence, by multiplicative closure of α, $\leq \alpha^2$ many steps. Computing the Σ_2-statement takes $< \alpha$ many steps, and verifying whether the statement belongs to $\Sigma_2 - \text{Th}(L_\gamma, \alpha + 1)$ is done by running once through α, and thus takes $\leq \alpha$ many steps. Since α is multiplicatively closed, it is also additively closed, and thus this part of the procedure takes place in time $\leq \alpha$. In total, computing a level code from $\Sigma_2 - \text{Th}(L_\gamma, \alpha+1)$ thus takes $\leq \alpha^2 + \alpha < \alpha^3$ many steps. □

Lemma 3.6.27.

(i) *There is an α-ITTM-program $P_{\alpha\text{-nl}}$ ('next level') such that, for all $\tau \in \text{On}$, $P_{\alpha\text{-nl}}$ computes an α-code for $L_{\tau+1}$ from an α-code for L_τ in $< \alpha^\omega + \alpha^3$ many steps.*

(ii) *There is an α-ITTM-program $P_{\alpha\text{-nll}}$ ('next limit level') such that, for all $\tau \in \text{On}$, $P_{\alpha\text{-nl}}$ computes an α-code for $L_{\tau+\omega}$ from an α-code for L_τ in $\leq \alpha^\omega \omega$ many steps.*

Proof. (i) Let c be an α-code for L_τ. We start by computing a code for the elementary diagram $\text{Diag}_{\text{el}}(L_\tau)$[13] in the sense that the tape cell with index $p(n, \gamma_0, \dots, \gamma_k)$ will contain 1 if and only if $L_\tau \models \phi_n(a_0, \dots, a_k)$, where γ_i codes a_i in the sense of c. When $\phi_n \in \Sigma_m$, this requires m nested searches through α, while evaluating quantifier-free formulas is possible in time $< \alpha$ by multiplicative closure of α. Thus, computing $T := \text{Diag}_{\text{el}}(L_\tau)$ is possible in time α^ω. Now, to compute an α-code for $L_{\tau+1}$ from T, we proceed as follows: Split the output tape into two disjoint portions. On one portion, a copy of c is written. The other is reserved for elements of $L_{\tau+1} \setminus L_\tau$; these can be named by pairs of the form $(n, \vec{\gamma})$, where $\vec{\gamma}$ is a finite sequence of elements of α and the element coded by that pair is $\{x \in L_\tau : L_\tau \models \phi_n(x, \vec{p})\}$, where $\vec{p}$ is the sequence of elements of L_τ coded by the elements of $\vec{\gamma}$ in the sense of c. Now determine, for any $\iota < \alpha$ and any pair $(n, \vec{\gamma})$, whether the element a_ι of L_τ coded by ι in the sense of c is equal to or an element of the element $a(n, \vec{\gamma})$ of $L_{\tau+1}$ named by $(n, \vec{\gamma})$. This is done by simply evaluating whether the formulas $\forall x(x \in a_\iota \leftrightarrow \phi_n(x, \vec{p}))$ and $\phi_n(a_\iota, \vec{p})$ hold in L_τ, which can be looked up in $\text{Diag}_{\text{el}}(L_\tau)$. For a given $\iota < \alpha$ and a given pair, this takes $< \alpha$ many steps by multiplicative closure of α. Doing this for every $\iota < \alpha$ and every such pair thus takes α^2 many steps. Now, when $a(n, \vec{\gamma})$ is not equal to any a_ι, then the $p(n, \vec{\gamma})$th element of the second tape portion is used to represent $a(n, \gamma)$; otherwise, the element is already present, and does not need to be coded again. The $\in$-relation is then updated according to the results of the routine mentioned above (note that no element of $L_{\tau+1} \setminus L_\tau$ can be element of another such element). This second step can be carried out in $< \alpha^3$ many steps. Thus, the whole computation takes $< \alpha^\omega + \alpha^3$ many steps.

13 For a first-order language $\mathcal{L}$, the elementary diagram of a $\mathcal{L}$-structure S is obtained by amending $\mathcal{L}$ with a constant symbol for each element of S and then taking the set of all sentences in the amended language that hold in S. See, e. g., Marker [136] for the notion of an elementary diagram.

(ii) To compute an α-code for $L_{\tau+\omega}$ from an α-code c for L_τ, we split the output tape into ω many disjoint portions and then iterate the above ω many times to successively write codes for $L_{\tau+i}$, $i \in \omega$, using the 0th portion for L_τ and the $(i+1)$th portion to encode elements of $L_{\tau+i+1} \setminus L_{\tau+i}$, for $i \in \omega$. This can be done in $\leq (\alpha^\omega + \alpha^3)\omega \leq \alpha^\omega \omega$ many steps. □

Lemma 3.6.28 (Cf. [34], Lemma 7). *There is an α-ITTM-program $P_{\alpha\text{-tfl}}$ ('theory from level') such that, for every $\tau \in$ On, $P_{\alpha\text{-lft}}$ computes $\Sigma_2(L_\tau, \alpha+1)$ from an α-code for L_τ in α^3 many steps.*

Proof. This is a simple application of P_{truth}. To evaluate a Σ_2-statement in a structure given by an α-code, the algorithm will need α^2 many steps. This needs to be done for any pair (i, y) of an index of a Σ_2-formula and a parameter, and these can be ordered in order type α. □

Lemma 3.6.29 (Cf. [34], Lemma 7). *There is an α-ITTM-program $P_{\alpha\text{-next}}$ such that, for any $\tau < \Sigma'(\alpha)$, $P_{\alpha\text{-next}}$ will compute $\Sigma_2 - \text{Th}(L_{\gamma+\omega}, \alpha+1)$ from $\Sigma_2 - \text{Th}(L_\gamma, \alpha+1)$ in $\leq \alpha^{\omega+1}$ many steps.*

Proof. This is an immediate consequence of combining Lemma 3.6.26, Lemma 3.6.27 and Lemma 3.6.28: First, compute an α-code for L_γ from $\Sigma_2 - \text{Th}(L_\gamma, \alpha+1)$ via $P_{\alpha\text{-lft}}$, then use $P_{\alpha\text{-nll}}$ to compute an α-code c for $L_{\gamma+\omega}$ from the code for L_γ, finally use P_{tfl} to read off $\Sigma_2 - \text{Th}(L_{\gamma+\omega}, \alpha+1)$ from c. By the running time estimates given in the lemmata, this takes $< \alpha^3 + \alpha^\omega\omega + \alpha^3 \leq \alpha^{\omega+1}$ many steps. □

3.6.3.2 The theory machine at limit levels

The theory machine almost works at limit levels: Namely, when τ is a limit of limit ordinals, then after successively writing $\Sigma_2 - \text{Th}(L_\iota, \alpha+1)$ to the output tape for all $\iota < \tau$, the output tape will contain a 1 on the cell corresponding to $\phi(\iota, \alpha)$ for $\phi \equiv \exists x \forall y \psi(u, v, x, y) \in \Sigma_2$, $\psi \in \Delta_0$, $\iota \in \alpha$ if and only if $\phi(\iota, \alpha)$ was true in all sufficiently large L_δ, $\delta < \tau$ a limit ordinal. That is, the output tape contents tells us whether the Σ_2-formula in question was true of the respective parameters from certain point on. Clearly, if $L_\tau \models \phi(\iota, \alpha)$ then the corresponding cell will contain a 1: for there must be some $a \in L_\tau$ such that $L_\tau \models \forall y(\iota, \alpha, a, y)$, and as τ is a limit of limit ordinals, there is a limit ordinal $\gamma < \tau$ such that $a \in L_\gamma$. But then, by downwards absoluteness of Π_1-formulas, we have $L_\delta \models \forall y \psi(\iota, \alpha, a, y)$, and thus $L_\delta \models \phi(\iota, \alpha)$ for all limit ordinals $\delta \in [\gamma, \tau)$. Thus, true Σ_2-statements will receive a 1 in the limit.

Unfortunately, the reverse implication is not true: As was pointed out in [62], if $\phi(\iota, \alpha)$ holds in all sufficiently large L_δ with $\delta < \tau$ a limit ordinal, it may still fail in L_τ. An easy example is the formula "there is a largest limit ordinal," i. e., 'there is a limit ordinal β such that $\gamma \leq \beta$ for every limit ordinal." This is clearly true in any level limit level of the L-hierarchy that is not a limit of limit levels, while it fails in every limit of limit levels and is thus the worst kind of counterexample one could think off. The problem is that the witness a for the existential quantifier changes: Every new limit level has a new largest limit ordinal, while the old witness ceases to work. We need to be able to fix a witness in a way that remains absolute between various L-levels. One way to do this would be to describe the witness as a certain value of the canonical Σ_1-Skolem function; this is the idea behind the proof of Lemma 3.6.31.

Definition 3.6.30. Let δ be a limit ordinal. Then the limit Σ_2-theory $\Sigma_2 - \text{limit}(L_\delta, \alpha+1)$ of L_δ with parameters in α is defined as the set of all Σ_2-statements with parameters in $\alpha+1$ that hold in all sufficiently large L-levels before L_δ; formally, the definition is as follows:

$$\Sigma_2 - \text{limit}(L_\delta, \alpha+1) := \{(i,\iota) \in \omega \times (\alpha+1) : \exists \beta < \delta \forall \gamma < \delta(\beta < \gamma \rightarrow L_\gamma \models \psi_i(\iota))\}.$$

We now come to the main lemma of Friedman and Welch on the extraction of theories from limit theories.

Lemma 3.6.31 (Friedman and Welch, see [62]). *Let $\delta < \Sigma'(\alpha)$ be a limit ordinal, $\vec{p} \subseteq \alpha+1$ finite, $\phi \equiv \exists x \forall y \psi(x,y,z) \in \Sigma_2$, where $\psi \in \Delta_0$. Then $L_\delta \models \phi(\vec{p})$ if and only if there are $n \in \omega$, $\iota \in \alpha$ and $\xi < \delta$ such that the following holds in L_β for all $\beta \in (\xi, \delta)$:*

There is $\hat{\xi} \in (S_\beta \cap \beta) \cup \{0\}$, $\hat{\xi} = 0$ or $\hat{\xi} > \alpha$ such that $L_{\hat{\xi}} \models \phi(\vec{p})$ or $L_\beta \models \forall y(\psi(h_1^{L_\beta}(n,(\iota,\alpha,\hat{\xi})),y,\vec{p}))$. (*)

Proof. '$\rightarrow$': Suppose that $L_\delta \models \phi(\vec{p})$. Let $a \in L_\delta$ such that $L_\delta \models \forall y \psi(a,y,\vec{p})$.

Case I: S_δ is unbounded in L_δ.

In this case, pick $\beta' \in S_\delta$ such that $\beta' > \alpha$ and $a \in L_{\beta'}$. Then $L_{\beta'} \prec_{\Sigma_1} L_\beta$ for all $\beta \in (\beta', \delta]$, and moreover, $L_{\beta'} \models \forall y \psi(a,y,\vec{p})$. Thus L_β believes in the existence of a Σ_1-stable ordinal γ such that $L_\gamma \models \phi(\vec{p})$ for every $\beta \in (\beta', \delta)$, so (*) holds for all $\xi > \beta'$.

Case II: $S_\delta \cap (\alpha, \delta)$ is bounded in L_δ (or empty).

In this case, it follows from Lemma 3.6.23 that S_δ has a maximal element μ. By Lemma 3.6.25(ii), we have $\Sigma_1^{L_\delta}(\alpha+1 \cup \{\mu\}) = L_\delta$. Thus, there must be $\iota \in \alpha$ and a Σ_1-formula ϕ' such that a is the $<_L$-minimal witness in L_δ for $\phi_i \equiv \phi'(\iota,\alpha,\beta,x)$, i.e., $L_\beta \models \forall y \psi(h_1^{L_\beta}(i,(\iota,\beta,\alpha)),y,\vec{p})$ for all $L_\beta \ni a$, hence (*) is again satisfied (with ξ large enough such that $a \in L_\xi$).

Since one of case *I* and case *II* must occur, this direction is proved.

'$\leftarrow$': Suppose that (*) holds, and let ξ witness this.

Case I: S_δ is unbounded in δ.

Let $\beta \in S_\delta$, $\beta > \xi$. Then, by assumption, one of the following cases holds:

Case Ia: Let $L_\beta \models \exists \gamma \in S_\delta \phi(\vec{p})^{L_\gamma}$.

Then $L_\gamma \prec_{\Sigma_1} L_\beta \prec_{\Sigma_1} L_\delta$, so $L_\gamma \prec_{\Sigma_1} L_\delta$. As $L_\gamma \models \phi(\vec{p})$, it follows from Lemma 3.6.3 that $L_\delta \models \phi(\vec{p})$.

Case Ib: $L_\beta \models \exists \gamma \in S_\delta \forall y \psi(h_1^{L_\beta}(n,(\iota,\alpha,\gamma)),y,\vec{p})$. Then in particular $L_\beta \models \phi(\vec{p})$ and as $L_\beta \prec_{\Sigma_1} L_\delta$, it follows again from Lemma 3.6.3 that $L_\delta \models \phi(\vec{p})$.

Case II: S_δ is bounded in L_δ. Then, by Lemma 3.6.23, it contains a maximal element μ. Denote by C_δ the set of ordinals γ such that, for all $\beta < \gamma$, $L_\beta \prec_{\Sigma_1} L_\gamma$ implies $\beta < \mu$.

Claim. *C_δ is unbounded in δ.*

To see this, note that by Lemma 3.6.25, we have $\Sigma_1^{L_\delta}(\alpha + 1 \cup \{\mu\}) = L_\delta$. Pick $\beta' < \delta$, then $\beta' = h_1^{L_\delta}(n, (\iota, \alpha, \mu))$ for appropriate $n \in \omega$, $\iota < \alpha$. Let γ be minimal such that $h_1^{L_\gamma}(n, (\iota, \alpha, \mu))$ is defined. Then $\gamma > \beta'$ and no level L_η with $\eta < \gamma$ such that $\mu \in L_\eta$ can be Σ_1-stable in L_γ. Thus $\gamma \in C_\delta$. As β' was arbitrary, the claim is proved.

By assumption, (*) holds in particular in L_γ for all $\gamma \in C_\delta$ with $\gamma > \xi, \mu$. Below, γ is always of this kind.

Case IIa: $L_\gamma \models \exists \gamma' \in S_\gamma \phi(\vec{p})^{L_{\gamma'}}$.

Then $\gamma' < \mu$ since $\gamma \in C_\delta$ and also $L_{\gamma'} \prec_{\Sigma_1} L_\gamma$, hence, as $\gamma > \mu$, also $L_{\gamma'} \prec_{\Sigma_1} L_\mu \prec_{\Sigma_1} L_\delta$. As $L_{\gamma'} \models \phi(\vec{p})$ by assumption, Lemma 3.6.3 implies that $L_\delta \models \phi(\vec{p})$.

Case IIb: $L_\gamma \models \exists \gamma' \in S_\gamma \forall y \psi(h_1^{L_\gamma}(n, (\iota, \alpha, \gamma')), y, \vec{p})$.

As $L_\mu \prec_{\Sigma_1} L_\delta$ and $\gamma < \delta$, we have $L_\mu \prec_{\Sigma_1} L_\gamma$, thus $h_1^{L_\gamma}(n, (\iota, \alpha, \gamma')) \in L_\mu$ and hence $h_1^{L_\gamma}(n, (\iota, \alpha, \gamma')) = h_1^{L_\mu}(n, (\iota, \alpha, \gamma'))$. But then $L_\mu \models \forall y \psi(h_1^{L_\mu}(n, (\iota, \alpha, \gamma')), y, \vec{p})$, so $L_\mu \models \phi(\vec{p})$, and hence $L_\delta \models \phi(\vec{p})$ by Lemma 3.6.3.

Case IIc: IIa and IIb occur for no $\gamma \in C_\delta$ with $\gamma > \xi, \mu$. Then, by (*), it holds for all such γ that $L_\gamma \models \forall y \psi(h_1^{L_\gamma}(n, (\iota, \alpha, \mu)), y, \vec{p})$. Moreover, $a := h_1^{L_\gamma}(n, (\iota, \alpha, \mu))$ has the same value for all such γ by upwards absoluteness of Σ_1-formulas. Now let $b \in L_\delta$. By assumption, there is $\gamma \in C_\delta, \gamma > \xi, \mu$, such that $\gamma \in L_\gamma$. As $L_\gamma \models \forall y \psi(a, y, \vec{p})$, we have in particular $L_\gamma \models \psi(a, b, \vec{p})$. Since b was arbitrary, we have $L_\delta \models \forall y \psi(a, y, \vec{p})$. Thus $L_\delta \models \exists x \forall y \psi(x, y, \vec{p})$, i. e., $L_\delta \models \phi(\vec{p})$. □

Of course, Lemma 3.6.31 only helps us when the statements used to characterize $L_\delta \models \phi(\vec{p})$ are actually Σ_2-expressible. But this is not hard to see.

Lemma 3.6.32 (Friedman and Welch, see [62]). *'There is $\beta' \in (S_\beta \cap \beta) \cup \{0\}, \beta' > \alpha$ such that $L_{\beta'} \models \phi(\vec{p})$ or $L_\beta \models \forall y(\psi(h_1^{L_\beta}(n, (\iota, \beta')), y, \vec{p}))$' is Σ_2 over L_β in the parameters $\vec{p}$, ι and α.*

Proof. The condition that β' is Σ_1-stable in L_β is Π_1 over L_β by Proposition 3.6.22.

The condition that $L_{\beta'} \models \phi(\vec{p})$ is Δ_0 in the parameters $L_{\beta'}$ and $\vec{p}$.

To express $L_\beta \models \forall y(\psi(h_1^{L_\beta}(n, (\iota, \beta')), y, \vec{p}))$ by a Σ_2-formula, let $\phi_{sf} \equiv \exists x \psi_{sf}(u, v, w, x)$ (with $\psi_{sf} \in \Delta_0$) be the defining Σ_1-formula for $h_1(u, v) = w$, $\psi_{sf} \in \Delta_0$; then the formula $\exists x, w \forall y(\psi_{sf}(n, (\iota, \beta'), w, x) \wedge \psi(w, y, \vec{p}))$ is as desired.

Combining these results, we see that the statement in question is indeed Σ_2. □

It remains to see that the criterion given by Lemma 3.6.31 can actually be read off from $\Sigma_2 - \text{limit}(L_\delta, \alpha)$.

Lemma 3.6.33 (Cf. [34], Lemma 3). *There is an α-ITTM-program $P_{\alpha\text{-tflt}}$ ("theory from limit theory") such that, for all limit of limit ordinals τ, $P_{\alpha\text{-tflt}}$ computes $\Sigma_2 - \mathrm{Th}(L_\tau, \alpha+1)$ from $\Sigma_2 - \mathrm{limit}(L_\tau, \alpha+1)$ in α many steps.*

Proof. The proof of Lemma 3.6.32 shows how to obtain effectively from $\phi \in \Sigma_2$ and $\vec{p} \subseteq (\alpha+1)$ a Σ_2-statement ϕ' such that $L_\tau \models \phi(\vec{p})$ if and only if $L_\beta \models \phi'(\vec{p})$ for all sufficiently large $\beta < \tau$. The latter condition can easily be looked up from $\Sigma_2 - \mathrm{limit}(L_\tau, \alpha+1)$ by running to the corresponding cell and checking the content, which takes $< \alpha$ many steps.

By additive closure of α, the whole procedure takes $< \alpha$ many steps. Doing this for all choices of a Σ_2-formula ϕ and a parameter $\iota < \alpha$ takes α many steps. □

3.6.3.3 The initial theory

We still need to show how to compute $\Sigma_2 - \mathrm{Th}(L_\alpha, \alpha)$ within the given time bound.

Lemma 3.6.34. *There is an α-ITTM-program P_{fl} ("first level") such that P_{fl} computes $\Sigma_2 - \mathrm{Th}(L_\alpha, \alpha)$ in $\leq \alpha^{\omega+2}$ many steps.*

Proof. We sketch the construction somewhat roughly, leaving the details to the reader. Split some scratch tape into α many disjoint portions. Now proceed as in the proof of part (ii) of Lemma 3.6.27: Start with a code for $L_0 = \emptyset$ on the 0th portion; then successively use the $(\iota+1)$th portion for amending the code of L_ι on the ιth portion to a code for $L_{\iota+1}$ as in the proof of Lemma 3.6.27. Each new level requires one application of $P_{\alpha\text{-nl}}$ to the code written last, which takes $\leq \alpha^\omega + \alpha^3$ many steps. At limit stages $\delta \leq \alpha$, the sequence $(c_\iota : \iota < \delta)$ of earlier level codes is arranged into a single code via pairing (i. e., $p(\iota, y)$ now represents the element represented by y in the ιth code).

Doing this α many times will take $\leq \alpha^{\omega+1}$ many steps. Finally, we take another α^4 many steps to rearrange the α-"code" for L_α thus obtained, in which some elements of α will not represent an element and in which some elements may be represented repeatedly, into a "proper" α-code based on a bijection between α and L_α. □

Exercise 3.6.35 (Friedman and Welch, [62]). Describe in detail an ITTM-program that halts after $< \omega^2$ many steps with a code for $L_{\omega+1}$ on its output tape.

3.6.3.4 Existence of the Σ_2-machine

Finally, we can prove Theorem 3.6.15:

Proof. The theory machine will mainly use two tapes T_t for storing Σ_2-theories of levels and T_l for storing level codes. It will also use a flag to indicate when the content of T_t has changed a limit number of times.

The theory machine starts by computing an α-code c_α for L_α and storing it on a tape T_l. By Lemma 3.6.34, this can be done in $\alpha^{\omega+2}$ many steps. Now, we compute $\Sigma_2 - \mathrm{Th}(L_\alpha, \alpha)$ from c and store it on a second tape T_t. We saw above that this can be done in $< \alpha^3$ many steps. Thus, we obtain $\Sigma_2 - \mathrm{Th}(L_\alpha, \alpha)$ on T_t in $< \alpha^{\omega+2} + \alpha^3$ many steps. In line with our desired time estimate, a code for L_α should be on the output tape at time $\alpha^{\omega+2}\alpha$; it is not hard to see that this ordinal is α-clockable, and moreover, that $\alpha^{\omega+2} = \alpha^{\omega+2}\alpha = \alpha^{\omega+2}(1+\alpha) = \alpha^{\omega+2}\alpha = \alpha^{\omega+3}$. Thus, running a program for clocking $\alpha^{\omega+3}$ after the computation of the code for L_α and its Σ_2-theory has finished and only continuing with the computation after that clock as stopped, the time estimate is respected.

The theory machine now proceeds in phases as follows: At the beginning of each phase, T_t and T_l have certain contents. If the flag indicates that the content of T_t has been changed a limit number δ of times, we use P_{tflt} from Lemma 3.6.33 to compute $\Sigma_2 - \mathrm{Th}(L_{\omega\delta}, \alpha+1)$ from the limit theory currently written on T_t and write it to T_t. After that or if the flag indicates a successor number of changes, we compute a code for the ωth successor level of the L-level coded by the content of T_l, which can be done by the program $P_{\alpha\text{-nll}}$ from Lemma 3.6.27 in $< \alpha^\omega\omega$ many steps and write it to T_l; then we use $P_{\alpha\text{-next}}$ from Lemma 3.6.29 to compute the Σ_2-theory with parameters in $\alpha+1$ for that level in $\alpha^{\omega+1}$ many steps.

Using the additive indecomposability of $\alpha^{\omega+1}$, a phase takes $\alpha^{\omega+1}$ many steps in total (or $\alpha^{\omega+1} \cdot 2$ when $\alpha = \omega$). By clocking $\alpha^{\omega+2}$ along the way and waiting for the required number of steps when necessary, we can achieve that it will take exactly $\alpha^{\omega+2}$ many steps.

Thus, when $\delta = \omega\tau < \Sigma'(\alpha)$ is a limit ordinal, an α-code for L_δ will be on the output tape at time $\alpha^{\omega+2} \cdot \tau$, as desired. □

Exercise 3.6.36. Does that fact that the theory machine loops from time $\zeta'(\alpha)$ on, repeating the sequence of configurations between times $\zeta'(\alpha)$ and $\Sigma'(\alpha)$ imply that, for all $\gamma, \beta \in \mathrm{On}$, $\Sigma_2 - \mathrm{Th}(L_{\Sigma'(\alpha)\gamma+\beta}, L_\alpha) = \Sigma_2 - \mathrm{Th}(L_{\zeta'(\alpha)+\beta}, L_\alpha)$?

3.6.4 Further consequences

Besides yielding the λ-ζ-Σ-theorem, the theory machine is a quite strong tool for a number of results.

Corollary 3.6.37 (Welch, cf. [181]). *$\mathcal{U}_\alpha$ has the same configuration c at times $\zeta(\alpha)$ and $\Sigma(\alpha)$. Moreover, $\Sigma(\alpha)$ is the first time at which the configuration from time $\zeta(\alpha)$ is repeated.*

Proof. The first claim follows from Corollary 3.6.10. By Lemma 3.6.9, every configuration arising after time $\zeta(\alpha)$ will weakly majorize c in every component. Thus, if c would occur at some time $\tau \in (\zeta(\alpha), \Sigma(\alpha))$, then $(\zeta(\alpha), \tau)$ would witness the looping of $\mathcal{U}_\alpha$. Then every α-aw subset of α would be accidentally written before time τ. As $\Sigma(\alpha)$ is a limit of admissible ordinals, $\tau^+ < \Sigma(\alpha)$ and so τ^+ is α-aw by Corollary 3.6.13. Hence τ^+ has a code in $L_{\tau+\omega}$, and thus in L_{τ^+}. As admissible sets contain the ordinals for which they contain codes by Lemma 3.1.32, we get $\tau^+ \in L_{\tau^+}$, a contradiction. □

Exercise 3.6.38. Show that the configuration of $\mathcal{U}_\alpha$ at time $\zeta(\alpha)$ did not occur before time $\zeta(\alpha)$.

Lemma 3.6.39 (Cf. Hamkins and Lewis, see [81]). *Let α be multiplicatively closed. Then $\Sigma(\alpha) = \Sigma'(\alpha)$ is not admissible.*

Proof. We show that there is a cofinal map $f : \omega \times \alpha \to \Sigma(\alpha)$ (and hence also such an $f : \alpha \to \Sigma(\alpha)$, as $\omega\alpha = \alpha$) which is Σ_1 over $L_{\Sigma(\alpha)}$, which contradicts admissibility. Recall that the configurations, in particular the tape contents of the universal α-ITTM $\mathcal{U}_\alpha$ are the same at times $\zeta(\alpha)$ and $\Sigma(\alpha)$ and that $\Sigma(\alpha)$ is the first time at which the configuration at time $\zeta(\alpha)$ is repeated. Also, we saw that, if the tape cell C_ι ($\iota \in \alpha$) contains 1 at time $\zeta(\alpha)$, it contains 1 at all times $\zeta(\alpha) \leq \xi \leq \Sigma(\alpha)$. It follows that there cannot be a limit ordinal $\delta \in (\zeta(\alpha), \Sigma(\alpha))$ such that all cells that contain 0 at time $\zeta(\alpha)$ contain 0 at time δ; otherwise, the configuration of $\mathcal{U}_\alpha$ at time $\zeta(\alpha)$ would already repeat at time $\delta < \Sigma(\alpha)$ (the head positions being dealt with in the usual manner). However, all of the cells that contain 0 at time $\zeta(\alpha)$ must contain 0 cofinally often in $\Sigma(\alpha)$ (otherwise, their content would be 1 at time $\Sigma(\alpha)$, hence also at time $\zeta(\alpha)$, by the liminf-rule). Now define $f : \omega\alpha \to \Sigma$ as follows: $f(0) = \zeta(\alpha)$ and, for all $i \in \omega$, $\nu \in \alpha$, $f(\alpha i + \nu)$ is the first time larger than all $f(\alpha k + \nu')$ with $\alpha k + \nu' < \alpha i + \nu$ at which the content of cell C_ν is 0. Clearly, f is Σ_1 over $L_{\Sigma(\alpha)}$. Consider $f[\omega \times \alpha] = \{f(\alpha i + \nu) : i \in \omega, \nu \in \alpha\}$. The supremum of this set is some limit ordinal $\delta \leq \Sigma(\alpha)$. If $\delta < \Sigma(\alpha)$, it would be a time strictly between $\zeta(\alpha)$ and $\Sigma(\alpha)$ at which the configuration of $\mathcal{U}_\alpha$ at time $\zeta(\alpha)$ reappears, a contradiction. Thus $\delta = \Sigma(\alpha)$, contradicting the admissibility of $\Sigma(\alpha)$. □

The following is a variant of Welch's "quick writing" theorem for ITTMs (see [184]; see also [34]).

Theorem 3.6.40 (Quick writing for α-ITTMs). *Suppose that $\beta > \alpha$ is an α-ITTM-clockable limit ordinal. Then an α-code for L_β is α-ITTM-writable in $\leq \alpha^{\omega+2}\beta$ many steps.*

Proof. Run the computation $Q(\iota)$ for clocking β along with $P_{\alpha\text{-level}}$ from Theorem 3.6.15 in the sense that, whenever $P_{\alpha\text{-level}}$ writes a new code for an L-level, one step of Q is carried out; when Q halts, halt. By definition of $P_{\alpha\text{-level}}$, this will halt after $\alpha^{\omega+2}\beta$ many steps with a code for L_β on the output tape. □

As a consequence, we obtain a characterization of the computability strength of (α, β)-Turing machines that was first given in [34].

Definition 3.6.41 (See [34]). For ordinals α, β, let $\text{gap}_\alpha(\beta)$ denote the ordinal $< \beta$ that starts an α-clockable gap containing β if such an ordinal exists; otherwise, let $\text{gap}_\alpha(\beta) = \beta$.

Theorem 3.6.42 (Cf. [34]). *Let α be multiplicatively closed, and let $\beta > \alpha$ be exponentially closed. Then $x \subseteq \alpha$ is (α, β)-TM-writable if and only if $x \in L_{\text{gap}_\alpha(\beta)}$.*

Proof. Let $x \subseteq \alpha$ be (α,β)-TM-writable, and let P be an α-ITTM-program, $\iota < \alpha$ such that $P(\iota)$ writes x in $\eta < \beta$ many steps. As $\eta < \beta$ is the halting time of some α-ITTM-computation, it follows that $\eta < \text{gap}_\alpha$. Then the computation of $P(\iota)$ is contained in $L_{\eta+\omega}$ and thus in $L_{\text{gap}_\alpha(\beta)}$, as gaps are always started by limit ordinals by the speedup-theorem from Chapter 2. By transitivity of $L_{\text{gap}_\alpha(\beta)}$, the output tape content x of the final configuration of $P(\iota)$ belongs to $L_{\text{gap}_\alpha(\beta)}$ as well.

Conversely, let $x \in L_{\text{gap}_\alpha(\beta)}$. Since $\text{gap}_\alpha(\beta)$ starts a gap, it is a limit ordinal, and thus, there is $\gamma < \text{gap}_\alpha(\beta)$ such that $x \in L_\gamma$. By Theorem 3.6.40, an α-code c for L_γ is α-ITTM-writable in $\alpha^{\omega+2}\gamma$ many steps. Since $\alpha, \omega+2, \gamma < \beta$, we have $\alpha^{\omega+2}\gamma < \beta$ by exponential closure of β, so c is writable in $< \beta$ many steps. From c, x can be obtained in $< \text{gap}_\alpha(\beta)$ in the usual way, also in $< \beta$ many steps. As β is additively closed, x is α-ITTM-writable in $< \beta$ many steps, and thus (α,β)-writable. □

Exercise 3.6.43. Let α be multiplicatively closed, $\beta > \alpha$.

(a) Show that, in the situation of Theorem 3.6.42, $x \in L_{\text{gap}_\alpha(\beta)} \cap \mathfrak{P}(\alpha)$ is indeed α-ITTM-writable in $< \text{gap}_\alpha(\beta)$ many steps.

(b) Show that Theorem 3.6.42 also works under the assumption that β is a power of α^{ω^2}.

(c) Show that, if β is exponentially closed, then so is $\text{gap}_\alpha(\beta)$. Generalize.

The next two exercises show how the sets of hyperarithmetic and arithmetic real numbers can be characterized as time-bounded versions of ITTM-computability.

Exercise 3.6.44 (Cf. Hamkins and Lewis, [81]). (a) Show that, given a code $c \subseteq \omega$ for an ordinal α, there is an ITTM-program P such that P^c runs for exactly $\alpha\omega$ many steps.[14]

(b) Conclude that every limit ordinal $< \omega_1^{\text{CK}}$ is ITTM-clockable.

(c) Show further that every ordinal $< \omega_1^{\text{CK}}$ is ITTM-clockable. Deduce that $\text{gap}_\omega(\omega_1^{\text{CK}}+1) = \omega_1^{\text{CK}}$.

(d) Conclude that, for $x \subseteq \omega$, we have $x \in L_{\omega_1^{\text{CK}}}$ if and only if x is ITTM-writable in $< \omega_1^{\text{CK}}$ many steps.

Exercise 3.6.45 (Cf. Hamkins and Lewis, [81]). Recall that a set $x \subseteq \omega$ is arithmetical if and only if there is an arithmetical formula ϕ such that $x = \{i \in \omega : \phi(i)\}$. This is equivalent to saying that, for some $n \in \omega$, x is recursive in the nth Turing jump $0^{(n)}$.

(a) Show that an arithmetical set is ITTM-writable in $< \omega^2$ many steps.[15]

(b) Show that the tape contents at time ω of an ITTM on input x that starts on input x are recursive in $x^{(2)}$, the second Turing jump of x.[16]

14 Hint: Adapt $P_{\text{stopwatch}}$ from Lemma 2.5.37.

15 It saves a lot of work to recall Section 2.5.5 on weak ITTMs.

16 Adapt the proof of Corollary 2.5.42.

(c) Show that the tape contents at time ωn of an ITTM that starts on input x are recursive in $x^{(2n)}$, for $n \in \omega$.

(d) Conclude that x is arithmetical if and only if it is ITTM-writable in $< \omega^2$ many steps.

3.6.5 Further results on α-ITTMs

Finally, we characterize the α-ITTM-semidecidable and the α-ITTM-decidable subsets of $\mathfrak{P}(\alpha)$; recall that $X \subseteq \mathfrak{P}(\alpha)$ is called α-ITTM-semidecidable if and only if there is some α-ITTM-program and some $\xi < \alpha$ such that, for all $y \subseteq \alpha$, $P^y(\xi)\downarrow = \chi_X(y)$. Again, we assume that α is multiplicatively closed.

Proposition 3.6.46 (Cf. Friedman and Welch, [62]). *Let $X \subseteq \mathfrak{P}(\alpha)$.*

(a) *X is α-ITTM-semidecidable if and only if there are a Σ_1-formula $\phi(x)$ and some $\xi < \alpha$ such that, for all $y \subseteq \alpha$, $L_{\lambda^y(\alpha)}[y] \models \phi(y, \xi)$ if and only if $y \in X$.*

(b) *X is ITTM-decidable if and only if it can be characterized in the sense of (a) both by a Σ_1 and by a Π_1-formula.*

Proof. (a) Suppose that X is α-ITTM-semidecidable. Let P be an α-ITTM-program and $\xi < \alpha$ such that, for all $y \subseteq \alpha$, $P^y(\xi)\downarrow$ if and only if $y \in X$. The condition that $P^y(\xi)\downarrow$ can be expressed as 'There is a halting P-computation in the oracle y and the parameter ξ', which is Σ_1 in the parameter ξ. Now, if $P^x(\xi)$ halts, it does so in $< \lambda^x(\alpha)$ many steps, so the computation will be contained in $L_{\lambda^x(\alpha)}[x]$, and thus the Σ_1-formula just described will hold in $L_{\lambda^x(\alpha)}[x]$ if and only if $x \in X$.

On the other hand, let ϕ be a Σ_1-formula and $\xi < \alpha$ such that $x \in X$ if and only if $L_{\lambda^x(\alpha)}[x] \models \phi(x, \xi)$, for all $x \subseteq \alpha$. Consider the α-ITTM-program Q that works as follows: Given the oracle x, Q runs the universal α-ITTM $\mathcal{U}_\alpha^x$ and computes grand sums. For every value ρ of the grand sum, it computes a code for $L_\rho[x]$ and checks whether $L_\rho[x] \models \phi(x, \xi)$ (alternatively, one can also use $P_{\alpha\text{-level}}$ here). If yes, then Q halts. Otherwise, the search is continued. This program Q semidecides X.

(b) This is clear from (a), as the ITTM-decidability of X is equivalent to the ITTM-semidecidability of X and $\bar{X}$. □

We conclude this section with a few exercises on results about ITTM-clockable ordinals, thus keeping our promise in Chapter 2 that we would return to clockable ordinals.

Exercise 3.6.47 (Welch, see [184]). In [184], Lemma 9, Welch proved his "quick writing theorem," which strengthens Theorem 3.6.40 above in the case $\alpha = \omega$: If α is ITTM-clockable, then codes for α and L_α are ITTM-writable in $\leq \alpha$ many steps. Moreover, it is shown in Friedman and Welch [62] that there is an ITTM-program Q which, at time $(\omega^2 + \omega) \cdot \alpha$, has a code for L_α on its output tape. Though we will not prove this theorem here, we will use it in this exercise, the point of which is to prove another theorem of Welch, a counterpart to Theorem 3.1.27 above: If α starts an ITTM-clockable gap, then α is admissible. Let us suppose otherwise: Thus α starts a gap and there are $\beta < \alpha$ and a total map $f : \beta \to \alpha$ which is defined by the Σ_1-formula ϕ in the parameter $\vec{p}$ over L_α and has unbounded range in α.

(a) Show that there is a clockable ordinal $\gamma \in [\beta, \alpha)$ such that $\vec{p} \in L_\gamma$.

(b) Use the speedup-theorem (Theorem 2.5.31, Chapter 2) to show that α is not of the form $\delta + \omega \cdot k$ for any $\delta \in \mathrm{On}$, $k \in \omega$. Conclude that α is a multiple of ω^2.

Moreover, use induction to show that α is a multiple of ω^n for every $n \in \omega$. Show that, if $\delta, \delta' < \alpha$, then $\delta + \delta' < \alpha$ and $\delta \cdot \delta' < \alpha$.

Now consider the ITTM-program P that works as follows: First, P computes a real code b for γ via a bijection $f : \omega \to \gamma$. Imagine b as stored on an extra tape. We also reserve another scratch tape T, initially filled with 0s. On this tape, we imagine a "pointer," which is initially at position 0. Also, we reserve two cells as a flag, the first of which initially contains 1, the other 0. Then start running the program Q mentioned above. Whenever Q produces the code for an L-level L_δ with $\delta > \beta$, check, for every $i \in \omega$, whether $L_\delta \models \exists y \phi(f(i), y, \vec{p})$; whenever the answer is positive, write 1 in the ith cell of T. After all elements of ω have been checked, move the pointer on T to the right up to the leftmost 0 on T; for each step the pointer makes to the right, flash the flag. Then continue with the run of Q. At each limit time, also check whether both flags contain 0. If this is the case, halt.

(c) Prove that P halts at time $\alpha + 2$. Use the speedup-theorem to deduce that α is clockable, a contradiction.

Exercise 3.6.48. (a) (Welch, see also [31]) Show that there is an admissible ordinal α which is properly contained in a clockable gap. That is, there are $\beta < \alpha$ and $\gamma > \alpha$ such that no ordinal in $[\beta, \gamma]$ is clockable. Conclude that there are gaps $[\alpha, \beta)$ in the clockable ordinals for which $\beta > \alpha \cdot 2$, $\beta > \alpha^\omega$ and $\beta > \alpha^\alpha$.

(b) (Cf. Carl, Durand, Lafitte and Ouazzani, [31]) Show that there is a gap $[\alpha, \beta)$ in the clockable ordinals that contains exactly one admissible ordinal besides α.

3.6.6 Without parameters

Above, we characterized writability, eventual writability and accidental writability for Turing machines with tape length α with finitely many parameters $< \alpha$ allowed. What happens when we drop the parameters? This question was first studied by Rin in [149] and then further by Carl, Rin and Schlicht in [35]. Here, we offer some exercises that guide you through some of the results. In this section, α is always multiplicatively closed.

It might seem at first glance that Welch's analysis that we explained above simply goes through. Unfortunately, this is not the case: The proof for the eventual writability of stabilization times $< \Sigma(\alpha)$ that was crucial in the argument requires the ability of the machine to “watch” the contents of a certain tape cell C_ι. However, if parameters are dropped, the machine might not be able to “identify” its ιth cell. This motivates the following definition.:

Definition 3.6.49 ([35]). Let $\alpha \in \mathrm{On}$. Then $\Sigma^*(\alpha)$, $\zeta^*(\alpha)$ and $\lambda^*(\alpha)$ denote the suprema of the ordinals coded by subsets of α that are writable/eventually writable/accidentally writable by an α-tape Turing machine without parameters.

(Rin, [149]) Also, let us say that $\beta < \alpha$ is “α-reachable” if and only if there is a Turing program that, when run on an α-tape Turing machine on the empty tape, halts with the read/write head on cell β.

Clearly then, we have $\lambda^*(\alpha) \leq \zeta^*(\alpha) \leq \Sigma^*(\alpha)$ and $\lambda^*(\alpha) \leq \lambda(\alpha)$, $\zeta^*(\alpha) \leq \zeta(\alpha)$ and $\Sigma^*(\alpha) \leq \Sigma(\alpha)$ for all $\alpha \in \mathrm{On}$.

From now on, we assume that α is multiplicatively closed, thus in particular closed under Cantor pairing.

Exercise 3.6.50 ([149, 35]). (a) Show that there are ordinals ι, α such that $\iota < \alpha$ and ι is not α-reachable. Show further that α can be taken to be countable.

(b) Show that there are ordinals α', β, γ, δ such that $\beta < \gamma < \delta < \alpha'$ such that β and δ are α'-writable without parameters, but γ is not: Thus there are gaps in the α-writable ordinals without parameters!

(c) Show that, if α and α' are the minimal ordinals with the respective properties in (a) and (b), then $\alpha = \alpha'$.

Exercise 3.6.51 ([35]). (a) By considering the program $\mathcal{U}'_\alpha$ that simultaneously simulates all Turing programs with all parameters $< \alpha$ on $\alpha\omega$ many disjoint portions of the tape, show that every subset of α that is accidentally α-writable with parameters is also accidentally α-writable without parameters. Conclude that $\Sigma(\alpha) = \Sigma'(\alpha)$.

(b) Show that, in contrast to the set of parameter-freely α-writable ordinals according to the last exercise, the set of parameter-freely accidentally α-writable ordinals is downwards closed.

(c) Using the corresponding fact for α-accidental writability with parameters obtained above, show that, for every $x \subseteq \alpha$ that is α-aw without parameters, there is $c \subseteq \alpha$ which codes an L-level L_β that contains x and such that c is α-aw without parameters.

Exercise 3.6.52 (P. Schlicht, see [35]). Let β be α-clockable by the program P with parameter $\vec{p} \subseteq \alpha$.

(a) Show that there is an ordinal $\gamma > \beta$ that is α-aw without parameters. (Hint: Use the corresponding fact with parameters and the last exercise.)

Now suppose that $\vec{p} = \emptyset$, i. e., β is α-clockable by P without parameters.

(b) Show that there is an ordinal $\gamma > \beta$ that is α-writable without parameters. (Hint: Use $\mathcal{U}'_\alpha$ from the last exercise to search for an accidentally α-aw ordinal γ such that P halts in $< \gamma$ many steps; once it has been found, halt and write it to the output tape.)

Exercise 3.6.53 ([35]). Let $x \subseteq \alpha$ be α-writable without parameters.

(a) Show that there is an α-writable code $c \subseteq \alpha$ of some L-level L_β such that $L_\beta \ni x$. (Hint: Conclude from the last exercises that such a c is α-aw without parameters. Use again $\mathcal{U}'_\alpha$ to search for such a code and halt once it has been found.)

(b) Show that, if $x \subseteq \omega$ is α-writable, then there is an α-writable real number c which codes an L-level that contains x. (Hint: Use the fact that, if δ is a limit ordinal such that $x \in L_\delta$, then L_δ will contain such a code and imitate the argument from (a).)

(c) Show that, for all $\alpha, \beta \in \mathrm{On}$, the set of real numbers that are α-writable without parameters is either a subset of a superset of the set of real numbers that are β-writable without parameters.

Exercise 3.6.54 ([35]). Let $x \subseteq \alpha$.

(a) Show that x is accidentally α-ITTM-writable without parameters if and only if $x \in L_{\Sigma(\alpha)}$.

(b) Show that x is α-ITTM-writable without parameters if and only if x belongs to the Σ_1-Skolem hull of $\{\alpha\}$ in $L_{\Sigma(\alpha)}$.

(c) Show that x is eventually α-ITTM-writable without parameters if and only if x belongs to the Σ_2-Skolem hull of $\{\alpha\}$ in $L_{\Sigma(\alpha)}$.

3.7 Accidental and eventual writability

Though originally only defined for ITTMs, the notions of eventual and accidental writability make perfect sense for other machine models. This is obvious for the tape models, but rather straightforward extensions to the register-based models also exist. In this section, we consider these notions.

3.7.1 Accidental and eventual writability for ITRMs

Definition 3.7.1. Let $x \subseteq \omega$.

$x \subseteq \omega$ is ITRM-ew if and only if there is a nonhalting ITRM-program P such that the following holds: During the computation of P, P writes outputs to the register R_1. We regard these outputs as codes for ordered pairs $(i, j) \in \omega \times \{0, 1\}$. For each $i \in \omega$ and each time α, a pair of the form (i, j) will be output after time α; furthermore, for each $i \in \omega$, there is an ordinal α_i such that, if (i, j) appears as output after time α_i, then $j = x(i)$.

x is ITRM-aw if and only if there is an ITRM-program P such that the following holds (with notation as in the definition of ITRM-ew): there is a time α such that, among the next ω many outputs after time α, each $i \in \omega$ appears exactly once as the first component and then the second component is 1 if $i \in x$ and otherwise 0.

Theorem 3.7.2. *Let $x \subseteq \omega$. The following are equivalent:*
1. *x is ITRM-computable.*
2. *x is ITRM-ew.*
3. *x is ITRM-aw.*

Proof. Clearly, (1) implies (2).

To see that (2) implies (3), let P be an ITRM-program that eventually writes x. We modify P to P' that works as follows: If P outputs (i, j) and no pair with first component i was written to the output register by P after the last limit time, then P' outputs (i, j). Otherwise, P' leaves the output unchanged. It is easy to see that one can obtain such a P' from P by storing the finitely many outputs since the last limit time in a stack. Now, let α_i be as in the definition of ITRM-ew, and let $\alpha = \sup_{i\in\omega} \alpha_i$. Then P' will accidentally write x from time α on.

Finally, we show that (3) implies (1): Suppose that P accidentally writes x, and suppose that P uses k many registers. We saw above that P halts in less than ω_{k+1}^{CK} many steps or loops from this time on. Hence, if x is accidentally written by P from time α on for the first time, we have $\alpha < \omega_{k+1}^{\text{CK}}$. So there is an ITRM-program Q that computes a code c for $L_{\omega_{k+1}^{\text{CK}}}$; in c, the ordinal α is coded by some natural number k.

Now, using the ability of ITRMs to evaluate truth predicates in coded structures, we evaluate, for each $i \in \omega$, the following statement ϕ_i in $L_{\omega_{k+1}^{\text{CK}}}$: "The first output of the form (i, j) by the program P after the time coded by k has $j = 1$." Clearly, this halts for every $i \in \omega$ and outputs 1 if and only if $i \in x$ and otherwise, it outputs 0. □

3.7.2 Accidental and eventual writability for OTMs

We now turn to OTMs. Here, we say that a set $X \subseteq \text{On}$ is OTM-ew if and only if there is an OTM-program P (without parameters), the output of which stabilizes at the characteristic function of X and X is OTM-aw if and only if there is such a program such that the characteristic function of X appears at some time on the output tape in the computation of P.

Theorem 3.7.3. *$x \subseteq \omega$ is accidentally OTM-writable if and only if $x \in L$.*

Proof. This is a mere reformulation of Lemma 3.5.3 above. □

Definition 3.7.4. A constructible real number a is a Σ_2-singleton in L if and only if there is a Σ_2-formula $\phi(x)$ with a free variable x such that $L \models \forall x(\phi(x) \leftrightarrow x = a)$.

Theorem 3.7.5 (Carl, Schlicht and Welch, see [39], Lemma 3.7). *Any OTM-ew real number is constructible; moreover, a real number a is OTM-eventually writable if and only if it is a Σ_2-singleton in L.*

Proof. Any OTM-ew real number a appears on the output tape during the computation of some OTM-program P; say this happens at time α for the first time. Then stopping the computation at this time will compute a. But this is clearly possible when α is used as a parameter. Thus any OTM-ew real number is computable by a parameter-OTM, and thus constructible by Lemma 3.5.2 above.

Now suppose that a is OTM-ew by the program P. Then $\phi(z) \equiv \exists\alpha \in \text{On}\forall C\forall\beta[(\beta > \alpha\wedge$'$C$ is the computation of P of length $\beta+1$ in the oracle z') $\rightarrow$ ('C has z on the output tape at time β')] is a Σ_2-statement that uniquely characterizes a in L.

On the other hand, suppose that $a \subseteq \omega$ is a Σ_2-singleton in L, as witnessed by the formula $\phi(z) \equiv \exists x\forall y\psi(x,y,z)$, where ψ is Δ_0. We show that a is OTM-ew. Consider the following procedure, whose main components are two nested runs of the OTM that 'enumerates L' from Lemma 3.5.3 above, called "inner" and "outer" enumeration. We start with the outer enumeration. Each output is a set S of ordinals, which we regard as coding a pair (x,A) with $x \subseteq \omega$, $A \subseteq \text{On}$ by letting $i \in x$ if and only if $i \in S$ for $i \in \omega$ and $\iota \in A$ if and only if $\omega + \iota \in A$ for $\iota \in \text{On}$. Whenever such an S coding (x,A) is produced, x is written to the output tape; it is then our current candidate for a, while A is the current candidate for a witness to the existential quantifier in ϕ. Now the inner enumeration starts. For every set $S' \subseteq \text{On}$ produced, it is tested, using evaluation of bounded truth predicates, whether the set X coded by A and the set X' coded by S' satisfy $\psi(X,X',x)$ (note that his is possible, as ψ is Δ_0). If not, the inner enumeration is reset to the initial configuration and the outer enumeration continues. Otherwise, the inner enumeration continues. Now, if a is the only witness to ϕ, then the outer simulation, enumerating all constructible sets, will eventually produce a pair (a,A) such that A codes a set X with $\forall y(X,y,a)$. After this has happened, all checks appearing in the inner simulation will have a positive answer and so a will remain on the output tape. Thus a is OTM-ew. □

By a similar argument, we can show that OTM-eventual writability is in fact characterized in a rather similar way to ITTM-eventual writability.

Definition 3.7.6. Let η be the minimal ordinal such that $L_\eta \prec_{\Sigma_2} L$.

Corollary 3.7.7 (Cf. [39], Lemma 3.7). *Let $x \subseteq \omega$. Then x is OTM-ew if and only if $x \in L_\eta$.*

Proof. Let X be the set of all sets that are coded by an OTM-ew real number y. We show that $X \prec_{\Sigma_2} L$ and that X is transitive; it then follows by minimality of η that $X \supseteq L_\eta$, and hence that all real elements of L_η are OTM-ew. The transitivity of X is clear from the definition. So let $\phi(v,\vec{u}) \equiv \exists y\forall z\psi(v,y,z,\vec{u})$, where $\psi \in \Delta_0$, and let $\vec{p} \subseteq X$ be finite. It follows that all elements of $\vec{p}$ have codes that are OTM-ew. For simplicity, suppose that $\vec{p}$ consists of a single element p, and let Q be an OTM-program that eventually writes a code for p. Now, as in the proof of Theorem 3.7.5, we use an OTM-program P to enumerate pairs (a,L_α), where $a \in L_\alpha \cap \mathfrak{P}(\omega)$. For every such pair, write a to the output tape. Along with P, we run Q. Suppose that P has written (a,L_α) and that the current output of Q is b. Then, as in Corollary 2.3.31, search for the element $p(b)$ of L_α that is coded by b, and also for the element $p(a)$ of L_α coded by a. If no such elements are found, continue with the next pair. Otherwise, test whether $L_\alpha \models \phi(p(a), p(b))$. If not, continue with the next pair. Otherwise, continue to write codes for L-levels L_β with the program from Lemma 3.5.3, and run Q in parallel. Whenever Q changes its output, start over with enumerating pairs and testing them. Whenever a new L-level L_β with $\beta > \alpha$ is written without a change of the output by Q, identify the elements $p(a)$, $p(b)$ of L_β coded by a and b and check whether $L_\beta \models \phi(p(a), p(b))$. If yes, continue writing L-levels and running Q; otherwise, continue with the next pair.

Now, if $\phi(v, \vec{p})$ has a witness in L, then this program will eventually write a real number coding such a witness. Thus $X \prec_{\Sigma_2} L$. By condensation and transitivity, X is equal to some L_γ; by minimality of η, we have $\gamma \geq \eta$, i. e., $L_\eta \subseteq X$.

On the other hand, suppose that x is OTM-ew. First note that, given any constructible $y \subseteq \omega$, we can use an OTM Q to enumerate the real numbers in L, waiting for a code c of some $L_\alpha \ni y$ to appear, and when it appears, write this code to the output tape and halt. Now, if P is an OTM-program that eventually writes x, we can run P, apply Q to the output every time the output changes and write the output of Q to the output tape. The output of P will eventually stabilize, and thus, a code for some L_α containing x will eventually appear and remain on the output tape. Thus c is OTM-eventually writable. As every real number in L_α is OTM-writable relative to c and OTM-eventual writability is closed under OTM-writability by the same proof as that for Theorem 2.5.10 in Chapter 2, every real number in L_α is OTM-ew. It follows that there must be some ordinal α such that the OTM-ew real numbers are exactly those in L_α.

Now, for every OTM-program P, the statement "P eventually writes a real number" is Σ_2, saying that there is some ordinal α such that for each two partial P-computations of successor length $> \alpha$, they have the same content in the first ω many cells of the output tape in their final configuration. Thus, by Σ_2-elementarity, P eventually writes a real number if and only if $L_\eta \models$ "P eventually writes a real number'. Let $\text{ew-}0'_{\text{OTM}} = \{i \in \omega : L_\eta \models$ 'P_i eventually writes a real number$\}$," then $\text{ew-}0'_{\text{OTM}}$ is definable over L_η and hence contained in $L_{\eta+1}$. We know from Theorem 3.7.5 that every OTM-ew real number is constructible, and it follows from our observation above that, if any constructible real number $y \notin L_\eta$ is OTM-ew, then so is every real number in $L_{\eta+1}$. But this includes $\text{ew-}0'_{\text{OTM}}$, which contradicts Theorem 2.6.4 from Chapter 2. □

Exercise 3.7.8. Define writability, eventual writability and accidental writability for OTMs with ordinal parameters in the obvious way. Show that $x \subseteq \text{On}$ is pOTM-writable if and only if it is pOTM-ew if and only if it is pOTM-aw.

Exercise 3.7.9. Generalizing the definition of OTM-ew to arbitrary sets of ordinals in the obvious way, show that $\{\sigma\}$ and $\{\omega_1^L\}$ are OTM-ew. Conclude that there are ordinals $\alpha < \beta < \gamma$ such that α and γ are OTM-ew, but β is not.

Exercise 3.7.10. Let $\text{OTM} - \mathcal{EW}$ be the set of OTM-ew real numbers.

(a) Show that $\text{OTM} - \mathcal{EW}$ is closed under OTM-computability.

(b) Show that OTM $-\mathcal{EW}$ is downwards closed with respect to $<_L$.

(c) Show that, if x is OTM-ew, then so is the $<_L$-minimal code for the minimal L-level that contains x as an element.

(d) Show that there is an L-level L_α such that OTM $-\mathcal{EW} = \mathcal{P}(\omega) \cap L_\alpha$.

(e) Try to characterize α.

3.8 Summary

Before we more on to consider some further models of transfinite computability that are based on somewhat different ideas than those considered in the preceding sections, we summarize the results obtained so far.

For infinitary Turing computability with parameters, the following is known (see [34]):[17]

	Tape length ω	**Tape length α multiplicatively closed**	**Tape length On**
Time ω	$\Delta_1(L_\omega) = \Delta_1^0$ (Folklore)	$\Delta_1(L_\omega) = \Delta_1^0$ (Folklore)	$\Delta_1(L_\omega) = \Delta_1^0$ (Folklore)
Time β admissible	$\mathfrak{P}(\omega) \cap L_{\mathrm{gap}_\alpha(\beta)}$ [4]	$\Delta_1(L_{\min(\alpha,\beta)})$ if $\beta \leq \alpha$ and α is exponentially closed [1] $L_{\mathrm{gap}_\alpha(\beta)} \cap \mathfrak{P}(\alpha)$ if $\beta > \alpha$ [4]	$\Delta_1(L_\beta)$ [1]
Time On	$\mathfrak{P}(\omega) \cap L_\lambda$ [2]	$\mathfrak{P}(\alpha) \cap L_{\lambda(\alpha)}$ [4]	$\mathfrak{P}(\omega) \cap L$ [3]

[1] Koepke and Seyfferth, see [112].
[2] Hamkins, Lewis and Welch, see [81, 184].
[3] Koepke, see [105].
[4] C., Ouazzani and Welch, see [34] using the ideas from [184] and [62].

For α-wITRMs with parameters, we saw the following:
- ω-wITRMs compute exactly those real numbers in $L_{\omega_1^{\mathrm{CK}}}$
- When α is exponentially closed, then (α, α)-wITRMs compute exactly those subsets of α that are $\boldsymbol{\Delta}_1(L_\alpha)$.
- When α is Σ_2-admissible, then α-wITRMs compute exactly those subsets of α that are $\boldsymbol{\Delta}_1(L_\alpha)$.
- In general, the subsets of α that are computable by an α-wITRM are contained in L_β, where β is the smallest Σ_2-admissible ordinal $> \alpha$.
- On-wITRMs compute exactly those sets of ordinals in L.

17 The table is reprinted with permission from Springer Nature from the article "Taming Koepke's Zoo" ([34]) by S. Ouazzani, P. Welch and the author, which appeared in the LNCS proceedings for the CiE 2018.

For α-ITRMs with parameters, we saw:

- ω-ITRMs compute exactly those real numbers in $L_{\omega_\omega^{\mathrm{CK}}}$
- When $\beta \in \{\omega_i^{\mathrm{CK}} : i \in \omega\}$, then (ω, β)-ITRMs compute exactly those real numbers contained in L_β.
- When $\kappa > \omega$ is a regular cardinal, then the κ-ITRM-computable subsets of κ are exactly those in $L_{\kappa+1}$. In fact, it suffices that $\kappa > \omega$ is regular in L_{κ^ω}.
- In general, the α-ITRM-computable subsets of α are contained in L_β, where β is the smallest Σ_2-admissible ordinal $> \alpha$.

3.9 Exercises

Exercise 3.9.1. Any ordinal is 0-admissible. For $\iota \in \mathrm{On}$, an ordinal is $(\iota+1)$-admissible if and only if it is an admissible limit of ι-admissible ordinals. If δ is a limit ordinal, then α is called δ-admissible if α is ι-admissible for each $\iota < \delta$. Show that σ is ω-admissible. See how much further you can extend the argument beyond ω.

Exercise 3.9.2. Let P be an ITTM-program.

(a) Show that P either halts or repeats some configuration before time λ.

(b) Show that, if $\mathcal{U}$ assumes some configuration at both times $\alpha < \beta < \lambda$, then (α, β) does not contain a clockable ordinal.

Exercise 3.9.3 (Cf. Matzner, [138], Section 5). A "bounded" ITTM (bITTM) works like an ITTM, but is only allowed to have finitely many 1s on the scratch tape at each time—if there happen to be infinitely many, the computation is undefined, similar to wITRMs with overflowing registers. (Note that this can only occur for the first time at a limit stage.)

(a) Show that bITTMs can simulate wITRMs. Conclude that all $x \subseteq \omega$ with $x \in L_{\omega_1^{\mathrm{CK}}}$ are bITTM-decidable in the sense that there is a bITTM-program P such that, for all $i \in \omega$, $P(i)\downarrow = 1$ if and only if $i \in x$, and otherwise $P(i)\downarrow = 0$.

(b) Show that, if $x \subseteq \omega$ is bITTM-decidable, then $x \in L_{\omega_1^{\mathrm{CK}}}$. Thus bITTMs are a tape model equivalent in computational power to wITRMs. (Hint: Proceed by analogy with the analysis of wITRMs given above.)

Exercise 3.9.4 (C., Schlicht). In analogy with the last exercise, this exercise will provide a tape model computationally equivalent to ITRMs. A "resetting bITTM" (rbITTM) is defined like a bITTM with a finite number of scratch tapes, but with a somewhat relaxed rule concerning the number of 1s on the tape: If, at a limit time δ, the liminf rule has the effect that some tape contains an infinite number of 1s, then this tape is simply erased (i. e., all cell contents on this tape are set to 0) and the computation is continued; in this case we say that the tape was "reset."

(a) Show that rbITTMs can simulate ITRMs. Conclude that all $x \subseteq \omega$ with $x \in L_{\omega_\omega^{CK}}$ are rbITTM-decidable in the sense that there is a bITTM-program P such that, for all $i \in \omega$, $P(i)\downarrow = 1$ if and only if $i \in x$ and otherwise $P(i)\downarrow = 0$.

For the following parts, we recommend reviewing the analysis of ITRMs given above.

(b) Show that, if an rbITTM-computation reaches a time of the form $\alpha + \omega_1^{CK}$ and no tape is reset at this time, then the computation is strongly looping (and thus will not stop).

(c) Show (e. g., by induction, similarly to Lemma 3.4.8): If P is an rbITTM-program less than n of the tapes of P have all cell contents equal to 0 at time $\alpha + \omega_n^{CK}$ for any $\alpha \in \mathrm{On}$, then P is strongly looping, and hence will not halt.

(d) Conclude that, if the rbITTM-program P uses n tapes, then P halts in $< \omega_{n+1}^{CK}$ many steps or does not halt at all.

(e) Show that, if $x \subseteq \omega$ is rbITTM-decidable, then $x \in L_{\omega_\omega^{CK}}$. Thus rbITTMs are computationally equivalent to ITRMs.

(f) Show that, for any $n \in \omega$, there is $f(n) \in \omega$ such that some rbITTM-program using $f(n)$ tapes can solve the halting problem for rbITTMs using n tapes.

Exercise 3.9.5 (Hamkins, Welch). (a) Show that the $<_L$-minimal code for Σ is ITTM-writable relative to every code for ζ. Hint: Use ζ to compute the configuration of $\mathcal{U}$ at time ζ, then wait for it to reoccur and use the fact that ITTM-clockable ordinals are ITTM-writable (Corollary 3.6.20).

(b) Show that, if ζ is ew in $x \subseteq \omega$, then so is Σ.

(c) Show that, if Σ is ew in $x \subseteq \omega$, then Σ is in fact ITTM-writable from x.

(d) Conclude that $\zeta \neq \lambda^x$ and $\Sigma \neq \lambda^x$ for every $x \subseteq \omega$.

(e) Can we have $\Sigma = \zeta^x$ for some $x \subseteq \omega$?

Exercise 3.9.6. (a) Show that $f : \omega \to \omega$ is wITRM-computable if and only if $\{p(i, f(i)) : i \in \omega\} \in L_{\omega_1^{CK}}$.

(b) Show that there is a universal wITRM-program $\mathcal{W}$ in the following sense: If $f : \omega \to \omega$ is wITRM-computable, then there is $i \in \omega$ such that, for all $j \in \omega$, $\mathcal{W}(j, i)\downarrow = f(i)$.[18] In particular, unlike ITRMs, wITRMs do not grow indefinitely in computational strength with the number of registers.

Exercise 3.9.7. In his Master's thesis [126], E. Lewis considered OTMs with infinite programs. An infinite OTM-program works like a finite one, but its inner states are indexed by ordinals rather than natural numbers.

(a) (Cf. [126]) Show that, for each set $x \subseteq \mathrm{On}$, there is an infinite OTM-program that decides x.

18 Note that it is not required that i can be effectively obtained from an index for a wITRM-program that computes f.

Thus, decidability of sets is not an interesting subject with infinite programs. Decidability of (proper) classes, on the other hand, is:

(b) (Cf. [126]) Let H be the set of sets of ordinals that code halting infinite time OTM-programs. Show that H is not decidable by an infinite OTM-program.

(c) (Cf. [126]) Show that a class $X \subseteq \mathrm{On}$ is decidable by an infinite OTM-program if and only if there are a set $y \subseteq \mathrm{On}$ and a finite OTM-program P such that P decides X in the oracle y.

(d) Show that, if $V = L$, then a class of ordinals is decidable by an infinite OTM-program if and only if it is decidable by an OTM-program.

4 Recognizability

In the last chapter, we determined the computability strength of various machine models of infinitary computability; there, we said that a set x of ordinals is computable if and only if some program can answer every question of the form "is α an element of x." With such a program, the set x can be produced. Intuitively, computability expresses the ability to **construct** a certain object.

Now, there is another natural way to approach objects: Unable as I am to construct the Eiffel tower, I can certainly tell whether what I have in front of me is the Eiffel tower. In other words, I can recognize the Eiffel tower. This intuitive idea of recognizability can easily be transferred to machine models of infinitary computability, leading to a new and more liberal concept of definability associated with these models, which corresponds to the logical concept of implicit definability. This chapter will deal with the recognizability strength, i. e., the realm of recognizable sets of ordinals, for various models of infinitary computability. The concept was originally introduced by Hamkins and Lewis for ITTMs in [81].

In the following, $\delta_{x,y}$ is the Kronecker symbol, i. e., $\delta_{x,y} = 1$ if and only if $x = y$, and otherwise $\delta_{x,y} = 0$.

Definition 4.0.1. Let $x \subseteq \omega$ be a real number. For a machine type M (such as ITRMs, ITTMs, OTMs, etc.), we say that x is M-recognizable if and only if there is an M-program P such that, for all real numbers y, $P^y(0)\downarrow = \delta_{xy}$.

If x, z are real numbers, then we say that x is M-recognizable relative to z if and only if there is a M-program P such that $P^y(0)\downarrow$ and $P^{y\oplus z}(0)\downarrow = \delta_{x,y}$ for all $y \subseteq \omega$. We will write $x \leq^r_M y$ to indicate that x is M-recognizable relative to y, where M is one of our machine types.

Remark 4.0.2. The notion of recognizability also makes sense for subsets of ordinals α when the machine type M is such that it can process subsets of α as its input. The main reason for focusing on subsets of ω is that these can be processed by all the machines we consider, so that machine types can be compared according to the real numbers they recognize.

Intuitively, the recognizability of a set x means that there is a procedure for identifying whether a "given" object is equal to x; and x is recognizable relative to y if and only if there is such a procedure that helps to identify x when one has z available as extra information.

4.1 Preliminaries

Two important theorems of set theory that we will frequently make use of are the following absoluteness results.

https://doi.org/10.1515/9783110496154-004

Theorem 4.1.1 (Shoenfield, [10], p. 189). *Let ϕ be a Σ^1_2-formula or a Σ_1-formula. Then ϕ holds in L if and only if ϕ holds in V.*

We saw in the last chapter that ITRMs compute up to the first limit of admissible ordinals. It is thus not surprising that the following result is particularly relevant when dealing with ITRMs.

Theorem 4.1.2 (Jensen and Karp, [95], pp. 162–163). *Let ϕ be a Σ_1-formula, α a limit of admissible ordinals, and suppose that ϕ holds in V_α. Then ϕ holds in L_α.*

Definition 4.1.3. Let $x \in L$ be transitive and countable in L. Then the $<_L$-minimal code $c \subseteq \omega$ for x (in the sense of Definition 2.3.18, Chapter 2) is called the "canonical code" for x, denoted by $c = \mathrm{cc}(x)$.

In the following, we will repeatedly make use of the fact that codes for L-levels cannot occur too early in the L-hierarchy.

Lemma 4.1.4. *Let $\alpha < \omega_1^L$ be a limit ordinal. Then $\mathrm{cc}(L_\alpha) \notin L_\alpha$.*

Proof. Suppose otherwise, and let α be a counterexample. Then $c = \mathrm{cc}(L_\alpha) \in L_\beta$ for some $\beta < \alpha$. Let $I_c := \{i \in \omega : \text{'}i \text{ codes a real number in } c\text{'}\}$. Clearly, I_c is definable over L_β in the parameter c. Denote by i_c the real number coded by i in the sense of c for $i \in I_c$. Now let $D_c := \{i \in I_c : i \notin i_c\}$. Then D_c is definable over L_β in the parameter c, and hence $D_c \in L_{\beta+1} \subseteq L_\alpha$. Thus, there is $j \in \omega$ such that $D_c = j_c$. But now $j \in j_c \leftrightarrow j \in D_c \leftrightarrow j \notin j_c$, a contradiction. □

4.1.1 Extended truth predicates

In Chapter 2, Theorem 2.3.28, we showed that wITRMs can evaluate truth predicates in $\in$-structures coded by real numbers. With respect to L-levels, we now go a step further: Suppose that $c \subseteq \omega$ is a code for L_α. Then, when asked the right way, wITRMs can use c to evaluate truth values of $\in$-sentences in $L_{\alpha+1}$, and in fact in $L_{\alpha+i}$ for all $i \in \omega$ (and more). The same thus holds for all stronger machine models. The role that this ability plays in the realm of recognizability is that it allows us in certain cases to identify $<_L$-minimal codes of L-levels: For example, if $c := \mathrm{cc}(L_\alpha) \in L_{\alpha+1}$, then one can test whether a given code of L_α is equal to c by using c to search $L_{\alpha+1}$ for an $<_L$-smaller code.[1]

Lemma 4.1.5. *Let $n \in \omega$. There is an ITRM (in fact a wITRM)-program $P_{\mathrm{truth}^{+n}}$ such that the following holds: If $c \subseteq \omega$ codes L_α via f and $i \in \omega$, then $P^c(i)\downarrow = 1$ if and only if $L_{\alpha+n} \models \phi_i$.*

1 This was first done for ITTMs in Hamkins and Lewis [81] and then for ITRMs in [32].

Proof. Given c, it is easy to ITRM-compute a code c' for α. From c', one obtains a code c'' for $\alpha + n$, e. g., by setting $c'' := \{p(i+n, j+n) : p(i,j) \in c'\} \cup \{p(i+n, j) : i \in \omega \wedge j \in n\}$. Now, using Lemma 3.4.26 from Chapter 3, we can compute a code c''' for $L_{\alpha+n}$ from c''. Finally, we run $P^{c'''}_{\text{truth}}(i)$. □

Exercise 4.1.6. It is instructive to prove Lemma 4.1.5 again without computing a code for $L_{\alpha+n}$, at least in the case $n = 1$. To this end, let c code L_α via $f : \omega \to L_\alpha$. Define a new surjection $g : \omega \to L_{\alpha+1}$ by $g(p(2i,j)) = f(j)$ and $g(p(2i+1, p(k, p_k(a_0, \ldots, a_{k-1})))) = \{x \in L_\alpha : L_\alpha \models \phi_i(x, f(a_0), \ldots, f(a_{k-1}))\}$ for all $i, j, k, a_0, \ldots, a_{k-1} \in \omega$. Now combine the ideas of Theorem 3.2.13 from Chapter 3 and Theorem 2.3.28 from Chapter 2 to evaluate the truth predicate for $\in$-formulas in $L_{\alpha+1}$.

Corollary 4.1.7. *For each $k \in \omega$, there is a wITRM-program $P_{k\text{-min}}$ with the following property: If $x \subseteq \omega$ is a code for L_α and $x \in L_{\alpha+k}$, then $P^x_{k\text{-min}}\downarrow = 1$ if and only if x is the $<_L$-minimal code of L_α, and otherwise $P^x_{k\text{-min}}\downarrow = 0$.*

Proof. First, use x to compute a code c for $L_{\alpha+k}$ as in the proof of Lemma 4.1.5. Suppose that c codes $L_{\alpha+k}$ via f. Now use P_{find} from Corollary 2.3.31 from Chapter 2 to identify the natural number i that codes x in $L_{\alpha+k}$, which exists by assumption. Now run through ω, check for each $j \in \omega$ whether $f(j) <_L f(i)$ using Lemma 4.1.5 and if the answer is positive, use P_{compare} from Lemma 2.3.29, Chapter 3 to check whether i and j code the same set. If yes, output 0. If not, continue with the search. When all j have been unsuccessfully tried, output 1. □

4.2 Lost melodies

A fascinating new phenomenon in infinitary computability that has no analogue in classical computability theory is the existence of real numbers that are recognizable, but not computable.

Definition 4.2.1 (Hamkins, Lewis [81]). For M a model of computation (i. e., wITRM, ITRM, ITTM, α-TM, OTM, ORM…) and $x \subseteq \omega$, we say that x is a lost melody for M if and only if x is M-recognizable, but not M-computable.

Remark 4.2.2. The somewhat poetic name, which originated in [81], is based on the similarity with the experience of having a melody in mind that one does not recall well enough to be able to, e. g., whistle it, but which one could immediately identify when one heard it.

We will now prove that lost melodies exist for ITTMs. A very similar proof works for ITRMs, so we will not repeat the details and only make some remarks with respect to the necessary adaptations.

Lemma 4.2.3. *L_Σ does not contain a triple (α, β, γ) of ordinals such that $L_\alpha \prec_{\Sigma_1} L_\beta \prec_{\Sigma_2} L_\gamma$.*

Proof. Note that this is not trivial: The characterization of (λ, ζ, Σ) as the "lexically minimal" triple (α, β, γ) satisfying $L_\alpha \prec_{\Sigma_1} L_\beta \prec_{\Sigma_2} L_\gamma$ is compatible with the existence of such ordinals (α, β, γ) with $\lambda < \alpha < \beta < \gamma < \Sigma$.

However, the existence of such a triple is a Σ_1-statement; let us call it ϕ. If such a triple existed in L_Σ, ϕ would hold in L_Σ, and hence, by Σ_1-elementarity, in L_λ. Thus such a triple would in fact already exist in L_λ, and thus have all three components $< \lambda$, which clearly contradicts the minimality of (λ, ζ, Σ). □

Theorem 4.2.4 (The lost melody theorem for ITTMs, Hamkins and Lewis [81], Theorem 4.9). *There is a real number r such that r is ITTM-recognizable, but not ITTM-writable.*

Proof. Let $r := \mathrm{cc}(L_{\Sigma+1})$. It is easy to see that r is not ITTM-writable: For if it was, we would have $r \in L_\lambda$, and thus a code for L_λ would be contained in L_λ, which contradicts Lemma 4.1.4.

We now show that r is ITTM-recognizable by describing a procedure for recognizing r. Let x be given in the oracle.

The first step is to test whether x codes a well-founded relation. This is possible by Theorem 2.3.25, Chapter 2. If x fails to code such a relation, we return "no." Otherwise, we know that x codes a well-founded relation and continue:

We test whether x codes an L-level. This can be done by checking whether the first-order formula $V = L$ holds in the structure coded by x, which is possible by Theorem 2.3.28 of Chapter 2. If not, we return "no," otherwise, we know that x codes some L_δ and continue.

Check whether there is a strictly increasing triple (α, β, γ) of ordinals in L_δ such that $L_\alpha \prec_{\Sigma_1} L_\beta \prec_{\Sigma_2} L_\gamma$. This is again expressible as a first-order formula, and thus a program for evaluating first-order statements in coded structures can perform this check. If not, we know that x does not code $L_{\Sigma+1}$, as $L_{\Sigma+1}$ contains the triple (λ, ζ, Σ) and return "no," Otherwise, we know that $\delta > \Sigma$ and continue.

Using exhaustive search and the truth predicate, determine whether L_δ contains a largest ordinal ε such that $\varepsilon = \gamma + 1$. Note that, by Lemma 4.2.3, $L_{\Sigma+1}$ contains only one ordinal γ' for which there are $\alpha' < \beta' < \gamma'$ satisfying $L_{\alpha'} \prec_{\Sigma_1} L_{\beta'} \prec_{\Sigma_2} L_{\gamma'}$. Hence, if $\delta = \Sigma + 1$, then the answer must be positive. If it is not, we halt and output "no." If it is, we know that δ is the successor of some ordinal γ such that L_γ has a proper Σ_2-elementary submodel L_β which in turn has some proper Σ_1-elementary submodel L_α.

It remains to determine whether $\gamma = \Sigma$. Clearly, we have $\gamma \geq \Sigma$. If $\gamma > \Sigma$, then L_γ contains $\alpha' < \beta' < \gamma'$ satisfying $L_{\alpha'} \prec_{\Sigma_1} L_{\beta'} \prec_{\Sigma_2} L_{\gamma'}$, which can again be checked by evaluating the truth predicate in L_γ.

At this point, we know that x codes $L_{\Sigma+1}$. We still need to determine whether x is $<_L$-minimal with this property, but this is possible by Corollary 4.1.7. □

Exercise 4.2.5. Show that $\mathrm{cc}(L_{\omega_\omega^{\mathrm{CK}}})$ is ITRM-recognizable, but not ITRM-computable.

Theorem 4.2.6 ([37], Lemma 11). *No element of $L_\Sigma \setminus L_\lambda$ is recognizable.*

Proof. Assume otherwise, and let $x \in L_\Sigma \setminus L_\lambda$ be recognizable by the ITTM-program P. Now consider the following program: Run the universal ITTM; whenever a new real number appears on the output tape, use P to test whether that real number is equal to x. If not, continue, otherwise halt.

Since x is accidentally writable, this program will eventually halt with x on the output tape. But this implies that x is writable, and thus an element of L_λ, a contradiction. □

Remark 4.2.7. Note that, as $0'_{\mathrm{ITTM}} \in L_{\lambda+1} \setminus L_\lambda$, this in particular implies that $0'_{\mathrm{ITTM}}$ is not ITTM-recognizable. This is in contrast with the case of ITRMs, for which we will see below that $0'_{\mathrm{ITRM}}$ is ITRM-recognizable.

A prominent role in definability theory and recursion theory is played by basis theorems; roughly, basis theorems say that nonempty sets that are in a certain way "nice" or "simple" must have some "nice" or "simple" elements. For ITTMs, we have the following basis theorem.

Theorem 4.2.8. *Let $X \subseteq \mathfrak{P}(\omega)$ be nonempty and ITTM-decidable. Then X contains an ITTM-recognizable element. In fact, if $x \in X$ is the $<_L$-minimal element of X, then x is ITTM-recognizable.*

Proof. Let X be ITTM-decidable and nonempty, and let P be an ITTM-program that decides X. Let α be the minimal element of $\{\lambda^x : x \in X\}$, and let $x \in X$ be such that $\lambda^x = \alpha$. Then $L_\alpha[x] \models P^x\downarrow = 1$, so $L_\alpha[x] \models \exists y P^y\downarrow = 1$. As α is a limit of admissible ordinals, it follows from Theorem 4.1.2 that actually $L_\alpha \models \exists y P^y\downarrow = 1$. Let z be the $<_L$-minimal real number such that $z \in L_\alpha$ and $P^z\downarrow = 1$, i. e., the $<_L$-minimal element of X. Since $z \in L_\alpha = L_{\lambda^x}$, z is ITTM-computable from x, so $\lambda^z \leq \lambda^x = \alpha$; on the other hand, as $P^z\downarrow = 1$, we have $z \in X$, so we have $\lambda^z \geq \alpha$ by minimality of α. Thus $\lambda^z = \alpha$.

We claim that z is ITTM-recognizable. To recognize z, we proceed as follows: Let a real number a be given in the oracle.

(1) First, run P^a. If this stops with output 0, then $a \neq z$, so we stop with output 0. Otherwise, we know that $a \in X$ and proceed.

(2) Let the universal ITTM run in the oracle a until a code for some L_β is found such that $L_\beta \models \exists y P^y\downarrow = 1$. We saw above that such a real number is contained in L_α, hence there is one in every L_{λ^b} with $b \in X$ by minimality of α, so in particular there is one in L_{λ^a}. Moreover, the length of the computation of P^a is bounded by λ^a, and hence there is some $\beta < \lambda^a$ such that L_β contains the whole computation. Thus, the universal ITTM will eventually produce a code for such a β. Since reals can be tested for coding well-founded $\in$-structures and also for coding structures that satisfy certain first-order-statements, this event can be detected when it occurs. From now on, we thus have a code c for some L_β with $L_\beta \models \exists y P^y\downarrow = 1$ available.

(3) Test whether $a \in L_\beta$. To do this, identify the natural numbers that code elements of ω in c and then search through L_β to check whether some element contains the same natural numbers as a. If not, then L_β contains an element x of X, but not a, so we must have $x <_L a$ and hence $a \neq z$ by the $<_L$-minimality of z. Hence, in this case, we stop with output 0. Otherwise, we know that $a \in L_\beta$ and proceed.

(4) Using c, search through L_β to check whether it contains an element r such that $r <_L a$ and $r \in X$. The first statement is a first-order statement over L_β, and thus can be checked, the second is just the claim that $P^r\downarrow = 1$. If this search is successful, then a is not the $<_L$-minimal element of X, so $a \neq z$, in which case we stop with output 0. Otherwise, we have $a = z$ and stop with output 1. □

Corollary 4.2.9. *The set of ITTM-recognizable real numbers is not ITTM-decidable.*

Proof. Otherwise, the complement would be decidable as well; as it is certainly nonempty, it would thus contain an ITTM-recognizable element by the last theorem, a contradiction. □

We can also show that the ITTM-recognizable real numbers are constructible, and in fact belong to a certain initial segment of L. The following theorem and its proof was suggested to us by P. Welch.

Theorem 4.2.10 (Cf. [19], Theorem 15). *Recall from Chapter 3 that σ is the smallest ordinal such that $L_\sigma \prec_{\Sigma_1} L$. Then every ITTM-recognizable real number x is an element of L_σ. Moreover, for every $\alpha < \sigma$, there is an ITTM-recognizable real number in $L_\sigma \setminus L_\alpha$, thus σ is minimal with this property.*

The same holds when one replaces ITTMs by ITRMs.

Proof. If x is ITTM-recognizable by the program P, then x is characterized as the unique $y \subseteq \omega$ such that P^y halts with output 1, which is expressible by a Σ_1-formula $\phi(y)$. Now $\exists y\phi(y)$ is also Σ_1, and it holds in V, as witnessed by x. By Shoenfield absoluteness, it must then hold in L. By definition of σ, L_σ is a Σ_1-elementary submodel of L, so $\exists y\phi(y)$ in fact holds in L_σ. So L_σ contains some real number y such that the computation P^y halts with output 1 in L_σ. By absoluteness of computations, the same holds in V, so P^y actually halts with output 1. But since P recognizes x, we must have $y = x$. Thus $x \in L_\sigma$, as desired.

For the second claim, let $\alpha < \sigma$. By definition of σ, there is some Σ_1-formula ψ such that $L_\sigma \models \psi$, but $L_\alpha \not\models \psi$. Let $\beta < \sigma$ be the smallest limit of admissible ordinals such that $L_\beta \models \psi$, and let c be the $<_L$-minimal code of L_β. (Note that, by upwards absoluteness of Σ_1-formulas, such a β must exist.) Now c is not an element of L_β by Lemma 4.1.4. On the other hand, c is ITTM-recognizable by a procedure very similar to the one used in the proof of Theorem 4.2.4.

The proof for ITRMs is analogous. □

ITTMs with parameters[2]

As we saw in Chapter 3, OTMs become considerably stronger when ordinal parameters are allowed. It thus seems natural to ask whether something similar happens for other machine models. However, it is at first sight not easy to make sense of general ordinal parameters for many machine models: All we can enter, e. g., in an ITRM are natural numbers, or real numbers for ITTMs.

In [72], Habic came up with an idea how to use ordinal parameters with ITTMs: Instead of entering them via the tape, enter them via time! This works as follows: When $\alpha \in \mathrm{On}$, or, more generally, $X \subseteq \mathrm{On}$, we can define an X-ITTM like an ordinary ITTM with a special reserved inner state $\bar{s}$ that is assumed at every point of computation time that is contained in X. Habic studied the case where X is the class of cardinals, which gives us "cardinality recognizing ITTMs," showing that they are considerably stronger than ITTMs (e. g., they can solve the halting problem for ITTMs).

In [37], the recognizability strength of ITTMs with general ordinal parameters was studied. We will explain some of the basic results.

Definition 4.2.11. Let $X \subseteq \mathrm{On}$, where X may be a proper class. An X-ITTM-program is an ITTM-program. We pick one special inner state $\bar{s}$. An X-ITTM-computation is defined like an ITTM-computation, with the following difference: Whenever $\tau \in X$, the inner state of the configuration at time τ is $\bar{s}$.

Remark 4.2.12 (Question). In the same way, one can define X-ITRMs, X-α-TMs etc. These models have not been considered to the best of our knowledge. They may well provide a worthwhile field of study, starting with their computational and recognizability strength.

We now give some results on the recognizability strength of X-ITTMs in the special case where X is the class of cardinals or contains only one element. These results come from [37], to which we also refer for a more detailed study. , P.0

Our first observation is that ITTM-recognizability with single parameters reaches beyond "plain" ITTM-recognizability:

Proposition 4.2.13 (Habic, [72], Proposition 6). *There is a real number x which is $\{\omega_1\}$-ITTM-writable, but not ITTM-recognizable.*

Proof. Let x code the contents of the tapes of the universal ITTM $\mathcal{U}$ at time ζ. Almost immediately by its definition, x is accidentally writable, and thus contained in L_Σ. On the other hand, it is not hard to see that x is not ITTM-computable: For, by Corollary 3.6.37 from Chapter 3, we know that $\mathcal{U}$ assumes the tape contents coded by x for the first time

---

2 The results in this section come from M. Carl and P. Schlicht. The Recognizability Strength of Infinite Time Turing Machines with Ordinal Parameters, LNCS 10307, Springer [37] with permission from Springer Nature. We thank Philipp Schlicht and Springer Nature for the kind permission to reuse material from this article here.

at time ζ. Thus, given x, we can use $\mathcal{U}$ to clock ζ. Hence, if x was ITTM-computable, then ζ would be ITTM-clockable; however, we know from Corollary 3.6.20 in Chapter 3 that it is not.

Now, by Theorem 4.2.6 above, all real numbers that are both ITTM-aw and ITTM-recognizable are also ITTM-writable. So x is not ITTM-recognizable. On the other hand, as Σ divides ω_1, the tape contents of $\mathcal{U}$ at times ζ, Σ and ω_1 are identical. Hence, running $\mathcal{U}$ up to time ω_1 and then writing a code for the tape contents to the output tape, we can write x. □

Proposition 4.2.14 ([37], Theorem 17). *Suppose that $X = \{\omega_1\}$, and let $x \subseteq \omega$ be recognizable by an X-ITTM. Then $x \in L_\sigma$.*

Proof. Let P an ITTM-program that recognizes x in the parameter ω_1. Let us imagine that P is run in the oracle $y \subseteq \omega$. Thus, by Corollaries 3.6.20 and 3.6.37 from Chapter 3, P^y either halts in $< \lambda^y$ many steps (and hence never assumes the special state) or loops from time Σ^y on, repeating the series of configurations between times ζ^y and Σ^y over and again. In the latter case, as ω_1 is divisible by every countable ordinal and Σ^y is countable, the configuration at time ω_1 will be the same as that at time Σ^y. From this point on, we have again an ordinary ITTM-computation, which starts in the special state with the same tape contents as the computation of P^y had at time Σ^y.

By these observations and the λ-ζ-Σ-theorem from Chapter 3, x can be characterized as the unique $z \subseteq \omega$ such that P^z halts with output 1 or such that there are ordinals α, β, γ such that $L_\alpha[z] \prec_{\Sigma_1} L_\beta[z] \prec_{\Sigma_2} L_\gamma[z]$, L_γ contains no further triples of this kind, and the ITTM computation starting in the special state with the tape contents that the computation of P^z had at time γ halts with output 1. This is expressible by a Σ_1-formula $\phi(z)$.

As $\phi(x)$ holds, the statement $\exists z \phi(z)$ is true in V. By Shoenfield's absoluteness theorem, it also holds in L. By Σ_1-elementarity, it thus holds in L_σ. By absoluteness of computations, it now follows that $x \in L_\sigma$, as desired. □

Exercise 4.2.15 ([37], Theorem 9). Adapt the proof of Proposition 4.2.14 to show that, whenever X is a class of cardinals closed under suprema and $x \subseteq \omega$ is recognizable by an X-ITTM, then $x \in L_\sigma$.

However, there are parameters that allow us to go beyond L_σ.

Proposition 4.2.16. *There are a real number x and a countable ordinal α such that $x \notin L_\sigma$, but x is $\{\alpha\}$-ITTM-recognizable. In fact, we can take $x = \mathrm{cc}(L_\sigma)$ and $\alpha = \sigma$.*

Proof. Recall from Chapter 3 that σ is admissible. We reserve a second scratch tape T'. Now, given the input $y \subseteq \omega$, we start by successively identifying the smallest natural number coding an ordinal in the structure coded M_y by y that is not yet written on T' and writing it to T'. When M_y contains ordinals not coded by any natural number on T', but no smallest such ordinal, this means that M_y is ill-founded and we output 0. Otherwise, the search for the smallest remaining ordinal will take $< \omega^\omega$ many steps.

If all natural numbers coding ordinals in M_y are written on T' before the special state indicates that time σ has been reached, the ordinal height of M_y is $< \sigma$ and we again output 0. If that special state is reached, but there are still ordinals in M_y not coded by any natural number written on T', then the ordinal height of M_y is $> \sigma$ and we also output 0.

If none of the above cases occurs, we know that y codes an $\in$-structure with ordinal height σ. In this case, we proceed as in the proof of Theorem 4.2.4 above to check whether y codes a well-founded L-level. If that is not the case, we output 0. If it is, then we know that y codes L_σ and we finally proceed as in the proof of Theorem 4.2.4 to test whether y is $<_L$-minimal with this property.

Thus $\mathrm{cc}(L_\sigma)$ is $\{\sigma\}$-ITTM-recognizable. By Lemma 4.1.4, $\mathrm{cc}(L_\sigma) \notin L_\sigma$. Thus $\mathrm{cc}(L_\sigma)$ and σ are indeed as desired. □

4.2.1 α-Machines

Lemma 4.2.17. *If α is a limit of admissible ordinals, then there is no lost melody for α-TMs.*

Proof. Suppose that $x \subset \omega$ is α-TM-recognizable, say by the program P. The computation of P^x is of length $< \alpha$ and is thus contained in $L_\alpha[x]$. Thus $L_\alpha[x] \models \exists y \subseteq \omega P^y \downarrow = 1$, which is a Σ_1-formula. As α is a limit of admissible ordinals, $L_\alpha \models \exists y \subseteq \omega P^y \downarrow = 1$ by the Jensen–Karp theorem. But then $x \in L_\alpha$ follows from absoluteness of computations as usual. Thus x is α-TM-computable by Theorem 3.5.5, since limits of admissible ordinals are clearly exponentially closed. □

Exercise 4.2.18. Show that there are $\alpha > \omega$, $x \subseteq \omega$ such that x is α-TM-computable, but not α-TM-recognizable.

4.2.2 Weak ITRMs[3]

Lost melodies are a rather typical phenomenon for models of infinitary computability but this does not mean that their existence can be taken for granted when considering such models. In the proof of the lost melody theorem for ITTMs, it was crucial to test real numbers for being codes of well-founded relations. It is not clear that wITRMs can also perform this test (they cannot, which will follow from the results of this section—though this is a bit of an overkill). Thus, it is not clear whether there are lost melodies

3 The following material is adapted from the article "The lost melody phenomenon" [30], which appeared in: S. Geschke, B. Löwe, P. Schlicht (eds.), Infinite, Computability and Metamathematics. Festschrift is celebrating the 60th birthdays of Peter Koepke and Philip Welch. We thank the editors for the kind permission to reuse parts of this work here.

for wITRMs. In this section, we show that there are none by showing that every real number recognizable by a wITRM is already computable by a wITRM. (It is also true that every wITRM-computable real number is wITRM-recognizable, though the argument requires a bit more care than in the case of ITRMs.)

This is particularly interesting as it shows that lost melodies can be absent even for models of computations that are unable to "enumerate their domain"[4]: If, for a certain machine type, there is a program P that goes on and on, eventually constructing every object that one could possibly give as an input, then one can compute every recognizable object x by running P and running the program for recognizing x on every object that P constructs. Thus, as we will see below, since OTMs can enumerate the constructible real numbers, there are no lost melodies for OTMs; likewise, as classical Turing machines can enumerate all natural numbers, there are no lost melodies for classical Turing machines. On the other hand, ITRMs and ITTMs can only enumerate small portions of the constructible real numbers, so that there is space left for lost melodies, which do in fact exist for both models. This fragment, however, is even smaller for wITRMs; it is thus worthwhile to note that the lost melody phenomenon does not occur here.

The results and proofs of this section come from [30].

What we will actually prove is a basis theorem for wITRMs, which as we will see in Chapter 7, is really a disguised version of a classical result in descriptive set theory, namely Kreisel's basis theorem. Our proof is an adaptation of the proof Kreisel's basis theorem (see, e. g., [135] or [153], p. 74) using wITRMs. Here is the theorem statement.

Theorem 4.2.19. *Let $X \subseteq \mathfrak{P}(\omega)$ be nonempty and wITRM-decidable. Then X has a wITRM-computable element.*

Note that this is quite different from the basis theorems for ITRMs and ITTMs given in this chapter, as we guarantee a **computable** instead of merely a **recognizable** element. The absence of lost melodies for wITRMs is an easy consequence of Theorem 4.2.19:

Corollary 4.2.20 (Cf. [30], Corollary 9). *Suppose that $x \subseteq \omega$ is wITRM-recognizable, then it is wITRM-computable. Hence, there are no lost melodies for weak ITRMs.*

Proof. By definition, the wITRM-recognizability of x is equivalent to the wITRM-decidability of $\{x\}$. Thus, if x is wITRM-recognizable, then $\{x\}$ is a nonempty wITRM-decidable set of real numbers, which has a wITRM-computable element by Theorem 4.2.19. As $\{x\}$ has only one element, namely x, x must be wITRM-computable. □

We now work toward a proof of Theorem 4.2.19.

4 See, e. g., http://www.3ammagazine.com/3am/playing-infinite-chess/

The following ingredient, known as König's lemma, is crucial in many areas of mathematics and logic. We refer the reader to Simpson, [163], Hirschfeldt [85] and Kaye, [98] for details.

Definition 4.2.21. Let S be a set. A "tree" over S is a set $\mathcal{T}$ of finite sequences of elements of S with the property that, if $s \in \mathcal{T}$, then all initial segments (including the empty sequence) are also elements of $\mathcal{T}$. If $S = \{0, 1\}$, $\mathcal{T}$ is called a "binary tree."

If $s, t \in \mathcal{T}$ and t arises from s by appending one element to the right, then t is an "immediate successor" of s in $\mathcal{T}$. If s is a proper initial segment of t, then t is a "successor" of s in $\mathcal{T}$. If $\mathcal{T}$ is such that every element of $\mathcal{T}$ has only finitely many immediate successors in $\mathcal{T}$, then $\mathcal{T}$ is called "finitely branching."

If $\mathcal{T}$ is a binary tree and $b \in^{\omega} \{0, 1\}$, then we say that b is an infinite branch of $\mathcal{T}$ if and only if $\mathcal{T}$ contains every finite initial segment of b.

Lemma 4.2.22 (König's lemma). *Let $\mathcal{T}$ be a finitely branching, infinite tree. Then $\mathcal{T}$ has an infinite branch.*

Proof. The proof works by an iterated application of the pigeonhole-principle. Let $b_0 = \langle\rangle$ be the empty sequence. As $\mathcal{T}$ is infinite, b_0 has infinitely many successors in $\mathcal{T}$, only finitely many of which can be immediate successors, as $\mathcal{T}$ is finitely branching. Thus there must be an immediate successor of b_0 with infinitely many successors in $\mathcal{T}$; pick one and call it b_1. Now the same reasoning applies to b_1, yielding an immediate successor b_2 of b_1 with infinitely many successors in $\mathcal{T}$. In general, having defined b_i, we obtain an immediate successor b_{i+1} of b_i with infinitely many successors in $\mathcal{T}$. Now let $b = (b_i : i \in \omega)$, then b is an infinite branch of $\mathcal{T}$. □

Corollary 4.2.23.
(i) *There is a wITRM-program P_{kl} such that the following holds: If c codes a finitely branching tree $\mathcal{T}$ via $f : \omega \to \mathcal{T}$ and $f(i) = t \in \mathcal{T}$, then $P^c_{\mathrm{kl}}(i)$ halts with output* 1 *if and only if t is the initial segment of an infinite branch of $\mathcal{T}$, and otherwise $P^c_{\mathrm{kl}}(i)$ halts with output* 0.
(ii) *Moreover, there is a wITRM-program P_{branch} such that the following holds: If c codes a finitely branching tree $\mathcal{T}$ via $f : \omega \to \mathcal{T}$ and $f(i) = t \in \mathcal{T}$ lies on an infinite branch of $\mathcal{T}$, then $P^c_{\mathrm{branch}}(i, j)$ outputs the jth bit of the leftmost such branch.*

Proof. (i) By König's lemma, t is an initial segment of an infinite branch of $\mathcal{T}$ if and only if t has infinitely many successors in $\mathcal{T}$. This can be determined on a wITRM as follows: We will perform an exhaustive search through all finite 0-1-sequences, coded as natural numbers, testing for each whether it extends t (which is clearly recursive) and belongs to $\mathcal{T}$ (which is possible using the oracle c). The natural number i that is currently being tested will be stored in a register R, which initially contains 0. Furthermore, we use two flags, F and F'. After it has been tested whether i codes a 0-1-sequence extending t and belonging to $\mathcal{T}$, we flash F and replace the content of R first by 0 and then by $(i + 1)$ (where the intermediate 0 is inserted to prevent an overflow of R at limit times). If the test was answered positively, we also flash F'. Now, using F, we will know when infinitely many natural numbers have been tried, which means that all natural numbers have been tried. At this point, F' will tell us whether infinitely many natural numbers were **successfully** tried, in which case we return "yes"; otherwise, we return "no."

(ii) This is done by iteratively applying (i). Let $n \in \omega$, and suppose we want to determine the nth bit of the leftmost infinite branch of $\mathcal{T}$. To do this, we iterate the following procedure n many times, starting with the empty sequence s_0 in the register R: Suppose we have so far computed (a code for) the sequence s of length k with infinitely many successors in $\mathcal{T}$, which is contained in R. Using P_{kl}, we test whether $s0 \in \mathcal{T}$. If yes, we replace the content of R by (a code for) $s0$. Otherwise, we replace the content of R by (a code for) $s1$. As soon as R contains a (code for a) sequence of length n, we output its last bit. This is clearly possible on a wITRM and computes the leftmost infinite branch of $\mathcal{T}$. Starting with t instead of the empty sequence, we find the leftmost infinite branch extending t. □

A nice consequence of König's lemma in recursion theory is the following.

Corollary 4.2.24 (Folklore, cf., e. g., Kaye [98], Chapter 1). *Let P be a (classical) Turing-program such that P^x halts for every $x \subseteq \omega$. Then there is $n \in \omega$ such that, for every $x \subseteq \omega$, P^x terminates in $< n$ many steps, and thus considers at most the first n many bits of x.*

Proof. We consider the binary tree $\mathcal{B}$ that consists of all finite 0-1-sequences s with the following property: If P is run in the oracle s, then P will eventually request some oracle bit with index bigger than the length of s; intuitively, s does not contain sufficient information to determine a run of P. Clearly, if s is an initial segment of t and t belongs to $\mathcal{B}$, then so does s, i. e., $\mathcal{B}$ really is a tree. Moreover, $\mathcal{B}$ is certainly finitely branching.

If the conclusion of our lemma fails, then $\mathcal{B}$ contains arbitrarily long sequences and is thus infinite. By König's lemma, it thus has an infinite branch b.

Now consider what happens when we run P in the oracle b. If this computation halts, it will do so within a finite number n of steps, considering at most the first n many bits of b. But this means that the sequence consisting of the first $(n + 1)$ many bits does not belong to $\mathcal{B}$, a contradiction. □

A set $X \subseteq \mathfrak{P}(\omega)$ is Σ^1_1 if and only if there is a Σ^1_1-formula ϕ such that $X = \{x \in \mathfrak{P}(\omega) : \phi(x)\}$. We will see below that the Σ^1_1-set and the wITRM-decidable sets coincide. It then suffices to show that every Σ^1_1-set contains a wITRM-computable element. To this end, we use a particularly nice form to write Σ^1_1-formulas (see, e. g., Chapter 3 of Mansfield and Weitkamp [135]).

Lemma 4.2.25. *Let $\phi(x)$ be Δ^0_1. Then there is a (classical) Turing program P such that P^z halts for all $z \subseteq \omega$ and $P^z\downarrow = 1$ if and only of $\phi(z)$, while $P^z\downarrow = 0$, otherwise.*

Proof (Sketch). Δ_0^0-statements can easily be decided by exhaustively searching the quantifier bounds. Thus, if ϕ is Σ_1^0, there is a program that halts with output 1 in the oracle x if and only if $\phi(x)$ (and otherwise does not halt) by trying out all possible witnesses. If ϕ is Δ_1^0, both ϕ and its negation are Σ_1^0, so there are such programs for both statements. Now just let them run in parallel and output 1 or 0, depending on which one halts. □

We note that Σ_1^1-formulas can be written in a particularly nice way.

Lemma 4.2.26 (Cf. Mansfield and Weitkamp [135], p. 32). *Let ϕ be a Σ_1^1-formula. Then there is a (classical) Turing-program P such that $P^z(i)$ halts for all $z \subseteq \omega$, $i \in \omega$ and for all x, $\phi(x)$ holds if and only if $\exists y \forall i P^{x \oplus y}(i)\downarrow = 1$.*

Exercise 4.2.27. Prove Lemma 4.2.26.[5]

Now suppose that $\phi(x)$ is a Σ_1^1-formula, P is the corresponding Turing-program in the sense of Lemma 4.2.26 and that there is at least one x such that $\phi(x)$ holds.

Our aim is to compute such an x on a wITRM. The idea is to search for such an x, in the following way: In fact, we will search for a real number $z(= x \oplus y)$ such that $\forall i P^z(i)\downarrow = 1$ holds. To obtain x, we will then only output the even bits of z. For each $i \in \omega$, we can regard the question whether $P^z(i)\downarrow = 1$ as a "test" whether z is as desired, and our aim is to determine on a wITRM a real number z that "passes" all of these tests. To this end, we shall use Corollary 4.2.23.

Let $\mathcal{T}$ be the set of all finite 0-1-sequences s with the property that for no $i \in \omega$ does $P^s(i)$ halt with output 0, i. e., either oracle bits beyond the length of s are requested at some point of the computation, or the computation terminates with output 1. Clearly, if $P^t(i)$ does not halt with output 0, then neither does $P^s(i)$ for any initial segment of s, so $\mathcal{T}$ is a tree, which is obviously finitely branching. Moreover, by our assumption, there is an infinite 0-1-sequence z such that $\forall i \in \omega P^z(i)\downarrow = 1$; hence all finite initial segments of z belong to $\mathcal{T}$, so $\mathcal{T}$ has an infinite branch. We show that $\mathcal{T}$ is in fact wITRM-decidable.

Proposition 4.2.28. *$\mathcal{T}$ is wITRM-decidable.*

Proof. We already know from Chapter 3 that a wITRM can compute all finite iterations of the Turing jump (see Exercise 2.8.7). Now, for given s and i, it is Turing-decidable relative to $0'$ whether $P^s(i)\downarrow = 0$. Consider the program H that uses the oracle $0'$ to successively decide, for each $i \in \omega$, whether $P^s(i)\downarrow = 0$ and halts when the answer is

5 Recall that Δ_0^0-sets are recursive. Use witness functions $f : \omega^k \to \omega$ to get rid of existential first-order quantifiers and find a way to encode such functions as subsets of ω in such a way that the decoding is recursive. Then use further codings to replace a finite sequence of universal first-order quantifiers by a single universal first-order quantifiers and a finite sequence of existential second-order quantifiers by a single existential second-order quantifier.

"yes." Then H halts if and only if $P^s(i)$ does not halt with output 0 for any $i \in \omega$, which hence can be computed in $0''$. So $\mathcal{T}$ is recursive in $0''$, and hence wITRM-decidable. □

We can now apply Corollary 4.2.23 to compute an infinite branch b of $\mathcal{T}$ on a wITRM. We show that $\forall i \in \omega P^b(i)\downarrow = 1$, so that b is as desired and the sequence b_e of its even bits satisfies $\phi(b_e)$ (and is clearly wITRM-computable by construction). By the assumption on P, we have that $P^b(i)$ will halt with output 0 or 1 for every $i \in \omega$. If there was some $i \in \omega$ such that $P^b(i)\downarrow = 0$, then in the course of this computation, only finitely many bits of b would be used, say the first n bits of b. It follows that $b \upharpoonright n$, the restriction of b to its first n many bits, satisfies $P^{b\upharpoonright n}(i)\downarrow = 0$, so that $b \upharpoonright n \notin \mathcal{T}$, which contradicts the fact that b is an infinite branch of $\mathcal{T}$. Thus we have indeed $\forall i P^b(i)\downarrow = 1$.

The above argument shows the following.

Theorem 4.2.29. *Every nonempty Σ^1_1-set of real numbers contains a wITRM-computable element.*

To complete the proof of Theorem 4.2.19, it now suffices to show that all wITRM-decidable sets of real numbers are Σ^1_1. This, however, is not entirely obvious: The tempting approach to show that the set X decided by the wITRM-program P is Σ^1_1 would be to say that $x \subseteq \omega$ belongs to X if and only if there is a real number c coding a P-computation in the oracle x that halts with output 1. However, to say that c codes a computation, we need to express well-foundedness, which is beyond the capabilities of Σ^1_1. Fortunately, we have a strong bound on the possibly halting times in the oracle x, namely $\omega_1^{\mathrm{CK},x}$, which saves the day for us.

Definition 4.2.30. Let $M \models \mathrm{KP}$. M is called an ω-model if and only if $\omega^M = \omega$, i. e., the natural numbers of M are the "actual" natural numbers.

Lemma 4.2.31 (Cf. [135], Proposition 5.6). *Let $x \subseteq \omega$ and let $M \ni x$ be an ω-model of KP. Then $\omega_1^{\mathrm{CK},x}$ is an initial segment of On^M.*

Proof. Admissible sets are closed under recursive functions; thus if $x \in M$, then every ordinal which is recursive relative to x, i. e., which is $< \omega_1^{\mathrm{CK},x}$, has a code in M. By Lemma 3.1.32 from Chapter 3, the corresponding ordinals are also contained in M, as desired. □

Lemma 4.2.32 (Cf. [135], Example 1.20). *There is a Σ^1_1-formula $\phi_\models$ such that, for all $c \subseteq \omega$ coding an $\in$-structure M and every $i \in \omega$, $\phi_\models(c, i)$ holds if and only if the ith $\in$-formula ϕ_i holds in $(M, \in)$.*

Lemma 4.2.33. *There is a Σ^1_1-formula ϕ_{KP} such that, for all $c \subseteq \omega$, $\phi(c)$ holds if and only if c codes a model of KP.*

Proof. It is not hard (though a bit cumbersome) to construct an arithmetical formula ψ such that, for all $i \in \omega$, $\psi(i)$ holds if and only if the ith $\in$-sentence ϕ_i is an axiom of

KP. Now let ϕ_{KP} be $\forall i(\psi(i) \to \phi_{\models}(c,i))$. For details, see [135], Example 1.21, where this is explained for ZF in place of KP. □

The idea is now to formulate, for a wITRM-program P and $x \subseteq \omega$, the claim that $P^x\downarrow = 1$ as "there is a real number c coding an ω-model $M \ni x$ of KP which believes that $P^x\downarrow = 1$." To do this, we need to be able to quantify over (coded) ω-models of KP. We already saw how to express coding a model of KP in the right way above, and it only remains to take care of the condition that $\omega^M = \omega$. The following observation makes things easier: For our purposes, it is not necessary to express "c codes an ω-model of KP" as a Σ^1_1-formula; rather, it suffices to find a Σ^1_1-formula that is only true of codes of ω-models of KP and such that every such model has a code for which the formula holds. This is the point of the following lemma.

Lemma 4.2.34 (Cf. [30], Lemma 7). *There is a Σ^1_1-statement $\phi_{\omega-\mathrm{KP}}(x)$ such that, for all $x \subseteq \omega$, $\phi(x)$ holds only if x codes an ω-model of KP and such that, for any countable ω-model M of KP, there is a code c for M such that $\phi(c)$ holds.*

Proof. Every countable ω-model M of KP can be coded by a real $c(M)$ in such a way that $i \in \omega$ is represented by $2i$ in c and ω is represented by 1. It thus suffices to characterize such codes by a Σ^1_1-formula.

Now, the demand that $2i$ represents i in c can be expressed by the arithmetical formula $\forall i,j(p(i,2j) \in c \leftrightarrow (2|i \land i < 2j))$; and that 1 represents ω in c is expressed by $\forall i(p(i,1) \in c \leftrightarrow (2|i))$. The conjunction of these arithmetical formulas with the Σ^1_1-formula ϕ_{KP} is still Σ^1_1, and this is our desired $\phi_{\omega-\mathrm{KP}}$. □

Lemma 4.2.35. *Let $X \subseteq \mathfrak{P}(\omega)$ be wITRM-decidable. Then X is Σ^1_1.*

Proof. Let P be a wITRM-program that decides X. We claim that, for all $x \subseteq \omega$, $P^x\downarrow = 1$ holds if and only if there is an ω-model $M \ni x$ of KP such that $M \models P^x\downarrow = 1$.

Using this claim, the lemma follows easily from Lemma 4.2.34 by simply adding the condition that $M \models P^x\downarrow = 1$, which is arithmetical in the code for M.

To see the claim, suppose first that M is such a model. It follows that M contains a set C such that M believes that C a halting computation of P^x with output 1. We need to show that C actually is such a computation, which means that its length is an ordinal (and not just an ordinal in the sense of M.) By the fact that P decides X, P^x terminates, and by the time bound on halting wITRM-computations obtained in Chapter 3, we know that the length α of the computation is $< \omega_1^{\mathrm{CK},x}$. Hence $\alpha + 1 \subseteq M$ by Lemma 4.2.31, so C is a halting computation of P^x with output 1 in V, and hence $P^x\downarrow = 1$, i. e., $x \in X$.

On the other hand, suppose that $x \in X$, i. e., $P^x\downarrow = 1$, and let C be the halting computation of P^x. As above, the length of C is $< \omega_1^{\mathrm{CK},x}$, so C is contained in $L_{\omega_1^{\mathrm{CK},x}}[x]$, which thus is an ω-model of KP modeling $P^x\downarrow = 1$. □

Exercise 4.2.36. $X \subseteq \mathfrak{P}(\omega)$ is wITRM-semidecidable if and only if there is a wITRM-program P such that, for all $x \subseteq \omega$, P^x terminates if and only if $x \in X$.

(a) Show that every wITRM-semidecidable set of real numbers is Σ^1_1.

(b) Show that every wITRM-semidecidable set of real numbers contains a wITRM-computable element.

Exercise 4.2.37. Conversely, show that every Σ^1_1-set is wITRM-semidecidable. (Hint: Use the "search" strategy described above, together with the nice presentation for Σ^1_1-formulas from Lemma 4.2.26.) Conclude that wITRM-decidability coincides with Σ^1_1.

4.3 Recognizability for infinite time register machines[6]

The real numbers recognizable by ITRMs have been extensively studied and are now rather well understood. We note that, by Exercise 4.2.5 above, the lost melody theorem holds for ITRMs. (Below, we will see that the set $0'_{\mathrm{ITRM}}$ of halting ITRM-programs is a lost melody for ITRMs.)

We consider ITRM-recognizability in more detail. We start with a crucial observation.

Lemma 4.3.1 ([21], Theorem 3.2). *Let x be ITRM-recognizable. Then $x \in L_{\omega_\omega^{\mathrm{CK},x}}$.*

Proof. This is a direct application of Theorem 4.1.2: Let P be an ITRM-program that recognizes x. By Corollary 3.4.12 from Chapter 3, the computation P^x takes less than $\omega_\omega^{\mathrm{CK},x}$ many steps and is thus contained in $V_{\omega_\omega^{\mathrm{CK},x}}$. So $V_{\omega_\omega^{\mathrm{CK},x}}$ believes that there is a real number y such that $P^y\downarrow = 1$, which is a Σ_1-statement. Furthermore, $\omega_\omega^{\mathrm{CK}}$ is certainly a limit of admissible ordinals. Hence, by Theorem 4.1.2, this statement is absolute between $V_{\omega_\omega^{\mathrm{CK},x}}$ and $L_{\omega_\omega^{\mathrm{CK},x}}$, and thus holds in $L_{\omega_\omega^{\mathrm{CK},x}}$. Consequently, there is some $y \subseteq \omega$ in $L_{\omega_\omega^{\mathrm{CK},x}}$ such that $L_{\omega_\omega^{\mathrm{CK},x}} \models P^y\downarrow = 1$. By absoluteness of computations, $P^y\downarrow = 1$ holds in the real world. But as P recognizes x, this means that $y = x$, so that $x \in L_{\omega_\omega^{\mathrm{CK},x}}$, as claimed. □

Note that this in particular implies that ITRM-recognizable real numbers are constructible, i. e., elements of L.

Corollary 4.3.2 ([21], Theorem 3.2). *If x is an ITRM-lost melody, then $\omega_\omega^{\mathrm{CK},x} > \omega_\omega^{\mathrm{CK}}$.*

Proof. By Lemma 4.3.1, we have $x \in L_{\omega_\omega^{\mathrm{CK},x}}$; as x is not ITRM-computable, we have $x \notin L_{\omega_\omega^{\mathrm{CK}}}$ by Corollary 3.4.12 from Chapter 3; thus, we must have $\omega_\omega^{\mathrm{CK},x} > \omega_\omega^{\mathrm{CK}}$. □

As for ITTMs, we have a basis theorem.

6 The results in this section come from the article "Optimal Results on ITRM-recognizability" [21], which appeared in the Journal of Symbolic Logic, vol. 80(4) in 2015. We thank the Association for Symbolic Logic for the kind permission to reproduce parts of that article here.

Theorem 4.3.3. *Every nonempty ITRM-decidable set of real numbers has a recognizable element.*

Proof. Let $X \neq \emptyset$ be ITRM-decidable, and let P be a program that decides X. Let $\Gamma := \{\omega_{\omega}^{\mathrm{CK},x} : x \in X\}$. Since X is nonempty, so is Γ; let μ be the minimal element of Γ and pick $x_y \in X$ such that $\omega_{\omega}^{\mathrm{CK},x_y} = \mu$. By Corollary 3.4.12 from Chapter 3, we thus have $P^{x_y}\downarrow = 1$ in less than μ many steps. By absoluteness of computations, it follows that $L_{\mu}[x_y] \models P^{x_y}\downarrow = 1$ and hence $L_{\mu}[x_y] \models \exists x P^{x}\downarrow = 1$. By the Jensen–Karp theorem 4.1.2, the Σ_1-formula $\exists x P^{x}\downarrow = 1$ is absolute between $L_{\mu}[x_y]$ and L_{μ}; hence $L_{\mu} \models \exists x P^{x}\downarrow = 1$. Let $\bar{x} \in L_{\mu}$ be the $<_L$-minimal element of L_{μ} such that $L_{\mu} \models P^{\bar{x}}\downarrow = 1$; thus $\bar{x} \in X$ since $P^{\bar{x}}\downarrow = 1$ by absoluteness of computations.

We claim that $\bar{x}$ is recognizable. To see this, first note that $\bar{x} \in L_{\omega_{\omega}^{\mathrm{CK},\bar{x}}}$, so that there is $k \in \omega$ with $\bar{x} \in L_{\omega_{k}^{\mathrm{CK},\bar{x}}}$. Suppose that P uses n registers and let $m = \max\{k, n+1\}$. Then the Σ_1-hull of $\{\bar{x}\}$ in $L_{\omega_{m}^{\mathrm{CK},\bar{x}}}$ will be isomorphic (via the collapsing map) to $L_{\omega_{m}^{\mathrm{CK},\bar{x}}}$, and so $L_{\omega_{m}^{\mathrm{CK},\bar{x}}+1}$ will contain a real number r coding $L_{\omega_{m}^{\mathrm{CK},\bar{x}}}$ by standard fine-structural arguments. Hence r is ITRM-computable from $\bar{x}$; let Q be an ITRM-program computing r from $\bar{x}$. To recognize $\bar{x}$, we now proceed as follows: Let y be given in the oracle. First, test whether $P^{y}\downarrow = 1$, which is possible by Corollary 3.4.33 from Chapter 3. If not, then $y \neq \bar{x}$. Otherwise, test, using Q and a halting problem solver for Q, whether Q^{y} computes a real number r coding an admissible L_{α} containing y. If not, then $y \neq \bar{x}$. If yes, then use r to search for an element z of L_{α} that is $<_L$-smaller than y and satisfies $P^{z}\downarrow = 1$. If the search is successful, then $y \neq \bar{x}$. On the other hand, if none is found, then y is the $<_L$-smallest element a of X with $a \in L_{\omega_{\omega}^{\mathrm{CK},a}}$, so $y = \bar{x}$. □

4.3.1 Gaps[7]

By our observation above, all ITRM-recognizable real numbers belong to L. We now turn our attention to the question where ITRM-recognizable reals appear in the L-hierarchy. Clearly, all ITRM-recognizable reals are also ITTM-recognizable, so that, by Theorem 4.2.10, all ITRM-recognizable reals belong to L_{σ}. The other half of this theorem also works for ITRMs, so that σ is again minimal with this property.

As we saw in Chapter 3, the ITRM-computable reals form an initial segment of the constructible reals in the sense of the canonical well-ordering $<_L$ of L. In particular, there are no "gaps" of non-ITRM-computable real numbers between the ITRM-computable real numbers. This, however, is not the case for the recognizable reals.

7 The material in this section is taken from the articles "The distribution of ITRM-recognizable reals" [19], which appeared in the Annals of Pure and Applied Logic in 2017 ©Elsevier 2017 and 'Optimal Results on ITRM-Recognizability' [21], which appeared in the Journal of Symbolic Logic in 2015 ©Association for Symbolic Logic 2015. We thank Elsevier and the Association for Symbolic Logic for the kind permission to reproduce parts of these articles here.

This is our next goal. We start by observing that, if α is a limit of admissible ordinals, then it is right about ITRM-recognizability for elements of L_α. Note that this is not true for arbitrary α: If, e. g., $x \in L_\alpha$ is recognizable by the ITRM-program P but P^x runs for more than α many steps, L_α will not notice that P recognizes x but believe that P will not halt on input x.

Lemma 4.3.4. *Suppose that α is a limit of admissible ordinals and $x \in L_\alpha$ is such that $L_\alpha \models$"x is not ITRM-recognizable." Then x is not ITRM-recognizable.*

Proof. Suppose for a contradiction that P recognizes x. Recall that, for $y \in L_\alpha$, we have $\omega_\omega^{\mathrm{CK},y} \leq \alpha$. Thus, for $y \in L_\alpha$, P^y will either halt in $< \alpha$ many steps or it will not halt at all. As P halts on every input by assumption, the first option must hold for all $y \in L_\alpha$. Furthermore, by absoluteness of computations, the computation of P^y in L_α is the same as the computation of P^y in V, and thus has the same output for every $y \in L_\alpha$, i. e., in L_α, P^y halts with output 1 if and only if $y = x$ and otherwise it halts with output 0. But then L_α believes that x is ITRM-recognizable after all, a contradiction. □

Theorem 4.3.5 ([19], Theorem 18). *There is a gap in the recognizable reals, i. e., there are constructible reals $x <_L y <_L z$ such that x and z are recognizable, but y is not.*

Proof. Let x be any real constructible real number that does not belong to L_σ (such as the $<_L$-minimal code of L_σ, or a number obtained from the countable many reals in L_σ by diagonalization). For each ITRM-program P_i, the fact that P_i does not recognize x is due to one of four possible reasons: (1) $P_i^x\uparrow$, (2) $P_i\downarrow$ with an output $\neq 1$, (3) there is some $y \subseteq \omega$ different from x such that $P_i^y\downarrow$ with some output different from 0 and (4) there is some $y \subseteq \omega$ such that $P_i^y\uparrow$.

Note that, in case (4), the statement that $P_i^y\uparrow$ is equivalent to the statement that P_i^y runs for at least $\omega_\omega^{\mathrm{CK},y}$ many steps, which is Σ_1-expressible in the parameter y, say by the formula $\phi(y)$. Thus $\exists y\phi(y)$ is Σ_1 as well, and hence if it holds in V, it holds in L by Shoenfield absoluteness. Thus if $P_i^y\uparrow$ for some $y \subseteq \omega$, then $P_i^y\uparrow$ for some constructible y. The same reasoning applies to (3). Thus, in cases (3) and (4), we find constructible witnesses for the non-recognizability of x; in each case, let w_i be the $<_L$-smallest such witness, if it exists.

Now consider $L_{\omega_1^L}$. It contains all w_i, along with x, so it is clear that $L_{\omega_1^L}$ will believe that x is not recognizable: After all, x is not recognizable in the real world, the possible reasons (1) and (2) are immediately reflected to $L_{\omega_1^L}$, and for (3) and (4) we know that the witnesses w_i are included in $L_{\omega_1^L}$. Also, ω_1^L is a limit or admissible ordinals. It follows that the Σ_1-formula 'There is some L_α which believes in the existence of a real number that is not ITRM-recognizable' is true in L, and hence in L_σ. Consequently, there is some L-countable limit of admissible ordinals α such that L_α believes in the existence of non-ITRM-recognizable real numbers. We pick the minimal such α.

Now by the same techniques used in the proof of Theorem 4.2.4, one shows that the $<_L$-minimal code of L_α is ITRM-recognizable: Being the code of an L-level whose height is a limit of admissible ordinals, which believes in the existence of a non-ITRM-recognizable real number and which is minimal with this property can be checked by testing for well-foundedness and evaluating truth predicates, while the minimality is obtained as explained at the end of the proof of Theorem 4.2.4. As usual, we cannot have $c \notin L_\alpha$.

Now let $x \in L_\alpha$ be such that L_α believes that x is not ITRM-recognizable. Then, by Lemma 4.3.4, x is in fact not ITRM-recognizable. Moreover, as x belongs to L_α while c does not, we clearly have $x <_L c$. Hence 0 and c are ITRM-recognizable, while x satisfies $0 <_L x <_L c$ and is not. Thus 0, x and c are as desired. □

Corollary 4.3.6 (Cf. [19]). *Relative ITRM-recognizability is not transitive. That is there are real numbers a, b, c such that a is ITRM-recognizable from b and b is ITRM-recognizable from c, but a is not ITRM-recognizable from c. In fact, we can take $c = 0$.*

Proof. Let $c = 0$, let b be the $<_L$-minimal code of L_α, where α is as in the proof of Theorem 4.3.5, and let a be some non-ITRM-recognizable element of L_α. □

We now consider those gaps in more detail. By Lemma 4.3.1 above, x being an element of $\in L_{\omega_\omega^{\mathrm{CK},x}}$ is a necessary condition for x being ITRM-recognizable. This motivates the following definition.

Definition 4.3.7. $x \subseteq \omega$ is ITRM-prerecognizable if and only if $x \in L_{\omega_\omega^{\mathrm{CK},x}}$.

It turns out that, among the ITRM-prerecognizable real numbers, the recognizable ones appear in large blocks.

Theorem 4.3.8. *Let $x, y \subseteq \omega$ be ITRM-prerecognizable such that $\omega_\omega^{\mathrm{CK},x} = \omega_\omega^{\mathrm{CK},y}$. Then x is ITRM-recognizable if and only if y is ITRM-recognizable.*

Proof. Suppose that x is ITRM-recognizable by the program P. As $x, y \in L_{\omega_\omega^{\mathrm{CK},x}} = L_{\omega_\omega^{\mathrm{CK},y}}$, there are ITRM-programs Q, R such that Q computes x in the oracle y and R computes y in the oracle x. By Corollary 2.3.13 from Chapter 2, we can assume that Q and R are safe, i. e., $Q^z(i)$ and $R^z(i)$ halt for every $z \subseteq \omega$ and every $i \in \omega$.

To recognize y, we proceed as follows: Given a real number z, first run Q^z; then use P to check whether the output is equal to x. If it is not, return 0. Otherwise, we have now computed x; apply R to x and compare the result (which must be y by the choice of R) to z. If they agree, output 1, otherwise, output 0. □

Corollary 4.3.9 (Cf. [21]). *For all $\alpha \in \mathrm{On}$, either all ITRM-prerecognizable real numbers in $L_{\alpha^+} \setminus L_\alpha$ are ITRM-recognizable or none is.*

Proof. Let $\hat{\alpha}$ denote the smallest limit of admissible ordinals $> \alpha$. If $x, y \in L_{\alpha^+} \setminus L_\alpha$ are ITRM-prerecognizable, then $\omega_\omega^{\mathrm{CK},x}$ and $\omega_\omega^{\mathrm{CK},y}$ both coincide with $\hat{\alpha}$ and thus with each other. Now apply Theorem 4.3.8. □

Exercise 4.3.10. Show that, for all $\alpha \in \mathrm{On}$, either all ITRM-prerecognizable real numbers in $L_{\alpha^{+i}} \setminus L_\alpha$ are ITRM-recognizable or none is.

Theorem 4.3.11 (Cf. [19]). *Let $\delta < \sigma$. Then there are real numbers x, y such that y is ITRM-recognizable, the set S of constructible ordinals that lie between x and y in the sense of $<_L$ has order type $\geq \delta$, but no element of S is ITRM-recognizable.*

Proof. Clearly, there is a $<_L$-interval in the constructible real numbers of length $\geq \delta$ that does not contain an ITRM-recognizable element, as all ITRM-recognizable real numbers are contained in L_σ and $\sigma < \omega_1^L$, while the indices of L-levels at which new real numbers appear are unbounded in ω_1^L. Now, by definition of δ, pick $\delta' \geq \delta$ and a Σ_1-statement ϕ (without parameters) such that δ' is minimal with the property that $L_{\delta'} \models \phi$.

Now the statement 'There are real numbers x, y and an ordinal δ' such that $L_{\delta'} \models \phi$, the $<_L$-interval (x, y) in the constructible real numbers has order type $\geq \delta'$ and does not contain an ITRM-recognizable real number' is Σ_1 and holds in L, and thus, by Σ_1-elementarity, it holds in L_σ. Pick $\bar{x}, \bar{y} \in L_\sigma$ such that this statement holds of $\bar{x}$ and $\bar{y}$ in L_σ. By Lemma 4.3.4 and the fact that σ is a limit of admissible ordinals, no constructible real number in the $<_L$-interval $(\bar{x}, \bar{y})$ is ITRM-recognizable. Moreover, by absoluteness of Δ_0-formulas, if L_σ believes that $L_{\delta'} \models \phi$, then it is true that $L_{\delta'} \models \phi$. Thus $\bar{x}$ and $\bar{y}$ are indeed as desired. □

4.3.2 Aristophanean pairs[8]

A real number x can contain information that helps in recognizing another real number y. For example, we can have $x \oplus y$ recognizable and x unrecognizable: Take x to be some nonrecognizable element of an L-level L_α with a recognizable code and let y be this code. Also, if x and y are both recognizable, then so is $x \oplus y$ by identifying the components separately. But can we have that neither x nor y is ITRM-recognizable, yet $x \oplus y$ is? This section is devoted to answering this question in the positive.

Definition 4.3.12. Two real numbers x, y form an "Aristophanean pair"[9] if and only if neither x nor y is ITRM-recognizable, but $x \oplus y$ is.

Do Aristophanean pairs exist? This was asked in [19] and answered by J. Hamkins, whose solution we now present.

---

8 The material in this section is taken from the article "The distribution of ITRM-recognizable reals" [19], which appeared in the Annals of Pure and Applied Logic in 2017 ©Elsevier 2017. We thank Elsevier for the kind permission to reproduce parts of this article here.

9 The name is motivated by the mythos about love reported by Aristophanes in Plato's "Symposium": Humans are incomplete halves of originally stronger 4-armed and 4-legged creatures. These stronger creatures were split in halves by the gods, and now everybody his looking for his or her other half to be complete again. Likewise, the members of Aristophanean pairs have nothing special about them when considered alone, while the pair can be uniquely identified by an ITRM.

Definition 4.3.13. For $x, y \subseteq \omega$, we denote by $x +_b y$ the bitwise sum of x and y, i. e., $x\Delta y$, the symmetric difference of x and y. If x is finite, then $x +_b y$ denotes $(x\Delta y) \cap \max\{x\} + 1$.

The construction below uses Cohen forcing over models of ZF^-. Here, ZF^- denotes Zermelo–Fraenkel set theory without the axiom of powerset; more precisely, we take the formulation of ZF^- given in Gitman and Hamkins [66]. The reader unfamiliar with forcing may consult Chapter 5 for a brief introduction and some references.

Lemma 4.3.14 ([19], Lemma 23). *Let M be a countable transitive model of ZF^-, let $(P, \leq, 1)$ be the notion of Cohen-forcing over M and let $x \subseteq \omega$ be arbitrary. Then there exists a real d such that both d and $d +_b x$ are P-generic over M.*

In other words, there are d, d' Cohen-generic over M such that $d +_b d' = x$.

Proof. Let $(D_i | i \in \omega)$ enumerate the dense subsets of P in M. Pick $p_0 \in D_0$ arbitrary. Let $q_0 \in D_0$ be such that $q_0 \leq p_0 +_b x$. Then we recursively let $p_{i+1} \in D_{i+1}$ be such that $p_{i+1} \leq q_i +_b x$ and $q_{i+1} \in D_{i+1}$ such that $q_{i+1} \leq p_{i+1} + x$. Such p_i and q_i can always be chosen by definition of density. Letting $p := \bigcup_{i\in\omega} p_i$, $q := \bigcup_{i\in\omega} q_i$, we get that both p and q are P-generic over M and $q = p +_b x$, as desired.

Now let $d' = d +_b x$. Then, by the rules of bitwise addition, $d +_b d' = x$, so d and d' are as desired. □

Theorem 4.3.15 ([19], Theorem 24). *There exists an Aristophanean pair (a, b) of ITRM-nonrecognizable constructible real numbers a and b such that $a \oplus b$ is ITRM-recognizable.*

Proof. Let $x = \mathrm{cc}(L_\gamma)$, where γ is minimal such that $L_\gamma \models \mathrm{ZF}^-$. Then $L_{\gamma+2}$ contains a bijection $f : \omega \to L_\gamma$. It is easy to see that the proof of Lemma 4.3.14 now shows that $L_{\gamma+3}$ contains a real number y such that both y and $y +_b x$ are Cohen-generic over L_γ. Let a be $<_L$-minimal with this property.

We first show that $z := a \oplus (a +_b x)$ is ITRM-recognizable. To see this, first note that x can easily be reconstructed from z and then be recognized as described in the proof of the lost melody theorem by checking whether the oracle codes a well-founded L-level in which all ZF^--axioms hold and whether this code is $<_L$-minimal. Furthermore, it is easy to see that, given x, one can decide whether a real number u given in the oracle is Cohen-generic over L_γ: Just run through ω, looking for codes of dense subsets, then check whether one of their elements is a finite subset of u. To see whether a is $<_L$-minimal, we use a straightforward generalization of the proof of the lost melody theorem to decide statements over $L_{\gamma+3}$ given x.

Now, we demonstrate that neither a nor $a +_b x$ are ITRM-recognizable. In fact, no Cohen-generic real over L_γ can be ITRM-recognizable: Suppose b was Cohen-generic over L_γ and suppose that P is an ITRM-program that recognizes b. Let $M = L_\gamma[G]$ be a generic extension of L_γ containing b. Then $M \models \mathrm{ZF}^-$. Furthermore, we have $L^M = L_\gamma$, and, as b is Cohen-generic over L_γ, we must have $b \in M - L_\gamma$.

Now, by Theorem 4.1.2, set-theoretical Σ_1-statements are absolute between L_γ and V_γ, hence between L_γ and M. By assumption, $M \models P^b\downarrow = 1$, so $M \models \exists y P^y \downarrow = 1$, and hence $L_\gamma \models \exists y P^y \downarrow = 1$. By absoluteness of computations, we get $b \in L_\gamma$, so that L_γ contains a real number that is generic over L_γ, a contradiction. □

Exercise 4.3.16 (Recognizable polyamory).

(a) Adapt the proof of Theorem 4.3.15 to show that there are real numbers x, y, z such that none of x, y, z, $x \oplus y$, $x \oplus z$, $y \oplus z$ is ITRM-recognizable, but $(x \oplus y) \oplus z$ is.

(b) Generalize (a) further to show that, for any $n \in \omega$, there are real numbers $x_1, \dots, x_n$ such that $\bigoplus_{i \in I} x_i$ is not ITRM-recognizable for any $I \subsetneq \{1, \dots, n\}$, but $\bigoplus_{i=1}^n x_i$ is ITRM-recognizable.

4.3.3 Recognizability of the halting numbers[10]

We obtain a very natural lost melody for ITRMs by showing that the halting number for ITRMs is in fact recognizable. Fix a natural well-ordering $(P_i \mid i \in \omega)$ of the ITRM-programs in order type ω by, e. g., sorting the programs lexicographically. The halting number $0'_{\text{ITRM}}$ for ITRMs is then defined as $0'_{\text{ITRM}} := \{i \in \omega \mid P_i \downarrow\}$. This real $0'_{\text{ITRM}}$ is natural insofar its definition is purely internal to ITRMs and it is also arguably the first non-ITRM-computable real number coming to mind.

Lemma 4.3.17 (Cf. [21]). $\omega_\omega^{\text{CK},0'_{\text{ITRM}}} > \omega_\omega^{\text{CK}}$.

Proof. First, note that using $0'_{\text{ITRM}}$, we can ITRM-decide the set of program indices i for which P_i computes a well-ordering. For each $i \in \omega$, use $0'_{\text{ITRM}}$ to check whether $P_i(k)$ halts for every input $k \in \omega$ with output 0 or 1, i. e., whether P_i computes a real number. If so, use the algorithm described in Theorem 2.3.25 in Chapter 2 to check whether the real number computed by P_i codes a well-ordering.

Now, by putting these codes together, we can ITRM-compute a code for a well-ordering of order type $\omega_\omega^{\text{CK}}$. More precisely, split ω into ω many portions via the usual bijection $p : \omega \times \omega \to \omega$. Now define a real number c as follows: For every $i \in \omega$, if P_i does not compute a code for a well-ordering, let $c(p(p(i,j), p(i,k))) = 1$ for all $j, k \in \omega$ with $j \le k$ and $c(p(p(i,j), p(i,k))) = 0$ for $j > k$, while if P_i computes a code x for a well-ordering, let $c(p(p(i,j), p(i,k))) = x(p(j,k))$. Finally, for all $i, j, k, l \in \omega$ with $i \ne j$ let $c(p(p(i,k), p(j,l))) = 1$ if $i < j$ and $c(p(p(i,k), p(j,l))) = 0$, otherwise. Then c codes a well-ordering the order type of which is a sum of a set of ordinals which includes all ordinals with an ITRM-computable code, i. e., all ordinals $< \omega_\omega^{\text{CK}}$ and is thus $\ge \omega_\omega^{\text{CK}}$.

As this ITRM-computation works in the oracle $0'_{\text{ITRM}}$, we have $\omega_\omega^{\text{CK},0'_{\text{ITRM}}} > \omega_\omega^{\text{CK}}$ as desired. □

10 The material in this section comes from the article "Optimal Results on ITRM-recognizability" [21], which appeared in the Journal of Symbolic Logic, vol. 80(4) in 2015. We thank the Association for Symbolic Logic for the kind permission to reproduce parts of that article here. ©Association for Symbolic Logic 2015.

Theorem 4.3.18 (Cf. [21], Theorem 4.6). *$0'_{\mathrm{ITRM}}$ is ITRM-recognizable.*

Proof. By Lemma 4.3.17, $\omega_\omega^{\mathrm{CK},0'_{\mathrm{ITRM}}} > \omega_\omega^{\mathrm{CK}}$.

Now, it is not hard to see that $L_{\omega_\omega^{\mathrm{CK}}+1}$ contains a real number coding $L_{\omega_\omega^{\mathrm{CK}}}$, and thus the $<_L$-minimal such code $c := \mathrm{cc}(L_{\omega_\omega^{\mathrm{CK}}})$: As $0'_{\mathrm{ITRM}}$ is definable over $L_{\omega_\omega^{\mathrm{CK}}}$, but not ITRM-computable, and thus not contained in $L_{\omega_\omega^{\mathrm{CK}}}$, $\omega_\omega^{\mathrm{CK}}$ is an index, so that the desired conclusion follows from Corollary 3.1.16.

Thus $c \in L_{\omega_\omega^{\mathrm{CK},0'_{\mathrm{ITRM}}}} \subseteq L_{\omega_\omega^{\mathrm{CK},0'_{\mathrm{ITRM}}}}[0'_{\mathrm{ITRM}}]$, so c is ITRM-computable from $0'_{\mathrm{ITRM}}$, say by program Q. Observe that the set of codes for $L_{\omega_\omega^{\mathrm{CK}}}$ is ITRM-decidable by (1) checking whether some given real number x codes a well-founded relation and (2) using evaluation of truth predicates to check whether x is a model of $V = L$, contains $L_{\omega_i^{\mathrm{CK}}}$ for all $i \in \omega$ and has no proper initial segment with this property. Let Q' be an ITRM-program that decides this set of codes. By solvability of the bounded halting problem (Corollary 3.4.33, Chapter 3), let H be an ITRM-program such that, for all $y \subseteq \omega$, $H^y\downarrow = 1$ if and only if $Q^y(i)$ halts for every $i \in \omega$ and otherwise $H^y\downarrow = 0$.

Now, putting this ingredients together, the following procedure recognizes $0'_{\mathrm{ITRM}}$: Given an oracle x, first run H^x. If $H^x\downarrow = 0$, this means that Q^x does not compute a real number. It follows that $x \neq 0'_{\mathrm{ITRM}}$, as Q was chosen to compute c from $0'_{\mathrm{ITRM}}$. Hence, in this case, we halt and output 0. Otherwise, if $H^x\downarrow = 1$, we use Q' to check whether Q^x computes a code for $L_{\omega_\omega^{\mathrm{CK}}}$. If not, then $x \neq 0'_{\mathrm{ITRM}}$ and we halt with output 0. Otherwise, we know that Q^x computes a code for $L_{\omega_\omega^{\mathrm{CK}}}$.

Now, for every $i \in \omega$, the statement that P_i halts is an $\in$-formula $\phi(i)$ that can easily be computed from i. Moreover, $\phi(i)$ is true if and only if it holds in $L_{\omega_\omega^{\mathrm{CK}}}$. Thus, we can use evaluation of truth predicates in the structure coded by x to compute all bits of $0'_{\mathrm{ITRM}}$ and compare them to x. If all bits match, then $x = 0'_{\mathrm{ITRM}}$ and we output 1, otherwise, we output 0. □

Exercise 4.3.19. Generalize the proof of Theorem 4.3.18 to show that, for all $x \subseteq \omega$, x'_{ITRM} is recognizable relative to x. Conclude that, for all ITRM-recognizable $x \subseteq \omega$, $x \oplus x'_{\mathrm{ITRM}}$ is also ITRM-recognizable. Conclude moreover that, if x is ITRM-recognizable, then so is x'_{ITRM}.

As a consequence, $0'_{\mathrm{ITRM}}$ is a minimal lost melody with respect to ITRM-reducibility.

Corollary 4.3.20 ([21], Corollary 4.7). *Let $x \subseteq \omega$ be ITRM-recognizable, but not ITRM-computable. Then there is an ITRM-program P that computes $0'_{\mathrm{ITRM}}$ from x.*

Proof. As x is an ITRM-lost melody, it follows from Theorem 4.3.2 that $\omega_\omega^{\mathrm{CK},x} > \omega_\omega^{\mathrm{CK}}$. Consequently, there is $i \in \omega$ such that $\omega_i^{\mathrm{CK},x} > \omega_\omega^{\mathrm{CK}}$. As $0'_{\mathrm{ITRM}}$ is definable over $L_{\omega_\omega^{\mathrm{CK}}}$, we have $0'_{\mathrm{ITRM}} \in L_{\omega_\omega^{\mathrm{CK}}+1}$, so $0'_{\mathrm{ITRM}} \in L_{\omega_i^{\mathrm{CK},x}} \subseteq L_{\omega_i^{\mathrm{CK},x}}[x] \subseteq L_{\omega_\omega^{\mathrm{CK},x}}[x]$. Hence $0'_{\mathrm{ITRM}}$ is ITRM-computable from x, as desired. □

We saw in Chapter 2 that the computability strength of ITRMs increases with the number of registers. Similar to the computability strength, the recognizability strength of ITRMs increases with the number of registers. We will now prove this using a forcing argument. The reader unfamiliar with forcing over models of KP should consult the preliminaries to Chapter 5 and the references given there. A forcing-free argument by diagonalization can be found in the exercises.

Theorem 4.3.21 ([19], Theorem 45). *For every $m \in \omega$, there are a natural number $n > m$ and a real number x such that x is ITRM-recognizable by an ITRM-program P using n registers, but not by an ITRM-program using m registers.*

Proof. Let x be the $<_L$-minimal Cohen-generic real number over $L_{\omega_{m+1}^{\mathrm{CK}}+1}$. We claim that x is as desired. Note that, by Corollary 5.2.18 from Chapter 5, we have $\omega_{m+1}^{\mathrm{CK},x} = \omega_{m+1}^{\mathrm{CK}}$, so that $L_{\omega_{m+1}^{\mathrm{CK},x}}[x] = L_{\omega_{m+1}^{\mathrm{CK}}}[x]$.

First, we claim that x is not recognizable by an ITRM using at most m registers. To see this, suppose for a contradiction that Q is such a program that recognizes x. Thus $Q^x\downarrow = 1$. As we saw in Chapter 3, this computation takes less than $\omega_{m+1}^{\mathrm{CK},x} = \omega_{m+1}^{\mathrm{CK}}$ many steps and is thus contained in $L_{\omega_{m+1}^{\mathrm{CK},x}}[x]$. Consequently, we have $L_{\omega_{m+1}^{\mathrm{CK},x}}[x] \models \exists x Q^x\downarrow = 1$. Let $\dot{x}$ be a name for x. By the forcing theorem, there is some condition $p \subseteq x$ such that $p \Vdash Q^{\dot{x}}\downarrow = 1$ over $L_{\omega_{m+1}^{\mathrm{CK}}}$. Again by the forcing theorem, we have $p \Vdash Q^y\downarrow = 1$ for every $p \subseteq y \subseteq \omega$ that is Cohen-generic over $L_{\omega_{m+1}^{\mathrm{CK}}+1}$. Hence, if we pick such a y different from x, we have $L_{\omega_{m+1}^{\mathrm{CK}}}[y] \models Q^y\downarrow = 1$; by absoluteness of computations, this implies that indeed $Q^y\downarrow = 1$. But this means that Q halts with output 1 for some oracle other than x, contradicting the assumption that Q recognizes x.

Now we show that x is ITRM-recognizable. To see this, first note that $L_{\omega_{m+1}^{\mathrm{CK}}+1}$ is countable in $L_{\omega_{m+1}^{\mathrm{CK}}+2}$, so that $L_{\omega_{m+1}^{\mathrm{CK}}+3}$ will contain Cohen-generics over $L_{\omega_{m+1}^{\mathrm{CK}}+1}$ by Lemma 5.2.5 in Chapter 5; in particular, we must have $x \in L_{\omega_{m+1}^{\mathrm{CK}}+3}$. But then, $x \in L_{\omega_{\omega}^{\mathrm{CK}}}$, so x is ITRM-computable, and hence also ITRM-recognizable. □

4.3.4 A machine-free characterization of ITRM-recognizability[11]

We can now characterize ITRM-recognizability in terms of definability, i. e., without reference to machine models. As similar characterization for ITTM-recognizability was obtained by Friedman and Welch in [62].

11 This section is adapted from the article "Optimal Results on ITRM-recognizability" [21], which appeared in the Journal of Symbolic Logic, vol. 80(4) in 2015. We thank the Association for Symbolic Logic for the kind permission to reproduce parts of that article here.

Theorem 4.3.22 ([21], Theorem 5.3). *Let $x \subseteq \omega$ be ITRM-prerecognizable. Then x is ITRM-recognizable if and only if there exists a Σ_1-formula ϕ without parameters such that x is the unique element a of $L_{\omega_\omega^{\mathrm{CK}}}$ with $L_{\omega_\omega^{\mathrm{CK}}} \models \phi(a)$.*

Proof. If x is ITRM-recognizable, then $x \in L_{\omega_\omega^{\mathrm{CK},x}}$ by Lemma 4.3.1 above. Now, if P is an ITRM-program that recognizes x, then $P^x\downarrow = 1$ is Σ_1-expressible over $L_{\omega_\omega^{\mathrm{CK},x}}$. By upwards preservation of Σ_1-statements, if $L_{\omega_\omega^{\mathrm{CK},x}} \models P^y\downarrow = 1$ for some $y \neq x$, then $P^y\downarrow = 1$ in the real world, which contradicts the assumption that P recognizes x.

On the other hand, if x is definable as above, then let L_γ be the first L-level containing x such that $L_\gamma \models \phi(x)$. Then $\gamma < \omega_\omega^{\mathrm{CK},x}$, so $c := \mathrm{cc}(L_\gamma)$ can be computed from x, say by the ITRM-program Q. Given a real y and c in the oracle, we can check whether $y \in L_\gamma$ and $L_\gamma \models \phi(y)$ hold.

Checking whether some oracle number y is equal to x then works as follows: Check whether Q^y computes a minimal code for an L-level containing y, then check whether $\phi(y)$ holds in that L-level and then whether it fails in all earlier L-levels. If all of this holds, then $y = x$ (since Σ_1 is preserved upwards), otherwise $y \neq x$. □

Exercise 4.3.23 (Friedman, Welch [62]). Find and prove a similar characterization for ITTM-recognizability.

4.4 Recognizability for ordinal Turing machines[12]

In this section, we consider recognizability for OTMs. By the simulation results of Chapters 2 and 3, what we say will also hold for ORMs.

A rather basic, yet important, observation is that the lost melody theorem fails for OTMs without parameters. It was first made by Dawson in his PhD thesis [45].

Theorem 4.4.1. *A real number $x \subseteq \omega$ is OTM-recognizable without parameters if and only if it is OTM-computable without parameters.*

Proof. Clearly, if x is OTM-computable, it is OTM-recognizable by simply computing it and comparing the output to the oracle.

Now suppose that x is OTM-recognizable, and let P be an OTM-program that recognizes x. The statement S:

'There is a real number y such that $P^y\downarrow = 1$' is Σ_1.

By Theorem 4.1.1, if S holds in V, then it holds in L. But as P recognizes x, S clearly holds in V, so ϕ is true in L and L contains some $y \subseteq \omega$ such that $L \models P^y\downarrow = 1$. By abso-

12 The material in this section is adapted from the article "Recognizable Sets and Woodin Cardinals: Computation beyond the constructible universe" [39] by P. Schlicht, P. Welch and the author, which appeared in the Annals of Pure and Applied Logic in 2017 ©Elsevier 2017. We thank our coauthors Philipp Schlicht and Philip Welch as well as the Annals of Pure and Applied Logic for the kind permission to reuse parts of this article here.

luteness of computations, $P^y\downarrow = 1$ must hold in V as well, and thus, as P recognizes x, we must have $x = y$.

To compute x with an OTM now works as follows: Use the program from Lemma 3.5.3, Chapter 3, to enumerate L, with a small modification: Whenever a new real number y is written on the tape, run P^y to check whether $x = y$. If P^y halts with output 0, continue with the enumeration, otherwise halt with output y. □

However, the picture changes drastically once ordinal parameters are allowed. Here, it turns out that real numbers that are far from being constructible are pOTM-recognizable (if they exist). The paradigmatical example, which we consider first, are sharps. The recognizability of $0^\sharp$ from ordinal parameters was first observed by B. Dawson in [45] and then independently in [30]. For the following, we assume familiarity with $0^\sharp$ and its relativizations; see, e. g., Jech, [92], Chapter 18 or Kanamori, [96]. We recall some basic points.

Lemma 4.4.2. *Σ^1_2-relations are absolute between transitive models of KP that contain ω_1.*

Proof. This was proved in Schlicht and Seyfferth, [160], Corollary 1 using OTMs. See also Theorem 8.3.3 in Chapter 8. □

Lemma 4.4.3. *Let $x, y \subseteq \omega$, and suppose that there is a Π^1_2-formula ϕ such that x is unique with the property that $\phi(x, y)$. Then x is OTM-recognizable relative to y in the parameter ω_1. In fact, this holds for every parameter $\alpha > \omega_1$ instead of ω_1.*

Proof. Let x and y be as in the theorem statement, and let $z \oplus y$ be given in the oracle. To determine whether $z = x$, we proceed as follows: First, run the OTM-program for enumerating L from Chapter 3 in the oracle $z \oplus y$ until a code for some relativized L-level $L_\gamma[z \oplus y]$ is written for which $\gamma > \alpha$. This can be done by using α along with P_{truth} to check, for each newly written code, whether it is as desired. Once such a code is produced, use it to evaluate whether $L_\gamma[z \oplus y] \models \phi(z, y)$ and return the result. By Lemma 4.4.2, this works as desired. □

Definition 4.4.4 (See, e. g., Jech, [92], Chapter 18). Let $\mathcal{L}^+_\in$ be the language of set theory together with countably many constant symbols $\{c_i : i \in \omega\}$. Then $0^\sharp$ is the set of (Gödel numbers of) $\mathcal{L}^+_\in$-formulas ϕ such that $L \models \phi$ under the assignment that assigns $\aleph^V_i$ to c_i for every $i \in \omega$.

Similarly, if $x \subseteq \omega$, then $x^\sharp$ is defined to be the set of $\mathcal{L}^+$-formulas ϕ that hold in $L[x]$ under the same assignment.

Remark 4.4.5. Since we cannot express truth in $L^+_\in$, the above definition is not a $\mathcal{L}^+_\in$-definition, and thus the ZFC-axioms do not immediately guarantee that $0^\sharp$ or any of the $x^\sharp$ exist. In fact, if $V = L$, then $0^\sharp$ does not exist, and large cardinal assumptions are necessary in order to guarantee that it does. By a theorem of Kunen, $x^\sharp$ exists if and only if there is an elementary embedding $\pi : L[x] \to L[x]$ other than the identity;

since such an elementary embedding must have a critical point (a smallest ordinal α such that $\pi(\alpha) > \alpha$), the existence of $0^\sharp$ or of $x^\sharp$ for all $x \subseteq \omega$ can be and is itself viewed as a large cardinal axiom. Often, $0^\sharp$ is used as a shorthand for the statement that $0^\sharp$ exists, and we will follow this convention below.

One can say that $0^\sharp$ provides a kind of dichotomy for the "closeness" or "similarity" of V to L. If $0^\sharp$ exists, then V looks very different from L: For example, all infinite cardinals in V, such as $\aleph_1^V$, are strongly inaccessible in L (Silver). On the other hand, if $0^\sharp$ does not exist, then Jensen's covering lemma guarantees that every set of ordinals in V is a subset of a set in L of the same cardinality, which means that V and L are combinatorially "close"; see, e. g., [92], Chapter 18.

Exercise 4.4.6. (The "information content" of the class of cardinals.) Use the last definition to show that there is an OTM-program P with the following property: If P is started on a tape on which every cell whose index is an infinite cardinal in V contains 1 while all other cells contain 0, then P halts with $0^\sharp$ written on the output tape. Show that the same holds for $(0^\sharp)^\sharp$, $((0^\sharp)^\sharp)^\sharp$, etc. More generally, show that the set S of real numbers that are pOTM-computable when the computation is started on such a tape is closed under sharps, i. e., $x \in S$ implies $x^\sharp \in S$.[13]

The following material is adapted or directly taken from [39]. We thank our coauthors Philipp Schlicht and Philip Welch as well as the Elsevier for the friendly permission to reuse it here.

We will use the following results.

Theorem 4.4.7. *There is Π^1_2-formula $\phi_\sharp$ such that, for all $x, y \subseteq \omega$, $\phi_\sharp(x, y)$ holds if and only if $x = y^\sharp$.*

In particular, there is a Π^1_2-formula $\phi_{0^\sharp}(x)$ in one free variable x such that, for all $a \subseteq \omega$, $\phi_{0^\sharp}(a)$ holds if and only if $a = 0^\sharp$.

Proof. See, e. g., Kaye [98], Theorem 14.11. □

Lemma 4.4.8 ([39], Lemma 3.8; independently (and earlier) Dawson, [45]). *If $x \subseteq \omega$ and $x^\sharp$ exists, then $x^\sharp$ is pOTM-recognizable from x.*

Moreover, for $x \subseteq \omega$, $x^\sharp$ (if it exists) is pOTM-recognizable from x in the parameter α if $\alpha \geq \omega_1$.[14]

Proof. By Lemma 4.4.3 and Theorem 4.4.7, $x^\sharp$ is pOTM-recognizable in the parameter α for any $\alpha \geq \omega_1$. □

13 This is considerably extended in up to now unpublished work by Philip Welch.

14 It can in fact be shown that x is pOTM-recognizable relative to $x^\sharp$ in the parameter α only if $\alpha \geq \omega_1$; see [39], Lemma 3.3.

Corollary 4.4.9. *If $0^\sharp$ exists, then it is independent of ZFC whether there are lost melodies for pOTMs.*

Proof. If $V = L$, then all real numbers are pOTM-computable by Lemma 3.5.4 of Chapter 3. So all real numbers are also pOTM-recognizable, and thus, trivially, no lost melodies exist. Thus, the nonexistence of lost melodies for pOTMs is consistent with ZFC.

On the other hand, suppose that $0^\sharp$ exists. We saw above that $0^\sharp$ is pOTM-recognizable, and we remarked that $0^\sharp \notin L$, so that $0^\sharp$ is not pOTM-computable. Hence, in $L[0^\sharp]$, ZFC holds together with the statement that there are lost melodies for pOTMs. Thus the existence of lost melodies for pOTMs is also consistent with ZFC. □

This strongly motivates the question how far OTM-recognizability with ordinal parameters can take us beyond the realm of Gödel's constructible universe L.

A particularly attractive answer would be that the pOTM-recognizable real numbers are exactly the elements of some inner model of ZFC. However, this is too much to ask: As we saw above in the discussion preceding Definition 4.3.12, the recognizable real numbers are not closed under very elementary operations like "taking every second bit." This motivates the following definitions.

Definition 4.4.10 ([39]). Let $X \subseteq \mathfrak{P}(\omega)$. Then $\mathrm{rcl}(X)$, the recognizable closure of X, is the closure of X under relativized pOTM-recognizability.

We let $\mathrm{rcl}(\emptyset) = \mathrm{RCL}$.

Though it might seem that closing the recognizable real numbers under relativized recognizability leads to a much richer class, this is not really the case:

Proposition 4.4.11 ([39], Lemma 2.7). *RCL coincides with the closure of the set of OTM-recognizable real numbers under primitive recursion.*

Proof. Let that $x_0, \ldots, x_n \subseteq \omega$ such that $x_0 \leq^r_{\mathrm{pOTM}} x_1 \leq^r_{\mathrm{pOTM}} \cdots \leq^r_{\mathrm{pOTM}} x_n$, where x_n is pOTM-recognizable. We claim that $x = (x_0 \oplus (x_1 \oplus (x_2 \oplus \cdots \oplus (x_{n-1} \oplus x_n) \ldots)))$ is pOTM-recognizable. To see this, let, for $i \in \{0, \ldots, n-1\}$, Q_i be a pOTM-program that recognizes x_i from x_{i+1} in the parameter α_i and let Q_n be a pOTM-program that recognizes x_n in the parameter α_n. Let $\alpha = p_{n+1}(\alpha_0, \ldots, \alpha_n)$.

Now suppose that some real number y is given in the oracle. We want to test, given the parameter α, whether $y = x$. To do this, first decompose y into $n + 1$ portions $y_0, \ldots, y_n$ so that $y = (y_0 \oplus (y_1 \oplus (y_2 \oplus \cdots \oplus (y_{n-1} \oplus y_n) \ldots)))$. Then decode α into its components $\alpha_0, \ldots, \alpha_n$. It is now easy to test the components of y successively: Run $Q_i^{y_i \oplus y_{i+1}}(\alpha_i)$ for $i \in \{0, \ldots, n-1\}$ and $Q_n^{y_n}(\alpha_n)$. If any of the outputs is 0, then $y \neq x$ and we halt with output 0, otherwise we halt with output 1. Thus x is pOTM-recognizable.

But now, x_0 is trivially primitive recursive in x, as it is simply the real number consisting of the bits of x with even index. □

Exercise 4.4.12 (Cf. [25]). We consider the following recognizable analogue of the jump operator: If $\alpha \in \mathrm{On}$, then RJ_α, the 'recognizable jump' with parameter α, is defined as the set of OTM-programs that recognize some real number in the parameter α, i. e., $\mathrm{RJ}_\alpha := \{i \in \omega : \exists x \subseteq \omega \forall z \subseteq \omega P_i^z \downarrow = \delta(x, z)\}$.

(a) For $i \in \omega$ and $\alpha \in \mathrm{On}$, let x_i be the real number recognized by P_i if $i \in \mathrm{RC}_\alpha$ and $x_i = 0$, otherwise. Show that, relative to RC_α, the join of the x_i (i. e., the real number that has the jth digit of x_i as its $p(i, j)$th digit) is OTM-recognizable in the parameter α.

(b) Use (a) to show that there is a real number y such that y is OTM-recognizable in the parameter α relative to RC_α and such that $y \oplus \mathrm{RC}_\alpha$ differs from each x_i in at least one digit.

(c) Conclude that RC_α is not OTM-recognizable in the parameter α for any $\alpha \in \mathrm{On}$.

(d) Suppose that $V = L$ and $\alpha < \omega_1$. Show that $\mathrm{RC}_{\omega_\alpha}$ is OTM-recognizable in the parameter $\omega_{\alpha+1}$. Conclude that it is consistent with ZFC that, for every $\alpha < \omega_1$, there is a real number r_α that is OTM-recognizable in the parameter $\omega_{\alpha+1}$, but not in the parameter ω_α.

4.4.1 The recognizable closure for OTMs

The task of this section is to determine RCL. Of course, this question is not well-posed: As we saw above, under $V = L$, RCL will just be $\mathfrak{P}^L(\omega)$, while under the existence of $0^\sharp$, we will have $0^\sharp \in \mathrm{RCL}$. We thus ask how large RCL can possibly become under appropriate set-theoretical extra assumptions. This question was considered and answered in [39]. The following material is adapted or directly taken from that article. We thank our coauthors Philipp Schlicht and Philip Welch as well as the Annals of Pure and Applied Logic for the friendly permission to reuse it here.[15]

This section requires a much stronger set-theoretical background than the rest of this book: In particular, we will use inner models for Woodin cardinals. It is hopeless to give an account of these here in an appropriate amount of space. Instead, we try to sketch the arguments in such a way that someone without a background in inner model theory may still find it illuminating. This unavoidably happens at the cost of detail and precision. The reader interested in full proofs is referred to [39]. Here, we merely briefly and roughly mention some definitions and facts, which we will use as "black boxes." We refer to Löwe and Steel [130] for a good introduction, and to Mitchell and Steel [141], Zeman [199] and the forthcoming book by Jensen [94] for a more detailed treatment.

Above we mentioned that $0^\sharp$ exists if and only if there is a nontrivial elementary embedding from L into itself, which expresses a certain "homogenity" of L. Similarly,

15 Moreover, we thank Philipp Schlicht for a careful proofreading of this section and suggesting various corrections and improvements to the presentation.

large cardinals can be defined as critical points of embeddings from the set-theoretical universe V to certain transitive classes. For example, a measurable cardinal can be defined as the critical point of V to some transitive class model M of ZFC. This can be strengthened by demanding that M is in some sense "close to V." By Kunen's inconsistency (see, e. g., [92], Theorem 17.7), the ultimate similarity $M = V$ is unattainable, as there is no elementary embedding from V to V. The definition of a Woodin cardinal, which we will state next, is in the same spirit. This is not the place to give more motivation for Woodin cardinals, and we restrict ourselves to the fact that it is a property that is often needed in set theory. For more support for Woodin cardinals, see Welch [186] or Löwe and Steel [130].

Definition 4.4.13 (Cf., e. g., [130], Definition 2.27). For ordinals $\kappa < \delta$ and $X \subseteq V_\delta$, κ is "X-reflecting in δ" if and only if, for all $\xi < \delta$, there are a class M and an elementary embedding $\sigma : V \to M$ with critical point κ and such that $\sigma(X) \cap V_\xi = X \cap V_\xi$.

Then $\delta \in \mathrm{On}$ is a **Woodin cardinal** if and only if every $A \subseteq V_\delta$ is reflected by some $\kappa < \delta$.

For studying the consequences of large cardinal axioms, an important tool are "core models," which are inner models of ZFC that are constructed in a similar way as L. The "building blocks" that replace the L_α-levels in the construction of such models are "premice," which are structures of the form $(L_\alpha[\vec{E}], \vec{E}, \in)$; here, $\vec{E}$ is a sequence of "extenders," which are set-sized codes for elementary embeddings; though the details are in fact a bit more subtle, one may think of a sequence of ultrafilters, which code the canonical embeddings to the respective ultrapower. (In fact, the sequence is allowed to contain merely "partial" ultrafilters that do not measure all sets in the structure. In that case, the ultrapower can only be formed when the structure is first "cut back" in order to make the partial ultrafilter a full one.) By forming the ultrapower of such a structure with one of the extenders on the sequence, one obtains a new structure of the same form, of which more ultrapowers can be formed, etc. This process is called an "iteration" of the premouse, and it can be extended to limit stages by forming direct limits. It is, however, not necessarily the last model to which an extender is applied; rather, that model is picked according to the iteration rules. Thus, one has various candidates for the direct limit at limit stages, and one uses an "iteration strategy" to pick out one of them. A structure generated in this way fails to be well-founded, in which case the process stops and we consider this attempt at iterating the premouse a failure. If that never happens, the premouse is called iterable; if it never happens for iterations of length $\leq \omega_1$, the premouse is called $\omega_1 + 1$-iterable. Iterability can also be weakened by only requiring that the iteration works for certain choices of the sequence of ultrafilters applied. For a precise account of premice, see Steel [169] or Mitchell and Steel [141].

Definition 4.4.14 (See, e. g., Schimmerling [154]; also cf. [39]). It can be shown that there is at most one premouse $M = (L_\alpha[\vec{E}], \in, \vec{E})$ with the following properties:

- There is a subset of ω which is Σ_1-definable over M, but not contained in M – i. e., the Σ_1-projectum of M is ω.
- If $\vec{p} \subseteq M$ is a finite set and $n \in \omega$ is such that a subset $x \notin M$ of ω is Σ_n-definable over M, then the Σ_n-hull of $\vec{p}$ in M coincides with M, i. e., M is ω-sound.
- M is iterable for $\omega_1 + 1$ many steps.
- There is $\alpha \in M$ such that $\models$ 'α is a Woodin cardinal'.
- M has a top measure U on some $\beta > \alpha$.
- If $\iota < \alpha$ and E_ι is an extender with critical point ρ, then $(L_\rho[\vec{E} \restriction \rho], \vec{E} \restriction \rho, \in) \models$ "There is no Woodin cardinal."

If it exists, this premouse is called $M_1^\sharp$.

Iterating U on $M_1^\sharp$ through the ordinals, we obtain a transitive class model of ZFC +"There is a Woodin cardinal," which is called M_1.

M_1 is equal to $L[\vec{E}]$ for an appropriate sequence $\vec{E}$ of extenders. Hence, like $M_1^\sharp$, it decomposes into "levels" or "stages"; we denote the αth level of M_1 by $M_1|\alpha$ for $\alpha \in \mathrm{On}$.

For the rest of this section, we work under the assumption that $M_1^\sharp$ exists; this follows, e. g., from the assumption that there is a measurable cardinal above a Woodin cardinal in V. That this is the right "largeness assumption" for the analysis of RCL is indicated by the following observation.

Lemma 4.4.15 ([39], Lemma 4.3). *Let $x \in$ RCL. Then $x \in M_1$.*

The proof uses the following theorem, known as Woodin's genericity iteration:

Theorem 4.4.16 (Cf. Steel, [169], Theorem 7.14). *There is a partial order Q in M_1 such that, for every real number $x \notin M_1$, there is a countable iterate N of M_1 such that x is $i_{M_1,N}(Q)$-generic over N, where $i_{M_1,N}$ denotes the iteration map. Moreover, the size of Q in M_1 is that of the unique Woodin cardinal of M_1.*

Proof. We now prove Lemma 4.4.15. Suppose for a contradiction that $x \subseteq \omega$ is such that some OTM-program P recognizes x relative to some ordinal parameter α, but $x \notin M_1$. By Woodin's genericity iteration, pick an iterate N of M_1 such that x is $i_{M_1,N}(Q) = Q'$-generic over N. As M_1 and N have the same real numbers, we have $x \notin N$. Now $N[x] \models P^{\dot{x}}(\check{\beta})\downarrow = 1$, where $\dot{x}$ is a name for x. By the forcing theorem, there is a condition $q \in Q'$ such that $q \Vdash P^{\dot{x}}(\check{\beta})\downarrow = 1$. As x does not belong to N, there must be two incompatible conditions $q', q'' \in Q'$. At least one of these is not above x, say it is q'. Now pick a Q'-generic real number y extending q'. Since $q' \leq_{Q'} p$, we have $P^y(\beta)\downarrow = 1$ in $N[y]$; by absoluteness of computation, the same holds in the real world. Moreover, as x is not above q' while y does, we have $x \neq y$. But this means that $P^z(\beta)\downarrow = 1$ for a value of z different from x, which contradicts the assumption that P recognizes x in the parameter β. □

This shows that RCL is bounded from above by $M_1 \cap \mathfrak{P}(\omega)$.

We now want to show that, under appropriate large cardinal assumptions, this bound can be attained. For this direction, we need to show that all real numbers in M_1 are pOTM-computable from some pOTM-recognizable real number.

Since all real numbers in M_1 are contained in $M_1|\omega_1^{M_1}$ and every real number in $M_1|\alpha$ is pOTM-computable from a real number coding $M_1|\alpha$ (by just picking out the right portion of the proof as in Chapter 3), it suffices to show the following.

Lemma 4.4.17. *There are cofinally in $\omega_1^{M_1}$ many ordinals α such that $M_1|\alpha$ has a pOTM-recognizable code.*

This part of the proof is much more involved, and we will merely sketch the idea, as we warned above.

Lemma 4.4.18. *There are cofinally in $\omega_1^{M_1}$ many α such that a real code for $M_1|\alpha$ is definable over $M_1|\alpha$.*

Proof. By fine structure in M_1, one sees that every M_1-index α, i. e., every α such that $\mathfrak{P}(\omega) \cap M_1|\alpha \subsetneq \mathfrak{P}(\omega) \cap M_1|(\alpha+1)$, is of this kind. □

Let α be an M_1-index. Of the real codes for $M_1|\alpha$ contained in $M_1|(\alpha+1)$, one is minimal in the sense of the lexicographic ordering sorting first by the index of the defining formula and second by the position of the parameter in the canonical well-ordering of M_1. Let us call this code $c(\alpha)$. We claim that $c(\alpha)$ is pOTM-recognizable from the parameters ω_1 and α, whenever α is an M_1-index.

To this end, we need a characterization of initial segments of M_1 due to Steel; see [168]. The crucial point is that the condition of iterability is replaced with Π^1_2-iterability. We will not define Π^1_2-iterability here, as doing so properly would take us too far afield; the crucial observation is that it is expressible by a Π^1_2-formula, and thus the set of reals coding a Π^1_2-iterable structure is pOTM-decidable in the parameter ω_1 as in the proof of Lemma 4.4.3.

Lemma 4.4.19 ([39], proof of Lemma 4.6). *$M_1|\alpha$ is uniquely characterized as a sound premouse M of height α that is Π^1_2-iterable and satisfies* $\mathrm{Def}(M) \cap \mathfrak{P}(\omega) \neq M \cap \mathfrak{P}(\omega)$, *together with some further first-order properties (see* [39]*).*

Lemma 4.4.20. *For every M_1-index α, $c(\alpha)$ is OTM-recognizable in the parameters α and ω_1 (i. e., in $p(\alpha, \omega_1)$).*

Proof. It is not hard to see that an OTM can test whether a given real number z codes a well-founded structure of the form $(L_\alpha[\vec{E}], \vec{E})$. It is also not hard to test whether the coded structure has projectum ω and is sound. Given that this is true, the minimality of z can be tested by computing from z a code for $L_{\alpha+1}[E]$ and searching through that structure for a smaller code.

It remains to test the Π^1_2-iterability of z. But Π^1_2-iterability is expressed by a Π^1_2-formula and given the parameter ω_1, OTMs can test Π^1_2-statements. □

Now suppose that $x \in M_1 \cap \mathfrak{P}(\omega)$. Then $x \in M_1|\omega_1^{M_1}$, and thus there is an M_1-index α such that $x \in M_1|\alpha$. In $c(\alpha)$, x is coded by some $i \in \omega$. Hence x is OTM-computable from $c(\alpha)$ (without parameters), and in particular recognizable from $c(\alpha)$. Thus $x \in C^{\infty}$.

Taking Lemma 4.4.15 and Lemma 4.4.17 together, we obtain the following.

Theorem 4.4.21. *Under the assumption that $M_1^{\sharp}$ exists, we have* $\mathrm{RCL} = M_1 \cap \mathfrak{P}(\omega)$.

4.4.1.1 Further results

Given the characterization of the recognizable closure for subsets of ω obtained above, a natural next step is to try to extend this to subsets of arbitrary ordinals. Here, we briefly summarize some of the results obtained in this direction.

Definition 4.4.22. For $\alpha \in \mathrm{On}$, let RCL_α be the pOTM-recognizable closure for subsets of α; i. e., RCL_α is the closure of the pOTM-recognizable subsets of α under relativized pOTM-recognizability (which, for the same reasons as for subsets of ω, coincides with the closure of the pOTM-recognizable subsets of α under pOTM-computability).

A first observation is that the recognizable closure for subsets of some $\alpha \in \mathrm{On}$ can be characterized in many different ways.

Theorem 4.4.23 ([39], Lemma 3.22). *Let $x \subseteq \alpha$. Then the following are equivalent: x is pOTM-computable from $y \subseteq \alpha$ such that:*
- *y is a pOTM-recognizable subset of α*
- *There are a Σ_1-formula $\phi(z,\beta)$ and an ordinal β such that $y \subseteq \alpha$ is unique with the property $L[y] \models \phi(y,\beta)$*
- *There is a formula $\phi(z,\beta)$ and an ordinal β such that $y \subseteq \alpha$ is unique with the property $L[y] \models \phi(y,\beta)$*
- *y is 'implicitly definable over L'[16]: There is a formula ϕ and an ordinal β such that y is unique with $\langle L, \in, y\rangle \models \phi(y,\beta)$*
- *y is an $(L_{\mathrm{On},0}^{(\alpha)} \cap L)$-definable subset of α, where $L_{\mathrm{On},0}^{(\alpha)}$ is the closure of the atomic statements $\gamma \in \dot{y}$ ($\gamma < \alpha$) and their negations under infinitary conjunctions and disjunctions.*

Recognizability is a concept of implicit definability; that an object x is recognizable means that there is a method for determining whether a given object happens to be x. In this sense, the recognizable closure RCL resembles the class HOD of hereditarily ordinal definable sets. It is in fact easy to see that $\mathrm{RCL} \subseteq \mathfrak{P}(\omega) \cap \mathrm{HOD}$. However, the other inclusion fails.

Theorem 4.4.24 ([39], Section 3.4). *It is consistent with ZFC that* $\mathrm{RCL} \neq \mathfrak{P}(\omega) \cap \mathrm{HOD}$.

16 The concept of implicit definability is due to J. Hamkins and C. Leahy; see [76].

For subsets of countable ordinals α, a complete characterization was obtained in [39].

Definition 4.4.25 (See, e. g., Mitchell and Steel, [141]). Let κ be the first measurable cardinal in M_1. For $\alpha \in \mathrm{On}$, let M_α be the αth iterated ultrapower of M_1 with a normal (closed under diagonal intersections) ultrafilter U on κ. Then $M^\infty := \bigcap_{\alpha\in\mathrm{On}} M^\alpha$.

Theorem 4.4.26 ([39], Theorem 4.7). *Suppose that $M_1^\sharp$ exists. If α is a countable ordinal, then $x \subseteq \alpha$ belongs to RCL if and only if $x \in M^\infty$*

For subsets of uncountable α, the following results are known.

Theorem 4.4.27 ([39], Theorem 4.8). *Under appropriate large cardinal assumptions, it is consistent with ZFC that* $\mathrm{RCL}_{\omega_1} \subseteq M^\infty$.

Passing from a model M of ZFC to a forcing extension $M[G]$, new subsets of an ordinal α may be added, which could have the effect that a subset $x \subseteq \alpha$ that is pOTM-recognizable in M ceases to be so in $M[G]$. It is therefore natural to "cancel out" this effect, thus obtaining "generic recognizability."

Definition 4.4.28 ([39], Definition 4.12). We say that $x \subseteq \alpha$ is "generically recognizable" if and only if there is $\beta \in \mathrm{On}$ such that, for all $\gamma > \beta$ and all filters G that are generic for the forcing collapsing γ to ω, x is recognizable in $V[G]$. The "generically recognizable closure," GRC, is the set of all sets of ordinals that are pOTM-computable from some generically recognizable set of ordinals.

Theorem 4.4.29 ([39], Theorem 4.13). *If $M_1^\sharp(X)$ exists for all sets $X \subseteq \mathrm{On}$, then GRC is the class of sets of ordinals in M^∞.*

4.5 Variants of recognizability

For the recognizability of a real number x by certain machine type, it is required that the set $\{x\}$ is decidable by a program for such a machine. Like decidability, this can be varied in several ways.

Definition 4.5.1 (Cf. [37]). Let $x \subseteq \omega$. We say that x is "semirecognizable" if and only if there is a program P such that, for all $y \subseteq \omega$, $P^y\downarrow$ if and only if $y = x$; that is, P semidecides $\{x\}$. x is "corecognizable" or "antirecognizable" if and only if there is a program P such that, for all $y \subseteq \omega$, $P^y\uparrow$ if and only if $y = x$, i. e., $\mathfrak{P}(\omega) \setminus \{x\}$ is semidecidable. x is called "eventually recognizable" or "limit recognizable" if and only if there is a program P such that, for all $y \subseteq \omega$, the output of P^y eventually stabilizes at $\delta(x, y)$ (but the computation may not halt).

Corollary 4.5.2. *Let $x \subseteq \omega$, and let $M \in \{\mathrm{ITRM}, \mathrm{ITTM}, \mathrm{OTM}, \mathrm{OTM}\}$.*
(i) *x is M-recognizable if and only if x M-semirecognizable and M-antirecognizable.*
(ii) *If x is M-semirecognizable or M-antirecognizable, then x is M-limit recognizable.*

For ITRMs, these notions are not very exciting.

Lemma 4.5.3. *For every $x \subseteq \omega$, the following are equivalent:*
(i) *x is ITRM-recognizable.*
(ii) *x is ITRM-semirecognizable.*
(iii) *x is ITRM-antirecognizable.*
(iv) *x is ITRM-limit recognizable.*

Proof. All of these claims follow easily from the solvability of the bounded halting problem for ITRMs. □

For ITTMs, the situation is a bit different.

Theorem 4.5.4. *There is a real number x that is ITTM-semirecognizable, but not ITTM-recognizable. In fact, the tape content of the universal ITTM $\mathcal{U}$ at time Σ has this property.*

Proof. As x is accidentally writable, but not eventually writable by Theorem 3.6.18 in Chapter 3, the $<_L$-minimal code c of $L_{\Sigma+1}$ is ITTM-writable in x. Let P be a program that does this. Now, in input y, run P^y. If it does not halt, then $y \neq x$ and things are as they should be. If it halts with output z, check whether $z = c$. If not, then $y \neq x$, so we enter an endless loop. Otherwise, x is coded in c by a certain natural number i that can be given to our program in advance, and we only need to check whether the number coded by i in c is equal to y. If not, then $y \neq x$ and we enter an endless loop, otherwise $y = x$ and we halt. □

Question 4.5.5. Is there a real number that is ITTM-antirecognizable, but not ITTM-recognizable? Is there a real number that is ITTM-limit recognizable, but neither ITTM-semirecognizable nor ITTM-antirecognizable?

Concerning OTMs, nothing changes with respect to semi-recognizability.

Lemma 4.5.6. *A real number x is OTM-semirecognizable if and only if it is OTM-recognizable (i. e., OTM-computable, see above).*

Proof. This works exactly as the proof that OTM-recognizable real numbers are OTM-computable: If x is OTM-semirecognizable by the program P, then $P^x\downarrow$, and $\exists y P^y\downarrow$ is still a Σ_1-statement which is thus absolute between V and L and further between L and L_σ. So $x \in L_\sigma$, i. e., x is OTM-computable and OTM-recognizable. □

However, changing to antirecognizability has a rather dramatic effect, even without parameters.

Theorem 4.5.7. $0^\sharp$ *is OTM-antirecognizable.*

Proof. As in the proof of Lemma 4.4.8, the relation $x = 0^\sharp$ is Π^1_2. Hence, the relation $x \neq 0^\sharp$ is Σ^1_2, and thus OTM-semidecidable as in the proof of Lemma 4.4.3, i. e., there is an OTM-program P such that P^x halts exactly when $x \neq 0^\sharp$, which means that P antirecognizes x. □

Question 4.5.8. What is the closure of the OTM-antirecognizable reals under primitive recursion?

4.6 Summary

We give an overview of the recognizable and computable objects for wITTMs, wITRMs, ITRMs, ITTMs and OTMs with and without parameters by summarizing them in the following picture, where M-COMP denotes the set of M-computable objects and M-RECOG the set of M-recognizable objects. All inclusions except the upper two are provably strict under ZFC; the upper two inclusions are strict when one additionally assumes that $0^\sharp$ exists. The equality on top of course depends on the existence of $M_1^\sharp$:

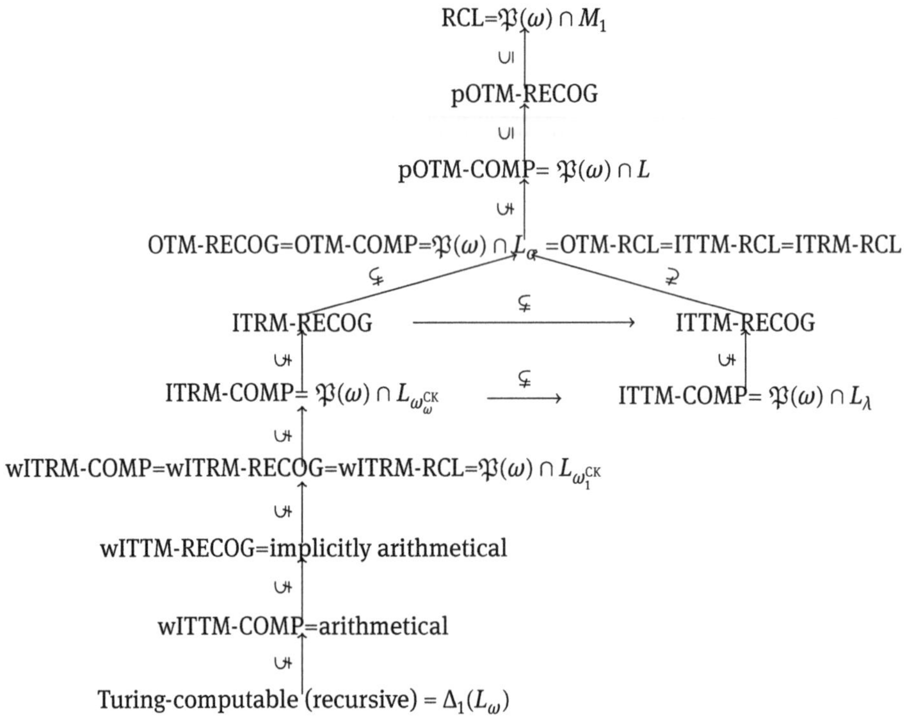

4.7 Exercises

Exercise 4.7.1. Show that, if x is wITRM-computable, then x is wITRM-recognizable.

Exercise 4.7.2. Show that, if x is wITRM-semirecognizable, then x is wITRM-computable.

Exercise 4.7.3. Show that there are real numbers $x <_L y <_L z$ such that x and z are lost melodies for ITTMs, but y is not.

Exercise 4.7.4. Let X be a nonempty set of reals that is decidable by a parameter-free OTM. Show that X contains an OTM-computable element.

Exercise 4.7.5. If M is one of ITRM, ITTM, OTM, the "recognizable jump" J_M consists of all natural numbers i such that the ith M-program recognizes a real number. Show that J_M is not M-recognizable.

Exercise 4.7.6 (Cf. [39]). Show that the recognizable closure for ITRMs, ITTMs and OTMs without parameters is $\mathfrak{P}(\omega) \cap L_\sigma$.

Exercise 4.7.7 (Cf. [19]). In this exercise, we will prove in a different way that, for every $n \in \omega$, there are $x \subseteq \omega$ and $m \in \omega$ (with $m > n$) such that x is not recognizable by an ITRM with n registers, but it is recognizable by an ITRM with m registers. Thus, like the computability strength, the recognizability strength of ITRMs increases with the number or registers.

Call a real number x "special" if and only if $x \in L_{\omega_1^{\mathrm{CK},x}}$.

(a) Show that the set of special real numbers is ITRM-decidable.

(b) Show that the set of special ITRM-recognizable real numbers is not ITRM-decidable. (Hint: If it was, then so would be its complement. Now use Theorem 4.3.3.)

For $n \in \omega$, let RECOG_n denote the set of real numbers that are recognizable by an ITRM-program using n registers and let $\mathrm{RECOG} = \bigcup_{i \in \omega} \mathrm{RECOG}_n$.

(c) Let $n \in \omega$. Show that the special elements of RECOG_n form a proper subset of the set of special elements of RECOG; deduce that RECOG_n does not contain all ITRM-recognizable real numbers. (Hint: Show that the set of special elements of RECOG_n is ITRM-decidable and use (b).)

Exercise 4.7.8. Recall from Chapter 3 the concept of a wITTM and the result that wITTMs can compute exactly the arithmetic subsets of ω.

(a) Show that $x \subseteq \omega$ is arithmetical if and only if $x \leq_T 0^{(n)}$ for some $n \in \omega$, where $0^{(n)}$ denotes the nth Turing jump of 0.

A real number $x \subseteq \omega$ is called "implicitly arithmetical" (see, e. g., Simpson [164]) if and only if there is an arithmetical formula ϕ using a new unary predicate symbol P such that ϕ is true if and only if P is interpreted as x.

(b) (Cf. Rogers, [152], Chapter 15, Theorem XII; also cf. Simpson [165]) Let $0^{(\omega)} := \{p(i,j) : i \in 0^{(j)}\}$. Show that $0^{(\omega)}$ is implicitly arithmetical.

(c) Show that, for each arithmetical formula ϕ in an extra unary predicate symbol P, there is a wITTM-program P such that, for all $x \subseteq \omega$, P^x halts with output 1 if and only if ϕ holds when P is interpreted as x. Conclude that, if x is implicitly arithmetical, then x is wITTM-recognizable.

(d) Show that $0^{(\omega)}$ is a lost melody for wITTMs.

Exercise 4.7.9. Recall from Chapter 3, exercise 3.4.37 the concept of a typed ITRM. Show that the $<_L$-minimal code for $L_{\omega_\omega^{\mathrm{CK}}}$ is recognizable by an ITRM of type $(1, n)$ for some $n \in \omega$. Thus, one overflowing register is enough for the existence of lost melodies!

Exercise 4.7.10. A natural analogue of Theorem 4.3.8 for ITTMs would be: If $\lambda^x = \lambda^y$ and $x, y \in L_{\lambda^x}$, then x is ITTM-recognizable if and only if y is ITTM-recognizable. In this exercise, we observe that this analogue fails.

(a) Let c be the tape content of the universal ITTM $\mathcal{U}$ at time ζ. Show that c is not ITTM-recognizable.

(b) Show that, relative to c, Σ is ITTM-clockable, and thus $\mathrm{cc}(\Sigma)$ is ITTM-writable.

(c) Show that $\lambda^c = \lambda^{\mathrm{cc}(\Sigma)}$.

(d) Show that $c \in L_{\lambda^c}$ and $\mathrm{cc}(\Sigma) \in L_{\lambda^c}$.

(e) Show that $\mathrm{cc}(\Sigma)$ is ITTM-recognizable. Conclude that the analogue of Theorem 4.3.8 for ITTMs fails.

Exercise 4.7.11. Show that, for parameter-OTMs and X a set of ordinals, the following are equivalent: (1) X is pOTM-recognizable, (2) X is pOTM-semirecognizable, (3) X is pOTM-antirecognizable.

Exercise 4.7.12. (Question) Let us say that an $\in$-structure $(A, \in)$ with A a set is ITRM-recognizable if and only if there is an ITRM-recognizable real number c that codes A. We call $(A, \in)$ "structurally ITRM-recognizable" if and only if the set of codes for $(A, \in)$ is ITRM-decidable.

(a) Show that, if $(A, \in)$ is structurally ITRM-recognizable, then $(A, \in)$ is ITRM-recognizable.

(b) (Open!) Is it true that conversely, the ITRM-recognizability of $(A, \in)$ implies the structural ITRM-recognizability of $(A, \in)$? What about ITTMs?

Question 4.7.13. Are there lost melodies for Infinite Time Blum–Shub–Smale machines (ITBSSMs)?

5 Randomness

5.1 Introduction

Customary as it is to use "random generators" in programming and working with randomized algorithms, computations relative to random information become a natural field of study for computability theory. In this chapter, we will consider infinitary computations in connection with randomness, along with its intimate connections to forcing.

Since forcing will play a prominent role in this chapter, we start with a short introduction, particularly to forcing over models of weak set theories.

5.2 Preliminaries on forcing and sets of real numbers

Below, we will be concerned with sets of real numbers that are intuitively "small." One classical formal notion of "smallness" is the following.

Definition 5.2.1. A set $X \subseteq \mathfrak{P}(\omega)$ is "meager" if and only if it is a countable union of nowhere dense sets.

Corollary 5.2.2. *A union of countably many meager sets is meager. Consequently, if* $X = \bigcup_{i\in\omega} X_i$ *is not meager, then at least one of the* X_i *is not meager.*

Proof. A union of countably many unions of countably many nowhere dense sets is still a countable union of nowhere dense sets. □

Forcing over admissible sets was worked out by I. Ershov [52] and then refined by A. Mathias in [137] and by Mathias and N. Bowler in [17]; our exposition is based on these papers, and also on Kunen [123]. This is not the place for a detailed introduction to forcing, for which we refer the reader to [123]. We will, however, recall some basic notions and facts and prove the results on forcing over admissible sets that we will need when their proof deviates from the proof in the usual setting.

A **forcing** is a partially ordered set $(P, \leq)$ with a maximal element 1_P. If $p, q \in P$ and $p \leq q$, we say that p is stronger than q or that p strengthens q. The idea is that the elements of P are approximations of some desired object, and that smaller elements contain more information. Therefore, the elements of P are called "conditions." A subset $X \subseteq P$ is **dense** if and only if, for every $p \in P$, there is $q \in X$ such that $q \leq p$. If $r \in P$ is such that this holds for every $p \leq q$, then X is called **dense below** r.

$F \subseteq P$ is called a **filter** if $1_P \in F$, F is upwards closed (i. e., for all $p, q \in P$, $p \leq q$ and $p \in F$ imply $q \in F$) and, for all $p, q \in F$, there is $r \in F$ such that $r \leq p$ and $r \leq q$. Intuitively, a filter is a set of conditions that has the potential of being the set of all conditions describing a certain (unknown) object. In particular, throwing away

https://doi.org/10.1515/9783110496154-005

information must not lead out of a filter and every two conditions must be consistent with each other, i. e., there must be a condition that comprises both.

If M is a set or a proper class and $F \cap D \neq \emptyset$ for every dense $D \in \mathfrak{P}(P) \cap M$, then F is called M-generic. If $F \cap D \neq \emptyset$ for every dense $D \subseteq P$ which is Σ_n-definable over M (i. e., there is a Σ_n-formula ϕ and some finite $\vec{p} \subseteq M$ such that $D = \{x \in M : (M, \in) \models \phi(x, \vec{p})\}$), then F is called Σ_n-generic over M. If this holds for all n, we call F "supergeneric" over M. (With the notation of Chapter 3, this means that F is generic over Def(M).)

Let $(P, \leq)$ be a forcing, and let F be a filter on $(P, \leq)$. A "name" is just an arbitrary set x. The F-evaluation x^F of x is defined recursively as $x^F := \{x^F : \exists p \in F(y, p) \in x\}$, which is a Δ_0-recursion.

Exercise 5.2.3. Verify that the above is indeed a Δ_0-recursion by writing out a Δ_0-definition.

If M is a set and F is a filter, then $M[F] := \{x^F : x \in M\}$ is the set of F-evaluations of the elements of M, called the forcing extension of M by F. We will presently see that, for sensible choices of M, $M[F]$ indeed extends M. (In the case that M is of the form L_α for some $\alpha \in \mathrm{On}$, this conflicts with our notation $L_\alpha[x]$ for relativized constructibility. We will show below that, in all cases that concern us, this conflict of notation is harmless, as the two sets coincide. In the meantime, we will write $L_\alpha[[x]]$ for a generic extension of an L-level L_α.) If F is generic over M, then $M[F]$ is called the generic extension of M by F.

Exercise 5.2.4. With notation as above, show the following:

(i) If M is transitive, then so is $M[F]$.

(ii) Show that, if M is closed under pairing, then so is $M[F]$. (Hint: For $\dot{x}, \dot{y} \in M$, consider $\{(\dot{x}, 1), (\dot{y}, 1)\}$.)

Now suppose that M is admissible.

(iii) For $x \in M$, define $\check{x}$ recursively as $\{(\check{y}, 1) : y \in x\}$. Show that $\check{x}^F = x$ and that $\check{x} \in M$. Conclude that $M \subseteq M[F]$.

The forcing we will mainly use in the following is Cohen-forcing, due to Cohen [41]. The conditions of Cohen-forcing are finite partial functions from ω to $\{0, 1\}$, ordered by $\supseteq$. Thus, what is approximated by Cohen forcing is a function f from ω to $\{0, 1\}$, which we regard as the characteristic function of a subset $x \subseteq \omega$; in the following, we will identify this function with the real number x. Moreover, we will associate with $f : \omega \to \{0, 1\}$ the filter $F_x := \{p : \omega \to \{0, 1\} | p \text{ is a partial function} \wedge |p| < \omega \wedge p \subseteq x\}$.

We will frequently need to compute filters (or, equivalently, real numbers) that are Cohen-(super)generic over a structure M for which a code c is given. This is possible and is guaranteed by the following lemma, which is an effective version of the Rasiowa–Sikorski-theorem (see, e. g., [92], Lemma 14.4).

Lemma 5.2.5. *There is a wITRM-program P_{rs} such that the following holds: If M is a countable transitive set, p is a Cohen-condition and c a real number coding M, then $P^{c}_{\mathrm{rs}}(p)$ computes a real number $x \supseteq p$ which is Cohen-generic over M.*

Similarly, there is a wITRM-program P_{sr^+} such that, under the same assumptions, $P^{c}_{\mathrm{sr}^+}(p)$ computes a real number $x \supseteq p$ which is Cohen-supergeneric over M.

Proof. To compute such an x, we proceed as follows: Suppose we are looking for the jth bit of x.

Start with $p_0 = p$. Now search ω for the minimal $i_0 \in \omega$ that codes a dense subset X_0 of Cohen-forcing in c, which is possible using P_{truth} from Chapter 2. As X_0 is dense, there is $p_1 \in X_0$ such that $p_1 \supseteq p_0$; pick the one with a minimal code in the sense of c. Now search for the minimal $i_1 \in \omega$ which is $> i_0$ and codes a dense subset X_1 of Cohen-forcing, and proceed in the same way to determine $p_2 \in X_1$ such that $p_2 \supseteq p_1$. Proceed in this manner until some p_k is generated that has j in its domain; then output $p_k(j)$. If the run through ω is completed and no such p_k was generated, output 0.

The second claim now follows since we can wITRM-compute a code for $\mathrm{Def}(M)$ from a code for M; see Chapter 3. □

A crucial concept in forcing is the forcing relation, which is a relation between conditions and formulas (which may use names as parameters); intuitively, that the condition p forces the truth of $\phi(\dot{x}_1, \dots, \dot{x}_n)$ means that, based on the approximation p alone, one can know for sure that, however the construction of the generic object G is continued, $\phi(\dot{x}_1{}^G, \dots, \dot{x}_n{}^G)$ will in the end hold in the generic extension. For example, in Cohen forcing, the condition 101111 forces that the generic real number x contains at least one prime number, but leaves open whether x contains a power of 7.

An important point is that the forcing relation admits a purely syntactical definition, which we now state. Afterwards, we will observe that the relation thus defined indeed has the above property.

Definition 5.2.6 (Cf., e. g., [123, 57]). Let $M \models \mathrm{KP}$, $(P, \leq)$ be a forcing, $p \in P$, and let ϕ, ψ be $\in$-formulas. The relation $\Vdash_M$ is defined by recursion as follows (we drop the subscript M):

- $p \Vdash \neg\phi$ if and only if there is no $q \leq p$ such that $q \Vdash \phi$.
- $p \Vdash \phi \wedge \psi$ if and only if $p \Vdash \phi$ and $p \Vdash \psi$.
- $p \Vdash \exists x \phi(x)$ if and only if the set of q for which there is a name $\dot{x}$ with $q \Vdash \phi(\dot{x})$ is dense below p.
- $p \Vdash \forall x \phi(x)$ if and only if for all $\dot{x}$, $p \Vdash \phi(\dot{x})$.

This leaves us with the atomic formulas, i. e., with the definition of $p \Vdash \dot{x} \in \dot{y}$ and $p \Vdash \dot{x} = \dot{y}$. This is a bit more complicated and is done by a simultaneous recursion. To simplify the definition, we also define $p \Vdash \dot{x} \subseteq \dot{y}$.

$p \Vdash \dot{x} \in \dot{y}$ if and only if the set of $q \in P$ for which there is $(\dot{z}, q') \in \dot{y}$ such that $q \leq q'$ and $q \Vdash \dot{x} = \dot{z}$ is dense in P below p.

$p \Vdash \dot{x} \subseteq \dot{y}$ if and only if, for every $(\dot{x}', p') \in \dot{x}$, the set of $q \in P$ which satisfy either that $q \nleq p'$ or for which there is $(\dot{z}, q') \in \dot{y}$ such that $q \leq q'$ and $q \Vdash \dot{x}' = \dot{z}$ is dense in P below p.

$p \Vdash \dot{x} = \dot{y}$ if and only if $p \Vdash \dot{x} \subseteq \dot{y} \wedge \dot{y} \subseteq \dot{x}$.

Exercise 5.2.7 (Monotonicity of Forcing, cf. [123]). Let $M \models$ KP, $(P, \leq)$ be a forcing, $p, q \in P$ with $p \leq q$, and ϕ an $\in$-formula (possibly with parameters) such that $q \Vdash \phi$. Show that $p \Vdash \phi$.

Exercise 5.2.8 (Definiteness of Forcing, cf. [123]). Let $M \models$ KP, $(P, \leq)$ be a forcing.

(i) Show that, for any $\in$-formula ϕ, the set $\{p \in P : p \Vdash \phi\} \cup \{p \in P : p \Vdash \neg\phi\}$ is dense in P.

(ii) Show that, if G is a P-generic filter over M, then for no ϕ there are $p, q \in G$ such that $p \Vdash \phi$ and $q \Vdash \neg\phi$.

Lemma 5.2.9 (Cf. Mathias [137], Ershov [52]; also see Carl and Schlicht [36], Lemma 31). *Let $M := L_\alpha$ be admissible, $(P, \leq) \in M$ be a forcing and let $(\phi_i : i \in \omega)$ enumerate the Δ_0-formulas in some natural way. Then $p \Vdash \phi(\dot{x})$ is Δ_1-expressible over M, i. e., there is a Δ_1-formula ψ such that, for $\dot{x} \in M$, $i \in \omega$ and $p \in P$, $M \models \psi(i, p, \dot{x})$ if and only if $p \Vdash \phi_i(\dot{x})$ over M. We say that the forcing relation for Δ_0-formulas is Δ_1.*

Moreover, the forcing relation for Σ_1-formulas is Σ_1 over M and the forcing relation for Σ_2-formulas is Σ_2 over M.

Proof. This can be done by induction on the construction of the formula ψ if ψ is not atomic, and by an $\in$-induction in the atomic case.

So suppose first that ϕ is $\neg\phi'$; by induction, the claim holds for ϕ', so $p \Vdash \phi'(\dot{x})$ is Δ_1 over M. But then $\neg\exists q \leq p p \Vdash \phi'(\dot{x})$ is also Δ_1, as Δ_1 is closed under negation.

If ϕ is $\phi_0 \wedge \phi_1$, then $p \Vdash \phi_0(\dot{x})$ and $p \Vdash \phi_1(\dot{x})$ are Δ_1 by induction, hence so is $p \Vdash \phi_0(\dot{x}) \wedge p \Vdash \phi_1(\dot{x})$, as Δ_1 is closed under conjunction.

If ϕ is $\dot{x} = \dot{y}$, then $p \Vdash \phi$ if and only if $p \Vdash \dot{x} \subseteq \dot{y}$ and $p \Vdash \dot{y} \subseteq \dot{x}$.

If ϕ is $\exists x \in y\phi'(x)$ and $\dot{a} \in M$, then $p \Vdash \phi$ holds if and only if, for all $q < p$ in P, there are $q' < q$ in P and $(\dot{x}, q'') \in \dot{a}$ with $q'' < q'$ and $q' \Vdash \phi'(\dot{x})$, which is Δ_1 over M, provided that $p \Vdash \phi'(\dot{x})$ is (which holds by the inductive assumption).

We now turn to the atomic cases $x \in y$ and $x \subseteq y$, which will be dealt with by a simultaneous recursion. To this end, we define the following partial ordering $\leq^*$ on pairs $(x, y) \in M \times M$: $(x, y) \leq^* (x', y')$ if and only if $x \in \text{tc}(x')$ and $y \in \text{tc}(y')$ or $x \in \text{tc}(x')$ and $y = y'$ or $x = x'$ and $y \in \text{tc}(y')$. It is easy to see that $\leq^*$ is a well-founded partial ordering. Moreover, if $x, y \in M$, then $(\text{tc}(\{x, y\}), \leq^*) \in M$.

Now, given a and b, we define a function $f : P \times \{\in, \subseteq\} \times \text{tc}(\{a, b\})^2 \to \{0, 1\}$ by the following recursion on $\leq^*$ on $\text{tc}(\{a, b\})$:

- $f(p, \in, (\dot{x}, \dot{y})) = 1$ if and only if $\{q \in P : \exists(\dot{z}, q') \in \dot{y}(q \leq q') \wedge f(q, =, (\dot{x}, \dot{z})) = 1\}$ is dense in P below p, and $f(p, \in, (\dot{x}, \dot{y})) = 0$, otherwise.

– $f(p, \subseteq, (\dot{x}, \dot{y})) = 1$ if and only if, for every $(\dot{x}', p') \in \dot{x}$, the set $\{q \in P : q \nleq p' \vee \exists(\dot{z}, q') \in \dot{y}(q \leq q' \wedge f(q, \subseteq, (\dot{x}', \dot{z})) = 1 \wedge f(q, \subseteq, (\dot{z}, \dot{x}')) = 1)\}$ is dense in P below p and $f(p, \subseteq, (\dot{x}, \dot{y})) = 0$, otherwise.

Both clauses are easily seen to be Δ_0. By Σ_1-recursion in KP-models (see Theorem 3.1.29 in Chapter 3; we leave it to the reader to adapt the relation $\leq^*$ to work on $P \times \{\in, \subseteq\} \times \mathrm{tc}(\{a, b\})^2$ instead of merely in its last component), there is exactly one function $f : P \times \{\in, \subseteq\} \times \mathrm{tc}(\{a, b\})^2 \to \{0, 1\}$ in M satisfying both clauses for all $\dot{x}$, $\dot{y}$ in $\mathrm{tc}(\{a, b\})$. Clearly, by the definition of $\Vdash$, we have $f(p, \in, (\dot{x}, \dot{y})) = 1$ if and only if $p \Vdash \dot{x} \in \dot{y}$ over M and $f(p, \subseteq, (\dot{x}, \dot{y})) = 1$ if and only if $p \Vdash \dot{x} \subseteq \dot{y}$ for all $p \in P$ and all $\dot{x}, \dot{y} \in \mathrm{tc}(\{a, b\})$.

Now $p \Vdash \dot{a} \in \dot{b}$ can be characterized in M in the following way: $p \Vdash \dot{a} \in \dot{b}$ if and only if there is a function $f : P \times \{\in, \subseteq\} \times \mathrm{tc}(\{a, b\})^2 \to \{0, 1\}$ such that for all $\dot{x}, \dot{y} \in \mathrm{tc}(\{\dot{a}, \dot{b}\})$, the two clauses above are satisfied and $f(p, \in, (\dot{a}, \dot{b})) = 1$ if and only if for all functions $f : P \times \{\in, \subseteq\} \times \mathrm{tc}(\{a, b\})^2 \to \{0, 1\}$ such that the two clauses above are satisfied, we have $f(p, \in, (\dot{a}, \dot{b})) = 1$. This gives us a Σ_1- and a Π_1-characterization of $p \Vdash \dot{a} \in \dot{b}$ over M; thus a Δ_1-characterization exists as desired. For $p \Vdash \dot{a} \subseteq \dot{b}$, we proceed in the obvious analogous way.

Together with the observations above, this completes the proof of the first claim.

Now, if $\phi \equiv \exists x \psi$ is Σ_1 where ψ is Δ_0, then, by definition, $p \Vdash \phi$ holds if and only if $\exists \dot{x} p \Vdash \psi(\dot{x})$; by the above, $p \Vdash \psi(\dot{x})$ is Δ_1, so this is a Σ_1-formula. By the same argument, $p \Vdash \psi$ is Π_1 when ψ is Π_1 and thus, applying the definition once again, $p \Vdash \phi$ is Σ_2 when ϕ is Σ_2. □

Exercise 5.2.10 (Uniform forcing definability). Let $(\psi_i : i \in \omega)$ be a natural enumeration of the Δ_0-formulas, $M = L_\alpha \models \mathrm{KP}$, $(P, \leq) \in M$ a forcing. Show that the relation $p \Vdash \psi_i(\dot{x}_0, \dots, \dot{x}_n)$ is Δ_1 over M.[1]

Exercise 5.2.11 (Forcing definability, cf. [137], [36]). Let $n \geq 1$, and let $(\sigma_i^n : i \in \omega)$ and $(\pi_i^n : i \in \omega)$ enumerate in some natural way the Σ_n-formulas and the Π_n-formulas, respectively. Suppose that $M = L_\alpha \models \mathrm{KP}$ is transitive and let $(P, \leq)$ be a forcing in M.

(a) Show that $\{(p, i, \dot{a}_1, \dots, \dot{a}_k) \in P \times \omega \times M^k : p \Vdash \sigma_i(\dot{a}_1, \dots \dot{a}_k)\}$ is Σ_n over M.

(b) Show that $\{(p, i, \dot{a}_1, \dots, \dot{a}_k) \in P \times \omega \times M^k : p \Vdash \pi_i(\dot{a}_1, \dots \dot{a}_k)\}$ is Π_n over M.

We now want to work toward the forcing theorem over admissible sets. Roughly, the forcing theorem says that a statement holds in the generic extension by a filter G if and only if it is forced by some element of G. The usual proof of the forcing theorem works over models of ZFC and crucially uses the fact that, if $(P, \leq)$ is a forcing, ϕ is an $\in$-formula and G is P-generic over some $M \models \mathrm{ZFC}$, then there is $p \in G$ such that $p \Vdash \phi$ or $p \Vdash \neg\phi$.

---

1 Hint: Recall that satisfaction for Δ_0-formulas is Δ_1-expressible by Lemma 3.1.11 in Chapter 3.

Exercise 5.2.12. Prove this, by using definiteness of forcing and the fact that ZFC-models satisfy full separation.

However, over a model M of KP, the situation is not that easy: For a general $\in$-formula ϕ, the definition of the set $D_\phi := \{p \in P : p \Vdash \phi\} \cup \{p \in P : p \Vdash \neg\phi\}$ via the forcing relation has at least the quantifier complexity of ϕ, and thus there is no reason to expect it to belong to M. But then, the M-genericity of a filter G no longer guarantees that $G \cap D_\phi \neq \emptyset$, and the statement of the last exercise may fail, and so will the proof of the forcing theorem building on it. It will, however, work again when we require G also to intersect sets definable over M. One can go into some detail here concerning how much genericity is to be required of G in order for the forcing theorem to hold for Σ_n-formulas. This is, e. g., done in [137] and [52] (see also the next exercise). However, for our purposes, the following rough variant is sufficient:

Lemma 5.2.13 (A. Mathias, [137]). *Let $M = L_\alpha$ be admissible, $(P, \leq)$ a forcing in M and G a P-supergeneric filter over M. Then there is $p \in G$ such that $p \Vdash \phi$ or such that $p \Vdash \neg\phi$.*

Proof. By definiteness of forcing, $D_\phi := \{p \in P : p \Vdash \phi\} \cup \{p \in P : p \Vdash \neg\phi\}$ is dense in P. By the exercise on forcing definability above, D_ϕ is definable over M, i. e., $D_\phi \in \mathrm{Def}(M)$. By the assumption on G, we have $G \cap D_\phi \neq \emptyset$, and any element of the intersection is as desired. □

Lemma 5.2.14 (A. Mathias, [137]). *If $M = L_\alpha$ is admissible, x is Cohen-supergeneric and ϕ is an $\in$-formula, then $M[x] \models \phi$ if and only if there is a Cohen-condition $p \subseteq x$ such that $p \Vdash \phi$ over M.*

Proof. This proof now works exactly as in the classical case (see, e. g., [123]); where it is used in that proof that, for every $\in$-formula ϕ, every generic filter contains an element that forces ϕ or its negation, Lemma 5.2.13 is used, but this is the only difference. □

Exercise 5.2.15 (A. Mathias, cf. [137]). Suppose that $M = L_\alpha$ is admissible and let $(P, \leq)$ be a forcing in M. Let $G \subseteq P$ be a filter. Show that, if G intersects every set that is Σ_1-definable or Π_1-definable over M, then the forcing theorem holds for Σ_1-formulas. Also show that, if G intersects every set that is Σ_n-definable or Π_n-definable over M, then the forcing theorem holds for Σ_n-formulas.

Lemma 5.2.16 (Cf. [137], Section 2). *Let α be admissible, and let x be Cohen-generic over L_α. Then $L_\alpha[[x]] = L_\alpha[x]$. The same holds when α is a limit of admissible ordinals and x is Cohen-generic over L_α.*

Proof. It is easy to see that $L_\alpha[x] \subseteq L_\alpha[[x]]$, as the recursive definition of the $L[x]$-hierarchy can be carried out inside the admissible set $L_\alpha[[x]]$, and hence, for every $\beta < \alpha$, we have $L_\beta[x] \in L_\alpha[[x]]$, which is transitive.

On the other hand, to see that $L_\alpha[[x]] \subseteq L_\alpha[x]$, first note that $x \in L_\alpha[x]$ implies $F_x \in L_\alpha[x]$, by admissibility of $L_\alpha[x]$. We need to show that, for $y \in L_\alpha[x]$, we have $y^{F_x} \in L_\alpha[x]$. But this follows from the recursive definition of y^{F_x} and the recursion theorem in admissible sets (see Theorem 3.1.29, Chapter 3), which is here applied to the well-founded partial ordering $(\mathrm{tc}(y), \in) \in L_\alpha[x]$.

To see the second claim, let $(\beta_\iota : \iota < \gamma)$ be a sequence of admissible ordinals $< \alpha$ with supremum α. If x is Cohen-generic over L_α, it is also Cohen-generic over $L_{\beta_\iota+1}$ for all $\iota < \gamma$. Thus, by the first claim, $L_{\beta_\iota}[[x]] = L_{\beta_\iota}[x]$ for all $\iota < \gamma$. But $L_\alpha[[x]] = \bigcup_{\iota<\gamma} L_{\beta_\iota}[[x]]$ and $L_\alpha[x] = \bigcup_{\iota<\gamma} L_{\beta_\iota}[x]$, so $L_\alpha[[x]] = L_\alpha[x]$. □

We are now justified in identifying $L_\alpha[x]$ and $L_\alpha[[x]]$ when α is admissible or a limit of admissible ordinals and will from now on denote both by $L_\alpha[x]$.

Lemma 5.2.17 (Cf. [52, 137]). *If α is admissible and x is Cohen-supergeneric over L_α, then $L_\alpha[x]$ is admissible.*

Moreover, if $x \subseteq \omega$, α is x-admissible and y is Cohen-supergeneric over $L_\alpha[x]$, then $(L_\alpha[x])[y]$ is admissible.

Proof. We only prove the unrelativized version, the relativization being straightforward.

Let $G := F_x$. Most axioms of KP are straightforward and left to the reader as an exercise. The only thing that needs some work is Δ_0-collection. So suppose that ϕ is a Δ_0-formula and $q, Y \in M[x]$ such that $M[x] \models \forall y \in Y \exists z \phi(y, z, \vec{q})$. By Lemma 5.2.14, there is a Cohen-condition p such that $p \Vdash \forall \dot{y} \subset \dot{Y} \exists \dot{z} \phi(\dot{y}, \dot{z}, \dot{q})$ over M.

In particular then, for all $(\dot{y}, r) \in \dot{Y}$ with $p \leq r$, there is $\dot{z} \in M$ such that $p \Vdash \phi(\dot{y}, \dot{z}, \dot{q})$. By Lemma 5.2.9, $p \Vdash \phi(\dot{y}, \dot{z}, \dot{q})$ is Σ_1-expressible over M. Thus, by Σ_1-collection in M, there is $Z' \subset M$ such that $\forall \dot{y} \subset \dot{Y} \exists \dot{z} \subset Z' p \Vdash \phi(\dot{y}, \dot{z}, \dot{q})$. Let $Z = (Z')^G$. Then $M[x] \models \forall y \in Y \exists z \in Z \phi(y, z, q)$, which is what we wanted. □

Corollary 5.2.18.
(i) *If x is Cohen-supergeneric over $L_{\omega_1^{\mathrm{CK}}}$, then $\omega_1^{\mathrm{CK},x} = \omega_1^{\mathrm{CK}}$.*
(ii) *If x is Cohen-generic over $L_{\omega_\omega^{\mathrm{CK}}}$, then $\omega_i^{\mathrm{CK},x} = \omega_i^{\mathrm{CK}}$ and $\omega_\omega^{\mathrm{CK},x} = \omega_\omega^{\mathrm{CK}}$. More generally, if α is a limit of admissible ordinals and x is Cohen-generic over L_α, then α is a limit of x-admissible ordinals.*

Proof. (i) By Lemma 5.2.17, $L_{\omega_1^{\mathrm{CK}}}[x]$ is admissible. So $\omega_1^{\mathrm{CK},x} \leq \omega_1^{\mathrm{CK}}$. Clearly, we have $\omega_1^{\mathrm{CK},x} \geq \omega_1^{\mathrm{CK}}$. Thus $\omega_1^{\mathrm{CK},x} = \omega_1^{\mathrm{CK}}$.

(ii) Using (i), one can see inductively that $\omega_i^{\mathrm{CK},x} = \omega_i^{\mathrm{CK}}$ for all $i \in \omega$. The suprema are equal now and follows trivially. The last claim is proved similarly. □

The link between Cohen-genericity and meagerness is given by the following observation.

Lemma 5.2.19 (Folklore). *If M is countable, then the set of real numbers that are Cohen-generic over* Def(M) *is comeager. Moreover, if p is a Cohen-condition, then the set of real numbers that are Cohen-generic over* Def(M) *is not meager.*

Exercise 5.2.20. Prove Lemma 5.2.19.

Below, we will frequently make use of the fact that, if $M = L_\alpha \models \mathrm{KP}$ and $x, y \subseteq \omega$ are such that y is Cohen-supergeneric over $M[x]$ and x is Cohen-supergeneric over $M[y]$, then $M[x] \cap M[y] = M$. To prove this, we need some facts about product forcing. As the only products we will need will be that of Cohen-forcing with itself, which will always be contained in the respective ground model M, we can ignore some technicalities arising otherwise.

Definition 5.2.21 (Cf. [123]). Let $(P, \leq)$ and $(Q, \leq')$ be forcings. Then the product forcing $(P, \leq) \times (Q, \leq)$ is given by $(P \times Q, \leq'')$, where $(p, q) \leq'' (p', q')$ if and only if $p \leq p'$ and $q \leq' q'$ for all $(p, q), (p', q') \in P \times Q$.

Lemma 5.2.22 (Cf. [123], Theorem VIII.1.4). *Let $(P, \leq)$ denote Cohen-forcing.*
(i) *Let α be admissible and let $x, y \subseteq \omega$ be Cohen-supergeneric over L_α. Then y is Cohen-supergeneric over $L_\alpha[x]$ if and only if x is Cohen-supergeneric over $L_\alpha[y]$ if and only if $x \oplus y$ is $P \times P$-supergeneric over L_α. Moreover, in these cases, we have $(L_\alpha[x])[y] = (L_\alpha[y])[x] = L_\alpha[x \oplus y]$.*
(ii) *Let α be a limit of admissible ordinals, and let x, y be Cohen-generic over L_α. Then x is Cohen-generic over $L_\alpha[y]$ if and only if $x \oplus y$ is $P \times P$-generic over L_α if and only if y is Cohen-generic over $L_\alpha[x]$.*

Proof. (i) The proof for forcings over models of ZFC given, e. g., in ([123], Chapter VIII, Theorem 1.4) adapts to this framework.

(ii) It is easy to see that $x \oplus y$ is $P \times P$-generic over L_α if and only if $y \oplus x$ is, so that it suffices to show the equivalence of the first two conditions. Note that, when $\beta < \alpha$, then x and y are Cohen-supergeneric over L_β.

First, let y be Cohen-generic over $L_\alpha[x]$, and let $D \in L_\alpha$ be dense in P. Hence there is an admissible $\beta < \alpha$ such that $D \in L_\beta$. As x and y are both Cohen-supergeneric over L_β, it follows from (i) that $x \oplus y$ is Cohen-supergeneric over L_β, and hence extends some condition $p \in D$. As D was arbitrary, $x \oplus y$ is Cohen-generic over L_α.

Now suppose that $x \oplus y$ is Cohen-generic over L_α; it is easy to see that this implies that x is Cohen-generic over L_α. We need to see that y is Cohen-generic over $L_\alpha[x]$. Let $D \in L_\alpha[x]$ be dense in P. So there is an admissible $\beta < \alpha$ such that $D \in L_\beta[x]$. As x and y are both Cohen-supergeneric over L_β, (i) implies that y is Cohen-supergeneric over $L_\beta[x]$ and thus extends some condition $p \in D$. Since D was arbitrary, y is Cohen-generic over $L_\alpha[x]$. □

Lemma 5.2.23 ([36], Lemma 28). *Let $M \models \mathrm{KP}$, and let $x, y \subseteq \omega$ be such that x is Cohen-supergeneric over M and y is Cohen-generic over $M[x]$. Then $M[x] \cap M[y] = M$.*

Proof. Call F_x, F_y be the filters corresponding to x and y, respectively, and denote by $\mathbb{P}$ the Cohen-forcing. Let $z \in M[x] \cap M[y]$, and let $\dot{z}_0, \dot{z}_1$ be such that $\dot{z}_0{}^{F_x} = \dot{z}_1{}^{F_y} = z$. Also, let $\dot{z}_0{}' := \{(a, (p, 0)) : (a, p) \in \dot{z}_0\}$ and $\dot{z}_1 = \{(b, (0, q)) : (b, q) \in \dot{z}_1\}$. Then $(\dot{z}_0{}')^{F_x \times F_y} = z$; hence, there is a condition $(p, q) \in F_x \times F_y$ such that $(p, q) \Vdash \dot{z}_0{}' = \dot{z}_1{}'$. Moreover, as

M is admissible, $\mathbb{P} \times \mathbb{P} \in M$ and the forcing relation for Δ_0-formulas is Σ_1 over M by Lemma 5.2.9.

Now suppose that there are $(p',q'), (p'',q'') \leq (p,q)$ such that, for some $i \in \omega$, $(p',q') \Vdash \check{i} \notin \dot{z}_0$. Then also $(p'',q'') \Vdash \check{i} \notin \dot{z}_1'$. Moreover, by definition of $\dot{z}_0', \dot{z}_1'$, $(p',0) \Vdash \check{i} \in \dot{z}_0'$, while $(0,q'') \Vdash \check{i} \notin \dot{z}_1'$. Thus $(p',q'') \Vdash \check{i} \in \dot{z}_0' \wedge \check{i} \notin \dot{z}_1'$, but $(p',q'') \leq (p,q) \Vdash \dot{z}_0' = \dot{z}_1'$, a contradiction.

It follows that, for every $i \in \omega$, every two conditions below (p,q) must agree on whether or not $\check{i} \in \dot{z}_0'$. Moreover, by the forcing theorem, for every $i \in \omega$ there is a condition $(p',q') \in F_x \times F_y$ that decides whether $\check{i} \in \dot{z}_0'$. As $F_x \times F_y$ is a filter, there is $(p'',q'') \in F_x \times F_y$ such that $(p'',q'') \leq (p',q')$ and $(p'',q'') \leq (p,q)$. Thus $i \in \dot{z}_0{}^{F_x \times F_y}$ if and only if there is $(p',q') \leq (p,q)$ such that $(p',q') \Vdash \check{i} \in \dot{z}_0'$. But this criterion is clearly Δ_1 in the parameters (p,q) and $\dot{z}_0'$. Thus x is Δ_1 over M, and as $M \models \mathrm{KP}$, we have $x \in M$ by Lemma 3.1.25 from Chapter 3, as desired. □

5.3 Infinitary analogues of Sack's theorem[2]

One should expect that throwing a dart at the unit interval and using the real number hit by the arrow as an oracle is not likely to help, e. g., in solving the halting problem. This intuition is made precise and supported by the following theorem of Sacks.

Theorem 5.3.1 (Cf. Downey and Hirschfeldt [49], Corollary 8.12.2). *Let Y be a set of reals of positive Lebesgue measure. Assume that $x \subseteq \omega$ is a real number and P is a Turing-program such that $P^y(i)\downarrow = \chi_x(i)$ for all $i \in \omega$ and every $y \in Y$, i. e., P computes x relative to any element of Y. Then x is recursive.*

Proof. For $X \subseteq \mathfrak{P}(\omega)$, let $\mu(X)$ denote the Lebesgue measure of X, if defined. By the Lebesgue density theorem (see, e. g., [49], Theorem 1.2.3), there is a finite sequence $t \in \{0,1\}^{<\omega}$ such that $\frac{\mu(\{y \in Y : t \subseteq y\})}{2^{|t|}} > \frac{1}{2}$; that is, x is Turing-reducible to more than half of all strings that extend t. Now let $i \in \omega$. By Corollary 4.2.24 from Chapter 4, there is $n(i) \in \omega$ such that, for all $y \subseteq \omega$, $P^y(i)$ considers at most the first $n(i)$ many bits of y. In particular then, $P^s(i)$ returns $\chi_x(i)$ for more than half of all 0-1-strings of length $n(i)$ that extend t. But now, $n(i)$ can be determined by an exhaustive search and then the finitely many 0-1-strings of length $n(i)$ that extend t can be completely enumerated and tried out; and then $\chi_x(i)$ is determined simply by which of the answers 0 and 1 shows up more often. □

2 The material in this section is adapted from the article "Infinite Computations with Random Oracles" [36] by P. Schlicht and the author, which originally appeared in the Notre Dame Journal of Formal Logic, Vol. 58:2, pages 249–270 ©201 University of Notre Dame. All rights reserved. Republished by permission of the copyright holder, and the present publisher, Duke University Press. We thank Philipp Schlicht and the Notre Dame Journal of Formal Logic for the kind permission to reuse parts of this article here.

Note that the above proof depends crucially on a "compactness" property of classical Turing machines: If the program P halts in the oracle x, then it will only consider finitely many bits of x before it halts. Clearly, this no longer holds for infinitary machine models, since these can run through the whole oracle many times and even uniquely identify certain oracles (see Chapter 4). Still, the idea that a random number should not help in solving an otherwise unsolvable problem seems very plausible also for infinitary machines. So does Sack's theorem still hold for ITRMs, ITTMs or OTMs? This is the question of the current section.

The "moral" message of Sacks' theorem is that, when x is recursive relative to all elements of a "large" sets of real numbers, then x is recursive, and in the above theorem, "large" was understood to mean "of positive Lebesgue measure." Here, we consider analogues where "large" is taken to mean "not meager." Analogues for measure can be obtained, but are only discussed briefly at the end of the chapter, as developing the methods involved would take us too far afield.

5.3.1 Sacks' theorem for wITRMs and ITRMs

Theorem 5.3.2 ([36], Section 4). *Let $Y \subseteq \mathfrak{P}(\omega)$, $x \subseteq \omega$ such that, for all $y \in Y$, we have $x \leq_{\mathrm{wITRM}} y$. Suppose further that Y is not meager. Then x is wITRM-computable.*

Proof. Suppose otherwise. Let C be the set of real numbers that are Cohen-generic over $L_{\omega_1^{\mathrm{CK}}+1}$. By Lemma 5.2.19, C is comeager, and thus must have non-empty intersection with Y. Let $y \in Y \cap C$; it follows that $x \in L_{\omega_1^{\mathrm{CK},y}}[y]$. As $\omega_1^{\mathrm{CK},y} = \omega_1^{\mathrm{CK}}$ by Corollary 5.2.18, $x \in L_{\omega_1^{\mathrm{CK}}}[y]$.

Now consider the set C' of real numbers that are Cohen-generic over $L_{\omega_1^{\mathrm{CK}}+1}[y]$. By Lemma 5.2.19, C' is comeager, and thus has a nonempty intersection with Y. Let $z \in C' \cap Y$. Then, as above, $x \in L_{\omega_1^{\mathrm{CK}}}[z]$. Moreover, by Lemma 5.2.22, y is Cohen-generic over $L_{\omega_1^{\mathrm{CK}}}[z]$. Now Lemma 5.2.23 implies that $L_{\omega_1^{\mathrm{CK}}}[y] \cap L_{\omega_1^{\mathrm{CK}}}[z] = L_{\omega_1^{\mathrm{CK}}}$.

Thus $x \in L_{\omega_1^{\mathrm{CK}}}$, i. e., x is wITRM-computable. □

Expanding this argument a bit, one obtains the analogue for ITRMs.

Theorem 5.3.3 ([36], Theorem. 47). *Let $Y \subseteq \mathfrak{P}(\omega)$, $x \subseteq \omega$ such that, for all $y \in Y$, we have $x \leq_{\mathrm{ITRM}} y$. Suppose further that Y is not meager. Then x is ITRM-computable.*

Proof. For each $y \in Y$, there is an ITRM-program P that computes x from y. For each ITRM-program P, let Y_P be the set of elements y of Y such that P computes x from y. Then Y is the union of the Y_P, and as there are only countably many ITRM-programs, it follows from Corollary 5.2.2 that at least one of the Y_P is not meager. We may thus assume without loss of generality that $Y = Y_P$ for some ITRM-program P. Let P use n many registers. Thus, by Corollary 3.4.9 from Chapter 3, if P computes x in the oracle y, then $x \in L_{\omega_{n+1}^{\mathrm{CK},y}}[y]$.

Let C be the set of real numbers that are Cohen-generic over $L_{\omega_{n+1}^{CK}+1}$; then C is comeager by Lemma 5.2.19, and thus has a nonempty intersection with Y. Pick $y \in Y \cap C$. By Lemma 5.2.18, $\omega_{n+1}^{CK,y} = \omega_{n+1}^{CK}$. By the same argument, there is a real number $z \in Y$ which is Cohen-generic over $L_{\omega_{n+1}^{CK,y}+1}[y]$ and satisfies $\omega_{n+1}^{CK,z} = \omega_{n+1}^{CK}$.

As $y, z \in Y$, P computes x from y and z, so $x \in L_{\omega_{n+1}^{CK}}[y] \cap L_{\omega_{n+1}^{CK}}[z]$, which is $L_{\omega_{n+1}^{CK}}$ by Lemma 5.2.23. Thus in particular, $x \in L_{\omega_{\omega}^{CK}}$, so x is ITRM-computable. □

5.3.2 Sacks' theorem for ITTMs

Sacks' theorem for ITTMs is slightly more complicated due to the fact that the characterizations of the characteristic ordinals λ, ζ and Σ are much more involved than those of ω_1^{CK} and ω_ω^{CK}.

Lemma 5.3.4 (Welch, [179], Lemma 3.3, Coskey and Hamkins [42], Theorem 4.8). *Suppose that α is admissible or a limit of admissible ordinals, let $n \in \mathbb{N}$, $L_\beta \prec_{\Sigma_n} L_\alpha$, and let $x \subseteq \omega$ be Cohen-generic over $L_{\alpha+1}$. Then $L_\beta[x] \prec_{\Sigma_n} L_\alpha[x]$.*

In particular, if x is Cohen-supergeneric over L_Σ, then $L_\lambda[x] \prec_{\Sigma_1} L_\zeta[x] \prec_{\Sigma_2} L_\Sigma[x]$.

Proof. Clearly, the second claim follows from the first.

Let x, α, β be as in the assumption, let $\vec{p} \subseteq L_\beta[x]$, and let ϕ be a Σ_n-formula such that $L_\alpha[x] \models \phi(\vec{p})$. We need to show that $L_\beta[x] \models \phi(\vec{p})$. If $\dot{\vec{p}}$ is a name for $\vec{p}$, it follows from Lemma 5.2.14 that there is a Cohen-condition q such that $q \Vdash \phi(\dot{\vec{p}})$ over L_α; in other words, the statement that $q \Vdash \phi(\dot{\vec{p}})$ holds in L_α.

By Lemma 5.2.9 (and the exercise following it), the statement $q \Vdash \phi(\dot{\vec{p}})$ is expressible as a Σ_n-formula. By assumption then, $q \Vdash \phi(\dot{\vec{p}})$ holds in L_β. But then, again by the forcing theorem, $L_\beta[x] \models \phi(\vec{p})$, as desired. □

Corollary 5.3.5 (Welch [179], Lemma 3.3, Coskey and Hamkins [42], Theorem 4.9). *If x is Cohen-supergeneric over L_Σ, then $\lambda^x = \lambda$, $\zeta^x = \zeta$ and $\Sigma^x = \Sigma$.*

Proof. We recall from Chapter 3 that, for every $x \subseteq \omega$, $(\lambda^x, \zeta^x, \Sigma^x)$ is the lexically minimal triple (α, β, γ) of ordinals such that $L_\alpha[x] \prec_{\Sigma_1} L_\beta[x] \prec_{\Sigma_2} L_\gamma[x]$. Clearly, if $L_\alpha[x] \prec_{\Sigma_1} L_\beta[x] \prec_{\Sigma_2} L_\gamma[x]$, then $L_\alpha \prec_{\Sigma_1} L_\beta \prec_{\Sigma_2} L_\gamma$ (as L_δ is Σ_1-definable in $L_\delta[x]$), so (λ, ζ, Σ) is lexically not larger than $(\lambda^x, \zeta^x, \Sigma^x)$ for any $x \subseteq \omega$.

It thus suffices to show that $L_\lambda[x] \prec_{\Sigma_1} L_\zeta[x] \prec_{\Sigma_2} L_\Sigma[x]$, as this implies that (λ, ζ, Σ) is lexically at least as large as $(\lambda^x, \zeta^x, \Sigma^x)$, but this is just Lemma 5.3.4. □

Exercise 5.3.6. The preservation of the characteristical ITTM-ordinals λ, ζ and Σ under Cohen-forcing can also be proved by a more computational argument due to Coskey and Hamkins, which appeared in [[42]]. This exercise guides you through the proof.

Let P be an ITTM-program that is intended to run in the oracle g, where g is Cohen-generic over $L_{\Sigma+1}$. Suppose that P uses the inner states $\{q_0, \dots, q_n\}$, where q_0 is the initial state, q_1 is the limit state and q_n is the halting state. To simplify things, we assume that P uses only one tape (the argument is easily adapted to the usual 3-tape architecture), so that the input is present as every other bit on the scratch tape. We introduce the following propositional variables: $S_0, \ldots, S_n$, $\{H_i : i \in \omega\}$ and $\{C_i^j : i \in \omega \wedge j \in \{0,1\}\}$. The intended interpretation of S_i is that i the index of the current inner state of the computation of P^g, while H_i signifies that the index of the cell at which the read/write-head is currently located is at least i and C_i^j expresses that the ith cell currently contains the symbol j.

Now, for every propositional variable A just introduced and every finite $p : \omega \to \{0,1\}$, we introduce another propositional variable A_p. In the following stagewise procedure, A_p at stage $\alpha \in \mathrm{On}$ is assigned the truth value 1 if and only if p forces that, at time α, A holds of the computation of $P^{\dot{g}}$ at time α. Thus for all finite $p : \omega \to \{0,1\}$, at stage 0, $S_{0,p} = 1$, $S_{i,p} = 0$ for $i \neq 0$, $H_{0,p} = 1$, $H_{i,p} = 0$ for $i \neq 0$, while $C^0_{2i,p} = 1$ for all $i \in \omega$ and $C^j_{2i+1,p} = 1$ if and only $i \in \mathrm{dom}(p)$ and $p(i) = j$.

We now want to show that the truth values of these propositional variables at every time can be computed on an ITTM-program Q. To this end, we imagine the truth values of the A_p written on the output tape of Q in some natural way.

(a) Show that there is an ITTM-program $P_{\mathrm{forcenext}}$ with the following property: Given as its input a certain truth assignment v to the A_p which describes the configuration of P at time τ, it can compute the corresponding truth assignment at time $\tau + 1$.

(b) Let δ be a limit ordinal. When A_p is one of the propositional variables introduced above, let A'_p be 1 if and only if there is $\alpha < \delta$ such that A_p had truth value 1 at all stages $\beta \in (\alpha, \delta)$ and otherwise let A'_p have truth value 0. Show that there is an ITTM-program $P_{\mathrm{forcelimit}}$ with the following property: Given the A'_p as its input, it can compute the correct values for each A_p at time δ.

(c) Combine (a) and (b) to show that there is an ITTM-program Q with an inner state q' such that, when Q is in state q' for the τth time, then the output tape of Q contains a 1 in the cell corresponding to A_p if and only if p forces that the computation of $P^{\dot{g}}$ satisfies A at time τ and otherwise contains 0.

Let us now assume that $g \subseteq \omega$ is Cohen-generic over $L_{\Sigma+1}$.

(d) Show that, if some proposition A (where A is one of the S_i, H_i or C_i^j) is true of the configuration of P^g at time τ, then there is a finite $p \subseteq g$ such that $A_p = 1$ at stage τ.

(e) Conversely, show that, if $A_p = 1$ at stage τ and $p \subseteq g$, then A holds of the configuration of P^g at time τ.

(f) Use the fact that Q works without oracle together with the preceding exercise parts to show that, from time Σ on, P^g loops, repeating over and again its behavior between time ζ and Σ. Conclude that $\zeta^g = \zeta$ and $\Sigma^g = \Sigma$.

(g) Suppose that $\tau \in \mathrm{On}$ is such that, for some Cohen-condition p, $p \Vdash$ '$P^{\dot{g}}\downarrow$ at time α'. Show that there is some clockable ordinal $> \alpha$. Conclude that α is writable, and thus $\alpha < \lambda$. Conclude further that $\lambda^g = \lambda$.

Theorem 5.3.7 ([36], Theorem 35). *Suppose that $Y \subseteq \mathfrak{P}(\omega)$ is not meager, and let $x \subseteq \omega$.*

(i) *If x is ITTM-writable relative to all $y \in Y$, then x is ITTM-writable.*
(ii) *If x is eventually ITTM-writable relative to all $y \in Y$, then x is eventually ITTM-writable.*
(iii) *If x is accidentally ITTM-writable relative to all $y \in Y$, then x is accidentally ITTM-writable.*

Proof. Let C be the set of all real numbers that are Cohen-generic over $L_{\Sigma+1}$. By Lemma 5.2.19, C is comeager, and thus has a nonempty intersection with Y. Let $y \in Y \cap C$; by the same argument, there is a real number $z \in Y$ which is Cohen-generic over $L_{\Sigma+1}[y]$.

By Corollary 5.3.5, we have $\lambda^y = \lambda^z = \lambda$, $\zeta^y = \zeta^z = \zeta$ and $\Sigma^y = \Sigma^z = \Sigma$. Consequently, we have $L_{\lambda^y}[y] = L_\lambda[y]$, $L_{\lambda^z}[z] = L_\lambda[z]$, $L_{\zeta^y}[y] = L_\zeta[y]$, $L_{\zeta^z}[z] = L_\zeta[z]$, $L_{\Sigma^y}[y] = L_\Sigma[y]$, $L_{\Sigma^z}[z] = L_\Sigma[z]$.

By Lemma 5.2.22 and Lemma 5.2.23, we have $L_\lambda[y] \cap L_\lambda[z] = L_\lambda$, $L_\zeta[y] \cap L_\zeta[z] = L_\zeta$ and $L_\Sigma[y] \cap L_\Sigma[z] = L_\Sigma$.

Now, if x is ITTM-writable relative to y and z (which follows from $y, z \in Y$), then $x \in L_\lambda[y] \cap L_\lambda[z] = L_\lambda$, so x is ITTM-writable.

Similarly, as x is eventually ITTM-writable relative to y and z, then $x \in L_\zeta[y] \cap L_\zeta[z] = L_\zeta$, so x is eventually ITTM-writable; and finally, if x is accidentally ITTM-writable relative to y and z, then $x \in L_\Sigma[y] \cap L_\Sigma[z] = L_\Sigma$, so x is accidentally ITTM-writable. □

Theorem 5.3.8. *Let α be multiplicatively closed, $x \subseteq \omega$. If x is α-ITTM-writable/ew/aw relative to all $y \in Y \subseteq \mathfrak{P}(\omega)$ and Y is not meager, then x is α-ITTM-writable/ew/aw.*

Exercise 5.3.9. Prove Theorem 5.3.8 by adapting the proof of Theorem 5.3.7.

Exercise 5.3.10. Show that, if X is ITTM-decidable and does not contain an ITTM-computable element, then $\lambda^x > \lambda$ for all $x \in X$. Conclude that X is meager.

Assuming without proof that $\{x \subseteq \omega : \lambda^x = \lambda\}$ has measure 1, show that X is also of measure 0.

5.3.3 Analogues for OTMs

We now consider Sacks' theorem for ordinal Turing machines with and without parameters.

5.3.3.1 Without parameters

In the constructible hierarchy, the intuition that arbitrary oracles should not add information fails in the strongest possible way: In fact, all but countably many real numbers allow it to solve the halting problem.

Theorem 5.3.11 ([36], Theorem 10). *Suppose that $V = L$. Then there is a co-countable (hence, in particular, comeager and measure 1) set X of real numbers such that $0'_{\mathrm{OTM}}$ is OTM-computable (without parameters) from every element of X.*

Proof. Let $x \in L$ be a real number that is not OTM-computable. We claim that $0'_{\mathrm{OTM}}$ is OTM-computable from x. As there are only countably many OTM-programs, and hence only countable many OTM-computable real numbers, it will then follow that the set of real numbers that are not OTM-computable is as desired.

To compute $0'_{\mathrm{OTM}}$ from x, we proceed as follows: By Lemma 3.5.3 from Chapter 3, there is an OTM-program that writes codes for every element of the constructible hierarchy to its output tape. Use this program, and test for each output whether it codes an L-level L_α that contains x. If not, the search is continued. Eventually, the search will terminate; say that (a code for) L_α is written on the output tape at this point.

Now, as x is not OTM-computable, we have $x \notin L_\sigma$, so we must have $\alpha > \sigma$. Thus α is larger than the supremum of the parameter-free OTM-halting times. It is now rather easy to determine, given $i \in \omega$, whether the ith OTM-program P_i will halt: Namely, if it does, it does so in $< \sigma$ many steps, so that the halting computation will be contained in L_σ and hence in L_α. On the other hand, if L_α contains a halting computation for P_i, then P_i actually halts by absoluteness of computations. It thus suffices to use the ability of OTMs to evaluate truth predicates (see Chapter 3) to check whether $L_\alpha \models P_i\downarrow$. □

Exercise 5.3.12. Exploit the above proof to show that the assumption $V = L$ can be replaced by the much weaker assumption that $\mathbb{R} = \mathbb{R}^L$.

The above observation implies that it is consistent with ZFC that the analogue of Sacks' theorem fails for OTMs without parameters. However, we will now see that the converse is also compatible with ZFC. The proof of this will follow the example of the Sacks analogue for ITTMs. There, the first thing to note was that, for x sufficiently generic, the "characteristic ordinals" of ITTMs, namely λ, ζ and Σ are preserved. For OTMs without parameters, the "characteristic ordinal" is σ.

Lemma 5.3.13 ([36], Lemma 8). *If x is Cohen-generic over L, then $\sigma^x = \sigma$.*

Proof. Suppose that P is an OTM-program such that P^x halts. Then there is a Cohen-condition $p \subseteq x$ such that $\exists \alpha p \Vdash$ '$P^{\dot{x}}$ halts in less than $\check{\alpha}$ many steps'. That P^x halts in less than $\check{\alpha}$ many steps is a Σ_1-statement, and so Lemma 5.2.9 above implies that so is the statement that p forces this. Hence, the whole statement that there is an α for which this holds is Σ_1 as well. As this statement holds in L, it holds in L_σ by definition. Thus, there is $\beta < \sigma$ such that $p \Vdash$ '$P^{\dot{x}}$ halts in less than $\check{\alpha}$ many steps'. By Lemma 5.2.14, it holds in $L_\sigma[x]$ that P^x halts in $< \beta$ many steps.

Thus, for all OTM-programs P such that P^x halts, this halting time is $< \sigma$, so $\sigma^x \leq \sigma$. But $\sigma \leq \sigma^x$ is obvious, and so we have $\sigma = \sigma^x$, as desired. □

Lemma 5.3.14 ([36], Section 2.1). *If x, y and z are real numbers such that x is OTM-computable from both y and z and y and z are mutually Cohen-generic over L, then x is OTM-computable.*

Proof. Clearly, y and z are in particular mutually generic over $L_{\sigma+1}$ and L_σ is admissible by Proposition 3.2.3 of Chapter 3. It thus follows from Lemma 5.2.23 above that $L_\sigma[y] \cap L_\sigma[z] = L_\sigma$. Moreover, as x is OTM-computable from y and z, we have $x \in L_{\sigma^y}[y] \cap L_{\sigma^z}[z] = L_\sigma[y] \cap L_\sigma[z] = L_\sigma$. Thus x is OTM-computable. □

Corollary 5.3.15 ([36], Section 2.1). *It is independent of ZFC whether any $x \subseteq \omega$ that is OTM-computable relative to all elements of a comeager set is OTM-computable.*

5.3.3.2 With parameters

With parameters, OTM-writability ("pOTM-writability") coincides with constructibility by Theorem 3.5.5 in Chapter 3. Similarly, relative OTM-writability with parameters corresponds to relative constructibility, a rather well-studied subject.

Proposition 5.3.16. *Suppose that $V = L$. Then every real number that is pOTM-writable relative to all elements of a nonmeager set Y is pOTM-writable.*

Proof. This holds trivially, as any real number in L is pOTM-writable by Theorem 3.5.5, Chapter 3. □

Showing that the analogue of Sacks' theorem can also fail requires a bit more work, and we skip the proof.

Lemma 5.3.17 ([36], Section 2.3). *There is a forcing $\mathbb{P} \in L$ such that, if G is $\mathbb{P}$-generic over L, then $L[G] \setminus L$ contains a real number x with the following properties:*
(i) *If $y \subseteq \omega$ is in $L[G] \setminus L$, then $x \in L[y]$*
(ii) *$\mathfrak{P}^{L[G]}(\omega) \setminus \mathfrak{P}^L(\omega)$ is comeager in $L[G]$.*

Consequently, in $L[G]$, x is an example of a real number which is not pOTM-computable, but pOTM-computable relative to all elements of a comeager set.

Proof. The forcing notion known as "Laver forcing" [see, e. g., Gray [70]] has property (i), which is shown in [70], and also property (ii), which is Theorem 7.3.28 of Bartoszynski and Judah [8]. □

5.3.3.3 Axioms from randomness?

One approach for finding new axioms to supplement ZFC is to find plausible statements that are independent of ZFC and add them. We note here that the intuition that one should not be able to extract useful information like the OTM-halting problem from a randomly chosen string is a rather strong one, so that the analogues of Sacks' theorem for OTMs with and without parameters are candidates for such statements. As was in Theorem 5.3.11, the former in particular refutes $V = L$.

5.4 Randomness and recognizability[3]

We will now combine the motive of Sacks' theorem with that of recognizability. More specifically, we ask: If x is recognizable relative to many other real numbers, does it follow that x is recognizable? This was the main topic of [25], where several instances of this question where answered positively.

Theorem 5.4.1 ([20]). *Let P be an ITRM-program such that P^z halts for every $z \subseteq \omega$ with output 0 or 1; furthermore, let $x \subseteq \omega$ and $Y \subseteq \mathfrak{P}(\omega)$ comeager such that, for all $y \in Y$ and all $z \subseteq \omega$, $P^{z \oplus y}\downarrow = \delta_{xz}$. Then x is ITRM-recognizable.*

Proof. Suppose that P uses k many registers.

The following is the crucial observation that allows us to replace the talk about the arbitrary comeager set Y with considerations of appropriate generics:

Claim. *Let $a \subseteq \omega$, and let $g \subseteq \omega$ be Cohen-generic over $L_{\omega_{k+1}^{\mathrm{CK},a}+1}[a]$. Then $P^{a \oplus g}\downarrow = \delta_{ax}$.*

To see this, first note that $\omega_{k+1}^{\mathrm{CK},a\oplus g} = \omega_{k+1}^{\mathrm{CK},a}$ by Lemma 5.2.18. As P^z halts for all $z \subseteq \omega$, $P^{a\oplus g}$ will halt, and hence do so in $< \omega_{k+1}^{\mathrm{CK},a\oplus g} < \omega_{\omega}^{\mathrm{CK},a}$ many steps. Now, if $P^{a\oplus g}\downarrow = i$, then this is a statement that holds in $L_{\omega_{k+1}^{\mathrm{CK},a}}[a \oplus g]$ and is hence forced by some Cohen-condition $p \subseteq g$ over $L_{\omega_{k+1}^{\mathrm{CK},a}+1}[a]$. It follows that $P^{a\oplus g'}\downarrow = i$ holds for every $g' \supseteq p$ which is Cohen-generic over $L_{\omega_{k+1}^{\mathrm{CK},a}+1}[a]$; as the set of these g' is not a meager set by Lemma 5.2.19, its intersection Z with Y must be nonempty; say it contains $\bar{g}$. Then $P^{a\oplus\bar{g}}\downarrow = \delta_{ax}$ since $\bar{g} \in Y$; on the other hand, the output of $P^{a\oplus\bar{g}}$ is the same as that of $P^{a\oplus g}$ by the choice of $\bar{g}$. It follows that $P^{a\oplus g}\downarrow = \delta_{ax}$, as desired.

By the claim, we can, for every $z \subseteq \omega$, use any Cohen-generic real number over $L_{\omega_{k+1}^{\mathrm{CK},z}+1}[z]$ to determine whether $z = x$. Thus, if such a real number was uniformly ITRM-computable from z, we could use z to compute the "extra information" necessary to decide whether $z = x$. However, computing such a generic real number from z is rather easy: First, given z, compute a code c for $L_{\omega_{k+2}^{\mathrm{CK},z}}[z]$, which is possible by

3 This section is adapted from the articles "Infinite Time Recognizability from Random Oracles and the Recognizable Jump Operator" [20], which appeared in Computability, vol. 5, no. 2, pages 181–196 ©2017, with permission from IOS Press, available at IOS Press through http://dx.doi.org/10.3233/COM-160061 and 'ITRM-Recognizability from Random Oracles' [22], which appeared in: A. Beckmann, V. Mitrana, M. Soskova (eds.), Evolving Computability, LNCS 9136 ©2015 with permission from Springer Nature. We thank Springer Nature and IOS Press for the kind permission to reuse parts of these articles here.

Theorem 3.4.31 of Chapter 3. Then a Cohen-generic real number over $L_{\omega_{k+2}^{\mathrm{CK},z}}[z]$ is ITRM-computable from z by Lemma 5.2.5. □

This theorem is the weakest variant of an analogue of Sack's theorem for ITRM-recognizability. Various strengthenings are possible.

Exercise 5.4.2. Show that, if $Y \subseteq \mathfrak{P}(\omega)$ is not meager and $x \subseteq \omega$ is such that, for some ITRM-program P, all $y \in Y$ and all $z \subseteq \omega$, $P^{z\oplus y}\downarrow = \delta_{xz}$, then x is ITRM-recognizable.

It is much harder to show the analogous statement for ITTMs, as the solvability of the bounded halting problem is absent in this case. It still holds, however, as was originally proved in [25]. The proof we give below is a simplified version, due to Philipp Schlicht (see [38]).

Theorem 5.4.3 (Cf. [25, 38]). *Let P be an ITTM-program, $x \subseteq \omega$ and $Y \subseteq \mathfrak{P}(\omega)$ comeager such that, for all $y \in Y$ and all $z \subseteq \omega$, $P^{z\oplus y}\downarrow = \delta_{xz}$. Then x is ITTM-recognizable.*

Proof. Let $z \subseteq \omega$ be given in the oracle.

Claim 1. *If g is Cohen-generic over $L_{\Sigma^z+1}[z]$, then $P^{z\oplus g}\downarrow = \delta_{xz}$ in $< \lambda^z$ many steps.*

To see this, first note that, by Corollary 5.3.5 above, if g is Cohen-generic over $L_{\Sigma^z+1}[z]$, then $\lambda^{z\oplus g} = \lambda^z$; as $P^{z\oplus g}\downarrow$ by the assumption on P, it will halt in $< \lambda^z$ many steps. Now, if $P^{z\oplus g}\downarrow = i \neq \delta_{xz}$, then, by Lemma 5.2.14, there is a condition p such that $p \Vdash P^{z\oplus \dot{g}}\downarrow = i$ over $L_{\Sigma^z+1}[z]$. It follows that $P^{z\oplus g'}\downarrow = i$ for every $g' \supseteq p$ that is Cohen-generic over $L_{\Sigma^z+1}[z]$; since $L_{\Sigma^z+1}[z]$ is countable, the set S of these g' is not meager and thus intersects Y. Let $g' \in S \cap Y$. Then $P^{z\oplus g'}\downarrow = i$ since $g' \in S$, and $P^{z\oplus g'}\downarrow = \delta_{xz}$ as $g' \in Y$. But this contradicts the assumption that $i \neq \delta_{xz}$. Hence the claim is proved.

Claim 2. *There is a triple $(L_{\alpha+1}[z], g, p) \in L_{\lambda^z}[z]$ such that $\alpha < \lambda^z$ is admissible, $g \subseteq \omega$ is generic over $L_{\alpha+1}[z]$ and $p \Vdash P^{z\oplus g}\downarrow$ over $L_{\alpha+1}[z]$. Moreover, for any such triple, we have that $P^{z\oplus g}\downarrow = \delta_{xz}$.*

Let g be Cohen-generic over $L_{\Sigma^z+1}[z]$; by Claim 1, the halting time τ of $P^{z\oplus g}$ is $< \lambda^z$. Let α be the smallest z-admissible ordinal $> \tau$, then $\alpha < \lambda^z$ since λ^z is a limit of z-admissible ordinals by Theorem 3.6.6 in Chapter 3. Then g is also generic over $L_{\alpha+1}[z]$, hence, by Lemma 5.2.14, there is a condition $p \subseteq g$ such that $p \Vdash P^{z\oplus \dot{g}}\downarrow$ over $L_{\alpha+1}[z]$. Pick $g' \subseteq p$ generic over $L_{\alpha+1}[z]$, then $(L_{\alpha+1}[z], g', p)$ is as desired.

For the second claim, suppose for a contradiction that $P^{z\oplus g}\downarrow = i \neq \delta_{xz}$. As in the proof of Claim 1, there is a condition q such that $q \Vdash P^{z\oplus \dot{g}}\downarrow = i$ over $L_{\alpha+1}[z]$, hence $P^{z\oplus g'}\downarrow = i$ for every $g' \supseteq q$ that is generic over $L_{\alpha+1}[z]$. But again, as $L_{\alpha+1}[z]$ is countable, the set S of these g' is not meager, and thus intersects Y; now we obtain a contradiction as in the proof of Claim 1.

The procedure for recognizing x now works as follows: Given z, let the universal ITTM $\mathcal{U}$ run in the oracle z, searching for a triple $(L_{\alpha+1}[z], g, p)$ satisfying the conditions from Claim 2. Such a triple is easily identifiable by an ITTM using well-foundedness tests and evaluation of truth predicates, and moreover contained in $L_{\lambda^z}[z]$ by Claim 2. Thus, it will eventually be found.

Once this is accomplished, run $P^{z\oplus g}$. By the second part of Claim 2, it will return δ_{xz}. Thus x is ITTM-recognizable. □

5.5 Some results on ITRM-genericity[4]

Intuitively, the sequence 1000111010 looks more "random" or more "arbitrary" than 1111111111 or 1010101010 (cf., e. g., Terwijn [173]). Probabilistically, however, each of these sequences has the same probability of showing up as the result of a sequence of 10 coin tosses. So how can we get precise grasp on the difference?

One attempt is the definition of Martin-Löf-randomness; see, e. g., Downey and Hirschfeldt [49]: Roughly, a real number x (i. e., an infinitary 0-1-sequence) is ML-random if and only if there is an "effective small set" $X \ni x$; more precisely, X is a set of Lebesgue measure 0 that can effectively be approximated by some Turing program.

This way of formalizing the idea that a random string x should not have any 'special properties' can be transferred to infinitary machines rather straightforwardly.

Definition 5.5.1. Let M be a machine type, such as ITRM, ITTM or OTM. A real number x is M-random if and only if there is no M-decidable set X of Lebesgue measure 0 such that $X \ni x$.

A real number x is M-generic if and only if there is no M-decidable meager set X such that $X \ni x$.

ITRM-genericity was studied in [22] and [23]. This third section of this chapter will give some results on this notion.

Definition 5.5.2. A real number x is ITRM-generic if and only if there is no ITRM-decidable meager set $X \ni x$.

One should not expect a random string to be informative. In particular, if y is random, the halting problem should not be solvable relative to y. For classical Turing machines and the notion of ML-randomness, however, this can happen: By the Kucera–Gacs-theorem (see [49], Theorem 8.3.2), every real number is Turing-reducible to some

4 This section is adapted from the articles "Algorithmic Randomness for Infinite Time Register Machines" [20], in: A. Beckmann, E. Csuhaj-Varju, K. Meer (eds.), Language, Life, Limits, LNCS 8493 ©Springer Nature 2014 with permission from Springer Nature and "Randomness and Degree Theory for Infinite Time Register Machines" [23], which appeared in Computability, vol. 5, no. 2, pages 181–196, ©IOS Press (2016), available at IOS Press through http://dx.doi.org/10.3233/COM-160055 with permission from IOS Press. We thank Springer Nature and IOS Press for the kind permission to reuse parts of these articles here.

ML-random real number. It can be seen as a virtue of ITRM-genericity that this cannot happen:

Theorem 5.5.3 ([22, 23]). *Let x be ITRM-generic. Then $0'_{\mathrm{ITRM}}$ is not ITRM-computable from x.*

Proof. Suppose otherwise, and let P be an ITRM-program such that P^x computes $0'_{\mathrm{ITRM}}$. By Corollary 2.3.13 in Chapter 2, we may assume without loss of generality that P is safe. By Theorem 4.3.18 from Chapter 4, let Q be an ITRM-program that recognizes $0'_{\mathrm{ITRM}}$. Now consider the following routine: Given $y \subseteq \omega$, run P in the oracle y, then run Q on the output (recall that, as ITRMs cannot store real numbers, what is really going on here is that Q is run, and whenever Q calls the oracle, P is used to determine the answer). This routine decides a set Y of real numbers y with the property that P computes $0'_{\mathrm{ITRM}}$ from y. By assumption, we have $x \in Y$. Moreover, by Theorem 5.3.3, Y is meager. Thus x is contained in an ITRM-decidable meager set, which contradicts the assumption that it is ITRM-generic. □

Exercise 5.5.4. More generally, show that, if x is a lost melody for ITRMs and y is ITRM-generic, then x is not ITRM-computable from y.

Definition 5.5.5. Let $x, y \subseteq \omega$. x is ITRM-generic relative to y if and only if there is no meager set $X \ni x$ such that X is ITRM-decidable in the oracle y. If x is ITRM-generic relative to y and y is ITRM-generic relative to x, then x and y are mutually ITRM-generic.

Intuitively, mutually generic objects should not share any nontrivial information. Again, this fails for ML-randomness and Turing computability, but it holds for ITRM-genericity.

Theorem 5.5.6 ([22], Theorem 11, [23]). *Let $y, z \subseteq \omega$ be mutually ITRM-generic and suppose that $x \leq_{\mathrm{ITRM}} y$ and $x \leq_{\mathrm{ITRM}} z$. Then x is ITRM-computable.*

Proof. Fix two ITRM-programs P and Q such that P^y and Q^z compute x. We may assume without loss of generality that P and Q are safe by Corollary 2.3.13 from Chapter 2.

Now, given y in the oracle, we can use P and Q to determine the set of all real numbers u such that Q^u computes x by simply applying Q to u and comparing the result to the result of applying P to y. So $Z := \{u \subseteq \omega : P^y(i)\downarrow\ = Q^u(i)\downarrow\}$ is ITRM-decidable relative to y. Moreover, as Q^z computes x by assumption, we have $z \in Z$.

But now, if x is not ITRM-computable but ITRM-computable relative to all elements of Z, it follows that Z is meager by Theorem 5.3.3. So z is contained in a meager set that is ITRM-decidable relative to y, and hence not ITRM-generic relative to y, a contradiction. Thus x must be ITRM-computable after all. □

We can also give a machine-independent characterization of ITRM-genericity.

Theorem 5.5.7 ([22], Corollary 16). *$x \subseteq \omega$ is ITRM-generic if and only if x is Cohen-generic over $L_{\omega_\omega^{\mathrm{CK}}}$.*

Proof. Suppose first that x is Cohen-generic over $L_{\omega_\omega^{CK}}$. Assume for a contradiction that x is not ITRM-generic and pick an ITRM-program P that decides a meager set $X \ni x$, and let P use n many registers. It follows from Lemma 5.2.18 that $\omega_{n+1}^{CK,x} = \omega_{n+1}^{CK}$, so that $P^x\downarrow$ in $< \omega_{n+1}^{CK}$ many steps. Hence $L_{\omega_{n+1}^{CK}}[x] \models P^x\downarrow = 1$. By the forcing theorem (Lemma 5.2.14), there is a Cohen condition p such that $p \Vdash P^{\dot{x}}\downarrow = 1$ over $L_{\omega_{n+1}^{CK}}$; hence, if $y \supseteq p$ is Cohen-generic over $L_{\omega_\omega^{CK}}$, then $P^y\downarrow = 1$. But by Lemma 5.2.19, the set of these y is not meager, so the set decided by P is not meager after all, a contradiction.

On the other hand, suppose that x is ITRM-generic, but not Cohen-generic over $L_{\omega_\omega^{CK}}$. Hence, there is a dense subset D of Cohen forcing such that finite $p \subseteq x$ is contained in D. As the set of Cohen-generics over $L_{\omega_\omega^{CK}}$ is comeager and all these extend some $q \in D$, the set of real numbers y with the property that no finite $p \subseteq x$ is contained in D is meager. Moreover, this set clearly contains x. To obtain a contradiction to the ITRM-genericity of x, it thus suffices to show that the set is ITRM-decidable. This can be seen as follows:

Let $m \in \omega$ be such that $D \in L_{\omega_m^{CK}}$, then there is an ITRM-computable real number c that codes $L_{\omega_m^{CK}}$ via some bijection $f : \omega \to L_{\omega_m^{CK}}$. Thus, there is some $i \in \omega$ such that $f(i) = D$. By Corollary 2.3.31 from Chapter 2, there is an ITRM-program Q such that Q^c determines, on input $p \subseteq \omega$ finite, some $j \in \omega$ with $f(j) = p$. Now, for a given real number y in the oracle, consider the ITRM-program P' that runs through all finite subsets p of ω, checking for each whether (1) $p \subseteq y$ and (2) whether $p \in D$; (2) can be done in the oracle c by first using P_{find} to determine the natural number coding p in the sense of c and then applying P_{truth}. If (1) or (2) fails, the next p is considered. If (1) and (2) both hold, the program halts with output 0. If all finite $p \subseteq \omega$ have been considered, the program halts with output 1. □

Exercise 5.5.8. Show that a real number x is ITRM-generic relative to a real number y if and only if x is Cohen-generic over $L_{\omega_\omega^{CK,y}}[y]$.

Exercise 5.5.9. Let us say that $x \subseteq \omega$ is wITRM-generic if and only there is no wITRM-decidable meager set $X \ni x$. Show that, if x is Cohen-supergeneric over $L_{\omega_1^{CK}}$, then x is wITRM-generic.

It is rather plausible that, for two "random" objects x and y, if x contains no information about y, then y contains no information about x; after all, a rough intuition tells us that, if y had information about x, then x would at least have information about that part of y that represents the information about x. In classical computability theory, this intuition is captured by van Lambalgen's theorem: If x is ML-random and y is ML-random relative to x, then $x \oplus y$ is also ML-random; as the reverse direction also holds (which is easy to see), it follows that x is also ML-random relative to y. For ITRMs, we have the following analogue.

Theorem 5.5.10 ([22], Theorem 17, [23]). *Let $x, y \subseteq \omega$. Then $x \oplus y$ is ITRM-generic if and only if x is ITRM-generic and y is ITRM-generic relative to x.*

Proof. By Theorem 5.5.7, the ITRM-genericity of x, y and $x \oplus y$ is equivalent to the Cohen-genericity of x, y and $x \oplus y$ over $L_{\omega_\omega^{\mathrm{CK}}}$, respectively. But then, the theorem statement translates into Lemma 5.2.22(ii). □

Exercise 5.5.11. In the proof above, we used general properties of forcing to prove van Lambalgen's theorem for ITRMs. In line with our general politics to stay as "computational" as possible, we feel the duty to also give a proof imitating the classical proof of van Lambalgen (see, e. g., [49], Corollary 6.9.3; the proof below first appeared in [22]). Let $x, y \subseteq \omega$. For the sake of this exercise, denote by $D(P, x)$ the set $\{y \subseteq \omega : P^{x \oplus y}\downarrow = 1\}$, where $x \subseteq \omega$ and P is an ITRM-program.

(a) Show that, if $x \oplus y$ is ITRM-generic, then x and y are ITRM-generic.

Let us assume that $x \oplus y$ is ITRM-generic, but y is not ITRM-generic relative to x. Let P be an ITRM-program that, in the oracle y, decides a meager set X containing x. Without loss of generality, we may assume that P is safe. Let n be the number of registers used by P.

(b) Let $z \subseteq \omega$, and let a be Cohen-generic over $L_{\omega_{n+1}^{\mathrm{CK},z}}[z]$ such that $P^{z \oplus a}\downarrow = 1$. Show that $D(P, z)$ is not meager.

(c) Conversely, show that, if $P^{z \oplus a}\downarrow = 0$ for every real number a that is Cohen-generic over $L_{\omega_{n+1}^{\mathrm{CK},z}}[z]$, then $D(P, z)$ is meager.

(d) Show that, in fact, $D(P, z)$ is meager r if and only if $P^{z \oplus a}\downarrow = 0$ for every real number a that is Cohen-generic over $L_{\omega_{n+1}^{\mathrm{CK},z}}[z]$ and contained in $L_{\omega_{n+2}^{\mathrm{CK},z}}[z]$.

(e) Use (d) to show that the set of $z \subseteq \omega$ for which $D(P, z)$ is meager is ITRM-decidable.

(f) Show that there is an ITRM-program P' such that for all $z \subseteq \omega$, if $D(P, z)$ is meager, then $D(P, z) = D(P', z)$, while if $D(P, z)$ is not meager, then $D(P', z) = \emptyset$.

(g) Show that P' decides a meager set that contains $x \oplus y$. Conclude that $x \oplus y$ is not ITRM-generic.

For the reverse direction, suppose that x is ITRM-generic and y is ITRM-generic relative to y. We assume that $x \oplus y$ is not ITRM-generic and work toward a contradiction. Let $X \ni x \oplus y$ be a meager and ITRM-decidable set, and let P be an ITRM-program that decides X.

(h) Show that the set S of $z \subseteq \omega$ for which $D(P, z)$ is not meager is ITRM-decidable.

We recall the Kuratowski–Ulam-theorem (see, e. g., [99], Theorem 8.41 or [143], exercise 5A.9): If $Z \subseteq [0, 1]$ is meager, then the set of $x \subseteq \omega$ for which $\{y : x \oplus y \in Z\}$ is meager is a comeager set.

(i) Use the Kuratowski–Ulam-theorem to show that S is meager.

(j) Show that, if $x \in S$, then x is not ITRM-generic.

(k) Show that, if $x \notin S$, then y is not ITRM-generic relative to y, contradicting our assumption. Conclude that the reverse direction holds as well.

5.6 OTMs and genericity

So far, we saw that Cohen-genericity over $L_{\omega_1^{\mathrm{CK}}+1}$ implies wITRM-genericity and that ITRM-genericity is equivalent to Cohen-genericity over $L_{\omega_\omega^{\mathrm{CK}}}$. Thus, there is a clear correspondence between genericity in the sense of forcing and genericity in the sense of infinitary machine models of computation. We supplement the picture on this correspondence by considering parameter-OTMs.

Theorem 5.6.1. *A real number $x \subseteq \omega$ is Cohen-generic over L if and only if it is pOTM-generic, i. e., not an element of a meager set that is decidable by an OTM with ordinal parameters.*

Proof. Suppose for a contradiction that x is Cohen-generic over L and that P is an OTM-program, $\xi \in \mathrm{On}$ such that $P(\xi)$ decides a meager set $X \ni x$. Thus $P^x(\xi)\downarrow = 1$, and hence, there is a condition $p \subseteq x$ such that $p \Vdash P^x(\xi)\downarrow = 1$ over L. Then the set of all $\tilde{x}$ that are Cohen-generic over L and extend p will be nonmeager and each such $\tilde{x}$ will satisfy $P^{\tilde{x}}(\xi)\downarrow = 1$, so that $P(\xi)$ does not decide a meager set, a contradiction.

In the other direction, if x is not Cohen-generic over L, then there is a dense set $D \in L$ such that x does not extend any element of D. Now, as $D \in L$, there are an OTM-program P and some $\xi \in \mathrm{On}$ such that $P(\xi)$ decides D (where the elements of D are coded as natural numbers in some natural manner). But then, the set of real numbers y that extend no element of D is meager, pOTM-decidable and contains x, so x is not pOTM-generic. □

5.7 Further results

We mention some further results, in particular, concerning the analogues of some of the above results when positive Lebesgue measure rather than non-meagerness is used as the underlying notion of largeness.

All variants of Sacks' theorem that we saw above also hold true for sets of oracles of positive Lebesgue measure rather than nonmeager sets.

Theorem 5.7.1 (Cf. Carl and Schlicht, [36]). *Let $x \subseteq \omega$, $Y \subseteq \mathfrak{P}(\omega)$ such that Y has positive Lebesgue measure.*
- *If x is wITRM-computable relative to every $y \in Y$, then x is wITRM-computable.*
- *If x is ITRM-computable relative to every $y \in Y$, then x is ITRM-computable.*
- *It is independent of ZFC whether the parameter-free OTM-writability of x relative to every $y \in Y$ implies the parameter-free OTM-writability of x.*
- *It is independent of ZFC whether the OTM-writability of x relative to every $y \in Y$ implies the OTM-writability of x.*

5.7.1 ITTMs and randomness

Finally, we mention some results ITTM-randomness that are analogous to those on ITRM-genericity we saw above. The forcing taking over the role of Cohen-forcing is random forcing (see, e. g., [92], Chapter 15); however, in contrast to Cohen forcing, the preorder of random forcing over an admissible set is in general a proper class, which complicates things and requires some modifications to the forcing technique due to Philipp Schlicht, which were done in [38]. Here, we only give some of the main results without proofs.

The notions and results in this subsection come from [38]. For a wealth of further results on ITTMs and randomness, see Monin and Angles D'Auriac [142].

Let us first note that we have the analogues of Sacks' theorem in this setting.

Theorem 5.7.2 (Cf. [38], Theorem 3.12, 3.16). *Let $x \subseteq \omega$, and let $Y \subseteq \mathfrak{P}(\omega)$ have positive Lebesgue measure.*
(i) *If x is ITTM-writable relative to all $y \in Y$, then x is ITTM-writable.*
(ii) *If x is eventually ITTM-writable relative to all $y \in Y$, then x is eventually ITTM-writable.*
(iii) *If x is accidentally ITTM-writable relative to all $y \in Y$, then x is accidentally ITTM-writable.*
(iv) *If x is ITTM-recognizable relative to all $y \in Y$, then x is ITTM-recognizable.*

Due to the solvability of the bounded halting problem, decidability and semidecidability coincide for ITRMs. For ITTMs, this is no longer the case, and thus, we have two notions of randomness.

Definition 5.7.3. Let $x, y \subseteq \omega$.

x is ITTM-random relative to y if and only if there is no $\mathfrak{P}(\omega) \supseteq X \ni x$ of Lebesgue measure 0 that is ITTM-semidecidable relative to y.

x is ITTM-decidably random relative to y if and only if there is no $\mathfrak{P}(\omega) \supseteq X \ni x$ of Lebesgue measure 0 that is ITTM-decidable relative to y.

With these notions, the analogues of the results that made ITRM-genericity appealing still hold true.

Theorem 5.7.4 ([38], Theorem 4.6 and Lemma 4.13). *Let $x, y, z \subseteq \omega$ such that $x \leq_{\mathrm{ITTM}} y$ and $x \leq_{\mathrm{ITTM}} z$.*
(i) *If y and z are mutually ITTM-random, then x is ITTM-writable.*
(ii) *If x is ITTM-recognizable and y is ITTM-random, then x is ITTM-writable.*

Finally, we have an analogue of van Lambalgen's theorem.

Lemma 5.7.5 ([38], Theorem 4.10). *Let $x, y \subseteq \omega$ be ITTM-decidably random. Then x is ITTM-decidably random relative to y if and only if y is ITTM-decidably random relative to x if and only if $x \oplus y$ is ITTM-decidably random.*

6 Degree theory

6.1 Preliminaries

Naively, any theory of computation will lead to a distinction between two kinds of objects: those that are computable and those that are not. However, there is a clear intuition that, among the noncomputable objects, some are "harder" than others. For example, one would certainly want to say that the double jump $0''$ is even "more remote from being computable" than the single jump $0'$.

This can be made precise by considering relativized computability: If x and y are two objects such that x is computable from y (or "x reduces to y") but not vice versa, then y is "harder" than x in the sense of the concept of computability considered. Likewise, if x is computable from y and y is computable from x, then x and y have the same hardness and we call them "equivalent," writing $x \equiv y$. For most sensible concepts of computability, the relation just described is indeed an equivalence relation (in particular, it is transitive), so that we may form equivalence classes. These equivalence classes are called the "degrees" of the respective theory of computability. In all cases that we will consider, the question whether x reduces to y only depends on the degrees of x and y, so that reducibility can be studied on the level of degrees without loss of granularity.

Once these concepts are defined, several questions come up rather naturally: Is "hardness," as one might conjecture, a linear ordering or are there "incomparable" problems x and y such that neither is reducible to the other? Are there incomputable objects that are strictly easier to compute than the jump? In general, what does the reducibility order look like? Is it dense, i. e., is there a degree strictly between x and y whenever x reduces to y, but not vice versa? These and other questions are studied in degree theory. In this chapter, we will consider them for transfinite models of computability.

6.2 Degree theory for ITTMs

The question whether there are incomputable real numbers strictly below $0'$ in the sense of classical Turing-reducibility is known as Post's problem and has been answered in the positive by Friedberg and Muchnik (see, e. g., Downey and Hirschfeldt [49], p. 32 or Theorem VI.1.2 of Soare [167]) with a proof that became the birthplace of priority constructions. Degree theory for ITTMs yields a rather different outcome and in general deviates from the classical theory in various respects. It was worked out mostly by Hamkins and Lewis in [77] and by Welch in [179], [180] and [193]. Our treatment will be brief. For many further results, we refer to these sources, where the following notions and results can also be found.

https://doi.org/10.1515/9783110496154-006

Definition 6.2.1. Let $x, y \subseteq \omega$.

We say that x is ITTM-reducible to y if and only if there is an ITTM-program P such that $P^x\downarrow = y$. In this case, we write $x \leq_{\text{ITTM}} y$. If $x \leq_{\text{ITTM}} y$ and $y \leq_{\text{ITTM}} x$, we write $x \equiv_{\text{ITTM}} y$ and say that x and y are ITTM-equivalent.

x is ITTM-eventually reducible to y if and only if there is an ITTM-program P such that P^y eventually writes x. In this case, we write $x \leq_{\text{ev}} y$.

Furthermore, we have a reducibility relation for sets of real numbers: If $X, Y \subseteq \mathfrak{P}(\omega)$, then X is ITTM-reducible to Y, written $X \leq^*_{\text{ITTM}} Y$, if and only if there is an ITTM-program P such that P^Y decides X.

$x \equiv_{\text{ev}} y$ and $X \equiv^* Y$ are now defined from $\leq_{\text{ev}}$ and $\leq^*$ as $\equiv_{\text{ITTM}}$ was from $\leq_{\text{ITTM}}$.

Proposition 6.2.2.
(i) $\leq_{\text{ITTM}}$, $\leq_{\text{ev}}$ *and* $\leq^*_{\text{ITTM}}$ *are transitive. In particular, when* $x, y \subseteq \omega$ *such that* y *is eventually writable and* $x \leq_{\text{ev}} y$, *then* x *is eventually writable.*
(ii) *If* x *and* y *are real numbers such that* $x \leq_{\text{ITTM}} y$, *then* $x \leq_{\text{ev}} y$.

Proof. (i) See Theorem 2.5.10 in Chapter 2.
(ii) Trivial. □

Definition 6.2.3. For $x \subseteq \omega$, $[x]_{\text{ITTM}}$ denotes $\{y \subseteq \omega : y \equiv_{\text{ITTM}} x\}$, while $[x]_{\text{ev}}$ denotes $\{y \subseteq \omega : y \equiv_{\text{ev}} x\}$ and for $X \subseteq \mathfrak{P}(\omega)$, $[X]^*_{\text{ITTM}}$ is $\{Y \subseteq \mathfrak{P}(\omega) : Y \equiv^* X\}$. By Proposition 6.2.2, $\mathfrak{P}(\omega)$ decomposes into disjoint equivalence classes with respect to $\equiv_{\text{ITTM}}$ and $\equiv_{\text{ev}}$, respectively, while $\mathfrak{P}(\mathfrak{P}(\omega))$ decomposes into equivalence classes with respect to $\equiv^*$.

An interesting observation is that all accidentally writable real numbers that are not eventually writable have the same ITTM-degree. As a consequence, ITTM-writability from an ITTM-aw real number does not imply accidental ITTM-writability.

Theorem 6.2.4 (Hamkins, Lewis [77]).
(i) *Let* x, y *be real numbers that are accidentally ITTM-writable, but not eventually ITTM-writable. Then* $x \equiv_{\text{ITTM}} y$.
(ii) *The accidentally ITTM-writable real numbers are not closed under* $\leq_{\text{ITTM}}$.

Proof. (i) Let x be accidentally ITTM-writable, but not eventually ITTM-writable. We claim that Σ is writable in x. By Lemma 3.4.26 in Chapter 3, it then follows that a code for L_Σ is writable in x, and hence so is every element of L_Σ. Thus every accidentally ITTM-writable real number is ITTM-reducible to x.

To see the claim, consider the following procedure: Let the level machine P_{level} from Theorem 3.6.15 in Chapter 3 run until a code $c \subseteq \omega$ for an L-level $L_\alpha \ni x$ is written to the output tape. Whether the output is of this form can be determined using P_{find} from Chapter 2. Since $x \in L_\alpha \setminus L_\zeta$, we have $\alpha > \zeta$. Thus, the configuration c_ζ of the universal ITTM-program $\mathcal{U}$ at time ζ is coded in c by some natural number i that can be given to the program as a parameter in advance.

Recall from Corollary 3.6.37 in Chapter 3 that C_ζ first appears at time ζ in the computation of $\mathcal{U}$ and that it first reappears at time Σ. It follows that, given c_ζ, we can clock Σ by simply running $\mathcal{U}$ and waiting for c_ζ to appear for the second time. By Corollary 3.6.20 from Chapter 3, Σ is ITTM-writable relative to c_ζ and hence relative to x.

(ii) Let x be accidentally ITTM-writable, but not eventually ITTM-writable. By (i), $\mathrm{cc}(\Sigma)$ is ITTM-writable in x, but $\mathrm{cc}(\Sigma)$ is not accidentally ITTM-writable, as it is the supremum of the accidentally ITTM-writable ordinals. □

Part (ii) of the last theorem seems a bit counterintuitive. After all, it seems rather easy to prove as follows that the ITTM-aw real numbers are closed under ITTM-writability: Suppose that y is aw, and let $x \leq_{\mathrm{ITTM}} y$; say P is an ITTM-program such that $P^y\downarrow = x$. Now we use the following procedure to accidentally write x: Run the universal ITTM $\mathcal{U}$. Whenever $\mathcal{U}$ has produced a new real number a on one of the simulated output tapes, run P^a and write the result to the output tape, then go on with the run of $\mathcal{U}$. Eventually, $\mathcal{U}$ will produce y, at which point x will be written to the output tape. Thus x is aw. (It is worthwhile to pause a moment to reflect what is wrong with this argument.)

The problem is that, though $P^y\downarrow = x$, P^a may not halt for some aw real number a. In this case, the above procedure will eventually stop to produce outputs and P might never be given the opportunity to run on the oracle y. In fact, part (ii) of the above theorem tells us that this actually must be the case for some aw y, some real number $x \leq_{\mathrm{ITTM}} y$ and any program P that computes x from y. The above idea would, however, work, if we could be sure that P^a halts for every $a \subseteq \omega$.

Definition 6.2.5. An ITTM-program P is "safe" if and only if P^a halts for all $a \subseteq \omega$. For real numbers x, y, x is "safely ITTM-computable from y," written $x \leq_{\mathrm{sITTM}} y$ if and only if there is a safe ITTM-program P such that $P^y\downarrow = x$.

Exercise 6.2.6. Show that, if $x \subseteq \omega$ is ITTM-aw and P is a safe ITTM-program such that P^x writes $y \subseteq \omega$, then y is ITTM-aw.

Corollary 6.2.7. *There are real numbers x, y such that $x \leq_{\mathrm{ITTM}} y$, but $x \nleq_{\mathrm{sITTM}} y$.*

Research Question. Develop the degree theory of safe ITTM-reducibility.

We are ready to attack Post's problem for ITTMs.

Proposition 6.2.8. *$0'_{\mathrm{ITTM}}$ is eventually writable. Hence $0'_{\mathrm{ITTM}} \in L_\zeta$.*

Proof. As the halting times of ITTM-programs are bounded by λ, $0'_{\mathrm{ITTM}}$ is definable over L_λ, so that $0'_{\mathrm{ITTM}} \in L_{\lambda+1} \subseteq L_\zeta$. (For a more algorithmic argument, run the universal ITTM and writing a 1 in the ith cell of the output tape when P_i halts. After running λ many steps, all programs that halt at all have halted (recall that λ is the supremum of the ITTM-clockable ordinals), so from this time on, the output tape will contain $0'_{\mathrm{ITTM}}$.) □

Lemma 6.2.9. *If x is eventually writable, but not writable, then $0'_{\text{ITTM}} \leq_{\text{ITTM}} x$.*

Proof. Let x be eventually writable, but not writable. Then $x \in L_\zeta$, and so there is an eventually writable ordinal ξ such that $x \in L_\xi$. Since x is not writable, we must have $\xi > \lambda$.

We claim that a code for such an L_ξ is ITTM-writable in the oracle x. To see this, run P_{level} from Chapter 3 and test for each output whether it codes an L-level L_α containing x, which is possible using P_{find} from Chapter 2. Once the answer is "yes," output the code for L_α on the output tape of P_{level}.

Now, since $\xi > \lambda$ and λ majorizes the clockable ordinals, we have that $L_\xi \models P_i\downarrow$ if and only if $L_\lambda \models P_i\downarrow$ if and only if P_i halts. The statement $P_i\downarrow$ is expressible as a first-order statement ψ_i, and from i, we can (Turing-)compute ψ_i. Hence, to decide whether $j \in 0'_{\text{ITTM}}$ for some $j \in \omega$, it suffices to compute (a code for) some L_ξ as above, and then decide, using P_{truth} from Chapter 2, whether $L_\xi \models \psi_i$. □

Theorem 6.2.10 (Hamkins, Lewis [77]). *There is no real number x such that $0 <_{\text{ITTM}} x <_{\text{ITTM}} 0'_{\text{ITTM}}$.*

Proof. By Proposition 6.2.8, $0'_{\text{ITTM}}$ is eventually writable. As $x \leq_{\text{ITTM}} 0'_{\text{ITTM}}$, we have $x \leq_{\text{ev}} 0'_{\text{ITTM}}$ by Proposition 6.2.2, hence, again by Proposition 6.2.2, x is eventually writable.

Now, if x is not writable, then, by Lemma 6.2.9, $0'_{\text{ITTM}} \leq_{\text{ITTM}} x$, contradicting our assumption that $x <_{\text{ITTM}} 0'_{\text{ITTM}}$. So x must be writable, but this contradicts the assumption that $0 <_{\text{ITTM}} x$. □

This argument can be extended.

Exercise 6.2.11 (Cf. Hamkins and Lewis, [77]). Let $0^{(1)}_{\text{ITTM}} = 0'_{\text{ITTM}}$, $0^{(n+1)}_{\text{ITTM}} = (0^{(n)}_{\text{ITTM}})'_{\text{ITTM}}$ for $n \in \omega$.

(a) Show (e. g., by induction) that, if $n \in \omega$, then $0^{(n)}_{\text{ITTM}} \in L_\sigma$.

(b) Show that, for all $n \in \omega$, we have $\lambda^{0^{(n)}}_{\text{ITTM}} < \zeta$.

(c) Show that, for all $n \in \omega$, we have $0^{(n)}_{\text{ITTM}} \in L_{\lambda^{0^{(n)}_{\text{ITTM}}}}$. Conclude that $L_{\lambda^{0^{(n)}_{\text{ITTM}}}}[\lambda^{0^{(n)}_{\text{ITTM}}}] = L_{\lambda^{0^{(n)}_{\text{ITTM}}}}$.

(d) Suppose that $0^{(n)}_{\text{ITTM}} <_{\text{ITTM}} x \leq_{\text{ITTM}} 0^{(n+1)}_{\text{ITTM}}$. Show that $x \in L_{\zeta^{0^{(n+1)}_{\text{ITTM}}}} \setminus L_{\lambda^{0^{(n)}_{\text{ITTM}}}}$.

(e) Suppose that $0^{(n)}_{\text{ITTM}} <_{\text{ITTM}} x \leq_{\text{ITTM}} 0^{(n+1)}_{\text{ITTM}}$. Show that some ordinal $\alpha > \lambda^{0^{(n)}_{\text{ITTM}}}$ is ITTM-clockable in the oracle x. Conclude that $\lambda^{0^{(n)}_{\text{ITTM}}}$ is ITTM-writable in the oracle x.

(f) Show that, if $0^{(n)}_{\text{ITTM}} <_{\text{ITTM}} x \leq_{\text{ITTM}} 0^{(n+1)}_{\text{ITTM}}$, then $x \equiv_{\text{ITTM}} 0^{(n+1)}_{\text{ITTM}}$. Conclude that there are no ITTM-degrees strictly between $0^{(n)}_{\text{ITTM}}$ and $0^{(n+1)}_{\text{ITTM}}$, for $n \in \omega$.

Considering sets of real numbers as oracles instead, the answer to Post's problem becomes positive, however.

Definition 6.2.12 (Cf. [81]). For $X \subseteq \mathfrak{P}(\omega)$, let X^*_{ITTM} denote the set $\{(i, y) : y \subseteq \omega, i \in \omega, P^X_i(y)\downarrow\}$, where P_i is the ith ITTM-program.

Theorem 6.2.13 (Hamkins, Lewis [77]). *There is a set A of real numbers such that* $\emptyset <^* A <^* \emptyset^*_{\text{ITTM}}$.

Proof. The proof follows the priority method of the original solution to Post's problem. It can be found in [77]. □

This positive solution opens the door for further degree-theoretical considerations about ITTMs concerning **sets** of reals rather than reals. We mention the following example:

In classical degree theory, we know that there are real numbers whose Turing-degree lies strictly between those of 0 (recursiveness) and the halting problem $0'$. A natural question to ask is then: Given such a real number x, how does its jump x' relate to $0'$? Clearly, x' is at least as informative as $0'$, since, if we want to know whether a program P halts in the empty oracle, we can (if necessary) effectively replace P with a program P' that works like P, but automatically answers every oracle call with 0. Then P^0 will halt if and only if $(P')^x$ halts, and this information is contained in x'. A natural conjecture would be that x' is in fact strictly stronger in terms of computational power than $0'$, i. e., that x' is not computable from $0'$. After all, how is $0'$ supposed to know what happens to programs in the oracle x? However, this is not true: By a result of Jockusch and Soare (see, e. g., Theorem 2.19.8 of [49]), every nonempty Π^0_1-set of real numbers contains a real number of 'low' degree for Turing machines, i. e., a real number x strictly between 0 and $0'$ in the sense of Turing reducibility that satisfies $x' = 0'$. A notion of lowness for ITTMs was considered by Hamkins and Lewis in [77].

Definition 6.2.14. A set A of real numbers A is 'ITTM-low' if and only if $A^*_{\text{ITTM}} \equiv^* \emptyset^*_{\text{ITTM}}$.

Theorem 6.2.15 (Hamkins, Lewis [77], Corollary 2.3). *If A is a set of real number such that* $A <_{\text{ITTM}} \emptyset^*_{\text{ITTM}}$, *then A is ITTM-low.*

Another important notion in classical degree theory is that of minimality: A nonrecursive real number x is "minimal" if and only if there is no nonrecursive real number y that strictly reduces to x. The same can be considered for ITTMs.

Definition 6.2.16. An ITTM-degree d is 'minimal' if every $x <_{\text{ITTM}} y$ is ITTM-writable.

Thus, as we just demonstrated, $0'_{\text{ITTM}}$ is minimal, so there is a minimal ITTM-degree; however, there are many more.

Theorem 6.2.17 (Welch, see [179]). *There are continuum many pairwise incomparable minimal ITTM-degrees, i. e., continuum many real numbers that are ITTM-minimal and such that none of them is ITTM-reducible to another.*

Proof. The proof uses methods that are beyond the scope of our exposition. We thus refer the reader to (Welch, [179], Theorem 3.9). □

6.2.1 Eventually writable degrees

Since ITTMs are associated with three notions of computability, namely writability, eventual writability and accidental writability, one potentially obtains three notions of reducibility and thus of degrees. We saw above that accidental writability does not lead to a very satisfying theory. This leaves us with eventual writability, which we briefly explore in this section. For a detailed treatment, we refer to Coskey and Hamkins ([42]) and Welch ([180] and [179]).

Definition 6.2.18. For $x, y \subseteq \omega$, if neither $x \leq_{\text{ev}}$ nor $y \leq_{\text{ev}} x$, then we call x and y **eventually incomparable**.

Lemma 6.2.19 (Coskey, Hamkins [42], Theorem 4.8). *Suppose that the real numbers x and y are mutually Cohen-generic over $L_{\Sigma+1}$. Then x and y are eventually incomparable.*

Proof. By genericity of x and y and Corollary 5.3.5 from Chapter 5, we have $\zeta^x = \zeta^y = \zeta$. Furthermore, by Lemma 5.2.23 from Chapter 5, we have $L_\zeta[x] \cap L_\zeta[y] = L_\zeta$. As neither $x \in L_\zeta$ nor $y \in L_\zeta$, it follows that neither $x \leq_{\text{ev}} y$ nor $y \leq_{\text{ev}} x$. □

Definition 6.2.20. A set X of real numbers is **perfect** if and only if it is closed and has no isolated points.

For T a binary tree (see Definition 4.2.21, Chapter 4), we denote by $[T]$ the set of all infinite paths through T. A tree T is **perfect** if and only if, for every $s \in T$, there are $t, t' \in T$ which both extend s, but none of which is an initial segment of the other.

Proposition 6.2.21 (Cf., e. g., [135], Proposition 2.11). *If T is a perfect tree, then $[T]$ is a perfect set.*

Theorem 6.2.22 (Welch, [179], [180]). *There is a nonempty perfect set of mutually Cohen-generic real numbers over $L_{\Sigma+1}$.*

Proof. Let $(D_i : i \in \omega)$ enumerate the dense subsets of the notion of Cohen-forcing contained in $L_{\Sigma+1}$ and let us construct a perfect binary tree T recursively as follows: We start with the root T_0. In step $i \in \omega$, we have constructed a finite binary tree T_i. For every path p through T that ends with a terminal node $n(p)$ of T, pick some extension $q \supseteq p$ with $q \in D_i$. Then extend p to q to obtain a finite binary tree T_i' all of whose branches will extend some element of D_i; no extend T_i' to T_{i+1} by extending each terminal node of T_i' with 0 and with 1 and let $T = \bigcup_{i\in\omega} T_i$. Thus T contains a branching node above each node and is hence perfect.

Clearly, $[T]$ is a nonempty set, and by Proposition 6.2.21, it is perfect. We show that each $x \in [T]$ is Cohen-generic over $L_{\Sigma+1}$. So let $D \in L_{\Sigma+1}$ be a dense subset of Cohen-forcing. Thus $D = D_i$ for some $i \in \omega$. Let $\bar{x}$ be the initial segment of x contained in T_i. Then every branch of T_i' extends some element of D_i by definition of T, and thus the same holds for x. As D was arbitrary, x is indeed Cohen-generic over $L_{\Sigma+1}$. □

Exercise 6.2.23 (Cf. Mansfield and Weitkamp [135], Proposition 2.13). Show that a perfect set is either empty or of cardinality $2^{\aleph_0}$.

Theorem 6.2.24 (Welch, [179]; also cf. Coskey, Hamkins [42], Theorem 4.8). *There is an uncountable perfect set X of real numbers such that, for all $x, y \in X$, x and y are $\leq_{\text{ev}}$-incomparable.*

Proof. This is now a direct consequence of Theorem 6.2.22 and Lemma 6.2.19. □

One of the main open questions in ITTM-degree theory is the following.

Question 6.2.25 (Welch, [179], [180], [193]). Does every countable set of (ev-)degrees have a $\leq_{\text{ev}}$-minimal upper bound?

6.3 Degree theory for OTMs and ORMs

Under the assumption of constructibility, the degree theory for OTMs is heavily simplified by the fact that the degree hierarchy is linear.

Lemma 6.3.1 (C. and Schlicht, [36]). *Suppose that $V = L$, and let $x, y \subseteq \omega$. Then $x \leq_{\text{OTM}} y$ if and only if $\sigma^x \leq \sigma^y$.*

Proof. Given $x \subseteq \omega$ in the oracle, a real number coding an L-level containing x is OTM-writable without parameters by running the OTM-program for enumerating L (see Chapter 3, Lemma 3.5.3) and waiting for level to turn up that contains x. Thus $x \in L_{\sigma^x}$ for each $x \in L \cap \mathfrak{P}(\omega)$.

Now, if $\sigma^x \leq \sigma^y$, then $x \in L_{\sigma^x} \subseteq L_{\sigma^y}$, so $x \in L_{\sigma^y} \subseteq L_{\sigma^y}[y]$, so $x \leq_{\text{OTM}} y$ by the relativization of Theorem 3.5.5 in Chapter 3.

On the other hand, if $x \leq_{\text{OTM}} y$, then clearly $\sigma^x \leq \sigma^y$. □

Corollary 6.3.2 ([36]). *Suppose that $V = L$, and let $x, y \subseteq \omega$. Then x is parameter-freely OTM-computable from y or y is parameter-freely OTM-computable from x.*

For OTMs, a similar argument as for ITTMs shows that there are no intermediate reals, and in fact, no intermediate subsets of any $\alpha < \sigma$.

Theorem 6.3.3 (Hamkins and Miller, [140]). *Let $\alpha < \sigma$, $x \subseteq \alpha$ such that $x <_{\text{OTM}} 0'_{\text{OTM}}$. Then x is OTM-computable.*

Proof. Since $0'_{\text{OTM}} \in L$, we have $x \in L$. Assume that x is not OTM-computable. Then $x \notin L_\sigma$. Hence, if γ is minimal such that $x \in L_\gamma$, we have $\gamma > \sigma$.

We show that $0'_{\text{OTM}} \leq_{\text{OTM}} x$, contradicting the strict reducibility assumption. To see this, let x be given on an extra tape and use the program from Lemma 3.5.3, Chapter 3, to enumerate all constructible sets of ordinals. Whenever a new set S of ordinals is written, check, using P_{truth} from Chapter 3, whether S codes an L-level L_β that contains x. If not, continue.

Since $x \in L$, this routine will eventually come up with a code C for some $L_\beta \ni x$. Now, as $\beta > \sigma$, an OTM-program P halts if and only if $L_\beta \models P\downarrow$. We can hence solve the halting problem for OTMs simply by evaluating the truth predicate in C. □

In fact, this argument can be exploited to yield much stronger conclusion.

Theorem 6.3.4. *When x is a constructible real number, then there is no real number y with $x <_{\text{OTM}} y <_{\text{OTM}} x'_{\text{OTM}}$.*

Proof. Relativize the argument above. □

However, when we consider subsets of ordinals that are not writable, we get incomparable sets, i. e., a positive solution to Post's problem, as proved in Hamkins and Miller [140].[1] We give the proof as a paradigmatic example of the use of a priority construction in ordinal computability.

Theorem 6.3.5 (Hamkins, Miller [140]). *There are OTM-incomparable, OTM-enumerable $x, y \subseteq \sigma$.*

Proof. We will produce two sets A and B of ordinals such that, for all OTM-programs P_i, there are OTM-writable ordinals α_i, β_i such that either $P_i^A(\alpha_i)\uparrow$ or $P_i^A(\alpha_i)\downarrow = 0$ while $\alpha_i \in B$ or $P_i^A(\alpha_i)\downarrow \neq 0$ while $\alpha_i \notin B$ and, furthermore, that either $P_i^B(\beta_i)\uparrow$ or $P_i^B(\beta_i)\downarrow = 0$ while $\beta_i \in A$ or $P_i^B(\beta_i)\downarrow \neq 0$ while $\beta_i \notin A$. In other words, we have, for every $i \in \omega$, to fulfill the following requirements:

$\mathcal{R}_i$: There is an ordinal α_i such that $P_i^A(\alpha_i)\uparrow$ or $P_i^A(\alpha_i)\downarrow = 0$ while $\alpha_i \in B$ or $P_i^A(\alpha_i)\downarrow \neq 0$ while $\alpha_i \notin B$

$\mathcal{S}_i$: There is an ordinal β_i such that $P_i^B(\beta_i)\uparrow$ or $P_i^B(\beta_i)\downarrow = 0$ while $\beta_i \in A$ or $P_i^B(\beta_i)\downarrow \neq 0$ while $\beta_i \notin A$

Note that we have dropped the "writability" condition for α_i and β_i. We will now give a method for OTM-enumerating A and B satisfying all of these criteria; we will then show that we get the writability of all elements of A and B for free.

For the enumeration method, we order the requirements by letting $\mathcal{R}_i, \mathcal{S}_i \geq \mathcal{R}_j, \mathcal{S}_j$ whenever $i \geq j$ and $\mathcal{R}_i > \mathcal{S}_i$ for all $i, j \in \omega$. The interpretation of the ordering is that requirements that are higher up in the ordering have higher 'priority' than those lower down.

The procedure now works as follows: At each time τ, an ordinal ρ_i is associated with $\mathcal{R}_i$ and an ordinal σ_i with $\mathcal{S}_i$; these ordinals will be called the "candidate ordinals" of the requirements; the starting value of σ_i and ρ_i will be i, and they will be updated

1 Actually, in [140], Hamkins and Miller took up the technical challenge of giving a direct solution to Post's problem on ORMs and then used this to solve the OTM-case. Due to the limited storing capabilities of the register models, this is much trickier than working with OTMs. Since our intention is merely to illustrate the general idea of priority constructions in the transfinite context, we give below a technically much less sophisticated adaptation of their argument to OTMs.

in such a way that $(\rho_i : i \in \omega)$ and $(\sigma_i : i \in \omega)$ will always form strictly increasing sequences. Moreover, σ_i and ρ_i will be ordinals of the form $p(i, \alpha)$, so that no requirement can ever assume a candidate ordinal that before was a candidate ordinal for a requirement with a different index. (The idea of the candidate ordinals is that they provide the current best guess at an ordinal that is rejected by a relativized P_i-computation in the oracle A or B and can hence be enumerated in the respective other set in order to fulfill the corresponding requirement.)

Furthermore, we will have approximations A_τ and B_τ to A and B at time τ. These approximations will behave monotonically, i. e., we will have $A_{\iota_1} \subseteq A_{\iota_2}$ and $B_{\iota_1} \subseteq B_{\iota_2}$ whenever $\iota_1 \leq \iota_2$; at limit times λ, we will have $A_\lambda = \bigcup_{\iota<\lambda} A_\iota$ and $B_\lambda = \bigcup_{\iota<\lambda} B_\iota$.

Now, by a suitable organization of the tape, we will simultaneously for all $i \in \omega$ simulate the computations of $P_i^{A_\tau}(\rho_i)$ and $P_i^{B_\tau}(\sigma_i)$. Whenever A_τ, B_τ, ρ_i or σ_i changes, the simulations start over with the new values. More specifically, one simulation step will consist in carrying out one computation step in each computation.

Furthermore, for each of these simulations, we will have an extra tape whose content represents the current space-usage of that computation. This is done by simply writing a 1 in cell α of the extra tape whenever the read/write-head on the other tape is at place α; the extra tape will then, at each point of time, contain a 1 exactly in those cells that have been used by the computation on the other tape. The candidate ordinals will be updated in such a way that ρ_i will always dominate the space-usage of all computations $P_j^{B_\tau}(\sigma_j)$ that have halted so far for $j < i$; and σ_i will always dominate the space-usage of $P_j^{A_\tau}(\rho_j)$ that have halted so far for $j \leq i$. At each time, every requirement will have either the status "satisfied" or "unsatisfied"; initially, this status will be "unsatisfied" for all requirements.

We say that the requirement $\mathcal{R}_i$ that is currently not satisfied "requires attention" at a certain stage τ of this simulation if and only if $P_i^{A_\tau}(\rho_i)\downarrow$ and no requirement of higher priority satisfies the same condition. Let us say that the computation has only used the first μ many tape cells; we then either have $P_i^{A_\tau}(\rho_i)\downarrow = 0$ or $P_i^{A_\tau}(\rho_i)\downarrow \neq 0$. In the first case, we enumerate ρ_i into B, replace σ_j with $p(j, \mu + \sigma_j + j)$ for all $j \geq i$, change the status of $\mathcal{R}_i$ to "satisfied" and that of any requirement $\mathcal{S}_k$ with $k \geq i$ to "unsatisfied." In the second case, we replace σ_j with $p(j, \mu + \sigma_j + j)$ for all $j \geq i$, change the status of $\mathcal{R}_i$ to "satisfied" and that of any requirement $\mathcal{S}_k$ with $k \geq i$ to "unsatisfied." In both cases, we say that $\mathcal{R}_i$ has "acted."

Similarly, $\mathcal{S}_i$, if currently not satisfied, "requires attention" at stage τ if and only if $P_i^{B_\tau}(\beta)\downarrow$ and no requirement of higher priority satisfies the same condition. Let us again say that the computation has only used the first μ many tape cells; we either have $P_i^{B_\tau}(\sigma_i)\downarrow = 0$ or $P_i^{B_\tau}(\sigma_i)\downarrow \neq 0$. In the first case, we enumerate σ_i into A, replace ρ_j with $p(j, \mu + \rho_j + j)$ for all $j > i$, change the status of $\mathcal{S}_i$ to "satisfied" and that of any requirement $\mathcal{R}_k$ with $k > i$ to "unsatisfied." In the second case, we replace ρ_j with $p(j, \mu + \rho_j + j)$ for all $j > i$, change the status of $\mathcal{S}_i$ to "satisfied" and that of any requirement $\mathcal{R}_k$ with $k > i$ to "unsatisfied." In both cases, we say that $\mathcal{S}_i$ has "acted."

This finishes the description of the algorithm.

An important observation is the following: Say $\mathcal{R}_i$ has acted, because $P_i^{A_\tau}(\rho_i)$ has halted. Then the computation, having used only μ many cell tapes, has at most used the first many bits of A_τ. The point in changing the candidate ordinals σ_j for $j \geq i$ into $\mu + \sigma_j + j$ is that this ensures that no further ordinal will be enumerated into A that affects this computation; hence, if $\mathcal{R}_i$ remains satisfied from this point on (i. e., no requirement of higher priority acts afterwards), the computation of $P_i^A(\rho_i)$ will be exactly the same as that of $P_i^{A_\tau}(\rho_i)$.

Claim 1. *Each requirement status is only changed finitely many times.*

Proof. A requirement only acts when it is unsatisfied; once satisfied, its status can only change back to "unsatisfied" when some requirement of higher priority acts, i. e., changes its status. Now, $\mathcal{R}_0$ will change its status at most once. The rest is a matter of induction: Given that all requirements of higher priority than, say, $\mathcal{S}_k$, only change their status finitely often, the same must hold for $\mathcal{S}_k$ itself—it changes its status at most once for each status change of a higher priority requirement, possibly plus one more initial time. □

Claim 2. *There is a stage γ such that no further elements will be enumerated into A and B after γ.*

Proof. New elements are only enumerated into A or B when a requirement acts. But by Claim 1, every requirement acts only finitely many times. Consequently, if γ is the supremum of the ordinal times at which some requirement acts, no further elements are added after time γ. (As a supremum of a countable sequence of ordinals, γ is clearly an ordinal.) □

Claim 3. *With γ as in Claim 2, the approximations A_γ and B_γ at stage γ are OTM-incomparable.*

Proof. Assume otherwise; without loss of generality, suppose that A_γ is OTM-reducible to B_γ, say by program P_i. In particular, this means that one of the following two cases occurs (where ρ_i is the candidate ordinal for $\mathcal{R}_i$ at time γ):

Case 1: $P_i^{A_\gamma}(\rho_i)\downarrow = 0$ and $\rho_i \notin B_\gamma$.

The case assumption implies that, unless $\mathcal{R}_i$ is already satisfied, it will eventually act and enumerate ρ_i into B, so that the procedure will indeed add an element to B_γ later on, which contradicts the definition of B_γ. (On the other hand, if $\mathcal{R}$ is satisfied, our observation above guarantees that P_i^A cannot compute B.)

Case 2: $P_i^{A_\gamma}(\rho_i)\downarrow \neq 0$ and $\rho_i \in B_\gamma$.

By the case assumption, ρ_i has at some point of the construction been enumerated into B_γ. However, ρ_i, which is of the form $p(i, \alpha)$, could only have been enumerated into B by an acting of $\mathcal{R}_i$, say at time τ. But this means that we had $P_i^{A_\tau}(\rho_i)\downarrow = 0$; now, if $\mathcal{R}_i$

had remained satisfied from this point on, the relevant part of A would have remained unchanged and we would have $P_i^A(\rho_i)\downarrow = 0$ as well; but if $\mathcal{R}_i$ would have changed its status to “unsatisfied” again afterwards, the value of ρ_i would have increased. In both cases, we have a contradiction. □

It remains to see that A_γ and B_γ are indeed subsets of σ, i. e., have only (OTM-)writable elements.

Claim 4. *All ordinals enumerated in this process are writable. Thus, $A_\gamma, B_\gamma \subseteq \sigma$.*

Proof. To see this, recall an ordinal is only enumerated when one of the requirements acts, and each requirements acts only a finite number of times. Now say that α is enumerated by the ith acting of the jth requirement in the priority ordering. Then α can be OTM-computed by simply letting the above procedure run until the jth requirement acts for the ith time and see what (if anything) it enumerates. A real code for α can then be written by enumerating L (see Lemma 3.5.3 in Chapter 3) and testing for all occurring reals whether they code α. □

Now let $A = A_\gamma$ and $B = B_\gamma$. Taken together, we have shown that A and B are OTM-enumerable, OTM-incomparable subsets of σ, as desired. □

Remark 6.3.6. The above proof is a prototypical use of a priority construction in infinitary computability theory. Priority constructions have been used for other machine models as well, such as α-Turing machines (see [112]). The details on these would take us beyond the scope of our exposition. We have given this argument as an example of how these arguments work.

Corollary 6.3.7 (Hamkins, Miller [140]). *There is a subset $X \subseteq \sigma$ such that $0 <_{\mathrm{OTM}} X <_{\mathrm{OTM}} 0'_{\mathrm{OTM}}$.*

Proof. We claim that the set $A \subseteq \sigma$ constructed in the proof of Theorem 6.3.5 is of the desired kind. First, A is clearly not OTM-decidable, since otherwise, it would be OTM-reducible to B.

Second, relative to the OTM-halting problem, we can decide for any given $\alpha < \sigma$ whether $\alpha \in A$ as follows: Run the universal OTM until some program P halts after writing some real code for α (it can easily be checked whether some written real number codes α). Now consider the program Q that first runs P to write a code for α, then enumerates A as described in the proof of Theorem 6.3.5 and, whenever a new element has been enumerated, checks whether it is equal to α, in which case it halts; otherwise, it keeps running. Clearly, Q halts if and only if α is eventually enumerated into A, i. e., if $\alpha \in A$. Hence A is decidable relative to the halting problem, so $A \leq_{\mathrm{OTM}} 0'_{\mathrm{OTM}}$.

To see that we must in fact have $A < 0'_{\mathrm{OTM}}$, it suffices to observe that the above argument also works to show that $B \leq_{\mathrm{OTM}} 0'_{\mathrm{OTM}}$, so that $0'_{\mathrm{OTM}} \leq_{\mathrm{OTM}} A$ would imply $B \leq_{\mathrm{OTM}} A$, contradicting the incomparability of A and B. □

By the general correspondence between OTMs and ORMs (see Chapter 3), we also obtain the following.

Corollary 6.3.8 (Hamkins, Miller [140]). *There are ORM-incomparable, ORM-enumerable $x, y \subseteq \sigma$.*

6.4 Degree theory for ITRMs[2]

We now deal with degree theory for infinite time register machines. The following results presuppose some of the results of Chapter 5 on ITRM-genericity.

For $x, y \subseteq \omega$, we define $x \equiv_{\mathrm{ITRM}} y$, $x <_{\mathrm{ITRM}} y$ and $[x]_{\mathrm{ITRM}}$ as we defined the corresponding notation for ITTMs in Definition 6.2.1. If neither $x \leq_{\mathrm{ITRM}} y$ nor $y \leq_{\mathrm{ITRM}} x$, we write $x \perp y$ and call x and y ITRM-incomparable.

Corollary 6.4.1 ([20], Corollary 22 and [23], Corollary 28). *If a is ITRM-generic, $a = a_0 \oplus a_1$, then $a_0 \perp_{\mathrm{ITRM}} a_1$.*

Proof. By Theorem 5.5.10 in Chapter 5, a_0 and a_1 are mutually ITRM-generic. If a_0 was ITRM-computable from a_1, then $\{a_0\}$ would be decidable in the oracle a_1, meager and contain a_0, so a_0 would not be ITRM-generic relative to a_1, a contradiction. By symmetry, the claim follows. □

Lemma 6.4.2 ([20], Lemma 23 and [23], Lemma 29). *There is an ITRM-generic real $x \leq_{\mathrm{ITRM}} 0'_{\mathrm{ITRM}}$.*

Proof. Given h, we can compute a code for $L_{\omega_\omega^{\mathrm{CK}}+1}$ as follows: By Theorem 4.3.18 from Chapter 4, $0'_{\mathrm{ITRM}}$ is ITRM-recognizable. Clearly, $0'_{\mathrm{ITRM}}$ is not ITRM-computable. Hence, by Lemma 4.3.17 of Chapter 4, we have $\omega_\omega^{\mathrm{CK},h} > \omega_\omega^{\mathrm{CK}}$. By Theorem 3.4.31 from Chapter 3, a real number r is ITRM-computable from $0'_{\mathrm{ITRM}}$ if and only if it is an element of $L_{\omega_\omega^{\mathrm{CK},0'_{\mathrm{ITRM}}}}[0'_{\mathrm{ITRM}}]$. As $L_{\omega_\omega^{\mathrm{CK}}+2}$ contains a code for $L_{\omega_\omega^{\mathrm{CK}}+1}$, such a code is contained in $L_{\omega_\omega^{\mathrm{CK},0'_{\mathrm{ITRM}}}} \subseteq L_{\omega_\omega^{\mathrm{CK},0'_{\mathrm{ITRM}}}}[0'_{\mathrm{ITRM}}]$ and is hence ITRM-computable from $0'_{\mathrm{ITRM}}$.

Now $L_{\omega_\omega^{\mathrm{CK}+1}}$ will contains a real number x which is Cohen-generic over $L_{\omega_\omega^{\mathrm{CK}}}$ by Lemma 5.2.5 from Chapter 5. Such an x can easily be ITRM-computed from a code for $L_{\omega_\omega^{\mathrm{CK}+2}}$ and consequently, x itself is ITRM-computable from $0'_{\mathrm{ITRM}}$. By Theorem 5.5.7 from Chapter 5, x is ITRM-generic. □

2 This section is adapted from the articles "Algorithmic Randomness for Infinite Time Register Machines" [20] which appeared in: A. Beckmann, E. Csuhaj-Varju, K. Meer (eds.), Language, Life, Limits, LNCS 8493 ©Springer Nature 2014 with permission from Springer Nature and "Randomness and Degree Theory for Infinite Time Register Machines" [23], which appeared in Computability, vol. 5, no. 2, pages 181–196, ©IOS Press (2016), available at IOS Press through http://dx.doi.org/10.3233/COM-160055 with permission from IOS Press. We thank Springer Nature and IOS Press for the kind permission to reuse parts of these articles here.

We can use this to solve Post's problem for ITRMs.

Corollary 6.4.3 ([20], Corollary 24 and [23], Corollary 30). *There are $x_0, x_1 \subseteq \omega$ such that we have $[0]_{\mathrm{ITRM}} <_{\mathrm{ITRM}} [x_0]_{\mathrm{ITRM}}, [x_1]_{\mathrm{ITRM}} \leq 0'_{\mathrm{ITRM}}$ and $x_0 \perp_{\mathrm{ITRM}} x_1$. In particular, there is a real number x_0 such that $[0]_{\mathrm{ITRM}} <_{\mathrm{ITRM}} [x_0]_{\mathrm{ITRM}} <_{\mathrm{ITRM}} 0'_{\mathrm{ITRM}}$.*

Proof. Pick x as in Lemma 6.4.2, let $x = x_0 \oplus x_1$, and use Corollary 6.4.1. □

Using the van Lambalgen for ITRMs, we can show much more.

Theorem 6.4.4 ([23], Theorem 31). *Let $\mathfrak{T}$ denote the full tree of finite 0-1-sequences, ordered by reverse inclusion. Then $\mathfrak{T}$ embeds in the ITRM-degrees between 0 and $0'$, i. e., there is an injection $f : \mathfrak{T} \to \mathfrak{P}(\omega)$ such that, for all $x, y \in \mathfrak{T}$, we have $0 <_{\mathrm{ITRM}} f(x), f(y) <_{\mathrm{ITRM}} 0'_{\mathrm{ITRM}}$ and $f(x) \leq_{\mathrm{ITRM}} f(y)$ if and only if $x \supseteq y$. In particular, there is an infinite descending sequence of ITRM-degrees between 0 and $0'_{\mathrm{ITRM}}$.*

Proof. We shall, for each $s \in^{<\omega} 2$, define a real number r_s in such a way that the set $\{[r_s]_{\mathrm{ITRM}} \mid s \in^{<\omega} 2\}$ has the desired property. Let $r = r_{()}$ be ITRM-generic such that $0 <_{\mathrm{ITRM}} r <_{\mathrm{ITRM}} 0'_{\mathrm{ITRM}}$. Assuming that r_s is already defined, let $r_s = x_0 \oplus x_1$ and set $r_{s0} = x_0$, $r_{s1} = x_1$. Clearly, if s is a proper initial segment of t, then $r_s <_{\mathrm{ITRM}} r_t$. We show that furthermore, when s and t are incompatible (i. e., none is an initial segment of the other), then $r_s \perp_{\mathrm{ITRM}} r_t$: Let $u = r_s \cap r_t$ be the longest common initial segment of r_s and r_t. Without loss of generality, assume that s starts with $u0$ and that t starts with $u1$. Then $r_u = r_{u0} \oplus r_{u1}$, so, by Theorem 5.5.10 in Chapter 5, $u0$ and $u1$ are mutually ITRM-generic. Now suppose that r_s and r_t are not mutually ITRM-generic, e. g., that r_s was not ITRM-generic relative to r_t. Then, as r_t is recursive in r_{u1}, r_s is not ITRM-generic relative to r_{u1}. Let U be a meager set containing r_s which is ITRM-decidable in the oracle r_{u1}.

Let s' be a binary string such that $u0s' = s$ and let l denote its length. We recursively define a sequence $(U_i : 0 \leq i \leq l)$ of subsets of $[0,1]$ as follows: Let $U_0 = 0$ and for $\leq i < n$ let $U_{i+1} := U_i \oplus [0,1]$ when the $(l-i)$th digit of l is 0 and otherwise $U_{i+1} := [0,1] \oplus U_i$. By an easy induction, we have that $r_{u0(s'|i)} \in U_{l-i}$ for $0 \leq i \leq l$, where $s'|i$ denotes the restriction of s' to its first i many bits. In particular, we have $r_{u0} \in U_l$. Moreover, when X is meager, then so are $X \oplus [0,1]$ and $[0,1] \oplus X$, so as $U_0 = U$ is meager, it follows by another induction that U_i is meager for $0 \leq i \leq l$ and in particular that U_l is meager. It is also easy to see inductively that, as $U_0 = U$ is ITRM-decidable in the oracle r_{u1}, so is U_i for all $0 \leq i \leq l$, and hence in particular for $i = l$. So U_l is meager, contains r_{u0} and is ITRM-decidable in the oracle r_{u1}, contradicting the fact that r_{u0} is ITRM-generic relative to r_{u1}. □

Corollary 6.4.5 ([23], Corollary 32). *The ITRM-degrees of ITRM-generic real numbers contain no $\leq_{\mathrm{ITRM}}$-minimal or maximal element.*

Proof. Let y is ITRM-generic, $y = y_1 \oplus y_2$.

Then first, by Theorem 5.5.10 in Chapter 5, y_1 is ITRM-generic and $y_1 <_{\text{ITRM}} y$. Hence there is no $<_{\text{ITRM}}$-minimal ITRM-generic $y \subseteq \omega$.

Furthermore, we have $\omega_{\omega}^{\text{CK},y} = \omega_{\omega}^{\text{CK}}$ by Corollary 5.2.18 in Chapter 5, so $L_{\omega_{\omega}^{\text{CK},y}}[y] = L_{\omega_{\omega}^{\text{CK}}}[y]$; Let x be Cohen-generic over $L_{\omega_{\omega}^{\text{CK}}}[y]$, then by the relativized version of Theorem 5.5.7 in Chapter 5, x is ITRM-generic relative to y. By Theorem 5.5.10 in Chapter 5, $z := x \oplus y$ is ITRM-generic, and we clearly have $y <_{\text{ITRM}} z$, so y is not maximal. □

Lemma 6.4.6 ([23], Theorem 36). *For $i \in \omega$, we have $\omega_{\omega}^{\text{CK},0_{\text{ITRM}}^{(i)}} = \omega_{\omega(i+1)}^{\text{CK}}$; consequently, $i+1$ is the minimal $n \in \omega$ such that $0_{\text{ITRM}}^{(i)} \in L_{\omega_{\omega n}^{\text{CK}}}$.*

Proof. For $i \in \omega$, let σ_i denote $\omega_{\omega}^{\text{CK},0_{\text{ITRM}}^{(i)}}$.

It is easy to see that $0_{\text{ITRM}}^{(i)}$ is ITRM-computable from $0_{\text{ITRM}}^{(i+1)}$ for all $i \in \omega$. By applying Theorem 4.3.18 in Chapter 5 (and the exercise below it) inductively, we see that, $0_{\text{ITRM}}^{(i)}$ is ITRM-recognizable for each $i \in \omega$. By Theorem 4.3.22 from Chapter 4 then, $0_{\text{ITRM}}^{(i)} \in L_{\sigma_i}$ for each $i \in \omega$; thus $L_{\sigma_i}[0_{\text{ITRM}}^{(i)}] = L_{\sigma_i}$. Hence a real number x is ITRM-computable from $0_{\text{ITRM}}^{(i)}$ if and only if $x \in L_{\sigma_i}$. As $0_{\text{ITRM}}^{(i+1)}$ is not ITRM-computable from $0_{\text{ITRM}}^{(i)}$, we must have $\omega_{\omega}^{\text{CK},0_{\text{ITRM}}^{(i+1)}} > \omega_{\omega}^{\text{CK},0_{\text{ITRM}}^{(i)}}$ for all $i \in \omega$; hence $(\sigma_i | i \in \omega)$ is a strictly increasing sequence of limits of admissible ordinals.

We now proceed inductively to show that $\sigma_i = \omega_{\omega(i+1)}^{\text{CK}}$:

As $0'_{\text{ITRM}}$ is not ITRM-computable, it is not an element of $L_{\omega_{\omega}^{\text{CK}}}$. As it is definable over $L_{\omega_{\omega}^{\text{CK}}}$, it is an element of $L_{\omega_{\omega}^{\text{CK}}+1}$, and hence of $L_{\omega_{\omega2}^{\text{CK}}}$. Hence $\omega_{\omega}^{\text{CK}} < \sigma_1 \leq \omega_{\omega2}^{\text{CK}}$; as σ_1 is a limit of admissible ordinals, we have $\sigma_1 = \omega_{\omega2}^{\text{CK}}$.

Similarly, assuming that $\sigma_i = \omega_{\omega(i+1)}^{\text{CK}}$, $0_{\text{ITRM}}^{(i+1)}$ is definable over $L_{\omega_{\omega(i+1)}^{\text{CK}}}$ and hence an element of $L_{\omega_{\omega(i+2)}^{\text{CK}}}$. Hence $\omega_{\omega(i+1)}^{\text{CK}} < \sigma_{i+1} \leq \omega_{\omega(i+2)}^{\text{CK}}$, so $\sigma_{i+1} = \omega_{\omega(i+2)}^{\text{CK}}$, as desired. □

We can now extend Theorem 6.4.4 to obtain intermediate degrees also for iterations of the jump operator:

Corollary 6.4.7 ([23], Corollary 37). *There is a real number x such that $0'_{\text{ITRM}} <_{\text{ITRM}} x <_{\text{ITRM}} 0''_{\text{ITRM}}$.*

Proof. Let $x \in L_{\omega_{\omega3}^{\text{CK}}} - L_{\omega_{\omega2}^{\text{CK}}}$ be Cohen-generic over $L_{\omega_{\omega2}^{\text{CK}}}$ (and hence ITRM-generic). (Such an x exists because $L_{\omega_{\omega2}^{\text{CK}}} = \Sigma_1^{L_{\omega_{\omega2}^{\text{CK}}}}\{\emptyset\}$, so $L_{\omega_{\omega2}^{\text{CK}}}$ is countable in $L_{\omega_{\omega2}^{\text{CK}}+1}$ and thus, by Lemma 5.2.5 from Chapter 5, $L_{\omega_{\omega2}^{\text{CK}}+1} \subseteq L_{\omega_{\omega3}^{\text{CK}}}$ contains a real number x that is Cohen-generic over $L_{\omega_{\omega2}^{\text{CK}}}$, and hence cannot be an element of $L_{\omega_{\omega2}^{\text{CK}}}$.)

So $x \in L_{\omega_{\omega3}^{\text{CK}}} = L_{\omega_{\omega}^{\text{CK},0''_{\text{ITRM}}}}[0''_{\text{ITRM}}]$ by Lemma 6.4.6, hence $x \leq_{\text{ITRM}} 0''_{\text{ITRM}}$.

We cannot have $x \in L_{\omega_{\omega}^{\text{CK},0'_{\text{ITRM}}}}[0'_{\text{ITRM}}] = L_{\omega_{\omega}^{\text{CK},0'_{\text{ITRM}}}}$ since $L_{\omega_{\omega}^{\text{CK},0'_{\text{ITRM}}}} = L_{\omega_{\omega2}^{\text{CK}}}$ by Lemma 6.4.6. Consequently, we have $x \leq 0''_{\text{ITRM}}$ and $x \nleq 0'_{\text{ITRM}}$.

As x is Cohen-generic over $L_{\omega_{\omega2}^{\text{CK}}}$ and $L_{\omega_{\omega2}^{\text{CK}}} = L_{\omega_{\omega}^{\text{CK},0'_{\text{ITRM}}}}[0'_{\text{ITRM}}]$ by Theorem 6.4.6, we have $\omega_{\omega}^{\text{CK},x\oplus 0'_{\text{ITRM}}} = \omega_{\omega}^{\text{CK},0'_{\text{ITRM}}} = \omega_{\omega2}^{\text{CK}} < \omega_{\omega3}^{\text{CK}} = \omega_{\omega}^{\text{CK},0''_{\text{ITRM}}}$. In particular, there is a real number coding a well-ordering of order type $\omega_{\omega 3}^{\mathrm{CK}}$ ITRM-computable from $0''_{\mathrm{ITRM}}$, but not from $x \oplus 0'_{\mathrm{ITRM}}$. Hence $0''_{\mathrm{ITRM}}$ cannot be ITRM-computable from $x \oplus 0'_{\mathrm{ITRM}}$, so that $x \oplus 0'_{\mathrm{ITRM}} \leq_{\mathrm{ITRM}} 0''_{\mathrm{ITRM}}$ implies $x \oplus 0'_{\mathrm{ITRM}} <_{\mathrm{ITRM}} 0''_{\mathrm{ITRM}}$.

But then, $0'_{\mathrm{ITRM}} <_{\mathrm{ITRM}} x \oplus 0'_{\mathrm{ITRM}} <_{\mathrm{ITRM}} 0''_{\mathrm{ITRM}}$, so $y := x \oplus 0'_{\mathrm{ITRM}}$ is as desired. □

By a similar argument, one gets intermediate degrees between any two successive (finite) iterations of the ITRM-jump.

Corollary 6.4.8 ([23], Corollary 38). *For every $i \in \omega$, there is a real x_i such that $0^{(i)}_{\mathrm{ITRM}} <_{\mathrm{ITRM}} x < 0^{(i+1)}_{\mathrm{ITRM}}$.*

Exercise 6.4.9. Prove Corollary 6.4.8.

6.4.1 Degrees of recognizables

Given Corollary 6.4.8, one may ask whether any of these intermediate degrees is in some natural sense "unique." In the Turing degrees, it is a striking experience that intermediate degrees (i. e., degrees lying properly between two iterations of the Turing jump) don't seem to come up "naturally," but need to be constructed on purpose (see, e. g., the discussion in Simpson [164]). Concerning ITRMs, we saw in Chapter 4, Corollary 4.3.20 that, to a certain extent, something similar is going on: Namely, reals of intermediate degree are never recognizable. We now generalize this result a bit. It turns out that, in this interpretation, no intermediate degree in the finite iterations of the ITRM-jump is "unique."

Theorem 6.4.10 ([23], Theorem 39). *Assume that $x \leq_{\mathrm{ITRM}} 0^{(i)}_{\mathrm{ITRM}}$ for some $i \in \omega$ and that x is ITRM-recognizable. Then there is $j \in \omega$ such that $x \in [0^{(j)}_{\mathrm{ITRM}}]_{\mathrm{ITRM}}$.*

Proof. We start by showing: If $x >_{\mathrm{ITRM}} 0^{(i)}_{\mathrm{ITRM}}$ is ITRM-recognizable, then $x \geq_{\mathrm{ITRM}} 0^{(i+1)}_{\mathrm{ITRM}}$. To see this, note that $x > 0^{(i)}_{\mathrm{ITRM}}$ implies that $x \notin L_{\omega^{\mathrm{CK}}_{\omega(i+1)}}$ by Theorem 6.4.6. As x is ITRM-recognizable, we have $x \in L_{\omega_{\omega}^{\mathrm{CK},x}}$ by Theorem 4.3.22 from Chapter5. Hence $\omega_{\omega}^{\mathrm{CK},x}$ is a limit of admissibles strictly bigger than $\omega^{\mathrm{CK}}_{\omega(i+1)}$, so $\omega_{\omega}^{\mathrm{CK},x} \geq \omega^{\mathrm{CK}}_{\omega(i+2)}$. As $0^{(i+1)}_{\mathrm{ITRM}} \in L_{\omega^{\mathrm{CK}}_{\omega(i+2)}}$, we have $0^{(i+1)}_{\mathrm{ITRM}} \in L_{\omega_{\omega}^{\mathrm{CK},x}}[x]$, and hence $0^{(i+1)}_{\mathrm{ITRM}} \leq x$. Now assume for a contradiction that x is as in the assumption of the theorem, but not ITRM-computably equivalent to some $0^{(i)}_{\mathrm{ITRM}}$. In particular then, x is not ITRM-computable (otherwise $x \in [0^{(0)}_{\mathrm{ITRM}}]_{\mathrm{ITRM}}$). So $x >_{\mathrm{ITRM}} 0^{(0)}_{\mathrm{ITRM}}$. As $x \leq_{\mathrm{ITRM}} 0^{(i)}_{\mathrm{ITRM}}$ for some $i \in \omega$, there is some $j \in \omega$ such that $x \not>_{\mathrm{ITRM}} 0^{(j)}$; without loss of generality, let j be minimal with this property. Then $j > 0$ and $x >_{\mathrm{ITRM}} 0^{(j-1)}_{\mathrm{ITRM}}$, so $x \geq_{\mathrm{ITRM}} 0^{(j)}_{\mathrm{ITRM}}$. As $x >_{\mathrm{ITRM}} 0^{(j)}_{\mathrm{ITRM}}$ is excluded, we have $x =_{\mathrm{ITRM}} 0^{(j)}_{\mathrm{ITRM}}$, so j is as desired. □

A nice phenomenon for ITRMs is that one can "jump into recognizability":

Theorem 6.4.11 ([25], Lemma 43). *There is a real number x which is not ITRM-recognizable, but such that x'_{ITRM} is ITRM-recognizable.*

Proof. Let x be the $<_L$-minimal real number which is Cohen-generic over $L_{\omega_\omega^{\mathrm{CK}}+1}$. We claim that x has the desired properties.

To see this, first note that x is not ITRM-recognizable: By Theorem 5.5.7 of Chapter 5, x is ITRM-generic and thus not contained in an ITRM-decidable meager set. But if x was ITRM-recognizable, then $\{x\}$ would be an ITRM-decidable, meager set containing x, a contradiction.

Now we show that x'_{ITRM} is ITRM-recognizable. To see this, first note that, by Theorem 6.4.6, $\omega_\omega^{\mathrm{CK},0'_{\mathrm{ITRM}}} = \omega_{\omega 2}^{\mathrm{CK},x} > \omega_\omega^{\mathrm{CK}} + 2$, so that the $<_L$-minimal code c for $L_{\omega_\omega^{\mathrm{CK}}+2}$ is ITRM-computable in the oracle x'. Let Q be an ITRM-program that computes c from x'; by Corollary 2.3.13 from Chapter 2, we may assume that Q is safe.

Now, by Lemma 5.2.5 of Chapter 5, we have $x \in L_{\omega_\omega^{\mathrm{CK}}+2}$. Thus, there is a natural number i_x that codes x in the sense of c. As i_x is a fixed natural number, we can use it freely in our algorithm (even though we do not know what it is—but after all, all we want to show is the **existence** of an algorithm for recognizing x'_{ITRM}). Similarly, there is $\hat{i} \in \omega$ that codes $L_{\omega_\omega^{\mathrm{CK}}+1}$ in the sense of c.

To recognize x'_{ITRM}, we can now proceed as follows: Given $y \subseteq \omega$ in the oracle, run Q^y. Suppose that Q computes z in the oracle y.

Now, using P_{truth} from Chapter 3 and a well-foundedness test, we can check whether z codes $L_{\omega_\omega^{\mathrm{CK}}+2}$ in such a way that $\hat{i}$ represents a code for $L_{\omega_\omega^{\mathrm{CK}}+1}$ and i_x is the $<_L$-minimal Cohen-generic real number over the set coded by $\hat{i}$ in z. If any of these fails, then $y \neq x'_{\mathrm{ITRM}}$ and we output 0.

Otherwise, we know that i_x represents x, so with the help of P_{find} from Chapter 2, we can ITRM-compute x from z. By exercise 4.3.19 from Chapter 4, x'_{ITRM} is ITRM-recognizable relative to x; let $\bar{Q}$ be a program that does this. Then simply run $\bar{Q}^{x\oplus z}$ and output the result, which will be δ_{xz}, as desired. □

Exercise 6.4.12. The point of this exercise is to generalize Theorem 6.4.11 to show that, for every $k \in \omega$, there is a real number x such that none of $x, x'_{\mathrm{ITRM}}, x^{(2)}_{\mathrm{ITRM}}, \ldots, x^{(k)}_{\mathrm{ITRM}}$ is ITRM-recognizable, but $x^{(k+1)}_{\mathrm{ITRM}}$ is ITRM-recognizable. Fix $k \in \omega$, and let x be the $<_L$-minimal real number that is Cohen-generic over $L_{\omega_{k+1}^{\mathrm{CK}}+1}$.

(a) Let $i \leq k$. Show that $\omega_{\omega(i+1)}^{\mathrm{CK},x} = \omega_{\omega(i+1)}^{\mathrm{CK}}$.

(b) Use Theorem 6.4.6 to show that for all $i \leq k$, we have $\omega_\omega^{\mathrm{CK},x^{(i)}_{\mathrm{ITRM}}} = \omega_{\omega(i+1)}^{\mathrm{CK}}$.

(c) Conclude that $x^{(i)}_{\mathrm{ITRM}} \notin L_{\omega_\omega^{\mathrm{CK},x^{(i)}_{\mathrm{ITRM}}}}$ for all $i \leq k$.

(d) Show that, for all $i \leq k$, $x^{(i)}_{\mathrm{ITRM}}$ is not ITRM-recognizable.

(e) Adapt the proof of Theorem 6.4.11 to show that $x^{(k+1)}_{\mathrm{ITRM}}$ is ITRM-recognizable.

We do not know whether an analogous statement holds for ITTMs. The following exercise, however, indicates that it might not.

Exercise 6.4.13. (a) Show that there is an ITTM-program P such that, for any $x \subseteq \omega$, $P^{x'_{\text{ITTM}}}$ halts with output x.

(b) Let $x \subseteq \omega$. Show that, if x'_{ITTM} is ITTM-recognizable, then x is ITTM-semirecognizable. (Hint: Recall that, for any $x \subseteq \omega$, x'_{ITTM} is ITTM-eventually writable relative to x and use (a).)

(c) Show that, if $x \subseteq \omega$ is such that x'_{OTM} is OTM-recognizable, then x is OTM-recognizable. Show that, if x'_{OTM} is OTM-semirecognizable, then x is OTM-semirecognizable.

A prominent result in the theory of Turing degrees is the existence of degrees that "join to the jump," i. e., $x, y \subseteq \omega$ such that $x, y <_{\text{T}} 0'$, but $x \oplus y \equiv_T 0'$.[3] Clearly, no such result can hold for ITTMs, simply due to the lack of non-writable real numbers strictly below $0'_{\text{ITTM}}$ at all. For ITRMs, however, we have a rich "zoo" of intermediate degrees, as we saw above. We will now show, by adapting Hamkins' trick in the proof of the existence of Aristophanean pairs in Chapter 4, that there is indeed an analogue of this theorem for ITRMs.

Theorem 6.4.14. *There are real numbers $x, y < 0'_{\text{ITRM}}$ such that $x \oplus y \equiv_{\text{ITRM}} 0'_{\text{ITRM}}$.*

Proof. We claim that there are $x, y \subseteq \omega$ such that x and y are Cohen-generic over $L_{\omega_\omega^{\text{CK}}}$, $x, y \leq_{\text{ITRM}} 0'_{\text{ITRM}}$ and such that the bitwise sum $x +_b y$ of x and y is $0'_{\text{ITRM}}$.

If this holds, then x and y are as desired: Being Cohen-generic over $L_{\omega_\omega^{\text{CK}}}$, x and y satisfy $\omega_\omega^{\text{CK},x} = \omega_\omega^{\text{CK},y} = \omega_\omega^{\text{CK}} < \omega_\omega^{\text{CK},0'_{\text{ITRM}}}$, so both x and y will be strictly $<_{\text{ITRM}}$-below $0'_{\text{ITRM}}$. Moreover, as $x, y \leq_{\text{ITRM}} 0'_{\text{ITRM}}$, we have $x \oplus y \leq_{\text{ITRM}} 0'_{\text{ITRM}}$, and as $0'_{\text{ITRM}} = x +_b y$ is clearly ITRM-computable from $x \oplus y$, we have $0'_{\text{ITRM}} \equiv_{\text{ITRM}} x \oplus y$.

To show that such real numbers x, y exist is an ITRM-effectivization of Lemma 4.3.14 from Chapter 4. Suppose that $0'_{\text{ITRM}}$ is given in the oracle. We can then ITRM-compute a code c for $L_{\omega_\omega^{\text{CK}}+\omega}$. This structure in particular contains a (code for an) enumeration $(D_i : i \in \omega)$ of the dense subsets of Cohen forcing contained in $L_{\omega_\omega^{\text{CK}}}$ in the order type ω. We now build two sequences $(x_i : i \in \omega)$, $(y_i : i \in \omega)$ of finite 0-1-strings recursively, starting with $x_0 = y_0 = \emptyset$, in such a way that $|x_n| = |y_n|$ and $x_n +_b y_n = 0'_{\text{ITRM}} \restriction |x_n|$ for all $n \in \omega$. In the nth step, let x_n and y_n be given and satisfy the properties just described. Then D_n, being dense, contains some extension of x_n; we pick the one with minimal index in c and call it p_x. Then $p_x +_b 0'_{\text{ITRM}} \restriction |x_n|$ is an extension of y_n by the assumptions about x_n and y_n. Again by the density of D_n, it contains an extension p_y of $p_x +_b 0'_{\text{ITRM}} \restriction |p_x|$. Now, if $|p_x| = |p_y|$, we let $x_{n+1} = p_x$ and $y_{n+1} = p_y$. If $|p_x| < |p_y|$, we let $y_{n+1} = p_y$ and define x_{n+1} by $x_{n+1}(i) = p_x$ for $i \leq |p_x|$ and $x_{n+1}(i) = p_y +_b 0'_{\text{ITRM}}(i)$ for $|p_x| < i \leq |p_y|$. This finishes the construction of $(x_i : i \in \omega)$ and $(y_i : i \in \omega)$. Finally, we let $x = \bigcup_{i\in\omega} x_i$ and $y = \bigcup_{i\in\omega} y_i$.

We claim that the x, y obtained in this way are as desired. Clearly, the construction just described is ITRM-effective in the oracle $0'_{\text{ITRM}}$ (in fact, the oracle is only needed

3 See, e. g., Lewis [125].

for computing $p_x +_b 0'_{\mathrm{ITRM}} \upharpoonright |x_n|$). Hence $x, y \leq_{\mathrm{ITRM}} 0'_{\mathrm{ITRM}}$. By construction, x and y each have an initial segment in D_n for every $n \in \omega$ and are thus Cohen-generic over $L_{\omega_\omega^{\mathrm{CK}}}$. It is also clear that $|x_n| = |y_n|$ for all $n \in \omega$. Finally, we have $x_n +_b y_n = 0'_{\mathrm{ITRM}} \upharpoonright |x_n|$ for all $n \in \omega$, so $x +_b y = 0'_{\mathrm{ITRM}}$, as desired. □

Exercise 6.4.15. Generalize the proof of Theorem 6.4.14 to show that, for any $n \in \omega$, there are real numbers $x_1, \dots, x_n$ such that $\bigoplus_{i \in I} x_i <_{\mathrm{ITRM}} 0'_{\mathrm{ITRM}}$ for every $I \subsetneq \{1, \dots, n\}$, but $\bigoplus_{i=1}^{n} x_i \equiv_{\mathrm{ITRM}} 0'_{\mathrm{ITRM}}$.

6.5 Some results on the degree theory of other models

For α-Turing machines, Post's problem has been worked out by Koepke and Seyfferth in [112]. Considering subsets of α instead of real numbers and using a priority construction, the following result is obtained.

Theorem 6.5.1 (Koepke, Seyfferth [112], Theorem 22). *Let α be admissible. Then there are α-TM-semidecidable subsets $x, y \subseteq \alpha$ such that x is not α-TM-computable from y and y is not α-TM-computable from x.*

As a consequence, we get the existence of intermediate degrees for α-Turing machines.

Corollary 6.5.2 (Koepke, Seyfferth [112]). *Let α be admissible. Define $0'_\alpha := \{p(i, \beta) : i \in \omega \wedge \beta < \alpha \wedge P_i(\beta)\downarrow\}$. Then there is a subset $x \subseteq \alpha$ such that x is not α-TM-computable, x is α-TM-computable from $0'_\alpha$, and $0'_\alpha$ is not α-TM-computable from x.*

Proof. Let x, y be as in Theorem 6.5.1. Then neither of them is α-TM-computable (otherwise, it would be reducible to every subset of α), and both are reducible to $0'_\alpha$ (as they are α-TM-semidecidable). As none reduces to the other, this implies that $0'_\alpha$ can reduce neither to x nor y; thus x and y are both as desired. □

6.6 Degrees of recognizability[4]

The concept of relativized recognizability makes it tempting to define "degrees of recognizability." This, however, is hindered by the observation made in Corollary 4.3.6

---

4 This section is adapted from the articles "Infinite Time Recognizability from Random Oracles and the Recognizable Jump Operator" [20], which appeared in Computability, vol. 5, no. 2, pages 181–196 ©2017, with permission from IOS Press, available at IOS Press through http://dx.doi.org/10.3233/COM-160061 and "ITRM-Recognizability from Random Oracles" [22], which appeared in: A. Beckmann, V. Mitrana, M. Soskova (eds.), Evolving Computability, LNCS 9136 ©2015 with permission from Springer Nature. We thank Springer Nature and IOS Press for the kind permission to reuse parts of these articles here.

from Chapter 4 that relativized recognizability is not transitive, not even for ITRMs. There are two ways around this: One can either give up on having degrees and merely study the reducibility relation on single real numbers, or one can replace relativized recognizability with its transitive closure to make it an equivalence relation. We shall take the second route here and give some results on this notion.

Definition 6.6.1 (Cf. [25], Definition 48, 49). Let M be ITRM, ITTM or OTM.

We let $x \preceq_M y$ if and only if x belongs to the M-recognizable closure of y, i. e., if there is a finite sequence $x = z_0, z_1, \ldots, z_n = y$ such that $z_i \leq^r_M z_{i+1}$ for $i \in \{0, \ldots, n-1\}$. If $x \preceq_M y$, we say that x is hereditarily M-recognizable from y. (Cf. Definition 4.4.10 in Chapter 4.) As in Proposition 4.4.11 in Chapter 4, a two-step iteration actually suffices to give the whole transitive closure.

x and y are called M-recognizably equivalent, written $x \asymp_M y$ if and only if $x \preceq_M y$ and $y \preceq_M x$. It is easy to see that M-recognizable equivalence is indeed an equivalence relation. $[x]^r_M$ will denote the $\asymp_M$-equivalence class of x, called M-recognizability degree of x.

One can now formulate and solve a recognizable analogue for Post's problem.

Theorem 6.6.2 (Cf. [25], Theorem 50). *For* $M \in \{\text{ITRM}, \text{ITTM}\}$, *there is* x *such that* $[0]^r_M \prec_M [x]^r_M \prec_M [0^r_M]^r_M$.

We conclude by observing that the structure of M-recognizability degrees turns out to depend heavily on the set-theoretical background:

Theorem 6.6.3 (Cf. [25], Theorem 51). *Let* $M \in \{\text{ITRM}, \text{ITTM}, \text{OTM}\}$. *Assume that* $V = L$. *Then the* M*-recognizability degrees are linearly ordered by the ordering induced by the canonical well-ordering* $<_L$ *of* L.

Proof. Relative to $x \subseteq$, we can recognize the $<_L$-minimal code of the minimal L-level L_{α_x} containing x. From a code for L_{α_x}, every real $z \in L_{\alpha_x}$ is computable and hence recognizable. Hence every real in L_{α_x} is hereditarily recognizable from x.

Now, for any $x, y \in L$, we have either $\alpha_x \leq \alpha_y$ or $\alpha_x > \alpha_y$, i. e., $x \in L_{\alpha_y}$ or $y \in L_{\alpha_x}$. Hence one of them is hereditarily M-recognizable from the other. Moreover, if $x \leq_L y$, then $\alpha_x \leq \alpha_y$, so x is hereditarily M-recognizable from y. □

On the other hand, we have the following.

Proposition 6.6.4 (Cf. [25], Proposition 52). *Let* $M \in \{\text{ITRM}, \text{ITTM}, \text{OTM}\}$. *If* x, y *are mutually Cohen-generic over* L, *then neither* $x \preceq_M y$ *nor* $y \preceq_M x$.

Proof. Assume for a contradiction that $x \preceq_M y$. Let P be a program that recognizes a real z relative to y such that x is primitive recursive in y which exists by Proposition 4.4.11 from Chapter4. Then x is Σ_1-definable from y, and hence, by Shoenfield absoluteness, we get $x \in L[y]$, which contradicts the assumption that x is Cohen-generic over $L[y]$. □

Consequently, the existence of $\preceq_M$-incomparable M-degrees of recognizability is independent of ZFC for $M \in \{\text{ITRM}, \text{ITTM}, \text{OTM}\}$. The study of the structure of the degrees of recognizability will hence need to work under set-theoretical extra assumptions to obtain substantial results.

7 Complexity

7.1 Introduction

In the classical theory of computability, we study the solvability of problems (modeled, e. g., as sets of finite binary sequences) by computational methods without regards to the required resources, like memory (e. g., the number of tape cells used for storing during "scratch work," not counting those used for representing the in and output) and time, i. e., the length of computations. These issues are dealt with in complexity theory. For a comprehensive source on classical complexity theory, see Arora and Barak [6]. Here, we give a very brief introduction to motivate the following sections.

In this chapter, when f is a function mapping an ordinal α to the set $\{0,1\}$, $|x|$ denotes the length of x, i. e., the domain of f.

In complexity theory, one determines the complexity of a problem $X \subseteq \{0,1\}^{<\omega}$ by looking for functions f and g such that some Turing-program P answers "$x \in X$?" for every x in $\leq f(|x|)$ many steps, using $\leq g(|x|)$ many tape cells of working memory (not counting the in and output tape), and ideally, one also wants to show that "much smaller" functions will not work.

An important observation is the following.

Theorem 7.1.1 (Cf., e. g., Hromkovic, [89]).

Speedup Theorem 1: *If $X \subseteq \{0,1\}^{<\omega}$ is a problem, $f, g : \omega \to \omega$ are at least linear (i. e., there are $a \in \mathbb{R}^+$, $b \in \mathbb{Z}$ with $f(x) > ax + b$ and $g(x) > ax + b$ for all $x \in \omega$) and P is Turing program that solves X with time bound f and space bound g, and $\varepsilon > 0$, then there is a Turing program Q which solves X with time bound εf and space bound εg.*

This motivates the introduction of complexity classes: A problem X is $\mathcal{O}(f)$ if and only if there are a Turing program P and a real number K that P determines whether $x \in X$ in $\leq Kf(|x|)$ many steps. The "trick" to prove Theorem 7.1.1 is to use Turing programs with (much) larger alphabets. Using only the alphabet $\{0,1\}$, we still have the following.

Theorem 7.1.2 (Cf., e. g., Cutland, [44], Chapter 12).

Speedup Theorem 2: *There is a problem X such that, for every program P using the alphabet $\{0,1\}$ that solves X, there is another program Q using the alphabet $\{0,1\}$ that solves X and the running time of which on each instance x is at most the (rounded down) logarithm of the running time that P uses. The logarithm can be replaced by any recursive function.*

This is a conceptually important theorem, as it shows that, in general, one cannot talk about "the fastest" algorithm, not even canceling out constant factors. Still,

https://doi.org/10.1515/9783110496154-007

there are some important classes of complexities that frequently apply to "real-world" problems.

Definition 7.1.3 (Cf., e. g., [6]). P is the class of problems X for which there are a polynomial p and a Turing-program P such that P decides X and $P(x)$ runs for $\leq p(|x|)$ many steps on input x.

NP is the class of problems X for which there are a polynomial p a non-deterministic[1] Turing-program P such that, for every finite binary string x, there is a computation of P on input x that halts with output 1 if and only if $x \in X$ and, moreover, in this case, there is such a computation of length $\leq p(|x|)$.

co-NP is the set of complements of problems in NP.

PSPACE is the class of problems X for which there are a polynomial p and a Turing-program P such that P decides X and $P(x)$ uses at most the first $p(|x|)$ many cells of the working tape.

NPSPACE is the class of problems X for which there are a polynomial p a nondeterministic Turing-program P such that, for all finite binary string x, there is a computation of P on input x that halts with output 1 if and only if $x \in X$ and, moreover, in this case, there is such a computation using at most the first $p(|x|)$ many cells of the working tape.

An important alternative characterization of NP is the following: X is NP if and only if there are a Turing program P and a polynomial p such that P runs for $\leq p(|x|)$ many steps on input $(x, y) \in \{0,1\}^{<\omega} \times \{0,1\}^{<\omega}$ and such that, for every $x \in X$, there is $y \in \{0,1\}^{<\omega}$ such that $P(x, y)$ halts with output 1, while for $x \notin X$, there is no such y. It is this characterization, rather than non-determinism, on which the infinitary analogues considered below will be based.[2]

Given two problems X and Y, we say that X is **polynomially reducible** to Y if and only if there are a Turing program P and a polynomial p such that, for all binary strings x, (1) P runs for $\leq p(|x|)$ many steps on the input x and (2) if y is the output of the computation of P on the input x, then $y \in Y$ if and only if $x \in X$. Thus, polynomial reducibility means that one can quickly "translate" instances of the program X into instances of the problem Y.

Definition 7.1.4 (Cf., e. g., [89]). The problem X is NP-hard if and only if every problem in NP is polynomially reducible to X. If X is both NP-hard and an element of NP, X is called NP-complete.

1 See, e. g., [6] for the notion of nondeterministic Turing machines.

2 There is no problem with introducing non-deterministic variants of infinitary machines. For OTMs, this is, e. g., done in Schlicht and Seyfferth [160], Löwe [129] and Winter, [196] and [195]. Our reason for going the other route is simply that we want to work around the necessary formalism for nondeterminism.

Intuitively, NP-complete problems are the hardest problems in NP: Were we given access to one of them for free, then every problem in NP could be solved in polynomial time. Actually, it is unknown whether the problems in NP are solvable in polynomial time. This is the content of the biggest open question of complexity theory:

$$\text{IS P} = \text{NP?}$$

It is generally expected that the answer is "no." Indeed, being able to quickly verify a solution with appropriate extra information seems to be something weaker than being able to quickly come up with a solution. In spite of this, there seems to be little hope that this problem will be solved in the near future. On the bright side, we will see below that the infinitary analogues are much easier. Interestingly, it is known that, if P≠NP, then there are problems in NP that are neither in P nor NP-complete; this fact is known as Ladner's theorem (see, e. g., [6], Theorem 3.4).

The paradigmatical example of an NP-complete problem is the problem of satisfiability (SAT); SAT is the set of (binary strings coding) propositional formulas in disjunctive normal form that are satisfiable, i. e., that become true for at least one assignment of truth values to the propositional variables. The fact that SAT is NP-complete is known as the Cook–Levin theorem (see, e. g., [6], Theorem 2.10). Infinitary analogues will be considered below.

7.2 Complexity theory for ITTMs

The first work on complexity theory for infinitary machines concerned ITTMs and was carried out by Schindler in [155] and then continued by Deolalikar, Hamkins and Schindler in [46]. We will here give a brief summary of the basic notions and results. Some of the proofs should be accessible to attentive readers of the preceding chapters, and we thus give them in the form of guided exercises, referring to [155] and [46] for complete proofs.

In the framework of [155], problems are sets not of finite, but infinite strings of length ω over the alphabet $\{0,1\}$, i. e., sets of real numbers. For such a set $X \subseteq \mathfrak{P}(\omega)$, we want to consider the time necessary for an ITTM-program to decide, for any $x \subseteq \omega$, whether $x \in X$. An immediate observation is that, since all strings in question are of length ω, the complexity bounds will not, as in the classical context, depend on the length of the input, but simply be ordinal constants.

Definition 7.2.1 (Schindler, see [155]). Let $\alpha \in \mathrm{On}$. Then $X \subseteq \mathfrak{P}(\omega)$ belongs to $\mathrm{TIME}_{\alpha}^{\mathrm{ITTM}}$ if and only if there are $\beta < \alpha$ and an ITTM-program P such that, for any $x \subseteq \omega$, P^x halts in $< \beta$ many steps and outputs 1 if and only if $x \in X$ and otherwise outputs 0.

The analogue of the class of polynomial time computable functions for ITTMs is then the following.

Definition 7.2.2 (Schindler, see [155]). $X \subseteq \mathfrak{P}(\omega)$ belongs to ITTM-P if and only if there is $n \in \omega$ such that X belongs to $\text{TIME}^{\text{ITTM}}_{\omega^n}$. In other words, ITTM-P $= \text{TIME}^{\text{ITTM}}_{\omega^\omega}$.

Adapting the characterization of NP via witnesses, one is led to the ITTM-analogue of nondeterministic complexity classes.

Definition 7.2.3 (Schindler, see [155]). Let $\alpha \in \text{On}$. Then $X \subseteq \mathfrak{P}(\omega)$ belongs to ITTM-NTIME$_\alpha$ if and only if there are $\beta < \alpha$ and an ITTM-program P such that P^z halts in time $< \beta$ on every input $z \subseteq \omega$ and, moreover, for any $x \subseteq \omega$, there is $y \subseteq \omega$ such that $P^{x\oplus y}\downarrow = 1$ if and only if $x \in X$. Also, ITTM-co-NTIME$_\alpha$ is the set of complements of elements of ITTM-NTIME$_\alpha$ in $\mathfrak{P}(\mathfrak{P}(\omega))$.

Moreover, X belongs to ITTM-NP if and only if there is $n \in \omega$ such that X belongs to ITTM-NTIME$_{\omega^n}$; in other words, ITTM-NP = ITTM-NTIME$_{\omega^\omega}$.

Finally, ITTM-co-NP is the set of complements of elements of ITTM-NP in $\mathfrak{P}(\mathfrak{P}(\omega))$.

Exercise 7.2.4 (See Schindler [155] and Deolalikar, Hamkins and Schindler, [46]). (a) Show that, for all $\alpha < \beta \in \text{On}$, we have $\text{TIME}^{\text{ITTM}}_\alpha \subseteq \text{TIME}^{\text{ITTM}}_\beta$ and $\text{NTIME}^{\text{ITTM}}_\alpha \subseteq \text{NTIME}^{\text{ITTM}}_\beta$

(b) Show that, for all $\alpha \in \text{On}$, $\text{TIME}^{\text{ITTM}}_\alpha \subseteq$ ITTM-NTIME$_\alpha \cap$ ITTM-co-NTIME$_\alpha$.

With these definitions, the P versus NP-question for ITTMs turns into a question of descriptive set theory, see [155]. Using this connection, one obtains the following.

Theorem 7.2.5.
(i) (Schindler, see [155]) ITTM-P $\subsetneq$ ITTM-NP.
(ii) (Deolalikar, Hamkins and Schindler, see [46]) ITTM-P $\subsetneq$ ITTM-NP $\cap$ ITTM-co-NP.

Exercise 7.2.6 (See [46], Lemma 5.2, Corollary 5.3; also cf. [195]). For $\alpha \in \text{On}$, let h^{ITTM}_α be the set of ITTM-programs that halt in time $< \alpha$. We can regard h_α as a subset of $\mathfrak{P}(\omega)$ by identifying natural numbers with subsets of ω in some natural way. For example, we can let $x \in h_\alpha$ if and only if x contains a 0 and if i is the length of the initial segment of x up to the first 0, then P_i halts in time $< \alpha$.

(a) Adapt the proof of the unsolvability of the halting problem to show that $h^{\text{ITTM}}_\alpha \notin \text{TIME}^{\text{ITTM}}_\alpha$.

(b) Exploit the proof in (a) to show that actually $h^{\text{ITTM}}_\alpha \notin \text{TIME}^{\text{ITTM}}_{\alpha+1}$.

(c) Show that, if α is ITTM-clockable, then $h_\alpha \in \text{TIME}^{\text{ITTM}}_{\alpha+2}$. Conclude that, for α clockable, we have $\text{TIME}^{\text{ITTM}}_{\alpha+1} \subsetneq \text{TIME}^{\text{ITTM}}_{\alpha+2}$.

Exercise 7.2.7 (See [46]). This exercise guides you through the proof of Theorem 7.2.5 as given in [46]. We reuse the notation from the last exercise.

(a) Show that $h_{\omega^\omega} \notin$ ITTM-P.

(b) Show that there is a (finite time) Turing computable real number x which codes a well-ordering of length $\omega^\omega + \omega$.

(c) Combine (b) with the results on evaluating bounded truth predicates from Chapter 2 to show that there are a real number y coding $L_{\omega^\omega+\omega}$, a natural number n and an ITTM-program P such that P^z halts in $< \omega^n$ many steps for all $z \subseteq \omega$ and such that P recognizes y.

(d) Show that there are an ITTM-program Q and a natural number n such that, for all real numbers z, Q^z halts in $< \omega^n$ many steps and Q decides $\{a \oplus y : a \in h_{\omega^\omega}\}$ (with y as in part (c)).

(e) Show that there are an ITTM-program Q' and a natural number n such that, for all real numbers z, Q'^z halts in $< \omega^n$ many steps and Q' decides $\{a \oplus y : a \notin h_{\omega^\omega}\}$ (with y as in part (c)).

(f) Deduce that $h_{\omega^\omega} \in$ (ITTM-NP $\cap$ ITTM-co-NP) $\setminus$ ITTM-P. Conclude that ITTM-P $\subsetneq$ ITTM-NP $\cap$ ITTM-co-NP.

In fact, there is a concise characterization of ITTM-NP$\cap$ITTM-co-NP (see Chapter 3 for the definition of Δ^1_1):

Theorem 7.2.8 (See Deolalikar, Hamkins and Schindler, [46], Theorem 3.2). ITTM-NP $\cap$ ITTM-co-NP $= \Delta^1_1$.

The next two exercises guide you through the proof, which is adapted from Deolalikar, Hamkins and Schindler [46], page 581, to our terminology.

Exercise 7.2.9 (See [46]). We want to show that ITTM-NP $\cap$ ITTM-co-NP $\subseteq \Delta^1_1$. If a set is Σ^1_1, then its complement is Π^1_1. It thus suffices to show that ITTM-NP $= \Sigma^1_1$.

(a) Show that there is a wITRM-computable function $f : \mathfrak{P}(\omega) \to \mathfrak{P}(\omega)$ such that, for all $x \subseteq \omega$, $f(x)$ codes $L_{\omega^\omega+\omega}[x]$.

(b) Conclude that there is wITRM-decidable set X of real numbers such that (1) for all $x \subseteq \omega$, there is $y \subseteq \omega$ such that $x \oplus y \in X$ and (2) for all $x \oplus y \in X$, y codes $L_{\omega^\omega+\omega}[x]$. Use Lemma 4.2.35 from Chapter 4 to conclude that X is Σ^1_1.

Now let Y be ITTM-NP, and let P be an ITTM-program witnessing this.

(c) Show that, with X as in (a), the set $Y' := \{(x \oplus y) \oplus z \in X : P^{x\oplus y}\downarrow = 1\}$ is wITRM-decidable. (Express $P^{x\oplus y}\downarrow = 1$ as a first-order formula and use evaluation of bounded truth predicates.) Conclude that Y' is Σ^1_1.

(d) Now show that $Y'' := \{x \subseteq \omega : \exists y, z(x \oplus y) \oplus z \in Y'\}$ is Σ^1_1.

Exercise 7.2.10 (See [46]). Now we show that $\Delta^1_1 \subseteq$ ITTM-NP$\cap$ITTM-co-NP. So let X be Δ^1_1, and let ϕ be a Σ^1_1-formula and ψ a Π^1_1-formula such that $X = \{x \subseteq \omega : \phi(x)\} = \{x \subseteq \omega : \psi(x)\}$. Further, let $\phi = \exists x\phi'$, $\psi = \exists x\psi'$, where ϕ' and ψ' are arithmetical formulas. As Δ^1_1 is by definition closed under complements, it suffices to show that $\Delta^1_1 \subseteq$ ITTM-NP.

(a) Let σ be an $\in$-formula. Show that there are $n \in \omega$ and an ITTM-program P with the following property: Given $x, y \subseteq \omega$ such that y codes $L_\alpha[x]$ for some $\omega < \alpha \in$ On, $P^{x\oplus y}$ halts in $< \omega^n$ many steps and $P^{x\oplus y}\downarrow = 1$ if and only if $L_\alpha[x] \models \sigma(x)$ and otherwise $P^{x\oplus y}\downarrow = 0$.

(b) Show that, for all $x \subseteq \omega$, $\phi'(x)$ holds if and only if $L_{\omega^\omega+\omega} \models \phi'(x)$.

(c) Conclude that $\Sigma^1_1 \subseteq$ ITTM-NP (which is stronger than what we wanted). Deduce that $\Delta^1_1 \subseteq$ ITTM-NP $\cap$ ITTM-co-NP.

7.2.1 ITTM-complexity with dependency on the input

Though the complexity classes defined above lead to a convenient complexity theory, the independence of the time complexity on the input may seem a bit unsatisfactory. Clearly, we cannot expect a dependency on the length of the input, as all inputs have the same length in this context. Thus, in [155], Schindler considered some ways to let the time bound depend on the "complexity" of the input x; here, we briefly describe one of them.

The idea is that the complexity of a real number x can be measured by the supremum of the computable ordinals relative to x, i. e., $\omega_1^{\mathrm{CK},x}$. Taking this measure to replace the length used in classical complexity theory, one obtains the following.

Definition 7.2.11 (Schindler, [155]). A set X of real numbers is ITTM-P$^+$ if and only if there is an ITTM-program P such that, for all $x \subseteq \omega$, P^x halts in $< \omega_1^{\mathrm{CK},x}$ many steps and outputs 1 if and only if $x \in X$ and otherwise outputs 0.

Likewise, X is ITTM-NP$^+$ if and only if there is an ITTM-program P such that, for all $x \subseteq \omega$, there is $y \subseteq \omega$ with $P^{x\oplus y}\downarrow = 1$ in $< \omega_1^{\mathrm{CK},x}$ many steps if and only if $x \in X$.

Concerning these complexity classes, we have the following.

Theorem 7.2.12.
(i) (Schindler, see [155], Corollary 2.14) ITTM-P$^+ \subsetneq$ ITTM-NP$^+$.
(ii) (Deolalikar, Hamkins, Schindler, see [46], p. 591) ITTM-P$^+$ = ITTM-NP$^+$ $\cap$ ITTM-co-NP$^+$.

It is shown in [46] how to generalize further: The guiding idea above is that, being deprived of using the length of the input x as the basis of bounding the length of the computation, we use some function f depending on x. However, not all functions are equally meaningful choices here: It would, e. g., be strange if we would allow less time on input $0'_{\mathrm{ITTM}}$ than on input 0. Thus, we say that $f : \mathfrak{P}(\omega) \to \mathrm{On}$ is a **suitable** function if $f(x) \geq \omega + 1$ for all $x \in \omega$ and $f(x) \geq f(y)$ whenever y is (finite time) Turing reducible to x. Then the classes ITTM-P^f, ITTM-NP^f and ITTM-co-NP^f can be defined by the obvious analogy with Definition 7.2.11. We conclude this section by mentioning without proofs the following results on this notion of complexity.

Theorem 7.2.13. *Suppose that f is a suitable function such that $f(x) > \Sigma^x$ for all $x \subseteq \omega$.*
(i) (Hamkins, Welch [192], Theorem 1.7), *then* ITTM-P$^f \neq$ ITTM-NP^f.
(ii) (Deolalikar, Hamkins, Schindler [46], Corollary 6.2), *then* ITTM-P$^f \subsetneq$ ITTM-NP^f $\cap$ ITTM-co-NP^f.

7.2.2 Space complexity for ITTMs

While it is obvious how to measure the computational time of an ITTM-program, the required space resources are not as easily quantified: By exercise 2.8.1 from Chapter 2, ITTMs that use only a proper initial segment of their working tape are rather trivial, so that all interesting ITTM-computations will use all tape cells, and thus have the same space complexity by the naive space complexity measure. For this reason, Löwe proposed a different approach in [129]: For a finite Turing machine, the length of a finite sequence of elements of an alphabet can be viewed as a measure for the complexity for that sequence. Similarly, one can measure the space usage of an ITTM-computation not by the length of the tape contents, but by the complexity of the occurring snapshots.

Definition 7.2.14 (Löwe, [129]). Let $C = (c_\iota : \iota < \delta)$ be a terminating ITTM-computation. Then the space usage of C is the minimal ordinal η such that $C \subseteq L_\eta[x]$, for any input x.

For $\alpha \in \mathrm{On}$, we denote by ITTM-PSPACE$_\alpha$ the set of all sets or real numbers that are decidable by an ITTM-program with computations with space usage $< \alpha$. Also, we let ITTM-PSPACE = ITTM-PSPACE$_{\omega^\omega}$.

Remark 7.2.15. An alternative proposal in [129] is to take the supremum of $\{\omega_1^{\mathrm{CK},c_\iota} : \iota < \delta\}$ as a measure for the space required by a certain tape content.

Except for some observations by Löwe in [129] and Winter in [195] and [196], this concept has not been studied much so far.

Exercise 7.2.16 (Winter, [195] Proposition 4.7, [196] Proposition 3.6). Show that, if $\alpha > \omega$, then $\mathrm{TIME}_\alpha^{\mathrm{ITTM}} \subseteq$ ITTM-PSPACE$_\alpha$.

The question is then whether $\mathrm{TIME}_\alpha^{\mathrm{ITTM}}$ is in general a proper subset ITTM-PSPACE$_\alpha$ (in classical complexity theory, the question whether P = PSPACE is open). Here, we mention the following results; for further results, we refer to [129, 195] and [196]:

Theorem 7.2.17 (Winter, [195, 196]).
(i) ([195], Theorem 5.12) *If* $\alpha \in [\omega + 2, \omega^2]$, *then* $\mathrm{TIME}_\alpha^{\mathrm{ITTM}} \subsetneq$ ITTM-PSPACE$_{\omega+2} \subseteq$ ITTM-PSPACE$_\alpha$.
(ii) ([195], Theorem 5.14) *If* $\alpha < \omega_1^{\mathrm{CK}}$, *we have* $\mathrm{TIME}_{\alpha+1}^{\mathrm{ITTM}} \subsetneq$ ITTM-PSPACE$_{\alpha+1}$.
(iii) ([195], Theorem 5.26) *If* β *is ITTM-clockable and* $\alpha \in (\beta + 3, \beta^+)$, *then* $\mathrm{TIME}_{\alpha+1}^{\mathrm{ITTM}} \subsetneq$ ITTM-PSPACE$_{\alpha+1}$.

One general conjecture in [129] is that sets decidable by computations that consist entirely of "simple" snapshots can also be decided in short time, which would mean that time and space complexity for ITTMs are strongly related. Note that exercise 2.8.1 and exercise 3.4.7 can be seen as somewhat encouraging results in this direction by providing time bounds on ITTM-computations in which every snapshot has only a fixed finite number of bits or only finitely many cells containing 1. However, we will now disprove this conjecture by showing that, for any $\alpha < \sigma$, there are sets of real numbers that are ITTM-decidable with extremely simple snapshots, but not with time bounded by α.[3]

Definition 7.2.18 (Cf. Löwe, [129]). A set $X \subseteq \mathfrak{P}(\omega)$ is ITTM-decidable with recursive snapshots if and only if there is an ITTM-program P that decides X and such that, for any $x \subseteq \omega$, all snapshots in the computation of P^x are recursive (in the classical sense).

In [129], it was asked whether any X that is ITTM-decidable with recursive snapshots belongs to ITTM-P, i. e., is ITTM-decidable with time bounded by ω^ω. We will now show that this even fails for any $\alpha < \sigma$ in place of ω^ω and much stricter simplicity criteria on the snapshots. The proof provides a nice example of how the study of one model of computability can help understanding another, and also of an application of the notion of recognizability.

Lemma 7.2.19. *If $x \subseteq \omega$ is ITRM-recognizable, then $\{x\}$ is ITTM-decidable with recursive snapshots.*

Proof. Recall from Chapter 2 that any ITRM-program using $n \in \omega$ many registers can be simulated by an ITTM using $(n+1)$ tape portions, the first n of which store the register contents, while the last one stores the active program line, where a natural number k is stored by writing k many 1s on the tape, followed by 0s.

Clearly, the snapshots of such an ITTM will be extremely simple (each of the $(n+1)$ tape portions either has finitely many 1s followed by 0s or, in the case of a register overflow, is entirely filled with 1s); in particular, they are recursive.

Now, if P is an ITRM-program that recognizes x, then the ITTM-program that simulates P in this way will decide $\{x\}$ with recursive snapshots. □

Lemma 7.2.20. *Suppose that $x \subseteq \omega$ is such that $\{x\}$ is ITTM-decidable by a program P such that P^x halts in α many steps. Then $x \in L_{\alpha^{+\omega}}$.*[4]

Thus, if $x \notin L_{\alpha^{+\omega}}$, then $\{x\}$ does not belong to $\mathrm{TIME}_\alpha^{\mathrm{ITTM}}$.

Proof. The statement $\exists y \subseteq \omega P^y\downarrow = 1$ is Σ_1 and, under the assumptions of the lemma, holds in $V_{\alpha^{+\omega}}$. Hence, by the Jensen–Karp theorem (Theorem 4.1.2), it holds in $L_{\alpha^{+\omega}}$, so $x \in L_{\alpha^{+\omega}}$.

For the second statement, observe that, if $\{x\}$ belongs to $\mathrm{TIME}_\alpha^{\mathrm{ITTM}}$ and this is witnessed by the program P, then P^x halts in $< \alpha$ many steps, and thus $x \in L_{\alpha^{+\omega}}$ by the first statement, contradicting the assumption. □

Theorem 7.2.21. *For every $\alpha < \sigma$, there is a set $X \subseteq \mathfrak{P}(\omega)$ that is ITTM-decidable with recursive snapshots, but not in $\mathrm{TIME}_\alpha^{\mathrm{ITTM}}$.*

---

3 Recall from Chapter 3 that σ is minimal with the property that $L_\sigma \prec_{\Sigma_1} L$.

4 Recall from Chapter 3 that $\alpha^{+\omega}$ denotes the smallest limit of admissible ordinals $> \alpha$.

Proof. Let $\alpha < \sigma$. As "there is a limit of admissible ordinals $> \alpha$" holds in L, it holds in L_σ, and thus $\alpha^{+\omega} < \sigma$. By Theorem 4.2.10, pick an ITRM-recognizable real number $x \in L_\sigma \setminus L_{\alpha^{+\omega}}$. By Lemma 7.2.20, $\{x\}$ does not belong to $\mathrm{TIME}_\alpha^{\mathrm{ITTM}}$. However, as x is ITRM-recognizable, $\{x\}$ is ITTM-decidable with recursive snapshots by Lemma 7.2.19. □

Corollary 7.2.22. *For any $\alpha < \sigma$, we have* $\mathrm{TIME}_\alpha^{\mathrm{ITTM}} \not\supseteq \text{ITTM-PSPACE}_{\omega+1}$.

Proof. It suffices to observe that all recursive real numbers belong to $L_{\omega+1}$. The claim then follows from Theorem 7.2.21. □

7.3 Complexity theory for OTMs[5]

As we discussed above, the most natural measure for space complexity, namely the supremum of the indices of the tape cells used by a computation, is not available for ITTMs. What is more, we also cannot measure time complexity in dependence from the length of the input, as all inputs have the same length ω. However, when we turn to OTMs, everything is well again: Inputs can have arbitrary ordinal length α, and we can consider how the computation time and space usage depends on α. To avoid technical coding details, we will for the purpose of complexity theory imagine OTMs to have a separate working tape besides an input and output tape, and the space usage will be the supremum of the indices of cells of the working tape used (i. e., visited by the read/write-head) in the course of the computation.

A warning is in order here: Already in classical complexity theory, the complexity of a problem crucially depends on the way one encodes the input. For example, the most natural way to test for primality—checking all smaller numbers for being divisors—is of polynomial complexity when encodes the natural number n as a string of n many 1s, and one can press it down further by picking even more inefficient codings. This is also the case for infinitary computability: We could reorganize any input of length α into one of length $\mathrm{card}(\alpha)$ or in fact any ordinal between $\mathrm{card}(\alpha)$ and $\mathrm{card}(\alpha)^+$. Thus, when coding is needed to make a problem accessible to an OTM—as is the case, e. g., for the SAT^∞—problem below—we need to say a few words on this issue.

In this section, we write OTMs for OTMs without parameters and pOTMs when parameters are admitted. The material in this section partly comes from [33], in particular Theorem 7.3.8 and Theorem 7.3.9 as well as the definitions and intermediate results leading there. I thank Benedikt Löwe, Benjamin Rin and Springer for the kind permission to reuse portions of this article here.

5 The material in this section comes partly from the article "Koepke Machines and Satisfiability for Infinitary Propositional Languages" [33] by B. Löwe, B. Rin and the author, which appeared in: J. Kari, F. Manea, I. Petre (eds.): Unveiling Dynamics and Complexity, LNCS 10307 ©Springer Nature 2017, with permission from Springer Nature. We thank our coauthors B. Rin and B. Löwe as well as Springer Nature for the kind permission to reuse it here.

We start by defining complexity classes for OTMs.

Definition 7.3.1 (Cf. [33]). An ordinal polynomial is an expression of the form $\alpha_0 x^{\gamma_0} + \alpha_1 x^{\gamma_1} + \cdots + \alpha_n x^{\gamma_n}$ with $n \in \omega$, $\alpha_0, \ldots, \alpha_n, \gamma_0, \ldots, \gamma_n \in \mathrm{On}$ and $\gamma_0 < \gamma_1 < \cdots . < \gamma_n$.

Definition 7.3.2. A "problem" or a "language" is a subclass of $\{0,1\}^{**}$, the set of ordinal-length 0-1-sequences.

A problem X is P^{∞} if and only if there are an OTM-program P and an ordinal polynomial p such that, for all $x \in \{0,1\}^{**}$, $P(x) \downarrow \chi_X(x)$ in $< p(|x|)$ many steps.[6]

A problem X is NP^{∞} if and only if there is are an P^{∞}-problem Y and an ordinal polynomial p such that $X = \{x \in \{0,1\}^{**} : \exists y \in \{0,1\}^{**}(|y| < p(|x|) \wedge x \oplus y \in X)\}$.

7.3.1 NP-completeness for OTMs

For OTMs, we have the following concept of NP-completeness.

Definition 7.3.3 (Cf. [33]). Let $X, Y \subseteq \{0,1\}^{**}$ be problems. We say that X is ∞-polytime reducible to Y or "∞-polynomially reduces to Y" if and only if there are an OTM-program P and an ordinal polynomial p such that, for all $x \in \{0,1\}^{**}$, $P^{Y}(x)\downarrow = \chi_X(x)$ in $< p(|x|)$ many steps.

In classical complexity theory, the NP-complete problem SAT of deciding the satisfiability of propositional formulas stands out as many other proofs of NP-completeness work by reducing SAT to the problem in question. We therefore strive for an infinitary analogue of SAT, which was obtain by Rin and Löwe; see [33].

Definition 7.3.4 (Cf. [33]). Let $\mathrm{Var} := (A_\iota : \iota \in \mathrm{On})$ be a proper class of propositional variables. The class IPF of infinite propositional formulas is the smallest class with the following properties:

- $\mathrm{Var} \subseteq \mathrm{IPF}$
- If $\phi \in \mathrm{IPF}$, then $\neg\phi \in \mathrm{IPF}$
- If $\Phi \subseteq \mathrm{IPF}$ is a set, then $\bigvee \Phi \in \mathrm{IPF}$

We will use $\bigwedge \Phi$ as an abbreviation for $\neg \bigvee \{\neg\phi : \phi \in \Phi\}$.

Definition 7.3.5. Let ϕ be an IPF.

If $\phi \in \mathrm{Var}$, then ϕ has no immediate subformulas. If $\phi \equiv \neg\psi$, then ψ is the only immediate subformula of ϕ. If $\phi \equiv \bigvee_{\psi \in X} \psi$, then X is the set of immediate subformulas of ϕ.

6 Note that we cannot say that X is "in" P^{∞}, as X is in general a proper class, and thus cannot be contained in another class.

Now, recursively, ψ is a subformula of ϕ if either $\psi \equiv \phi$ or ψ is a subformula of an immediate subformula of ϕ.

Let $S(\phi)$ denote the set of subformulas of ϕ.

Definition 7.3.6. Let ϕ be an IPF, and let f be a map from the propositional variables (elements of Var) appearing in ϕ to $\{0,1\}$. Such a map is called an "assignment." Then the valuation v_f of ϕ induced by f is the unique map from $S(\phi)$ to $\{0,1\}$ which satisfies the following clauses:

- $v_f(\xi) = f(\xi)$ for every $\xi \in \text{Var} \cap S(\phi)$
- If $\xi \equiv \bigvee_{\psi\in X} \psi$, then $v_f(\xi) = \max\{f(y) : y \in X\}$.
- If $\xi \equiv \neg\psi$, then $v_f(\xi) = 1 - v_f(\psi)$

ϕ is called "satisfiable" if and only if there is an assignment f such that $v_f(\phi) = 1$.

Let SAT^{∞} denote the class of satisfiable IPFs.

By the length of an IPF ϕ, we mean the order type of ϕ as a symbol string when written from left to right. We take this as the length of ϕ as an input, ignoring further coding issues (e. g., regarding coding the indices of the propositional variables).

For the proof of the next results, Theorem 7.3.8 and Theorem 7.3.9, we need the idea of coding an $\in$-structure with infinitary propositional logic. As the method is quite similar to the idea of coding $\in$-structure as sets of ordinals, we keep the description brief.

Let $(X, \in)$ be an $\in$-structure, let $\alpha \in \text{On}$ and $h : \alpha \to X$ bijective. Let $\mathcal{A}_\alpha := (A_{\iota,\xi} : \iota, \xi \in \alpha)$ be a set of propositional variables. Then the following assignment $f : \mathcal{A} \to \{0,1\}$ codes $(X, \in)$ via h: $f(A_{\iota,\xi}) = 1$ if and only if $h(\iota) \in h(\xi)$, and $f(A_{\iota,\xi}) = 0$, otherwise.

We now show how to express first-order properties of $\in$-structures as IPFs.

Lemma 7.3.7. *There is an OTM-program P_{ipf} with the following property: If $(X, \in)$ is an $\in$-structure, ϕ is an $\in$-formula (with parameters from X), $h : \alpha \to X$ is bijective and f is the assignment that codes $(X, \in)$ via h, then P_{ipf}, when run in the input (α, f, ϕ), halts and outputs an IPF $\overline{\phi}$ such that $v_f(\overline{\phi}) = 1$ if and only if $(X, \in) \models \phi$. Moreover, the length of $\overline{\phi}$ as a string will be $< \alpha^{\omega}$.*

Proof. $\overline{\phi}$ is obtained from ϕ by the following recursive rules, which can clearly be implemented on an OTM:

- If ϕ is $f(\iota) \in f(y)$, then $\overline{\phi}$ is $A_{\iota,y}$.
- If ϕ is $\neg\phi'$, then $\overline{\phi}$ is $\neg\overline{\phi'}$.
- If ϕ is $\phi_0 \vee \phi_1$, then $\overline{\phi}$ is $\overline{\phi_0} \vee \overline{\phi_1}$.
- If ϕ is $\exists x\phi'$, then $\overline{\phi}$ is $\bigvee_{\iota\in\alpha} \overline{\phi'[\frac{x}{f(\iota)}]}$.
- $=, \to, \wedge, \forall$ are defined in terms of $\in, \neg, \vee, \exists$ and then treated by the above rules.

The estimate on the length of $\overline{\phi}$ is now a matter of a simple induction on the construction of ϕ. □

We are ready to prove the ordinal analogue of the Cook–Levin theorem.

Theorem 7.3.8 (Rin and Löwe, [33], Theorem 4 and Theorem 5). SAT^∞ *is* NP^∞*-complete.*

Proof. It is easy to see that SAT∞ in fact belongs to NP^∞, as an assignment f satisfying a given IPF ϕ can clearly be verified in time ∞-polynomially bounded in the length of ϕ by an OTM.

We now aim to show that every NP^∞-problem is ∞-polynomially reducible to SAT^∞. (We will be a bit sketchy, as the next proof will elaborate more on the idea of coding structures in IPFs.) To this end, let $X \subseteq \{0,1\}^{**}$ be a problem that is NP^∞, and let P be an OTM-program and p an ordinal polynomial witnessing this. Thus, for $x \in \{0,1\}^{**}$, we have $x \in X$ if and only if there is $y \in \{0,1\}^{**}$ such that $P^{x\oplus y}\downarrow = 1$ in $< p(|x|)$ many steps. Now $P^{x\oplus y}$ halts with output 1 in $< p(|x|)$ many steps if and only if $L_{p(|x|)+\omega}[x \oplus y]$ believes that it does. It thus suffices to find, given x, an IPF ϕ_x that is satisfiable if and only if there is such a y in ∞-polynomial time. For all but boundedly many values of $|x|$, the cardinalities of $|x|$ and $p(|x|) + \omega$ will be the same, and we suppose without loss of generality that this is the case. Let $\rho := (p(|x|) + \omega)^\omega$. Thus, if a relativized L-level of the desired kind exists, it can be mapped bijectively via a map f onto ρ (since infinite L-levels always have the same cardinality as their index). Note that ρ is closed under ordinal multiplication. We assume without loss of generality that f maps ι to $\omega\iota$. As in Lemma 7.3.7, we fix $\rho \times \rho$ many propositional letters $A_{\iota\xi}$, with the intended meaning $f^{-1}(\iota) \in f^{-1}(\xi)$, and we additionally fix propositional letters A^x_ι and A^y_ι for all $\iota < \rho$. The intended meaning of A^x_ι is that $f^{-1}(\iota) \in x$, which is the case if and only if $\iota = \omega\xi < p(|x|) + \omega$ and $\xi \in x$, and likewise with A^y_ι. Given the condition on f just fixed, it is easy to encode x and y correctly by IPFs. Now the statement that the $\in$-structure M encoded in this way satisfies $V = L_{p(|x|)+\omega}[x \oplus y]$ and $P^{x\oplus y}\downarrow = 1$ is easily expressible by a short IPF using Lemma 7.3.7. Thus, we have reduced the question whether $x \in X$ to an instance of SAT^∞ in ∞-polynomial time, which is what we wanted. □

Theorem 7.3.9 (See [33]). SAT^∞ *is not OTM-decidable.*

Proof. The basic idea of the infinitary Cook–Levin theorem was to encode OTM-computations via IPFs. We try to reuse this idea encode, for a given OTM-program P, the statement that P halts in an appropriate IPF. If the encoding works effectively, this would reduce the OTM-halting problem to SAT^∞, showing that the latter is not OTM-decidable.

The approach is then this: P halts if and only if there is a countable L-level L_α that believes that P halts. So we can try to encode such an L-level. Thus, imagine that L_α is such a level and pick $g : \omega \to L_\alpha$ bijective.

Also, pick countably many propositional variables $(A_{i,j} : i,j \in \omega)$, with the intended interpretation of $A_{i,j}$ being "The set coded by i is an element of the set coded by j", i. e., $f(i) \in f(j)$. By Lemma 7.3.7, it is straightforward to express first-order sentences in L_α as IPFs. Thus we can, given P, compute an IPF $\phi(P)$ in the propositional variables $\{A_{i,j} : i,j \in \omega\}$ such that $\phi(P)$ is true under the valuation v_f if and only if the structure (ω, E_f) coded by the assignment f is a model of $V = L + P\downarrow$.

However, this does not help much unless we know that (ω, E_f) is isomorphic to an "actual" $\in$-structure, i. e., E_f is well-founded: For $(X, E) \models \mathrm{V} = \mathrm{L}$ only guarantees that (X, E) is isomorphic to an L-level when (X, E) is (isomorphic to) a transitive set!

We thus also have to express the well-foundedness of E_f. Now, for any particular sequence $s := (k_i : i \in \omega)$ of natural numbers, we can express "$(f(k_i) : i \in \omega)$ is not an infinite descending sequence in E_f" by $\bigvee_{i\in\omega} \neg A_{k_i,k_{i+1}}$. If we could do this for every constructible sequence s, this would be sufficient: For L, being a model of ZFC, would then contain a well-ordering of ω extending E_f, so there could be no infinite descending E_f-sequences at all.

The trouble is then that the set of all constructible sequences of natural numbers is not OTM-computable. This could be solved by adding the parameter ω_1^L: Given this parameter, one can compute (a code for) $L_{\omega_1^L}$, and from this obtain a list of all constructible sequences of natural numbers. However, the OTM-halting problem is also decidable in this parameter.

So let us work instead with the OTM-halting problem $H_{\omega_1^L} := \{i \in \omega : P_i(\omega_1^L)\downarrow\}$ relative to the parameter ω_1^L instead and try to encode that in IPF-formulas. Thus, we now want to say "there is $\alpha < \omega_2^L$ such that $L_\alpha \models P(\omega_1^L)\downarrow$."

To this end, we use ω_1^L many propositional variables $(A_{\iota,\xi} : \iota, \xi \in \omega_1^L)$ (again with the intended interpretation of $A_{\iota,\xi}$ being that $f(\iota) \in f(\xi)$ for some bijective $f : \omega_1^L \to L_\alpha$) and encode "the $A_{\iota,\xi}$ code an L-level L_α with $\alpha < \omega_2^L$ such that $L_\alpha \models P(\omega_1^L)\downarrow$."

The first-order properties that the $\in$-structure coded by the $A_{\iota,\xi}$ satisfies $V = L$ and $P(\omega_1^L)\downarrow$ are straightforward as above.

But now, all constructible sequences $(\alpha_i : i \in \omega)$ of ordinals $< \omega_1^L$ are contained in $L_{\omega_1^L}$.[7]

Thus, we can compute the set of all of these by using the parameter ω_1^L to compute a code for $L_{\omega_1^L}$ and then computing the subset of all sequences of ordinals that are countable from the point of view of $L_{\omega_1^L}$.

For each of these, we express that it does not correspond to a descending $\in$-sequence as explained above. In total, we obtain the desired IPF by an OTM-program in the parameter ω_1^L.

Now, if SAT^∞ was OTM-decidable, we would have an OTM-program P which, in the parameter ω_1^L, would solve the halting problem for OTMs in the parameter ω_1^L:

7 This follows from the condensation lemma; see Lemma 3.1.13 in Chapter 3. To see this, consider the transitive collapse of $\Sigma_1^{L_{\omega_1^L}}\{\sup_{i\in\omega} \alpha_i \cup \{(\alpha_i : i \in \omega)\}\}$.

Given Q, first construct the corresponding ϕ, then check its satisfiability. But this is a contradiction.

Thus SAT^{∞} is OTM-undecidable. □

Since every P^{∞}-problem is OTM-decidable by definition, we get the following.

Corollary 7.3.10. $\mathrm{P}^{\infty} \neq \mathrm{NP}^{\infty}$.

Exercise 7.3.11. (a) Adapt the above proof to show that there is no $\alpha \in \mathrm{On}$ such that SAT^{∞} is OTM-decidable in the parameter α. (Hint: The role of ω_1^L in the preceding proof can be played here by the next infinite L-cardinal after α.)

(b) By a further adaptation, show that there is no set S of ordinals such that SAT^{∞} is OTM-decidable when S is given as an oracle.

Exercise 7.3.12 (Cf. [27]). Let us say that a class $X \subseteq \{0,1\}^{**}$ is $\mathrm{EXPTIME}^{\infty}$ if and only if there are an OTM-program P, an ordinal α and an ordinal polynomial p such that P decides X and works for $\leq \alpha^{p(\beta)}$ many steps on an input of length β. Exploit the above argument to show that $\mathrm{EXPTIME}^{\infty}$ is properly contained in NP^{∞}. Generalize.

7.3.2 Structural properties of NP^{∞}[8]

An important theorem in classical complexity theory is the following.

Theorem 7.3.13 (Ladner's theorem; see, e. g., [6], Theorem 3.4). *If* $\mathrm{P} \neq \mathrm{NP}$, *then there is a language* $X \in \mathrm{NP}$ *which is neither in P nor* NP*-complete.*

Benjamin Rin asked in [33] whether the analogous statement holds for OTMs. We will show that it does. This was first done in [27]. Since we already saw above that $\mathrm{P}^{\infty} \neq \mathrm{NP}^{\infty}$, we formulate the statement as follows.

Theorem 7.3.14 ([27], Theorem 2). *There is an* NP^{∞}*-problem* X *which is neither* P^{∞} *nor* NP^{∞}*-complete.*

Proof. The idea is to define X by diagonalization against all OTM-programs with polynomial running time that decide a class. Now, every ordinal polynomial p is globally majorized by an ordinal polynomial of the form $\alpha \mapsto \alpha^{\beta}\beta$ for some $\beta \in \mathrm{On}$; it thus suffices to avoid decidability by programs having such a polynomial as their running time bound.

8 The material in this subsection comes partly from the article "Some Observations on Infinitary Complexity" [27], which appeared in: F. Manea, R. Miller, D. Nowotka (eds.), "Sailing Routes in the World of Computation," LNCS 10936, ©Springer Nature 2018, with permission from Springer Nature. We thank Springer Nature for the kind permission to reuse parts of this article here.

Also, for $\alpha \in \mathrm{On}$, $\alpha = \omega\beta + n$ with $n \in \omega$, we write α_0 for n and α_1 for β. In this way, we can regard a single ordinal α as an ordered pair consisting of an ordinal and a natural number.

We will write $P(x)\downarrow[\alpha] = y$ to express that the OTM-program on input x halts in $< \alpha$ many steps with output y and $P(x)\uparrow[\alpha]$ to mean that $P(x)$ does not halt in $< \alpha$ many steps.

So let us set $X := \{x \in \{0,1\}^{**} : P_{|x|_0}(x)\uparrow[|x|_1^{|x|_1}|x|_1] \vee P_{|x|_0}\downarrow = 0[|x|_1^{|x|_1}|x|_1]\} = \{x \in \{0,1\}^{**} : \neg P_{|x|_0}(x)\downarrow = 1[|x|_1^{|x|_1}|x|_1]\}$.

It is quite obvious that X is OTM-decidable: Given $x \in \{0,1\}^{**}$, calculate its length $|x|$ and the components $|x|_0$ and $|x|_1$ along with the bound $\beta := |x|_1^{|x|_1}|x|_1$, then simulate $P_{|x|_0}$ for β many steps on input x and answer 0 if this has halted with output 1 and otherwise answer 1.

It is also not hard to see that X is not P^{∞}: For suppose that P_k was an OTM-program that decides X in polynomial time, say with upper time bound $\alpha \mapsto \alpha^{\beta}\beta$. Let $x \in \{0,1\}^{**}$ be such that $|x|_0 = k$ and $|x|_1 > \beta$. Then $x \in X \leftrightarrow P_k(x)\downarrow = 1[|x|_1^{|x|_1}|x|_1] \leftrightarrow P_{|x|_0}(x)\downarrow = 1[|x|_1^{|x|_1}|x|_1] \leftrightarrow x \notin X$, a contradiction.

The interesting part of the proof is then to show that X is NP^{∞}. Thus, we need a quickly identifiable way of witnessing $x \in X$. Here is the idea:

Let $X' := \{(x,y) \in \{0,1\}^{**} \times \{0,1\}^{**}$: '$y \subseteq |x|$ codes a KP-model $M' \wedge |x| \in \mathrm{wf}(M) \wedge M \vDash \neg P_{|x|_0}(x)\downarrow = 1[|x|_1^{|x|_1}|x|_1]\}$.

Claim 1. *There are an OTM-algorithm Q and an ordinal polynomial q such that, for $(x,y) \in \{0,1\}^{**} \times \{0,1\}^{**}$ with $|x| = |y|$, Q decides whether or not $(x,y) \in X'$ in time $\leq q(|x|)$.*

Proof. By adapting the program P_{truth} from Chapter 3, it is not hard to see that, if ϕ is a first-order statement, then the question whether ϕ holds in the structure coded by $y \subseteq \alpha$ is decidable with time bounded by α^{ω}, uniformly in ϕ. As KP is a countable set of such statements, the question whether or not y codes a model of KP is decidable in time bounded by $\alpha^{\omega}\omega$.

Testing whether $|x|$ belongs to the well-founded part of the structure coded by y works by recursively constructing an embedding of $|x|$ to an initial segment of the ordinals of that structure: Suppose we have constructed such an embedding for $\beta < |x|$, then an exhaustive search through α reveals whether the embedding can be continued to one of $\beta + 1$. Thus, for every $\beta < |x|$, we need to decide a first-order statement about the coded structure, which is possible in ordinal polynomial time in α. □

Claim 2. $X = \{x \in \{0,1\}^{**} : \exists y(x,y) \in X'\}$.

Proof. For every $x \in \{0,1\}^{**}$, there is a KP-model $M \ni x$ such that $\mathrm{card}(M) = \mathrm{card}(|x|)$ and $|x| \in \mathrm{wf}(M)$. To see this, let κ be the cardinal successor of α and form the transitive collapse of the elementary hull of $|x| + 1 \cup \{x\}$ in $L_{\kappa}[x]$.

Moreover, if $M \models$ KP such that $|x| \in \mathrm{wf}(M)$, then $|x|_1^{|x|_1}|x|_1$: This follows from the fact that KP-models are closed under ordinal exponentiation and multiplication (as they are closed under Σ_1-recursion by Chapter 3), together with the fact that well-founded parts of KP-models are again KP-models (Lemma 3.1.39 of Chapter 3).

Now, if $x \in X$, then let M be such a KP-model and let $y \subseteq |x| \times |x|$ be a code for M, which exists as $\mathrm{card}(M) = \mathrm{card}(|x|)$. It follows that $(x, y) \in X'$. Hence, if $x \in X$, then there is y such that $(x, y) \in X'$.

On the other hand, suppose $(x, y) \in X'$ and let M be the KP-model coded by y. If $M \models \neg P_{|x|_0}(x)\downarrow = 1[|x|_1^{|x|_1}|x|_1]$, then $\neg P_{|x|_0}(x)\downarrow = 1[|x|_1^{|x|_1}|x|_1]$, as the computation of $P_{|x|_0}$ on input x of length $|x|_1^{|x|_1}|x|_1$ inside M belongs to the well-founded part of M, and thus coincides with this computation in the real world. Thus, $(x, y) \in X'$ implies $x \in X$. □

Together, Claim 1 and 2 yield that X is indeed NP^∞. Thus, X is NP^∞-intermediate, as desired. □

With a similar argument, we can show something stronger. Let us say two classes $X, Y \subseteq \{0,1\}^{**}$ are ∞-polytime-incomparable if and only if there is no OTM-program P that decides (without parameters) X in the oracle Y in time bounded by some ordinal polynomial or that decides Y in the oracle X in time bounded by some ordinal polynomial.

Theorem 7.3.15. *There are infinitely many classes* $(A_i : i \in \omega)$ *of ordinals*[9] *that are* NP^∞ *and that are pairwise* ∞*-polytime-incomparable; in particular, none of the* A_i *is* NP^∞*-complete.*

Proof. We use a priority argument, similar to the solution of Post's problem for OTMs in Chapter 6 (see Theorem 6.5.1). For $i \in \omega$, $i = p_3(j, k, l)$, $(j, k, l) \in \omega^3$, let $t_i = (j, k, l)$. Below, we will identify a class X of ordinals with the subset of $\{0,1\}^{**}$ containing, for each $\alpha \in X$, the string consisting of α many 0s followed by a 1 and no other elements.

We imagine an OTM-program Q working on ω many tapes $(B_i : i \in \omega)$, where B_i will be used to store the characteristical function of A_i; it is easy to simulate the workings of such a machine on an OTM with a single tape. Some further scratch work will be done on finitely many further tapes. We also have an ordinal β_i^ι associated with B_i at any time ι of the computation for $i \in \omega$; initially, we have $\beta_i^0 = 0$ for all $i \in \omega$. After that, the sequences $(\beta_i^\iota : \iota \in \mathrm{On})$ will be increasing. The significance of β_i^ι is that, from step ι on, the first β_i^ι many cells of B_i are "blocked," i. e., their content cannot be modified subsequently.

Q will work in On many phases, where each phase consists of ω many subphases. Suppose that the ith subphase of the αth phase starts at time $\tau \in \mathrm{On}$ and that $t_i =$

9 Strictly speaking, we cannot form a sequence of proper classes in ZFC. We trust our reader to nevertheless understand this expression and give a formal treatment of it within ZFC, should she desire so.

$(j, k, l) \in \omega^3$. Then we do the following: Let $\xi(\alpha, i, k) = \beta_k^T$ be the first cell of B_k that is not blocked. We will ensure that the jth OTM-program P_j does not reduce the content of B_k to that of B_l in time bounded by the ordinal polynomial $x \mapsto \alpha x^\alpha$. (Ensuring this for all $\alpha \in \mathrm{On}$ guarantees that P_j does not reduce the content of B_k to that of B_l in ∞-polynomial time at all.)

So let P_j run for $\xi(\alpha, i, k) \cdot \xi(\alpha, i, k)^{\xi(\alpha,i,k)}$ many steps in the input $\xi(\alpha, i, k)$ with the current content of B_l as its oracle and write 1 in the $\xi(\alpha, i, k)$th cell of B_k if and only if this computation does not terminate with output 1; otherwise, write 0 in this cell. Then, increase β_k and β_l to block the $\xi(\alpha, i, k)$th cell of B_k as well as all cells of B_l used in the computation.

By construction, for every $\iota \in \mathrm{On}$ and every $i \in \omega$, we will eventually have $\beta_i^T > \iota$, i. e., the ιth cell of B_i will eventually stop changing. In fact, every β_i is increased at least by ω in each phase, so that the αth cell of all B_i will be blocked after the αth phase for all $i \in \omega$.

Let $\iota \in A_i$ if and only if B_i contains a 1 at the time the ιth cell is blocked.

The classes A_i thus obtained are not OTM-reducible to each other in ∞-polynomial time by construction.

It remains to see that each A_i is NP^∞. Suppose that the ith subphase of the αth phase is finished after $g(\alpha, i)$ many steps. Then all simulations so far have taken $< g(\alpha, i)$ many steps, and thus never moved any head behind position $g(\alpha, i)$; so β_j is below $g(\alpha, i)$ for all $j \in \omega$ after the subphase has finished. Now the $(i + 1)$th phase will run for $< g(\alpha, i) \cdot g(\alpha, i)^{g(\alpha,i)}$ many steps, so $g(\alpha, i + 1) < g(\alpha, i) \cdot g(\alpha, i)^{g(\alpha,i)}$. The αth phase then finishes after $\sup_{i\in\omega} g(\alpha, i)$ many steps. This recursion can be used to define a Σ_1-function mapping $\alpha \in \mathrm{On}$ to an upper bound $h(\alpha)$ on the time it takes Q to finish the αth phase. Moreover, by induction, we will have $h(\alpha) < \alpha^+$. It follows that, if M is a model of KP that contains α in its well-founded part, then $\alpha \in A_i$ if and only if M contains a computation of Q up to a time when the αth cell of B_i is blocked and such that this cell contains 1 at this time, which is a first-order statement. Thus, as in the proof of Theorem 7.3.14, we can add an α-code for a KP-model containing α in its well-founded part to see that A_i is in fact NP^∞.

Finally, if any of the A_i was NP^∞-complete, then the others would have to be ∞-polytime-reducible to A_i, a contradiction. □

Exercise 7.3.16. Modify the proof to Theorem 7.3.15 to show the following:

(a) There are infinitely many classes $(A_i : i \in \omega)$ of ordinals that are NP^∞ such that none of them is pOTM-reducible in ∞-polytime (defined as above) to any other.

(b) For any cardinal κ, there are κ many classes $(A_\iota : \iota < \kappa)$ that are NP^∞ such that none of them is OTM-reducible (without parameters) in ∞-polytime to any other.

(c) Combine (a) and (b).

Exercise 7.3.17 ([27], Theorem 3). Let $S := \{w \in \{0, 1\}^{**} : \neg P_{w(n)}(w)\downarrow = 1[|w| \cdot 2]\}$ be the class of elements w of $\{0, 1\}^{**}$ that have the following property: Let $n(w) = i$ when w starts with $i \in \omega$ many 1s, followed by a 0, and let $n(w) = 0$ when w starts with

ω many 1s. Let $w \in S$ if and only if the $n(w)$th OTM-program $P_{n(w)}$ does not output 1 on input w in $< |w| \cdot 2$ many steps. Show that there is a natural number n such that S is decidable by an OTM-program (with multiple, but finitely many tapes) in running time $\omega \cdot n$, but not in running time $\omega \cdot 2$. Conclude that the analogue of the speedup theorem 1 Theorem 7.1.1 for OTMs fails.

Exercise 7.3.18 ([27], Corollary 2). For $\beta \in \mathrm{On}$, let us say that a problem X is P_β if and only if X is decidable by an OTM-program with running time bounded by $\alpha \mapsto \alpha^\beta\beta$. Show that, whenever $\alpha < \beta$ are OTM-clockable ordinals, then P_α is properly included in P^β.

7.3.3 Strong space bounds and regularity

In Chapter 2, we introduced deterministic ordinal automata (DOAs) and REG^∞-languages. A nice characterization of regular languages (in the classical sense) in terms of complexity theory is the following (see, e. g., Hromkovic [89] or Fortnow and Gasarch [58]).

Theorem 7.3.19. *Suppose that the language $\mathcal{L}$ is decidable by a Turing machine P that uses only the first* $\mathrm{loglog}(|x|)$ *many cells of its working tape on the input x. Then $\mathcal{L}$ is regular.*

A similar result holds for OTMs. The proof is too long to be given here in detail and can be found in [28].

Definition 7.3.20. An OTM-program P is "strongly space-bounded" if and only if there is an ordinal α such that, for all $x \in \{0,1\}^{**}$ with $|x| > \alpha$, the space usage of P on the input x is strictly smaller than $\mathrm{card}^L(|x|)$.

The result is then the following.

Theorem 7.3.21. *If X is decidable by a strongly space-bounded OTM, then X is* REG^∞.

8 Applications and interactions

8.1 Introduction

Originally, Turing computability was introduced to have a convincing formal model for the intuitive concept of computability to serve as a basis for a proof of the unsolvability of Hilbert's Entscheidungsproblem. Subsequently, Turing computability was applied to many areas of mathematics, yielding fruitful interactions: To only name a few, there is computable analysis (see, e. g., Weihrauch, [178]), computable algebra (see, e. g., Rabin, [147]) and computable model theory (see, e. g., Harizanov, [82]). Also, there are proofs for results in other areas of mathematics that use computability concepts, such as the existence proof for infinitely mans primes using Kolmogorov complexity (see, e. g., Hromkovic, [89]), which can be developed into a reasonable estimate for the density of prime numbers.

The question whether infinitary models of computability can be viewed as a "convincing formal model" of any intuitive concept will be taken up in the next chapter. In any case, infinitary computability can be applied in realms where Turing computability with its restriction to countable objects (or objects representable in some "countable way," such as separable spaces; see Simpson [163]) cannot. One can therefore hope to arrive at some interesting bits of mathematics by introducing concepts of "infinite time effectivity" into other areas of mathematics and set theory and by trying to reprove known results using these concepts. It is the purpose of this chapter to give an overview of some such interactions, more specifically using ordinal computability for reproving results in constructible and descriptive set theory (the fields studying Gödel's L and definable subsets of the real line from a set-theoretical point of view) by Koepke, Schlicht and Seyfferth, the approach of Coskey and Hamkins to replace Borel reducibility with ITTM-reducibility, the steps taken into computable model theory by Hamkins, Miller, Seabold and Warner and an ordinal-time variant of Weihrauch reducibility that can, e. g., be applied to formulations of the axiom of choice.

Brief surveys of applications of ITTMs covering some of the topics below can also be found in Coskey and Hamkins [43] and Hamkins [74].

8.2 Ordinal Turing machines and the constructible hierarchy

Soon after these ordinal Turing machines had been introduced, it turned out that proofs from constructible set theory (the theory of Gödel's L) and descriptive set theory could often be recast in the language of ordinal computations, arguably making them simpler or more directly accessible, at least to some readers. Though this may ultimately be a matter of taste or personal preferences, it is nonetheless an advantage to have such presentations available for those who find ordinal computability intuitive. In this section, we consider the first such example, namely Koepke's proof of the

https://doi.org/10.1515/9783110496154-008

generalized continuum hypothesis in L in Section 7 of [104], which elegantly avoids making use of the full power of the condensation lemma used in the usual proof.

Definition 8.2.1. The continuum hypothesis (CH) is the claim that $2^{\aleph_0} = \aleph_1$. More generally, the generalized continuum hypothesis (GCH) states that $2^\kappa = \kappa^+$ whenever κ is an infinite cardinal (where κ^+ denotes the cardinal successor of κ).

Both CH and GCH are independent of the axioms of ZFC. One half of this result was Gödel's proof that GCH holds in L (the other came from Cohen's introduction of the forcing technique). We will now prove this result with Koepke's argument in [104], which uses the characterization of L via ordinal Turing machines from Chapter 3.

Theorem 8.2.2 ([104], Theorem 7.3). *If κ is a cardinal in L, then $(2^\kappa)^L = (\kappa^+)^L$.*

Proof. We show the following: If $x \subseteq \kappa$ is OTM-computable in the parameter $\gamma \in \mathrm{On}$, then it is OTM-computable in some parameter $\alpha < (\kappa^+)^L$. (*)

To see that this suffices, note that, by Theorem 3.5.5 in Chapter 3, every constructible set of ordinals is OTM-computable in some ordinal parameter α. Now, there are only ω many programs and, by (*), only $(\kappa^+)^L$ many choices of the parameter that need to be taken into account. Since each program computes at most one subset of κ for each parameter, there are in total at most $\omega(\kappa^+)^L = (\kappa^+)^L$ many constructible subsets of κ. By Cantor's theorem, there must also be at least that many, and so $(\kappa^+)^L = |\mathfrak{P}^L(\kappa)|^L = (2^\kappa)^L$.

To prove (*), let P be an OTM-program and $\alpha \in \mathrm{On}$ be such that P computes $x \subseteq \kappa$ in the parameter α. Pick $\beta \in \mathrm{On}$ sufficiently large so that the computation C of P in the parameter α is contained in L_β. In particular then, we will have $x \in L_\beta$. Now let $\mathcal{H}$ be the elementary hull of $\kappa \cup \{\alpha, C\}$ in L_β, and let M be its transitive collapse.[1]

Now, by simple combinatorics, $|M|^L = |\mathcal{H}|^L = \kappa \cdot \aleph_0 = \kappa$. As the collapsing map π is an isomorphism, it will send ordinals to ordinals, so let $\bar{\alpha} \in \mathrm{On}$ be the image of α. Then $|\bar{\alpha}|^L \leq \kappa$, so $\bar{\alpha} < (\kappa^+)^L$. As an isomorphism, π is elementary, so $\pi(C)$ will in M be the computation of $\pi(P)$ in the parameter $\bar{\alpha}$ with output $\pi(x)$.

But $x \subseteq \kappa \subseteq \mathcal{H}$, so $\pi(\iota) = \iota$ for all $\iota \in \kappa$, and hence $\pi(x) = x$. Thus, in M, $\pi(C)$ is an OTM-computation on the parameter $\bar{\alpha} < (\kappa^+)^L$ with output x. As M is transitive, the same holds in V by absoluteness of computations. Thus we have found a parameter $< (\kappa^+)^L$ from which x is OTM-computable, as desired. □

8.3 Infinitary computations and descriptive set theory

We already saw an application of ordinal computability in descriptive set theory in Chapter 4, where in order to show the failure of the lost melody theorem for wITRMs,

1 Note that we do **not** require that $M = L_\gamma$ for some $\gamma \in \mathrm{On}$, as is done in Gödel's proof based on the condensation lemma—all that we use is the Mostowski collapse.

we proved that (i) every nonempty Σ_1^1-set $X \subseteq \mathfrak{P}(\omega)$ contains a wITRM-computable element (Theorem 4.2.29) and that every wITRM-computable set of real numbers is Σ_1^1 (see exercise 4.2.36). As the latter holds in particular for Σ_1^1-singletons, we have, by a wITRM-effective variant of the original argument, reproved a statement known as Kreisel's basis theorem (see, e. g., Mansfield and Solovay [135] or Sacks [153], p. 74).

Theorem 8.3.1. *Every nonempty Σ_1^1-set of real numbers contains a Σ_1^1-singleton.*

For another example, we turn to a variant of the Jensen–Karp-theorem (see Theorem 4.1.2 in Chapter 4) that was frequently used in Chapter 4 that we prove using ITRMs. Though the statement that we prove below is weaker than the full Jensen–Karp theorem, it suffices for all of our applications in Chapter 4.

Theorem 8.3.2 (Cf. Jensen and Karp [95], pp. 162–163). *Let $x \subseteq \omega$ and ϕ a Σ_1-formula such that $V_{\omega_\omega^{\mathrm{CK},x}} \models \phi(x)$. Then $L_{\omega_\omega^{\mathrm{CK},x}}[x] \models \phi(x)$.*

Proof. The proof is basically an ITRM-effective variant of that by Jensen and Karp in [95], which is based on primitive recursive set functions.

Let $x \subseteq \omega$ and $\phi(x) \equiv \exists y \psi(x, y)$ with ψ a Δ_0 formula such that $V_{\omega_\omega^{\mathrm{CK},x}} \models \phi(x)$.

The idea is to compute with an ITRM, using the oracle x, a (code for a) transitive set $M \ni x$ such that $M \models \phi(x)$. To this end, we extend $\phi(x)$ to an appropriate complete theory in which every existential statement is witnessed by a constant; then we use the Henkin model construction, which can be effectivized on an ITRM.

Once this is done, the claim follows: As c is ITRM-computable from x, we have $c \in L_{\omega_\omega^{\mathrm{CK},x}}[x]$. It is then not hard to conclude that $M \in L_{\omega_\omega^{\mathrm{CK},x}}[x]$. As $M \models \phi(x)$, there is some $y \in M$ such that $M \models \psi(x, y)$. As M is transitive, $\psi(x, y)$ holds in the real world. Finally, as $y \in M \in L_{\omega_\omega^{\mathrm{CK},x}}[x]$ and the latter is transitive, we have $y \in L_{\omega_\omega^{\mathrm{CK},x}}[x]$, so $L_{\omega_\omega^{\mathrm{CK},x}}[x] \models \phi(x)$, as desired.

We could of course attempt to compute a Henkin-model for $\phi(x)$ right away on an ITRM in the oracle x. The problem is that this model might fail to be transitive (or to be isomorphic to a transitive model), so that the witness it contains for $\phi(x)$ may fail to witness this statement in V. Therefore—and this is the crucial point of the Jensen–Karp proof—it has to be ensured that the model will be well-founded, and hence isomorphic to a transitive model. This is the point of the following construction.

As $L_{\omega_\omega^{\mathrm{CK},x}}[x] \models \phi(x)$, there is $\beta < \omega_\omega^{\mathrm{CK},x}$ such that $L_\beta[x] \models \phi(x)$; we pick such a β, say the minimal one. Then a code r_β for β is ITRM-computable from x. We may thus use β freely in the following construction.

We introduce constant symbols $(c_\iota : \iota < \beta)$ and another constant symbol c_x and form the theory $\mathcal{T}$ using the binary relation symbol $\in$ and the unary function symbol rk (intended to denote the rank function) which will contain the following statements: $\phi(c_x)$, $c_i \in c_x$ for $i \in x$, $c_i \notin c_x$ for $i \notin x$, $c_\iota \in c_{\iota'}$ for $\iota < \iota' \leq \beta$, $c_\iota \notin c_{\iota'}$ for $\iota' \leq \iota \leq \beta$ and $\mathrm{On}(c_\iota)$ for $\iota \leq \beta$, which we use as an abbreviation for $c_\iota \in \mathrm{On}$. Moreover, $\mathcal{T}$ contains the axiom of extensionality and statements giving the relevant information about the rank function, namely $\forall x, y(x \in y \rightarrow \mathrm{rk}(x) \in \mathrm{rk}(y))$, $\forall x(\mathrm{rk}(x) \in \mathrm{On})$ and $\forall \alpha \in \mathrm{On}(\mathrm{rk}(\alpha) = \alpha)$.

Clearly, this theory $\mathcal{T}$ is ITRM-computable from x; moreover, $\mathcal{T}$ is consistent, as $V_{\beta+1}$ with the standard interpretations of $\in$ and rk is a model of $\mathcal{T}$.

We now extend, in an ITRM-effective manner, $\mathcal{T}$ to a suitable complete consistent theory $\mathcal{T}'$ in which every existential statement is witnessed by one of ω many new constants $\{d_i : i \in \omega\}$, and which moreover has the property that, if $\mathcal{T}'$ contains the statement $\mathrm{On}(a)$ for some a, then it also contains the statement $a = c_\iota$ for some $\iota \leq \beta$. This latter condition ensures that the model we obtain in the end will only have "actual" ordinals, and thus be well-founded.

Let $\mathcal{L}$ be the first-order language using the function, relation and constant symbols just described. We enumerate the $\mathcal{L}$-formulas with one free variable in the order type ω as $(\psi_i : i \in \omega)$. We now build a tree $\mathcal{B}$ on the set of the closed $\mathcal{L}$-formulas. For all $i \in \omega$, at the $3i$th level, every node of $\mathcal{B}$ will have ω many immediate successors labeled $\exists z\psi_i(z) \rightarrow \psi_i(d_j)$, $j \in \omega$. At the $(3i + 1)$th level, we will again have ω many immediate successors for each node, labeled $\mathrm{On}(d_i) \rightarrow (d_i = c_\iota)$, $\iota \leq \beta$; these can be arranged in order type ω as we have a real number coding β available. Finally, at level $3i + 2$, we have the ith closed formula of $\mathcal{L}$. A finite sequence $\vec{s}$ of formulas will enter $\mathcal{B}$ if and only if $\mathcal{T} + \bigwedge \vec{s}$ is consistent, which can easily be tested with an ITRM. Thus, $\mathcal{B}$ is ITRM-computable from x.

Any infinite branch of $\mathcal{B}$ will be a complete consistent extension of $\mathcal{T}$. We know that such an extension exists, as we can form the elementary hull of β in $V_{\beta+1}$, which will result in a countable structure in which d_i can be interpreted as the ith element and all other symbols can be interpreted in the obvious way. To find such a branch with an ITRM given $\mathcal{B}$, we use the ITRM-algorithm P_{wf} to test for well-foundedness of binary relations (see Theorem 2.3.25 from Chapter 2): $\mathcal{B}$ has an infinite branch if and only if the binary relation defined by $\eta < \eta'$ if and only if η is a successor of η' in $\mathcal{B}$ is not well-founded. Moreover, this algorithm can also be used to test whether a given sequence of formulas extends to an infinite branch of $\mathcal{B}$. In this way, always selecting the lexically minimal (leftmost) continuation of a branch that still extends to an infinite branch of $\mathcal{B}$, we can ITRM-compute an infinite branch of $\mathcal{B}$ from x.[2]

Now let $\mathcal{T}'$ be a complete and consistent extension of $\mathcal{T}$ as described and consider the model M built from the constant symbols of $\mathcal{L}$ in the usual way: We form equivalence classes of constants depending on whether $\mathcal{T}'$ contains the statement that they are equal and, for two such classes $[d]$, $[d']$, we let $\mathrm{rk}([d]) = [d']$ if and only if the formula $\mathrm{rk}(d) = d'$ is contained in $\mathcal{T}'$ and $[d] \in [d']$ if and only if $(d \in d')$ is contained in $\mathcal{T}'$. Clearly, the resulting structure S has no infinite descending $\in$-sequences, as these would translate into an infinite descending $\in$-sequence among the c_ι, which would in turn induce an infinite descending sequence in their indices, i. e., in the ordinals, a contradiction. Moreover, taking the transitive collapse M of S, c_ι will be mapped to ι

2 In Corollary 4.2.23 in Chapter 4, we saw with a somewhat more subtle argument that the task of finding the leftmost infinite branch of a given tree is actually possible already for a wITRM.

for all $\iota \leq \beta$, so in particular, c_i will be mapped to i for $i \in \omega$, and thus, by extensionality in S, c_x will be mapped to x. As S satisfies $\phi(x)$ by definition, the same holds for M. Thus M is as desired.

We can now ITRM-compute a code for M from $\mathcal{T}'$ by implementing the above construction on an ITRM, which is easily possible. □

An important absoluteness result in set theory which we frequently used in this book is Shoenfield's absoluteness theorem; see, e. g., ([10], p. 189). A proof using OTMs was given by P. Schlicht and B. Seyfferth in [160]. We give here an argument in the spirit of that in [160], adapting the approach in the last proof.

Theorem 8.3.3 (Cf. [160], Corollary 1). *Let M be a transitive model of KP such that $\omega_1 \in M$, $x \in M$ a real number and let $\phi(x)$ be Σ_1. Then $\phi(x)$ holds in M if and only if it holds in V. In particular, if $x \in L$ and $\phi(x)$ holds in V, then $\phi(x)$ holds in L.*

Proof. Using an algorithm similar to that used in the Jensen–Karp variant above, we use an OTM to compute a transitive model of $\phi(x)$ in the parameter ω_1 on the input x. The construction is similar to the one above, and we merely sketch it for the case that $x = \emptyset$.

Let β be such that $V_\beta \models \phi$, let H be the elementary hull of $\emptyset$ in V_β and let $\pi : H \to \bar{M}$ be the collapsing map. Then M is a transitive, L-countable set with $\bar{M} \models \phi$. Moreover, $\alpha = \bar{M} \cap \mathrm{On} \in \omega_1$ and every set in $\bar{M}$ has its rank in α.

We now want to compute a complete theory $\mathcal{T}$ extending ϕ and a model thereof as above, using again the d_i as the underlying set of constants. The ordinal height of the desired model is the countable ordinal α, which we can give to the machine as a parameter. Thus, any halting OTM-computation in the parameter α will halt in a countable number of steps. It follows that the whole computation, and thus also its output, will be contained in $L_\alpha[x]$, which is a subset of M as $\alpha < \omega_1 \in M$ and the L-hierarchy is definable in models KP by Lemma 3.1.40 from Chapter 3. So M contains a code for a set witnessing the truth of $\phi(x)$. As $M \models$ KP and models of KP can decode such codes by Lemma 3.1.32 from Chapter 3, such a witness is contained in M, so $M \models \phi(x)$.

The second claim now follows trivially, as L is a transitive model of KP that contains ω_1. □

8.4 Infinite time computability and Borel reducibility

An interesting problem associated with any kind of structure appearing in mathematics, such as groups, fields, modules, vector spaces, linear orderings, etc. is their classification, i. e., of telling whether they are in some essential way the same, e. g., isomorphic.

When the structures in question are countable, they can be coded by real numbers in some natural way and equivalence relations like, e. g., isomorphy of groups are thereby translated into equivalence relations on real numbers. Thus, our general problem is: Given an equivalence relation $\sim$ on (some subset of) $\mathfrak{P}(\omega)$, we want to determine, for given $x, y \subseteq \omega$, whether $x \sim y$. Let us call this the "classification problem" associated with $\sim$.

In some cases, it seems that one such problem P is "easier" than another such problem Q in the sense that, provided one gets the solutions to Q for free, one would know how to solve P. In other words, there is a way to "translate" P into Q. As a very basic example, it is easy to classify countable groups up to isomorphism when one knows how to do it for arbitrary countable sets with a binary function.

How do we make of "translating" one classification problem into another precise? The obvious idea would be to say that, given two equivalence relations $\sim$ and $\sim'$ on (subsets of) $\mathcal{P}(\omega)$, $\sim$ is simpler than $\sim'$ if and only if there is a function $f : \mathfrak{P}(\omega) \to \mathfrak{P}(\omega)$ such that, for all $x, y \subseteq \omega$, we have $x \sim y$ if and only if $f(x) \sim' f(y)$.

This, of course, is too liberal: If we allow f to be **any** function, this only means that there is an injection from the $\sim$-equivalence classes into the $\sim'$-equivalence classes. Though this may indeed count as a "complexity measure," it is clearly too coarse and hardly captures the intuition that "one would know $\sim$ if one knew $\sim'$." What is missing is some kind of simplicity or effectivity condition concerning f. Friedman and Stanley had the idea of demanding that f is Borel (see [61]), which started the fruitful field of Borel reducibility. Later on, Coskey and Hamkins proposed to use ITTM-computable functions instead of Borel functions (see [42]); this section will give a brief introduction to this subject. We do not go into any depth here; for a wealth of results, see [42].[3]

8.4.1 ITTM-reducibility and Borel reducibility

In general, Borel reducibility does the job of serving as a conceptual basis for comparing the complexities of classification problems quite well. However, as Coskey and Hamkins point out in [42], there are some rather natural equivalence relations that one might well regard as "effective," but which are not Borel. As an example, they mention the relation on $\mathfrak{P}(\omega)$ where two real numbers x and y are equivalent if and only if $\omega_1^{\mathrm{CK},x} = \omega_1^{\mathrm{CK},y}$.

Thus, let us consider the following reducibility relation.

3 We note that Borel-reducibility can be characterized as a bounded version of ITTM-reducibility in a rather natural way, see Coskey and Hamkins [42], Theorem 2.1: Namely, a function $f : \mathfrak{P}(\omega) \to \mathfrak{P}(\omega)$ is Borel if and only if there are an ITTM-program P and a countable ordinal α such that, for all $x \subseteq \omega$, $P^x\downarrow = f(x)$ in $< \alpha$ many steps. Thus, Borel reducibility is a "uniformly time-bounded" version of ITTM-reducibility.

Definition 8.4.1 (Cf. [42], Section 3). Let $\sim$ and $\sim'$ be two equivalence relations on (subsets of) $\mathfrak{P}(\omega)$. We say that $\sim$ is ITTM-reducible to $\sim'$ if and only if there are a real number a and a function $f : \mathfrak{P}(\omega) \to \mathfrak{P}(\omega)$ that is ITTM-computable in the oracle a such that, for all $x, y \subseteq \omega$, we have $x \sim y$ if and only if $f(x) \sim' f(y)$.

ITTM-reducibility incorporates all the positive results of Borel reducibility, as all Borel functions are ITTM-computable relative to some real oracle; see [42]. On the other hand, Hamkins and Coskey observe that many proofs of non-Borel reducibility actually show non-ITTM-reducibility as well, a phenomenon which they call "overshooting." Moreover, they point out that ITTM-reducibility has the advantage that ITTM-programs often have rather intuitive informal descriptions which can be used in proofs of reducibility. One may say that "proof by Church–Turing thesis" is more readily available for ITTM-reducibility than for Borel reducibility; see Chapter 9.

Let us consider an example of "overshooting."

Definition 8.4.2. $=$ is the equality relation on $\mathfrak{P}(\omega)$. E_0 is the "equal but for finitely many places" relation on $\mathfrak{P}(\omega)$, i. e., xE_0y if and only if $|(x \setminus y) \cup (y \setminus x)| < \omega$.

A classical result is that equality is strictly Borel reducible to E_0; see [61]. The same holds in the ITTM-context.

Proposition 8.4.3 (Coskey and Hamkins, [42], Proposition 3.2). $=$ *is ITTM-reducible to* E_0*, but not vice versa.*

Proof. First, let us show that $=$ is ITTM-reducible to E_0. To see this, we need an ITTM-program P and a real number z such that, for all $x, y \subseteq \omega$, P computes real numbers r_x and r_y in the oracles $x \oplus z$ and $y \oplus z$, respectively, and such that r_x and r_y differ in infinitely many places when $x \neq y$. This is easily achieved: For $a \subseteq \omega$, let $f(a) = \{p(i,j) : j \in \omega \wedge i \in a\}$ (thus $f(a)$ codes infinitely many copies of a). Then $f(a)$ and $f(b)$ differ in infinitely many places when $a \neq b$ and moreover, f is clearly ITTM-computable.

Now we show that E_0 is not ITTM-reducible to $=$. Let us assume for a contradiction that P is an ITTM-program and $z \subseteq \omega$ such that, for all $x, y \subseteq \omega$, $P^{x\oplus z}$ and $P^{y\oplus z}$ compute real numbers r_x and r_y, respectively, and that $r_x = x_y$ if and only if xE_0y.

As $L_{\Sigma^z+1}[z]$ is countable, there are $x, y \subseteq \omega$ mutually Cohen-generic over $L_{\Sigma^z+1}[z]$. In particular, we have $\neg xE_0y$, so by assumption on P and z there are real numbers $P^{x\oplus z}\downarrow = r_x$, $P^{y\oplus z}\downarrow = r_y$ with $r_x \neq r_y$. Let $i \in r_x \Delta r_y$, without loss of generality let $i \in r_x \setminus r_y$. Thus $P^{x\oplus z}(i)\downarrow = 1$ and $P^{y\oplus z}\downarrow = 0$. By Corollary 5.3.5 from Chapter 5, $P^{x\oplus z}$ and $P^{y\oplus z}$ both terminate in $< \lambda^z$ many steps.

By Lemma 5.2.14 from Chapter 5, there are Cohen-conditions $p \subseteq x$ and $q \subseteq y$ such that $p \Vdash P^{\dot{x}\oplus z}\downarrow = 1$ and $q \Vdash P^{\dot{y}\oplus z}\downarrow = 0$ over $L_{\Sigma^z+1}[z]$. Now let $\tilde{y}$ be defined as follows: For $j \in \mathrm{dom}(p)$, we have $j \in \tilde{y}$ if and only if $p(j) = 1$. For $j \notin \mathrm{dom}(p)$, we have $j \in \tilde{y}$ if and only if $j \in y$. Thus, $\tilde{y}$ differs from y in only finitely many places and extends p. It is easy to check that $\tilde{y}$ is still Cohen-generic over $L_{\Sigma^z+1}[z]$. Thus, again by the forcing theorem, we must have $P^{\tilde{y}\oplus z}(i)\downarrow = 1$, as p forces this. On the other hand, since $\tilde{y}$ differs only in finitely many places from y, we must have $P^{\tilde{y}\oplus z}(i)\downarrow = 0$, a contradiction. □

A case where Borel reducibility and ITTM-reducibility differ is the following.

Definition 8.4.4 (Cf. [42], Section 3). Let $x, y \subseteq \omega$. We define a relation E_{CK} by $xE_{\mathrm{CK}}y$ if and only if $\omega_1^{\mathrm{CK},x} = \omega_1^{\mathrm{CK},y}$. Moreover, if x and y code ordinals α_x and α_y, we say that $xE_{\mathrm{WO}}y$ if and only if $\alpha_x = \alpha_y$.

Theorem 8.4.5 (Coskey and Hamkins, [42], Proposition 3.6). *With respect to Borel reducibility, neither of E_{CK} and E_{WO} reduces to the other. However, with respect to ITTM-reducibility, each reduces to the other.*

The point mentioned above about having informal program descriptions available to prove ITTM-reducibility can be seen by doing the following exercise:

Exercise 8.4.6. Show that E_{CK} is ITTM-reducible to E_{WO}.

Exercise 8.4.7. Let us define OTM-reducibility with and without parameters for equivalence relations on subsets of $\mathfrak{P}(\omega)$ by the obvious analogy with Definition 8.4.1.

(a) Show that, if $V = L$, then E_0 is OTM-reducible to $=$ without parameters.

Now assume that for every $x \subseteq \omega$, there is a Cohen-generic real number over $L[x]$.

(b) Adapt the proof of Proposition 8.4.3 to show that $=$ is OTM-reducible without parameters to E_0, but not vice versa.

(c) Adapt the proof of Proposition 8.4.3 to show that $=$ is OTM-reducible with parameters to E_0, but not vice versa.

8.5 Infinite time computable model theory

Computable model theory studies model theory under the aspect of effectivity; typical questions are whether certain objects that are proven to exist in model theory can also be computed or whether certain methods can be effectivized. This can yield to beautiful and interesting results; as an example, though it is easy to prove that Peano-arithmetic has nonstandard models, Tennenbaum's theorem (see, e. g., Kaye [97]) says that no such model can be recursive in the sense that its domain are the natural numbers and that its multiplication and addition, considered as sets of triples of natural numbers, are recursive. On the other hand, every recursive field has a recursive algebraic closure (see, e. g., Rabin [147], Chapter 2). For an introduction to the classical area of computable model theory using Turing machines, see Harizanov [82].

The degree up to which Turing computability can be applied to model theory is somewhat limited by the fact that Turing machines can only work on subsets of ω, i. e., codes for countable structures. Moreover, some important model-theoretical operations that one might regard as effective, such as passing from a given structure to its theory, cannot be carried out on Turing machines due to lack of time. This motivates the consideration of infinite time computable model theory, which replaces Turing machines with stronger models of computability.

So far, this has only been done for ITTMs, in work by Hamkins, Miller, Seabold and Warner (see [79]); see also the survey articles [43] and [74].

As ITTMs can also decide subsets, rather than mere elements, of $\mathfrak{P}(\omega)$, one can talk about decidable structures of size up to $2^{\aleph_0}$. One can then, e. g., ask whether Gödel's completeness theorem—the statement that every consistent set of first-order propositions in a language of first-order logic is satisfied in some structure—is ITTM-effective in the sense that when the language and the theory can be appropriately presented to an ITTM (in particular, when both are finite or countable), then a structure, coded as a set of real numbers, satisfying it is ITTM-computable. In the infinitary context, several versions of this claim can be formulated, some of which are provable; the main result, however, is that the most general version of this claim is independent of ZFC; see [79], Section 3.

Given that ITTMs are still restricted to objects of cardinality $\leq 2^{\aleph_0}$, it is natural to consider computable model theory with stronger models like OTMs, which can work on objects of any cardinality, or to consider objects of size κ on κ-TMs. So far, no work on this topic is known to us.

8.6 Generalized effective reducibility[4]

Given a concept of effectivity, such as Turing computability or OTM-computability, one can ask whether theorems of the form $\forall x \exists y \phi(x, y)$ are "effective" in the sense that there is an effective procedure which, given some x, produces some y such that $\phi(x, y)$ holds. We have already seen this motive at work in the section on infinite time computable model theory; in full generality, it can, e. g., be found in Shore [162]. Moreover, one can also ask about effective reducibility: Assuming that a mysterious device D, i. e., an "oracle" producing appropriate y's from x's for ϕ in the sense above is given, can is $\forall x \exists y \psi(x, y)$ effective "relative" to ϕ, i. e., is there an effective procedure for turning effectivizations of ϕ into effectivizations of ψ? Such questions are pursued by reverse mathematics and Weihrauch reducibility with Turing computability as the underlying notion of effectivity. For generalized effectivity, this is a largely unexplored field. There is the classical paper by Hodges [87] based on the primitive recursive set functions of

4 This section is adapted from the article "Generalized Effective Reducibility" [24], which appeared in: F. Manea, R. Miller, D. Nowotka (eds.), "Sailing Routes in the World of Computation," LNCS 10936 ©Springer Nature 2018, with permission from Springer Nature. We thank Springer Nature for the kind permission to reuse parts of that article here. Moreover, some material of this section has been used in [26] and we thank IOS press (the publisher of the journal Computability) for the kind permission to reuse it here and also de Gruyter for the kind permission to use material from this chapter in [27].

Jensen and Karp as the underlying concept of effectivity and the more recent work of Shore [162] using α-recursion theory.

In this section, we will take "effective" to mean computable by a parameter-free OTM-computability and consider effectivity and reducibility in this setting. Our approach is inspired by that of Hodges in [87].

To apply OTMs to general mathematical constructions, we need to represent arbitrary sets as an input for OTMs, i. e., as sets of ordinals. We saw in Chapter 2, Definition 2.3.18, how this can be done. Following this way of coding, we write $\text{rep}(A, x)$ to indicate that $A \subseteq \text{On}$ codes x. By a certain abuse of notation, if x is a set, we will sometimes write $c(x)$ for an "arbitrary" code for x. Note that $|c(x)| = |\text{tc}(x)|$ if $\text{tc}(x)$ is infinite.

We can now talk about OTM-computability of arbitrary functions from V to V:

Definition 8.6.1 (Cf. [24], Definition 2). Let $F : V \to V$ be a functional class.[5] We say that F is OTM-computable if and only if there is an OTM-program P such that, for every set x and every tape content τ, if $\text{rep}(\tau, x)$, then $P(\tau)$ converges to output σ such that $\text{rep}(\sigma, F(x))$, i. e., P takes representations of x to representations of $F(x)$.

By this definition, the representation of a set x will depend on the choice of a well-ordering of $\text{tc}(x)$. The output of a computation on input x may hence depend on the choice of the representation of x. This is fine as long as only the output, but not the object coded by the output, depends on the choice of the input representation.

This allows us to make our notion of "effectivity" precise.

Definition 8.6.2 (Cf. [24], Definition 3). A subclass $R \subseteq V \times V$ is a called a "construction problem." A "solution" or "canonification" of R is a functional class $F : V \to V$ such that $\forall x(\exists y R(x, y) \to R(x, F(x)))$.

Let $R \subseteq V \times V$ be a construction problem. Then R is effectively solvable if and only if there is an OTM-computable solution F for R. Moreover, a set-theoretical Π_2-statement $\forall x \exists y \phi(x, y)$ (where ϕ is Δ_0) is effective if and only if the construction problem $\{(x, y) \in V \times V : \phi(x, y)\}$ is effectively solvable.[6] We write R_x for $\{y : (x, y) \in R\}$.

One may now inquire whether various well-known construction problems and Π_2-statements are effective. Such questions were studied by Hodges in [87], though with a different notion of effectivity based on Jensen and Karp's primitive recursive set functions. We note here that the two methods Hodges uses also work for our model, which allows us to carry over his results.

The following lemma corresponds to Hodges' "cardinality method," i. e., Lemma 3.2 of [87].

---

5 A class $F \subseteq V \times V$ is "functional" if and only if, for any $x \in V$, there is at most one $y \in V$ such that $(x, y) \in F$. We will use the name "class function" for functional classes.

6 In particular, this implies that an effective Π_2-statement must be true.

Lemma 8.6.3 (Cf. [24], Lemma 1). *Let $\alpha \in$ On, and let $R \subseteq V \times V$ be such that, for some cardinal $\kappa > \alpha$, there is $x \in V$ such that $|\text{tc}(x)| = \kappa$, $R_x \neq \emptyset$ and $\forall y \in R_x\ |y| > \kappa$. Then no canonification for R is OTM-computable in the parameter α.*

Consequently, if R is such that there are such κ and x for every $\alpha \in$ On, then no canonification for R is computable by an OTM with parameters.

In particular, if, for some transitive x of infinite cardinality, $R_x \neq \emptyset$ and $\forall y \in R_x\ |y| > |x|$ then no canonification for R is OTM-computable without parameters.

In fact, it suffices that this holds in some transitive class model M of ZFC.

Proof. Clearly, in less than κ^+ many steps, the machine cannot write a code of a structure of cardinality $> \kappa$.

It hence suffices to show that, when P is an OTM-program and P is given a (code c of a) set x of size $\kappa \geq \omega$ for input and the computation halts, then the output of the computation will be of size $\leq \kappa$. This follows if we can show that the computation will take less than κ^+ many steps, since P can write at most α many symbols in α many steps. Suppose for a contradiction that P takes $\lambda > \kappa$ many steps, and let δ be the smallest cardinal $> \lambda$. Let H be the Σ_1-Skolem hull of $\kappa \cup \{c\}$ in $L_\delta[c]$ and let M denote the transitive collapse of H. We may assume without loss of generality that $c \subseteq \kappa$, so that we have $c \in M$; as $L_\delta[c]$ contains the computation of P in the input c, so does H, and hence there is $S \in M$ such that M believes that S is the computation of P with input c. By transitivity of M and absoluteness of computations, S is actually the computation of P with input c. Since S is contained in a transitive set of cardinality κ, $|S| \leq \kappa$, so the length of the computation is $< \kappa^+$, as desired.

The last statement is obtained by relativizing the proof to M. □

There is also an analogue of the "forcing method" (Lemma 3.7 of [87]), which is given in Lemma 8.6.19 below.

For many of the following results, we will need the existence of generic filters for various partial orderings in L and some of its (symmetric) extensions. To avoid technical complications, we use as a shortcut an extra assumption that guarantees the existence of such filters. $0^\sharp$ is more than enough for our purposes, and we assume from now on that it exists.[7] See, e. g., [92] for a proof that all uncountable cardinals in $L[0^\sharp]$ are strongly inaccessible in L, which implies that all the sets that we add to L by various forcing constructions below are still well in the countable realm in $L[0^\sharp]$.

These lemmata can be seen as expressing the intuition that neither the power set operation on infinite sets nor the use of the axiom of choice are "effective," not even in a very idealized sense. We note some sample applications, based on those by Hodges

7 For some of the following results, this assumption is actually necessary: It is, e. g., not hard to check that all choice principles considered in section 4 are effective (and hence trivially reducible to each other) if $V = L$.

in [87], where a variety of mathematical constructions common to the field of algebra is considered.

Lemma 8.6.4. *None of the following construction problems is effectively solvable:*
(1) *Linear ordering to its completions*
(2) *Set to its (constructible) power set*
(3) *Set to its well-orderings*

Proof. (1) and (2) are easy applications of Lemma 8.6.3 as, e. g., the completion of the countable linear ordering $\mathbb{Q}$ is of cardinality $2^{\aleph_0}$ and the power set of an infinite set always of larger cardinality than the set itself; relativizing the argument to L, we obtain the nonsolvability of the constructible power set operation. (3) follows from Lemma 8.6.19 below. □

It is, on the other hand, not hard to see that, e. g., the construction problem of taking a ring to its quotient field is effectively solvable as in [87]. The intuitions captured by Hodges' approach are hence preserved in our framework.

There are certainly various interesting questions to be asked about the effectivity, or otherwise, of various construction problems or Π_2-statements. However, we want to take the analogy with Turing computability a bit further: Instead of merely asking what problems are solvable, we want to consider what problems/statements are effectively reducible to which others in the sense that, given access to a solution to one as an "oracle," one can effectively solve the other. A quite straightforward way to make this idea precise is the following.

Definition 8.6.5 (Cf. [24], Definition 4). Assume that the OTM is equipped with an extra "miracle tape." Let F be a class function taking sets of ordinals to sets of ordinals. A miracle-OTM-program is defined like an OTM-program, but with an extra "miracle" command. When this command is carried out, the set X of ordinals on the miracle tape is replaced (in one step) by $F(X)$. We write P^F to indicate that P is run and whenever the miracle command is applied to X, it is replaced by $F(X)$.[8]

Definition 8.6.6 (Cf. [24], Definition 5). Let C_1 and C_2 be construction problems. Then C_1 is reducible to C_2, written $C_1 \leq C_2$, if and only if there is some miracle-OTM-program P such that the following holds: Whenever F is a canonification of C_2 and whenever $G : V \to V$ is a class function taking each code for a set x to some code y for $F(x)$ and x is a set and c a code for x, we have $P^G(c)\downarrow = d$, where d is a code for a set z such that $(x, z) \in C_1$.

8 We thus make the implicit assumption that the miracle tape behaves deterministically, i. e., that whenever the miracle command is applied to some X, the outcome will be the same. However, this property is not used anywhere in the arguments below. One may thus drop it, at the price of some extra formal complications.

Remark 8.6.7. Note that we do not demand in the conditions on G that $G(c)$ depends only on x when c is a code for x. By demanding that the same reduction works for every G, we rule out the possibility of coding extra information into the input representations.

Concerning this concept of reducibility, we observe that certainly a cardinality-raising construction is not reducible to one that is not.

Lemma 8.6.8 (Cf. [24], Lemma 14). *Let C_1, C_2 be construction problems. Assume that there are some canonification F of C_2 and some infinite set x such that, for all sets y,*
(1) *if $C_1(x, y)$, then $|y| > |\mathrm{tc}(x)|$ and*
(2) *if y is infinite, then $|F(y)| \leq |\mathrm{tc}(y)|$*

Then $C_1 \nleq C_2$.

Proof. As in the proof of Lemma 8.6.3 above, OTM-computable functions cannot raise infinite cardinalities. By assumption, the miracle operation will also not raise the cardinality. Hence the output of a program P with a C_2-miracle will (for infinite input) always have at most the cardinality of the input, and thus cannot in any case witness C_1. □

Exercise 8.6.9. Show that taking a linear ordering to its completions is not reducible to the problem of taking a set to its well-orderings.

The above definition captures the idea that one construction "helps" carrying out another. There is also a much more restrictive intuitive notion of reducibility between problems, namely that instances of one (construction) problem can be effectively "translated" to particular instances of another: Given an instance of a problem C_1, we can first effectively turn it into an instance of a problem C_2 and then effectively turn the solution to C_2 into a solution to C_1. Another way to view this is that C_2 may only be used once in solving C_2. Thus, we define the following.

Definition 8.6.10 (Cf. [24], Definition 6). Let C_1, C_2 be construction problems. Then C_1 is ordinal Weihrauch[9] reducible to C_2, written $C_1 \leq_{\mathrm{oW}} C_2$, if and only if there are OTM-programs P and Q such that the following holds for all sets x in the domain of C_1, every code c for x and every canonification F of C_2:
(1) $Q(c)$ halts with output c', where c' is a code for a set y
(2) For every code c'' of $F(y)$, $P(c'')$ halts with output c''', where c''' is a code for a set z
(3) We have $C_1(x, z)$

9 In [24], this relation was called "generalized Weihrauch reducibility." In his review to this paper, D. Dzhafarov pointed out that this name was already in use and proposed "ordinal Weihrauch reducible' as an alternative.

If these clauses hold, we say that (P, Q) witnesses the oW-reducibility of C_1 to C_2. Also, when F is a canonification, P and Q are OTM-programs and x is a set, we write $[P, F, Q](x)$ for the z obtained by the procedure just described.

If $C_1 \leq_{\text{oW}} C_2$ and $C_2 \leq_{\text{oW}} C_1$, we write $C_1 \equiv_{\text{oW}} C_2$.

Remark 8.6.11. The name of the reducibility relation is due to its obvious resemblance with Weihrauch reducibility, which is an analogous notion for classical computability. For some results on classical Weihrauch reducibility, see e. g., Brattka et al. [18]. Another notion of Weihrauch reducibility on higher cardinalities based on continuous transformations on generalized Baire spaces is introduced in Galeotti [64]. So far, no connection between the two approaches has been made.[10]

We note that reducibility notions satisfy the general order-theoretic properties of reducibility relations:

Lemma 8.6.12 (Cf. [24], Lemma 4). *Both $\leq$ and $\leq_{\text{oW}}$ are transitive and reflexive. Consequently, $\equiv_{\text{oW}}$ and $\equiv$ are equivalence relations.*

Exercise 8.6.13. Show that taking a linear ordering to its completions is oW-reducible to taking a set to its power set.

Definition 8.6.14 (Cf. [24], Cf. 7). Let C be a construction problem. Then $[C]$ denotes the $\equiv$-equivalence class of C and $[C]_{\text{oW}}$ denotes the $\equiv_{\text{oW}}$-equivalence class of C.

Our next goal is to develop a method for showing that a construction problem is not oW-reducible to another.

Definition 8.6.15. A class $X \subseteq V$ is "closed under OTM-computability" or "OTM-closed" if and only if, for every OTM-computable class function F and any $x \in X$, we have $F(x) \in X$.

The basic idea is now given by the following lemma.

Lemma 8.6.16. *Suppose that C and C' are construction problems such that there are a canonification F' of C' and a class M closed under F' and under OTM-computability with an element $x \in M$ such that $\neg\exists v \in M(x, v) \in C$. Then $C \nleq_{\text{oW}} C'$.*

Proof. Suppose otherwise, and let (P, Q) witness that $C \leq_{\text{oW}} C'$. Suppose that $Q(c(x))$ halts with output $c(y)$; then $y \in M$ as M is OTM-closed. Then we have $z := F'(y) \in M$ since M is closed under F'. Again by OTM-closure of M, if $c(u)$ is the output $c(u)$ of $P(z)$, then $u \in M$. But $u = [P, F', Q]$ satisfies $(u, x) \in C$ as (P, Q) witnesses that $C \leq_{\text{oW}} C'$, contradicting the assumption that $\neg\exists v \in M(x, v) \in C$. □

10 The generalization of uniform reducibility introduced in Hirschfeldt and Jockusch [86] based on (Turing-) computable strategies goes into a different direction, though it may be compatible with our setting. We thank one of our anonymous referees for pointing out this paper as well as the work of Galeotti to us.

The classes M we will consider will be transitive class models of KP. Clearly, if $A \in \mathfrak{P}(\mathrm{On}) \cap M$ and P^A halts for some OTM-program P, then this suffices to guarantee that the computation of P^A, and thus its output, along with the set coded by it, belong to M.

However, if $x \in M$ is merely a set, M may not contain a well-ordering of x and thus no code for x. In this case, even though $x \in M$, any OTM-computation on the input x is "external" to M, i. e., it takes place in some extension of M in which x is well-ordered, and thus codable as an OTM-input. In order for M to be OTM-closed, we should thus demand that such an extension exists. But even then, the computation takes place in that extension and not in M, and thus there is no guarantee that the set coded by its output will belong to M. The point of the next theorem is that, under certain circumstances, this can still be achieved.

Lemma 8.6.17 (Cf. [24], Lemma 5). *Let $M \models \mathrm{KP}$ be transitive and suppose that $x \in M$. Then $\mathbb{P}_x := \{f : \omega \to x : |f| < \omega \wedge f \text{ injective}\}$ is a set in M.*

Proof. Let $y := x \times \omega$. For each $n \in \omega$, we have $y^n \in M$ and the function $F : \omega \to M$ that maps n to y^n is Σ_1-definable over M. By Σ_1-collection and union, we have $A := \bigcup\{y^n : n \in \omega\} \in M$. Now P_x can be obtained from A via Δ_0-separation. □

Theorem 8.6.18 (Cf. [24], Lemma 5). *Let F be a computable class function, $M \models \mathrm{KP}$ transitive such that $\mathrm{On}^M = \mathrm{On}$.*[11] *Assume moreover that $x \in M$ is such that there are (in V) two mutually generic $\mathbb{P}_x$-generic filters G_1 and G_2 over M. Then $F(x) \in M$.*

Proof. Note that, as M is a transitive class, we have $L[a] \subseteq M$ for every $a \in M$ by definability of L in KP (relativize Lemma 3.1.40 of Chapter 3).

Let P be an OTM-program witnessing the computability of F. Let $x \in M$ be as in the assumption of the theorem. By passing to $\mathrm{tc}(x)$ if necessary, we may assume without loss of generality that x is transitive. Let G_1, G_2 be mutually $\mathbb{P}_x$-generic filters over M which exist by assumption. In $M[G_1]$ and $M[G_2]$, x is well-ordered in order type α by $\bigcup G_1$ and $\bigcup G_2$, respectively. Hence both $M[G_1]$ and $M[G_2]$ contain tape contents τ_1 and τ_2 coding x. Let α_1, α_2 be the halting times of $P(\tau_1)$ and $P(\tau_2)$, respectively, and let $\beta = \max\{\alpha_1, \alpha_2\}^{+,\tau_1 \oplus \tau_2}$ be the next $\tau_1 \oplus \tau_2$-admissible ordinal greater than α_1 and α_2. Then G_1 and G_2 are in particular $\mathbb{P}_x$-generic over $L_{\beta+1}[\tau_1] \subseteq M[G_1]$ and $L_{\beta+1}[\tau_2] \subseteq M[G_2]$ and thus, $(L_\beta[\tau_1])[G_1]$ and $(L_\beta[\tau_2])[G_2]$ will still be admissible by a slight variant of Lemma 5.2.17 in Chapter 5.

It follows that the computations of $P(\tau_1)$ and $P(\tau_2)$ are contained in $(L_\beta[\tau_1])[G_1]$ and $(L_\beta[\tau_2])[G_2]$, respectively, and hence so is the set $F(x)$ coded by their final output.

---

11 Note that some condition on the height of M is required to ensure that the convergence of programs is absolute between V and M. In particular, if there is some α with $L_\alpha \models \mathrm{ZFC}$, then the parameter-free OTM-program that enumerates L until it encounters such an L-level will run for more than β many steps, where β is minimal with $L_\beta \models \mathrm{ZFC}$.

Consequently, we have $F(x) \in (L_\beta[\tau_1])[G_1] \cap (L_\beta[\tau_2])[G_2]$, and hence $x \in M[G_1] \cap M[G_2]$. But since G_1 and G_2 are mutually generic over M, we have $M[G_1] \cap M[G_2] = M$ by a variant of Lemma 5.2.23 in Chapter 5. Thus, we have $F(x) \in M$, as desired. □

This suggests a general method for realizing the method indicated by Lemma 8.6.16 above.

Lemma 8.6.19 (Cf. [24], Lemma 6). *Let C_1, C_2 be construction problems. Assume that there are a canonification F of C_2 and a transitive class-sized $M \models \mathrm{KP}$ and some $x \in M \cap \mathrm{dom}(C_1)$ such that M is closed under F, but $\{y : C_1(x,y)\} \cap M = \emptyset$. Assume moreover that x is such that there are (in V) two mutually generic $\mathbb{P}_x$-generic filters G_1 and G_2 over M. Then $C_1 \not\leq_{\mathrm{oW}} C_2$.*

Proof. Assume otherwise, and let P and Q be OTM-programs such that (P, Q) witnesses the oW-reducibility of C_1 to C_2. Pick F, M and x as in the statement of the lemma. Then Q computes, for every code of x as an input, a code for some (unique) set y. By Theorem 8.6.18, we have $y \in M$. As M is closed under F, we have $F(y) \in M$. Now, for every code of $F(y)$ as an input, P computes a code for some (unique) set z. Again by Theorem 8.6.18, $z \in M$. Also, by the choice of P and Q, we have $C_1(x,z)$. So $z \in \{y : C_1(x,y)\} \cap M$ and so the latter is not empty, a contradiction. □

8.6.1 An application: versions of the axiom of choice

As a sample application of the notions and methods developed above, we consider variants of the axiom of choice (see e.g. [118]) with respect to effective reducibility.

Definition 8.6.20 (Cf., e. g., Devlin [48]). Denote by ZL ("Zorn's lemma") the claim that, for any partially ordered set $(X, \leq)$ such that every linearly ordered subset $Y \subseteq X$ has an upper bound in X possesses a maximal element, i. e., an element $x \in X$ such that $x \not\leq y$ for every $y \in X$.

Denote by HMP ("Hausdorff's maximality principle") the claim that, for any partially ordered set, there is a linearly ordered subset $Y \subseteq X$ such that there is no proper linearly ordered superset $Y \subsetneq Y' \subseteq X$.

Denote by AC the axiom granting the existence of choice functions, i. e., the statement that for all sets x, there is a function f such that $f(\emptyset) = \emptyset$ and for $y \in x$, if $y \neq \emptyset$, then $f(y) \in y$.

Denote by AC′ the axiom granting the existence of systems of representatives, i. e., the statement that for all sets x whose elements are nonempty and mutually disjoint, there is a set r such that $|r \cap y| = 1$ for all $y \in x$.

Denote by WO the well-ordering principle, i. e., the statement that for every set x, there is an ordinal α and a bijection $f : \alpha \leftrightarrow x$.

In addition, we consider the following principle.

Definition 8.6.21. Denote by PP ("picking principle") the claim that, for every nonempty set x, there is $y \in x$.

PP is, of course, not a variant of the axiom of choice. It expresses the utter triviality that a nonempty set has a member. Still, as we will see in Corollary 8.6.23 below, it is still not effective.

It is easy to turn the usual proof for the equivalence of AC and AC′ over ZF (see, e. g., [48]) into two oW-reductions, which shows that AC $\equiv_{\mathrm{oW}}$ AC′. Moreover, we can see that AC ≡ PP; clearly, PP reduces to AC′ by mapping $\{x\}$ to $\{\{x\}\}$; in the other direction, given a set x of nonempty sets, we can use a canonification of PP to simply pick an element of each element of x.

Theorem 8.6.22. *Suppose that* $0^\sharp$ *exists. Then* AC′ *is not effectively solvable.*

Proof. By the assumption that $0^\sharp$ exists, there is a symmetric extension M of L with a set $X \subseteq \mathfrak{P}(\mathfrak{P}(\omega))$ of disjoint nonempty sets of real numbers without a choice function, see, e. g., Chapter 15 of [92]. M will be a model of ZF and X will be countable in $L[0^\sharp]$, so that the assumptions of Theorem 8.6.18 are satisfied. Thus, if F was an OTM-computable canonification of AC′, we would have $F(X) \in M$, a contradiction. □

As a consequence, it is not even possible to canonically pick a single element of a nonempty set in an effective way.

Corollary 8.6.23. *Suppose that* $0^\sharp$ *exists. Then* PP *is not effectively solvable.*

Proof. If PP was effectively solvable, then so was AC ≡ PP, and hence so was AC′, contradicting Theorem 8.6.22. □

Corollary 8.6.24. *If* $0^\sharp$ *exists, then* AC $\not\leq_{\mathrm{oW}}$ PP.

Proof. Suppose for a contradiction that AC $\leq_{\mathrm{oW}}$ PP, and let P and Q be OTM-programs that witness this. Let M be as in the proof of Theorem 8.6.22, and let $F : M \to M$ be a class function such that $F(x) \in x$ for all $x \in M$. Let $X \in M$ be a set that has no choice function in M. Then Q, when applied to X, will compute a (code for a) set $Y \in M$. By assumption, we have $F(Y) \in M$, and P will compute a code for a choice function f of X when applied to $F(Y)$. Thus $f \in M$, so M contains a choice function for X after all, a contradiction. □

By effectivizing the classical proof, it is also easy to see in the same way that AC $\leq_{\mathrm{oW}}$ WO. However, the opposite direction is not quite as obvious: To deduce WO from AC, one applies AC not to the set X to be well-ordered, but to its powerset. We saw above in Lemma 8.6.4 that the function mapping a set to its powerset is not OTM-effective. Of course if, e. g., $V = L$, then WO is trivially effective (and so are all of the other principles in the last definition): Given a set X, one simply enumerates L until the first well-ordering for X shows up, which one returns. If V and L are very different, this is not quite as obvious. Indeed, if $0^\sharp$ exists, the reduction no longer works.

Theorem 8.6.25 (Cf. [24], Theorem 10). *If* $0^{\sharp}$ *exists, then* $\mathrm{WO} \not\leq_{\mathrm{oW}} \mathrm{AC}$.

Proof. (Sketch) In Theorem D.-A. C. of [198], it is shown how to construct a transitive model of[12] $\mathrm{ZF}^{-}+\mathrm{AC}+\neg\mathrm{WO}$ as a union of an ascending chain of symmetric extensions of a transitive ground model M of ZF^{-}. Starting with $M = L$, it is easily checked that, under the assumption that $0^{\sharp}$ exists, the construction leads to a definable transitive class model N of $\mathrm{ZF}^{-} + \mathrm{AC}$ such that some set $A \in N$ that is nonwell-orderable in N is countable in V and, moreover, $\mathbb{P}_A$ is countable, and thus has two mutually generic filters over N. Hence the assumptions of Lemma 8.6.19 are satisfied and the nonreducibility follows. □

In terms of ordinal Weihrauch reducibility, WO is quite strong. In fact, it is universal among the Π_2-consequences of ZFC.

Theorem 8.6.26. *Let* ϕ *be a* Π_2*-formula.*
(i) *If* ϕ *is a consequence of ZFC, then* $\phi \leq_{\mathrm{oW}} \mathrm{WO}$.
(ii) *If* ϕ *is a consequence of ZF, then* $\phi \leq_{\mathrm{oW}} \mathrm{PP}$.

Proof. Suppose ϕ is $\forall x \exists y \psi(x, y)$, where ψ is Δ_0.

(i) A reduction of ϕ to WO works as follows: Let F be a canonification of WO. Given a set x, determine $F(x)$. Then, since $F(x)$ is well-ordered, $L[F(x)]$ will be a model of ZFC and thus satisfy ϕ; in particular, it will contain some b such that $\psi(x, b)$ holds in $L[F(x)]$ and thus, by Δ_0-absoluteness, in the real world. We use the program from Lemma 3.5.3 in Chapter 3 to search through $L[F(x)]$ for the $<_{L[F(x)]}$-minimal such b and write it to the output tape once it has been found.

(ii) As $L[x] \models \mathrm{ZF}$, there is $y \in L[x]$ such that $L[x] \models \psi(x,y)$; as ψ is Δ_0, $\psi(x,y)$ also holds in V. Let F be a canonification of PP, and let a set x be given. By using the relativized version of the algorithm from Lemma 3.5.3, successively write codes for the levels $L_\alpha[x]$ to the tape. For each α, search through $L_\alpha[x]$ for a code of some such y to determine the smallest α such that $L_{\alpha+1}[x] \setminus L_\alpha[x]$ contains such a y. Then compute a code c for the set $\{z \in L_{\alpha+1}[x] : L_{\alpha+1}[x] \models \psi(x, z)\}$. Applying F to c will yield the desired witness. □

Exercise 8.6.27. Working in ZFC (i. e., in particular assuming that ZL is true), show that $\mathrm{ZL} \equiv_{\mathrm{oW}} \mathrm{PP}$.

Exercise 8.6.28. Show that $\mathrm{HMP} \leq_{\mathrm{oW}} \mathrm{WO}$.

Exercise 8.6.29. Let POT denote the problem of mapping a set x to its power set. Show that neither $\mathrm{POT} \leq_{\mathrm{oW}} \mathrm{WO}$ nor $\mathrm{WO} \leq_{\mathrm{oW}} \mathrm{POT}$.

12 Recall from Chapter 4 that ZF^{-} is ZF without the power set axiom in a proper formulation, see [66]. Clearly, ZF^{-} is a (strong) extension of KP.

This leaves the position of HMP among the choice principles mentioned above with respect to ordinal Weihrauch reducibility. In particular, we do not know the answer to the following question:

Question 8.6.30. Is HMP $<_{\text{oW}}$ WO? Is HMP $\leq_{\text{oW}}$ AC?

Remark 8.6.31. Further results on ordinal Weihrauch reducibility between set-theoretical choice principles can be found in [26].

8.7 Further prospects

We close by mentioning some further possible areas of application for OTMs that are not treated above, either because they are treated elsewhere or because they do not yet exist.[13]

- Complexity of structures:
  Roughly, Kolmogorov complexity measures the complexity of a binary string—and then, naturally, of an object coded by such a string—by the minimal length of a Turing program that can write that string. Using OTMs with binary strings of ordinal length as inputs (or equivalently, using the OTMs with infinite programs introduced by E. Lewis in his Master's thesis [126], see exercise 3.9.7 in Chapter 2 for a brief discussion), this can be extended to ordinal strings of arbitrary length: For a set of even a proper class $s \subseteq \text{On}$, let us define ∞-KK(s) as the minimal length α of a string $t \in \{0,1\}^{**}$ such that, for some OTM-program P, P^t decides s. This defines a complexity measure on arbitrary classes or ordinals, and, via coding, on arbitrary sets that has not been considered so far at all.
- Classification:
  The results arising from Coskey's and Hamkins' approach to use ITTM-reducibility instead of Bore reducibility motivate a closer look at the concept of classification. To avoid repetitions, see Chapter 9 for a further comment on this topic.
- Proofs as certificates: See Chapter 9.
- Connections to the philosophy of mathematics:
  In various positions on the foundations and the philosophy of mathematics, mathematical objects are viewed as constructions by or being observed by idealized cognitive agents. Such a view on set theory is, e. g., entertained by H. Wang [176] or P. Kitcher [101] and is also present in various remarks by Gödel (see, e. g., 7.1.17, 7.1.18 and 7.1.19 on p. 220 of [176]). It is plausible that the notion of an idealized cognitive agent can be expressed via infinitary machines in a similar way that

13 A very recent application of ITTMs in reproving a classical result in descriptive set theory, namely that, for any countable admissible ordinal α, there is a real number x such that $\alpha = \omega_1^{\text{CK},x}$, along with strengthening thereof due to Jensen, is the paper by Durand and Lafitte [50], which unfortunately appeared too late to be discussed in any detail in this book.

a Turing machine can be used as a surrogate for a finite agent. In this view, the computability of an object expresses its constructibility by such an "agent," while recognizability corresponds to a more "platonic" picture, in which an idealized "observer" encounters "objects." As P. Maddy writes in her review [134] of [101], it would be particularly interesting to see what consequences such a view has for the possibility of extensions of ZFC and the rationality of certain candidates. Modelling idealized agents via infinitary machines could help in developing such a connection and evaluate the range of such agent-based views. An example of a statement independent of ZFC, but apparently natural and well-motivated through a computational perspective is Sacks' theorem for OTMs (see Chapter 5), which in particular contradicts $V = L$. We will come back to this in Chapter 9.
- Finally, we mention the close connection between ITTM-computability and the revision theory of truth that was discovered by B. Löwe and led to a fruitful interaction of the areas, a sketch of which can be found in Chapter 9.

9 Philosophical aspects

9.1 Infinitary constructions in mathematics

In this chapter, we want to consider some connections between ordinal computability and the foundations and philosophy of mathematics.[1] The literature on this connection is currently somewhat sparse, but it includes [29] (from which parts of this chapter are adapted) as well as the contributions of B. Rin in [151] and [150]. Later on, we want to argue that ordinal Turing machines provide a formal counterpart to the intuitive concept of a "transfinite mechanical procedure" in the same way as Turing machines successfully formalize the intuitive concept of a "finite mechanical procedure." In fact, we will attempt to provide an analysis similar to Turing's analysis of a finite human being's calculating activity, leading to the concept of an OTM.

But does such an "intuitive concept of a transfinite mechanical procedure" even exist? Is there an informal concept to which our formalization could or could not be adequate? If there was not, our analysis would be pointless.

We believe that such a concept is indeed present in mathematical thinking and intuition. As a first indicator, consider the following list of statements:

1. In principle, I could check each number below $a = 10^{10^{10^{10!}!}!}!$ to see whether there is a counterexample to the Goldbach conjecture below a; it is only that time does not allow it.
2. In principle, I could check each natural number separately to see whether there is a counterexample to the Goldbach conjecture, only time does not allow it. When I was done, I would know whether the Goldbach conjecture was true.
3. In principle, I could check each natural number n separately to see whether $n - 1$ and $n + 1$ are both prime; it is only that time does not allow it. When I was done, I would know whether the prime twin conjecture was true.
4. In principle, I could check all constructible complex numbers separately to see whether one of them is a counterexample to the Riemann hypothesis, only time does not allow it. When I was done, I would know whether the Riemann hypothesis was true in L.
5. In principle, I could check all complex numbers separately to see whether one of them is a counterexample to the Riemann hypothesis, only time does not allow it. When I was done, I would know whether the Riemann hypothesis was true.

---

1 A philosophically potentially relevant aspect of ordinal computability, or more general, of hypercomputability that we will not consider here is the question of its physical realizability. It has been argued, e. g., in Earman and Norton [51] that relativistic physics allows for physical hypercomputations, which was criticized by Aaronson in [1]; see also Welch [183, 182]. We thank M. Bianchetti for pointing out the philosophical interest and some relevant references to us.

https://doi.org/10.1515/9783110496154-009

6. In principle, I could check all subsets of $\mathbb{R}$ to see whether one of them is a counterexample to the continuum hypothesis, only time does not allow it. When I was done, I would know whether the continuum hypothesis was true.
7. In principle, I could, for all infinite cardinals κ, check all subsets of $\mathfrak{P}(\kappa)$ to see whether one of them is a counterexample to the generalized continuum hypothesis. It is only that time does not allow it. When I was done, I would know whether the generalized continuum hypothesis was true.

Clearly, these statements are vague; everything seems to depend on the "in principle." Leaving out "in principle," all of these statements are plain wrong. However, experience shows that many mathematicians still see differences in plausibility between these statements, usually ranking them in the order given above.

Moreover, the degree to which these statements are considered as plausible seems to be connected to the degree to which the conjectures in question are perceived as having a definite truth value: If one can decide, even if only "in principle," whether the Goldbach conjecture has counterexamples below a given finite bound, then this question must have an answer, and likewise, the "in principle" possibility to search through the natural numbers (possibly even performing several nested searches) testifies the impression that statements like the Goldbach or the twin prime conjecture must have a truth value. Even if one of these should turn out to be independent of ZFC, this would be more of a reason to look for extensions of ZFC than to conclude that the questions are meaningless.[2] On the other hand, impossible as it is to search through the set-theoretical universe, there seems to be less trust in a "realistic content" of, e. g., the continuum hypothesis that ensures the existence of a definite truth value. The Riemann hypothesis, on the other hand, is usually still perceived as meaningful and, therefore, possessing a definite truth value; in this respect, it may be interesting to note that the Riemann hypothesis can in fact be reformulated in number-theoretical terms; see Lagarias [124].

Whatever one thinks of the preceding thought experiment, there is another indicator that we do indeed have an intuitive concept of a transfinite effective procedure, since occasionally, such a concept explicitly appears in the mathematical literature. Consider, e. g., the following quotations from an article by W. Hodges (see [87], p. 1, my emphases):

> Since the 1930s (Post, Turing) we have known exactly what it is for a function of natural numbers (...) to be *effectively or algorithmically computable*. (...) Now **every mathematician is at least**

2 Recently, I found that this point was also made by Kripke [120], page 37: "We do not know whether Goldbach's conjecture is true or false. So right now we certainly do not know anything *a priori* about it. Perhaps it will be alleged that we *can* in principle know *a priori* whether it is true. Well, maybe we can. Of course an infinite mind which can search through all the numbers can or could."

> **vaguely aware of another quite different notion of "effective function," which has nothing at all to do with denumerable sets.**
> (Hodges, [87])

As examples of effective functions, Hodges mentions the function F_0 taking an integral domain to its field of fractions and the function F_1 taking each ring R with identity to a maximal ideal of R. He then goes on as follows ([87], my emphases):

> In the sense which concerns us (...), function F_0 is *effective*, function F_1 is (apparently) highly noneffective (...) To prove theorems, **we have to replace this intuitive notion of effectiveness with something more precise** (...).
> (Hodges, [87])

Indeed, a heuristical idea of an infinitary extension of the concept of "effectivity" is present in mathematics in various places. We mention four examples:

1. Effective existence proofs:
   In his [87] cited above, Hodges gives various examples of algebraic constructions, some of which are intuitively "effective," while others are not. (The "more precise something" that Hodges uses is a variant of Jensen and Karps primitive recursive set functions; see [95].) One reason why such considerations are relevant is that effective proofs of existence are often more informative than indirect proofs: For example, given an explicit construction of an object, one can prove properties of that object by induction on the construction. To take a drastic example: Even if one could prove indirectly the existence of a $\subseteq$-minimal transitive class model of ZFC, the recursive definition of L would certainly still be strongly advantageous. A precise concept of generalized effectivity can thus help us to find out when it makes sense to look for such effective constructions and when it does not, since none exists.
2. Reducibility and effective equivalence:
   In various places in mathematical discourse, one finds claims that certain theorems are "equivalent." Logically, this does not make much sense, since every two true statements are equivalent in the sense of material biimplication. Nevertheless, there is an intuition that supports these claims. What is meant, rather than material biimplication, is that one can obtain one theorem "easily" from the other. Reverse mathematics and Weihrauch reducibility are two approaches to give a precise meaning to this intuition in the finite and countable realm. In particular, Weihrauch reducibility concerns the "effective reducibility" of one statement to another. Since a great deal of mathematics takes place beyond the countable, the question arises whether these approaches can be extended to structures of arbitrary size? One idea would be to make the objects countable via forcing and apply the classical notions; this approach is, e. g., taken in [91, 103]. Another would be to take the objects at face value, but apply a stronger notion of effectivity. An approach in this spirit can be found in [162]. Here again, models of infinitary com-

putability can be used to render precisely the idea of "effectivity." Some steps in this direction have been taken in Chapter 8.

3. Classification:
   In order to "classify" a family of structures, e. g., up to isomorphism in a satisfying way, it is not sufficient to use the axiom of choice to chose a representative for each equivalence class and call the result a "classification." What is missing in such a "classification" is apparently some "effective" way of passing from an object to its class and possibly vice versa, from a class to some designated representative therein. A precise notion of "effectivity" would then allow negative results about classifiability. Such an approach is pursued by G. Melles in [139], where this point is made as follows ([139], p. 294):

   > In order to prove nonclassification, we need an exact definition of the intuitive notion of effective construction within set theory.[3,4]

4. Proofs as certificates:
   According to a common view, mathematical proofs serve two purposes: They guarantee the truth of a statement and they explain the reasons for this statement's being true. From an abstract point of view, the "guarantee" aspect of a proof of a statement ϕ can be expressed thus: A proof of ϕ is a mathematical object X such that, if X exists, then ϕ must be true. Again, this alone is insufficient: If there was a largest prime p such that $p-2$ is also prime, it would witness the finiteness of the number of prime twins; but p alone should not count as a proof, for one cannot see that it has the required property. What is missing is apparently some "effective" connection between X and ϕ.[5]
   As Barwise puts it ([10], p. 92):

   > We would like to prove a generalization of the ordinary completeness theorem for L by showing that $\models \phi$ iff $\exists P[P$ is a proof of $\phi]$ for some notion of proof. For such a result to be of any use, there must be something effective about the notion of proof (...).[6]

---

3 For a different discussion of classification see, e. g., Hodges [88]. Here, as well, notions of generalized effectivity are manifestly present in the background. For example, on page 232: "Lachlan suggested that a satisfactory structure theory would consist of a way of **computing** the truth or otherwise of any statement of the form (...). It would be interesting to see similar suggestions made for classes containing arbitrarily large structure," or [88], page 233: "When is an operation Γ that takes structures to structures 'concrete and explicit' (...)?" or page 234: "A structure theory for a class K consists of an **explicit operation** which reduces each structure A in K to an object ΓA (...)" (my emphases).

4 How well, e. g., OTM-computability captures an intuitive sense of "classifiable" and how well it would agree with the approaches of Melles and Hodges has not been pursued so far.

5 A somewhat related very general concept of "proof" is also present in Woodin's Ω-logic [7].

6 Taking this "effective connection" to mean computability by some transfinite machine model, this may point to an "infinite time proof theory." Such a connection was suggested to us by R. Miller during CiE 2014.

Thus, there seems to be sufficient evidence to suspect that there is a potentially useful concept of "effective transfinite procedure" present at least in the informal mathematical discourse. It is natural to suspect that machine models of infinitary computability can serve as a basis for expressing this notion formally, allowing a mathematical study thereof, just like regular Turing machines started the study of finite effectivity in mathematics. To this end, Turing argued in [175] that a human being, following a precise description of a procedure is "equivalent to a Turing machines" (see [175]) when one abstracts from all too human properties like lack of concentration or a strictly bounded life span. In a similar respect, we want to argue that ordinal machines capture the idea of a calculating human being at a higher level of idealization. To this end, we will first have a look at idealized agents in the philosophy and the foundations of mathematics.

9.2 Idealized agents in the philosophy of mathematics

A tempting and widespread view on mathematical objects is that mathematical objects are mental objects, i. e., objects given in and perceived and created by thought. For example, the number 3 is realized by (mentally) counting to three. Mathematics is then a description of this mental activity.

Clearly, if such a view is supposed to have even a starting chance of incorporating a serious amount of contemporary mathematics, many of the restrictions that we encounter in our everyday mental activity must be rejected as mathematically inessential and irrelevant: The empirical fact that our memory, concentration and motivation are limited would otherwise prevent us from building up even tiny fragments of elementary arithmetic in this way. More importantly, mathematics would cease to be intersubjective, and even for a single person the question, e. g., what numbers there are will vary with time, age, mood, etc. What is relevant in such an approach is hence not what we are actually capable of, but "in principle."

This transition from "actually" to "in principle" can be viewed in at least two ways: First, one may consider the "in principle"-capabilities as an abstraction from mathematically irrelevant aspects of our cognition that one abstracts away. In this case, the mathematical agent is considered as an "idealized agent," which is the terminology of P. Kitcher [101]. On the other hand, we may regard our physical selves as (rather poor) approximations to this mathematical agent, which adequately captures the concept of cognition. We will remain neutral and silent with respect to this question. In the following, we will follow Kitcher in speaking of "idealized agents" without subscribing to either view.

The most famous example of a mathematical foundation based on the concept of an ideal mathematician is of course Brouwer's intuitionism [84]. The ideal mathematician of Brouwer has the ability to construct arbitrarily large finite objects, going on and on with this construction to approximate, e. g., infinite sequences of natural numbers (interpreted as real numbers) with arbitrary precision, but without ever finishing the construction. As it turns out, fruitful connections exist between intuitionism and recursion theory, via modeling the ideal mathematician by a Turing machine; see, e. g., Goodstein [68].

However, the restriction to the “arbitrarily large, but finite” implicit in this approach rules out many branches of mathematics, such as general algebra, general topology or set theory. The conclusion of a hard-nosed intuitionist is that these areas are nonsensual and should not be pursued. On the other hand, there are quite a few approaches to allow mathematical agency to enter the transfinite, e. g., granting the mathematical agent the ability to finish an infinite constructions and then go on constructing based on the result. We will now briefly consider three such approaches.

9.2.1 Idealized agents in G. Takeuti's foundational position

For the great proof theorist, G. Takeuti, a conception of idealized mathematical agency is at the core of modern mathematics and logic (see [171], p. 161):

> Let us, to begin with, adopt the standpoint of an ‘infinite mind,’ which we suppose can examine infinitely many objects one by one. (...) This approach is in its essence the way in which many working mathematicians conceive of sets. It is fair to say that in modern mathematics many of the arguments concerning sets are carried out along these lines, and the higher (finite) order predicate calculus is a formulation of such an approach to sets.
> (Takeuti, [171])

Thus, Takeuti claims that the idea of an infinite cognitive agent is in the background of a good part of mathematical practice. If he is right, one should certainly expect foundational merits from a clarification of this concept of agency.

A similar point is made again, with respect to the foundation of mathematics, in [172]:

> Foundational problems begin when we realize that we cannot examine infinitely many objects one by one. However, it is very easy for us to imagine an infinite mind which can do so. Actually by working in mathematics we have been building up our intuition on what an infinite mind can do. An infinite mind must be able to operate on infinitely many objects as freely as we operate on finitely many objects.
> (Takeuti, [172])

Finally, in [172], we also find the following punchline:

> Modern mathematics is the world of the infinite mind.
> (Takeuti, [172])

In [171], the idealized agent view of set theory, according to which sets are constructions of an “infinite mind,” is used critically: Takeuti argues that the axioms of ZFC are in part incompatible with such a view. Consider, for example, the axiom of comprehension. At first sight, this seems to be easy to justify on the basis of an idealized agent view: Given a set X and a property ϕ, our agent could "run through" X and pick out those objects that satisfy ϕ. However, if ϕ is given by a formula that quantifies over the whole set-theoretical universe, then in order for this story to work out, our agent would need access to the completed universe of sets while he is still busy constructing it. Thus, the proposed justification of the comprehension axiom fails, and it is not clear at all that the idealized agent view of set theory is even compatible with it. The only way the comprehension axiom could still hold in such a view seems to be that the sets arising from the use of comprehension are always "coincidentally" also constructed by the agent in some other way. For Takeuti, this is sufficient motivation to prefer higher-order logics as foundational systems for mathematics to set theory. A formalization of the idealized agent view via ordinal computability will allow us a closer look at this criticism.

9.2.2 Idealized agents and Philipp Kitcher's mathematical empiricism

A position in the philosophy of mathematics that is explicitly based on idealized agents and that also created the term "idealized agent" is that taken by P. Kitcher in [101], which is an attempt at an empiristically acceptable philosophy of mathematics. According to Kitcher, mathematical objects are the result of idealizations of mental constructions that we carry out in our everyday interaction with physical objects, such as sorting or grouping of objects. This allows him to empirically reconstruct arithmetic, but clearly, more is required when it comes to transfinite set theory. In this respect, Kitcher says ([101], p. 146):

> I see no bar to the supposition that the sequence of stages at which sets are formed is highly superdenumerable, that each of the stages corresponds to an instant in the life of the constructive subject, and that the subject's activity is carried out in a medium analogous to time, but far richer than time. (Call it "supertime".) (...) The view of the ideal subject as an idealization of ourselves does not lapse when we release the subject from the constraints of our time.

We see here an explicit reference to the concept of time, which any account infinitary agency will have to take into consideration. We will come back to this point below.

9.2.3 Idealized agents in Ramsey's foundations of mathematics

Another work in the foundations of mathematics where idealized cognitive agents play an important role is F. Ramsey's 1925 paper "The foundations of mathematics," [148]. As is well known, Russell had put forward the claim that the circularity of impredicative definitions, i. e., definitions that define an object in terms of a set containing that object lie at the root of Russell's antinomy and need to be ruled out in order to obtain consistent foundations for mathematics. The theory of types was his means to do so. However, this solution had its own difficulties, the possibly strongest one being that it forced Russell to accept the "axiom of reducibility," which can hardly count as self-evident.[7]

In his paper, Ramsey tries a different route: He points out that by far not all impredicative definitions lead to contradictions, and thus the challenge becomes to distinguish those that do from those that do not. His analysis is then, roughly, the following: The circularity that Russell observed is only present for some, but not all impredicative definitions. For the others, the apparent circularity is due to our inability to construct infinitely long logical expressions, an unlucky, yet irrelevant feature of our condition that one needs to abstract from in order to obtain a sensible foundational system. In Ramsey's words ([148], p. 64):

> The only difference is that, owing to our inability to write propositions of infinite length, which is logically a mere accident, $\forall(\phi).\phi(a)$ cannot, like $p \wedge q$, be elementarily expressed, but must be expressed as the logical product of a set of which it is also a member. If we had infinite resources and could express all atomic functions (...).

This analysis then forms the basis for Ramsey's foundational system that accepts infinite definitions and avoids some of the properties of the Principia Mathematica regarded as defects like the axiom of reducibility ([148], p. 65, my emphases):

> In this lies the great advantage of my method over that of `Principia Mathematica`. In `Principia` the range of ϕ is that of functions which can be elementarily expressed, and since $\forall\phi.f(\phi!\hat{z}, x)$ cannot be so expressed it cannot be a value of $\phi!$; but I define the values of ϕ not by how they can be expressed by us at all, let alone elementarily, **but only by a being with an infinite symbolic system**.

9.2.4 Remarks

We hope that the above examples, which are by far not an exhaustive listing of the appearance of idealized agency in the philosophy of mathematics, have made it apparent that the motive of infinitary idealized agency is present in various places in the foundations and the philosophy of mathematics.

It may be interesting to note the uses made of idealized agency in mathematics. So far, we have observed the following:

- In a **descriptive** use, idealized agents are used as a conceptual basis for describing mathematical activity and objects. An example of this kind of use is Kitcher's account of mathematics discussed above.

7 For a discussion of Russell's approach and Ramsey's reaction, see [67].

- In a **heuristical** use, idealized agents are used as a "mental picture" that helps in solving problems. For example, taking an agent-based look at Russell's antinomy, one gets the impression that an agent forming a set, no matter how "infinite," should never encounter the question whether the set he is currently forming should become a member of itself, as he can only form sets out of elements that are "already there." In this way, one is lead to the idea that sets should come in some kind of a hierarchy. Indeed, such motivations are given for the cumulative hierarchy of set theory, e. g., in Shoenfield [161].
- As a special case of the last point, idealized agency can also guide the **formation of new concepts**: For example, Turing computability was a result of Turing's analysis of finitary idealized agency.
- In a **didactical** use, idealized agency is used to explain proof steps, procedures and concepts, to motivate axioms, etc. For example, the picture of going through a family of sets, picking one element of each, and thus forming a system of representatives works rather well to explain the content of the axiom of choice and to convince people that it is true. And of course, proofs in natural language are full of declarations that "we do" something, or even imperatives like "form the elementary hull of $\emptyset$ in V_{ω_1}." The fact that such formulations seem to help understanding seems to confirm that thinking about mathematics in terms of idealized agency is quite natural.
- A **justificational** use is one in which the possibility to justify a mathematical practice, such as a set of axioms or a deduction scheme, via idealized agency, is used as an argument in favor of the consistency or even the correctness of this practice. Ramsey's use of idealized agency above is a strong example of this kind of use.
- A **foundational** use is one in which it is assumed that idealized agency, no matter how well it does as a means of describing mathematical practice, **should** be the backbone of legitimate mathematics and that, therefore, mathematics needs to be built up in a way compatible with it. Brouwer's intuitionism is a well-known example of this kind of use.
- Closely related to the last point is the **critical** way of using idealized agency: Assuming that idealized agency forms the only legitimate basis of mathematics, those parts of mathematics that are incompatible with it have to be regarded as untenable and either replaced by reformed versions going along with idealized agency or, when this is not possible, excluded from mathematics. Again, Brouwer's intuitionism is a strong example of this kind of use; a much softer one is Takeuti's criticism of impredicativity in set theory explained above.

9.3 Infinitary machines as formal models for idealized agents

By the above, it would be desirable to have formalizations for transfinite agency. Given that classical Turing machines were so successful in providing a formal counterpart to the intuitive concept of computability, one might hope that, similarly, machine models of ordinal computability can be used for modeling the concepts of transfinite effective computability that we indicated above.

Note that this point is not trivial: Even if one accepts that, beyond finiteness, clear degrees of idealization of our activity can be concretely captured, e. g., OTMs are not as canonical a model of it as, e. g. Turing machines are in the finite case. In the one direction, it does indeed seem plausible that the actions of an OTM are available to a transfinite idealized agent and that hence everything computable by an OTM should be computable by such an agent: The aspects of an OTM-computation going beyond classical computability consist in elementary limit operations like forming the inferior limit of a sequence of 0s and 1s. But the other direction is not as clear: For example, the limit rule of OTMs seems to be rather arbitrary. The intuition here is that other reasonable choices of limit rules will not change the class of computable objects, but it is exactly the intuition leading there that we want to capture here. We see no direct path from idealized agents to OTMs. Our approach is hence to develop a formal notion of a transfinitely computing agent modeled after our intuition and then see how it relates to OTMs. It turns out that it does indeed describe the same notion of computability, which we consider a good piece of evidence for the thesis that OTM-computability coincides with an intuitive concept of transfinite effective computability.

We thus need to argue that OTMs are suitable as a formal representation of an idealized agent like that described by Takeuti or Kitcher. In the finitary case, Turing achieved this by giving an informal analysis of a calculating human being, then working toward formalization. We will now attempt to give a similar analysis for transfinite agency. We do not claim, however, that this yields a definite and unique answer to the question what transfinite agency can mean. In fact, it seems that (and, e. g., the results of reverse mathematics strongly suggest) beyond finiteness, there is a spectrum of degrees of idealization corresponding to different areas and methodologies of mathematics. What we do claim, though is that (1) OTMs express some preformal intuitive concept and (2) that OTMs are an upper bound for transfinite computability, at least as far as "mechanical procedures" are concerned.[8]

9.3.1 The idealized agent of set theory

We now describe our picture of the idealized agent informally, then working toward a formalization.

The idealized agent performs acts in some order, in accordance with some given process description. Abstracting away the actions and only considering the order in

8 The outcome may be different when one follows Brouwer's example in admitting the cognitive subject to be "creative." We will briefly consider this below.

which they take place, we see that the agent is temporal, i. e., working along a linear ordering as his "time axis." At each point of time, the agent is in a certain state, and additionally he has a complete memory of his previous computational activity; on the basis of this memory, the given process description must determine what to do, i. e., which new state to assume. We call the transition from an initial segment of memorized computational activity to a new state an "elementary step."

We thus obtain three main components, and in order to give a formalization of idealized agency, we need to say what is an appropriate time axis, what is an elementary step (or an appropriate "process description") and what is a state.

In order not to lose generality, we will be very liberal about the states: A "state" is just any mathematical object. Since we are working in the context of set theory, this means that a "state" is simply a set.

9.3.2 Transfinite working time

We now deal with the concept of time. In the passage quoted in the first paragraph, Kitcher suggests that set theory can be considered as the outcome of the mental activity of an idealized agent working in "supertime", which is supposed to be "a medium analogous to time, but far richer than time." Here, we want to argue that the only sensible choice for such a medium are ordinals.

Certainly, we intend a notion of time as a medium of a deterministic computation to be a linear ordering, but we can say more. The computational activity has to start at some point. Every other state may depend on this earlier state and hence has to take place at a moment after the starting point. Hence, the "medium of computation" has to have a unique minimal element.

Whenever the agent has carried out a certain amount of computational activity, he has to know what to do next, i. e., there must be a unique next state for him to assume. This next state has to take place at some point of time. Hence, the medium in which he computes has to contain a unique next element after those through which the activity passed so far. Put differently: For every initial segment of time, there has to be a unique time point preceded by all moments in the initial segment and only by those.

This leads to the following notion of "supertime": A "supertime" is a linearly ordered set $(X, \leq)$ with a unique minimal element μ and such that, for every proper initial segment I of X, there is a $\leq$-minimal $x_I \in X$ such that $\forall t \in I(t < x_I)$. It is now easy to see that this means that all candidates for supertime are (isomorphic to) ordinals.[9]

9 Long after giving this definition of "supertime" in [29], we learned that this was actually Cantor's original—and thus the historically first—definition of well-foundedness; see, e. g., Hallett [73], page 52.

Proposition 9.3.1. *Let $(X, \leq)$ be a linearly ordered set such that, for every $I \subsetneq X$ which is downwards closed (i. e., $x < y \in I$ implies $x \in I$),*[10] *there is a minimal $x_I \in X$ such that $\forall t \in I(t < x_I)$. Then $(X, \leq)$ is isomorphic to an ordinal.*

Exercise 9.3.2. Prove Proposition 9.3.1.

We can further substantiate our analysis by comparing it to two mathematical methods that are of an inherently dynamical nature and thus at least implicitly appeal to a generalized understanding of time, namely (transfinite) recursion and induction. The following two exercises make this point; they show that well-foundedness is exactly the structural property required from a linear ordering that supports inductive proofs and recursive definitions. Their solutions are due to A. Blass and P. Komjath in their responses to a question of mine on MathOverflow.[11]

Exercise 9.3.3 (Folklore). Let $(X, \leq)$ be a linear ordering.

(a) Let $(X, \leq)$ be well-founded, and let ϕ be an $\in$-formula and $\vec{p}$ a finite sequence of parameters (i. e., sets). Suppose that, for all $x \in X$, we have that $\forall y < x\phi(x, \vec{p})$ implies that $\phi(x, \vec{p})$. Show that $\forall x \in X\phi(x, \vec{p})$.

(b) Suppose that, for all $\in$-formulas ϕ and all $\vec{p}$ as in (a), we have that $\forall x \in X$ $((\forall y < x\phi(y, \vec{p})) \rightarrow \phi(x, \vec{p}))$ implies $\forall x \in X\phi(x, \vec{p})$. Show that $(X, \leq)$ is well-founded.[12]

Exercise 9.3.4.[13] Again, let $(X, \leq)$ be a linear ordering. Furthermore, let ϕ be an $\in$-formula and $\vec{p}$ be a parameter such that, for all x, y, there is exactly one z such that $\phi(x, y, z, \vec{p})$.

(a) Show that, if $(X, \leq)$ is well-founded, then there is exactly one function f with $\mathrm{dom}(f) = X$ such that $\forall x \in X\phi(x, f \upharpoonright X_x, f(x), \vec{p})$.

(b) Suppose that, for all ϕ and $\vec{p}$ as above, there is at most one function f with $\mathrm{dom}(f) = X$ such that $\forall x \in X\phi(x, f \upharpoonright X_x, f(x), \vec{p})$. Show that $(X, \leq)$ is well-founded.[14]

(c) Now suppose that, for all ϕ and $\vec{p}$ as above, there is at least one function f with $\mathrm{dom}(f) = X$ such that $\forall x \in X\phi(x, f \upharpoonright X_x, f(x), \vec{p})$. Show that $(X, \leq)$ is well-founded.[15]

---

10 Note that this implies that the empty set is downwards closed.

11 https://mathoverflow.net/questions/96891/ill-founded-recursion-without-fund

12 Hint: Consider the formula "There is no infinite strictly $\leq$-decreasing sequence below x."

13 Blass, Komjath, cf. https://mathoverflow.net/questions/96891/ill-founded-recursion-without-fund

14 Hint: Suppose for a contradiction that Y is a subset of X without a smallest element. Then consider the function that assigns 1 to an element y of Y if and only if Y contains a smaller element y' that is assigned 1 and otherwise assigns 0 and work toward a contradiction.

15 Hint: Proceed as in the hint for (b), but now consider the function that assigns 1 to y if no smaller element has been assigned 1 and otherwise to 0.

9.3.3 Elementary steps

As said above, an elementary step in a transfinite computation means passing from any initial segment of computational activity to the next state. Since states are just sets and we found that our time axis is the class of ordinals, an "initial segment of computational activity" is just a sequence $(a_\iota : \iota < \delta)$ of sets for some $\delta \in \mathrm{On}$. Thus, the process description we are looking for will be a (description of) a class function mapping ordinal sequences of sets to sets.

Let us keep in view that we are looking for a concept of computability. This implies that the process description must be of such a nature that it can be followed "mechanically." Thus, in the finite case, we would not admit a command such as "given i on the tape, decide the ith formula of arithmetic and proceed with state 1 when it is true and with state 0, otherwise." So now we need to look a bit closer what a concept of "mechanical" appropriate for the transfinite might be. A minimal criterion seems to be that, in order for a process description to count as "mechanical," no information must be necessary in carrying it out that is "external" to the process (such as the truth values of arithmetical formulas above). Put in modal terms, following the description should yield the same result in all possible worlds. In our context, the natural interpretation of this demand would be that, given an initial amount of computational activity $A := (a_\iota : \iota < \delta)$, the next state only depends on A and not on any further properties of the set-theoretical universe, i. e., that the process is "absolute."

By making this demand, we work in the spirit of Barwises' discussion of "first-order logics," which we learned from a paper by J. Kennedy, M Magidor and J. Väänänen [100]:

> Imagine a logician k using T (an appropriate system of true set-theoretical statements [M. C.]) as his metatheory for defining the basic notions of a particular logic $\langle L^*, \models^* \rangle$. When is it reasonable for us, as outsiders looking on, to call L^* a "first-order" logic? If the words "first order" have any intuitive content it is that the truth or falsity of $\mathfrak{M} \models^* \phi$ should depend only on ϕ and $\mathfrak{M}$, not on what subsets of M may or may not exist in k's model of his set theory T. In other words, the relation $\models^*$ should be absolute for models of T.
> (Barwise, see [9])

Given this analysis, we need to decide for what class of models (i. e., $\in$-structures) we need to demand absoluteness. There is some leeway in the analysis here, but as the precise choice will turn out to have little influence on the final outcome, we will not make to much of it. Certainly, we only want to consider $\in$-structures (S, E) in which E can be interpreted as the "actual" $\in$-relation, i. e., those that are (isomorphic to ones that are) transitive. Moreover, the structures in which we evaluate our process descriptions should be chosen in such a way that they satisfy some basic properties of transfinite sequences, such as the possibility of concatenation. We follow Barwises proposal in [9] to use transitive models of KP for this purpose here.

Thus, the following approach toward formalization emerges.

Definition 9.3.5. An "elementary step operator" or a "process description" is a class function $F : V \to V$ mapping sequences of sets to sets, with the following property: If M, M' are transitive models of KP and $x \in M \cap M' \cap \mathrm{dom}(F)$, then $F^M(x) = F^{M'}(x)$.

Definition 9.3.6. Let F be an elementary step operator, $x \in V$, $\alpha \in \mathrm{On}$. Then the partial "agent computation" $C_F^\alpha(x)$ of F on input x of length α is defined by the following recursion:
- $C_F^0(x) = \{(0, x)\}$
- $C_F^{\alpha+1}(x) = C_F^\alpha(x) \cup \{(\alpha, F(C_F^\alpha(x)))\}$
- $C_F^\lambda(x) = \bigcup_{\iota<\lambda} C_F^\lambda(x)$, for λ a limit ordinal

We still need some way of determining whether the computation has come to an end. Clearly, there is a wealth of ways to achieve this formally; we could, e. g., take states to be ordered pairs in $\{0, 1\} \times V$ instead of mere sets, the first component indicating whether or not the computation has finished. However, as we have little use for such formalization details here, we will not pursue this further and simply assume that we have some easily recognizable way of telling whether a computation in the sense of Definition 9.3.6 if finished.

From this concept of computation, which emerged from a very liberal analysis of idealized computational activity and can thus count as an upper bound on the strength of transfinite computability, concepts like "agent-computable set" of "agent-computable function" derive in the obvious way.

9.3.4 Idealized agent machines and ordinal Turing machines

Having transformed the informal notion of computability by a transfinite mechanical agent into the formal concept of agent-computability, we will now see how agent computability relates to OTM-computability. It will turn out that both concepts coincide.

Lemma 9.3.7. *Let F be an elementary step operator. Then there is an OTM-program P_F such that, whenever x is a set and X is a set of ordinals coding x, then $P_F(X)$ terminates with output Y, where Y is a set of ordinals coding $F(x)$.*

Proof. We describe the actions of P_F informally. Let ϕ_F be an $\in$-formula defining P_F as a class function, satisfying the definition of an elementary step operator.

First, note that there is some $\alpha \in \mathrm{On}$ such that $L_\alpha[X] \models \mathrm{KP}$. This can, e. g., be seen as follows: Consider the $L[X]$-minimal cardinal κ such that $\kappa > \sup(X)$. It is easy to see that $L_\kappa[X] \models \mathrm{KP}$. Now form the elementary hull of $\sup(X) \cup \{X\}$ in $L_\kappa[X]$, which, by the condensation lemma, is isomorphic to some $L_{\bar{\kappa}}[X]$, which will again be a model of KP and is thus as desired.

P_F now works as follows: Given X, run the relativized version of the OTM-program P_L from Lemma 3.5.3 in Chapter 3 that successively writes codes for the constructible sets to the output tape in the input X until a code C for some $L_\alpha[X]$ with $L_\alpha[X] \models \mathrm{KP}$ is obtained. By Lemma 3.1.32 from Chapter 2, we have $x \in L_\alpha[X]$, and by running through the code and testing for isomorphy, we can effectively in X obtain an ordinal y that codes x in the sense of C. Since F is an elementary step operator, we have $F^{L_\alpha[X]}(x) = F^V(x)$. Now, searching through $L_\alpha[X]$ and using evaluation of bounded truth predicates, we find $y \in L_\alpha[X]$ such that $L_\alpha[X] \models \phi(x, y)$, which, by definition of an elementary step operator, implies that $F(x) = y$ holds in the real world. But now, a code for y can easily be read of from C and y, which is then written to the output. □

By iterating this lemma, we obtain the following.

Corollary 9.3.8.
(a) *If x is the final state of some agent-computation starting with $\emptyset$, then some code of x is OTM-computable.*
(b) *If $G : V \to V$ is an agent-computable class function, then the function G^c that sends every code c_x for a set x to a code c_y for $y = G(x)$ is OTM-computable.*

Proof. (a) Let F be the elementary step operator associated with the agent-computation. Start with a code for $\emptyset$. Now use P_F from the last lemma to successively obtain the $C_F^\alpha(\emptyset)$ and keep checking whether the first component is 1. Once this happens, output the second component, which will be a code for x.

(b) Like (a), starting with c_x instead of $\emptyset$ as the input. □

This now finishes our analysis, as we now have the following chain of inclusions:

$$\begin{aligned}\text{OTM-computable} &\subseteq \text{Computable by an idealized transfinite agent}\\ &\subseteq \text{agent-computable} \subseteq \text{OTM-computable}.\end{aligned}$$

Here, the first inclusion is intuitively clear—transfinite agents should be able to carry out OTM-programs, the second inclusion was established in the previous sections, and the last inclusion is a theorem of set theory. Taken together, these inclusions show that the intuitive concept of computability by an idealized transfinite agent coincides with OTM-computability.

9.3.5 Criteria for a defensible Church–Turing thesis for the transfinite

We argued above that OTMs realize the intuitive notion of a "transfinite mechanical procedure," and we hope that our analysis has somewhat contributed to the credibility of this claim. Still, it is open to challenges and discussion. We thus conclude by mentioning some criteria (one might also say "truth conditions") that our thesis should meet. It is then a matter of future developments to confirm or refute it. In this way, our discussion incidentally yields a rather extensive research agenda.

We begin with a quote from Hodges article [87] that we already discussed above. Hodges proposed the primitive recursive set functions, amended by a parameter and a bounded powerset operation for sets of cardinality $< \kappa$ as a reasonable model for transfinite "effectivity." The justification of this choice is discussed thus: [Hodges, [87], my emphases]

> In identifying 'effective' with 'Prim$(x, \mathcal{P}_{<\kappa})$' we are not proposing any kind of higher Church's thesis. Indeed we are not at all sure that this intuitive notion of effectiveness does answer to one precise mathematical formulation. **If the reader has a better explication of the notion, he will probably find that most of our results—both positive and negative—transfer to his version; there is plenty of leeway.** Our own reasons for choosing Prim$(x, \mathcal{P}_{<\kappa})$ are as follows.
> 1. It is not too generous. Every Prim$(x, \mathcal{P}_{<\kappa})$ function **can be computed according to an explicit and finite formula by a being who can search only a bounded distance beyond the contents of the function argument and the parameter** x.
> 2. It is not too narrow. In algebra at any rate, **the constructions most naturally regarded as effective are in fact** Prim$(x, \mathcal{P}_{<\kappa})$ **for appropriate** x **and** κ. The only way to show this is to examine a large number of cases, in the light of a general picture of what is going on. The present paper forms part of the case-examination.

With respect to our aim, the second sentence in this quote sounds somewhat discouraging at first. Yet, it is quite possible to keep up our thesis while still remaining doubtful that the "intuitive notion of effectiveness does answer to **one** precise mathematical formulation" [my emphasis]: It may well be that, e. g., different areas of mathematics such as number theory and transfinite set theory will require different degrees of idealization. But the need for varying degrees of remoteness from concrete feasibility also comes up in classical computability theory. This is why we have a zoo of complexity classes, which however can be defined naturally in terms of Turing machines. Indeed, we have only proposed OTMs as an **upper bound** for effectivity, i. e., as expressing a certain maximal degree of idealization. The new question is then: Can the degrees of idealization naturally arising in mathematics be reformulated as complexity-bounded OTMs, such as OTMs bounded in time, space or some other way?

Though Hodges has some clear reservations about the possibility of a thesis of the kind we are proposing, he also gives three criteria that a concept of transfinite effectivity should meet, and which we will now discuss (along with two others):

Criterion 1 ("No counterexamples"). Where we have an intuition about whether a certain construction should be regarded as "effective," our formal concept should agree with it. In the one direction (operations that are effective in our technical sense should be intuitively so), this is Hodge's first point and in the other (operations that are effective in the intuitive sense should be effective in the technical sense), this is Hodges second point. As Hodges says, "the only way to test this criterion seems to be to study a large amount of instances." Criterion 1 thus yields a research program, the outcome of which we can hardly anticipate.

Criterion 2 ("Stability"). Different formalizations of the concept of transfinite computability should lead to formally equivalent concepts. In the case of finite Turing machines, this is regarded as a strong evidence for the adequacy of Turing machines: The realm of computable objects does not depend on the details of formalization. Indeed, an impressive amount of proposed formalizations of the concept of (finite) effective computability have all led to the same extension of computable objects. So far, the same phenomenon can be observed for ordinal computability: The equivalence of ordinal Turing machines and ordinal register machines was obtained by P. Koepke (see Chapter 3). In his diploma thesis [53] where the idea of a Church–Turing thesis for ordinal computability was first put forward, T. Fischbach introduced an ordinal version of Kripke's equational calculus and showed that it is equivalent in power to OTMs and ORMs; he also introduced a kind of higher-type "programming language" for these machines. In [54] and [55], T. Fischbach and B. Seyfferth transferred λ-calculus to the ordinal realm and showed that, once again, one obtains the same computable functions as with OTMs. At the level of class-sized ordinal time and space, the stability criterion thus seems to be satisfied.

One objection to this is that many other models of ordinal computability like α-(w)ITRMs or α-ITTMs do not appear in this picture. Our answer to this is that these models can be regarded as complexity bounded versions of OTMs or ORMs, and thus correspond to what in finitary computability would be complexity classes.

Interestingly, OTM-computability is not the same as Hodges' Prim$(x, \mathcal{P}_{<\kappa})$-functions, for x is an arbitrary set parameter and we saw above that the power set operation is highly noneffective in the sense of OTMs. Here lies a potential challenge for our claim. The interesting question is thus whether there are operations for which there is an intuition about their effectiveness and for which Hodges' approach and the OTM-approach yield different answers. The program proposed in criterion 1 should show whether such cases exist. In the quote above, Hodges formulates the expectation that they do not (which would be an interesting outcome as well).

The following explicit challenge to the Church–Turing thesis for ordinal computability as formulated in [29] from the thesis of J. Long [127] also belongs under this subheading. In his thesis, he introduces ITTMs that change their own program and shows that they are equivalent to ITTMs. Since he regards this as a indication that ITTMs are a "natural" model of computation, the following subcriterion naturally arises, which also leads to a research question:

> (...) it would be natural to figure out ways to add self-modification to other models of infinite-time computation, and then determine if their underlying computational power is preserved or compromised. In particular, carrying out this study for Koepke's Ordinal Turing Machines (OTMs) would, depending on the outcome, affirm or dispute Carl's claim in [29] that the OTM model is in fact 'right'.
> (Long, [127])

Criterion 3 ("Naturalness of associated classes"). By a nowadays widespread view, computability theory is about definability, and programs and machine models, etc. are ways of defining sets, measuring their complexity, etc. Looking from this perspective, concepts of definability naturally associated with a concept of computability like OTMs should be "useful," i. e., frequently applicable for measuring the complexity of problems actually arising in mathematics. In this respect, finite computability and complexity theory are strongly confirmed. For the transfinite, there are some encouraging results: For example, several already known important realms like $L_{\omega_1^{\mathrm{CK}}}$, $L_{\omega_\omega^{\mathrm{CK}}}$, L_σ, L, M_1 were "reobtained"; SAT^∞, the paradigmatical example for an NP^∞-complete problem, is quite natural; and there is a strong correspondence between algorithmic randomness in the sense of infinitary machines and genericity in the sense of forcing. Still, much more work would be needed to clarify this point.

Criterion 4 ("Heuristic value"). Assuming that the idealized agent view of mathematics is right and that OTM-computability is a reasonable tool for formally expressing this view, OTM-computability would capture a key feature of mathematics and we should expect the study of OTM-computability to be mathematically instructive and productive. Hence, it should reveal something new: new concepts, new questions and new insights on mathematics that can be formulated without referring to OTMs. A possibly encouraging indicator here is the possibility to reprove some theorems from descriptive and constructive set theory with OTMs that we touched in Chapter 8.

Criterion 5 ("Discursive tenability"). Just like the Church–Turing thesis, our proposed analysis above, together with its conclusion, is open to debate. For example, have we really been liberal enough to be sure that our notion of agent-computability covers all cases that would intuitively be called "computable"? That Hodges' approach allows for parameters and a bounded power set operator casts a first shadow of doubt on it. Further alternative analyses could be given and compared with ours.

In [14], M. Bianchetti doubts that inferior limits should count as "effective," which would mean that OTMs are already too strong a model, and proposes to work with "eventually stable" limits instead; see the section on weak ITTMs in Chapter 2 for some results on this limit rule when applied to ITTMs. Behind this is the idea that "in principle computability" implies compatibility of a computation with the usual laws of physics, which is clearly an approach very different from the one presented above. A further objection communicated to us by M. Bianchetti is that it is "arbitrary" to grant the agent transfinite working time and space while keeping the program finite. Allowing infinite programs would correspond to OTMs with a set-sized input as studied in recent work by E. Lewis, which we briefly touched in Chapter 3 (Exercise 3.9.7); it would also be closer to Hodges' allowing arbitrary set parameters. (We will have a brief look at this idea in the next section.) One might reply that, as the agent is, in Kitcher's words, considered to be an "idealization of ourselves", the crucial point is that we can envisage carrying out a procedure for transfinitely many steps—we have frequently done so in this book—while an infinite program is completely inaccessible to us (unless it can be generated by a finite program, in which case the functioning of the infinite program could be simulated by a finite one that first writes the infinite program and then simulates it).[16] Whatever one thinks of this, it is a clear example that reasonable objections to the above approach can be put forward and should be taken into account.

Of course, to see how well our thesis fares in a critical discussion, it is required that such a discussion takes place. Hopefully, the preceding sections have helped to start it.

9.4 Infinitary Agency as modeled by OTMs

As an application of OTMs as a formalization of idealized agents in mathematics, let us see how this approach matches the criticism of certain ZFC axioms as nonconstructive, such as the one by Takeuti mentioned above. To this end, we propose the following extension of the notion of "effectivity" of a Π_2-statement from Chapter 8, which is Kleene's realizability interpretation of intuitionistic arithmetic (see Kleene [102]) transferred to ordinal computability:

Definition 9.4.1 (Cf. Kleene [102]). Let ϕ, ψ be $\in$-formulas, and let P be an OTM-program, $\alpha \in \mathrm{On}$, $a_0, \dots, a_n, b_0, \dots, b_m$ sets with codes $c(a_0), \dots, c(a_n), c(b_0), \dots, c(b_m)$ and R, R' be finite tuples.

1. If ϕ is quantifier-free, then (P, α) realizes $\phi(a_0, \dots, a_n)$ if and only if $\phi(a_0, \dots, a_n)$ is true (in any transitive sets containing $a_0, \dots, a_n$).
2. (R, R') realizes $(\phi(a_0, \dots, a_n) \wedge \psi(b_0, \dots, b_m))$ if and only if R realizes $\phi(a_0, \dots, a_n)$ and R' realizes $\psi(b_0, \dots, b_m)$.
3. (i, R) realizes $(\phi(a_0, \dots, a_n) \vee \psi(b_0, \dots, b_m))$ if and only if $i = 0$ and R realizes $\phi(a_0, \dots, a_n)$ or $i = 1$ and R realizes $\psi(b_0, \dots, b_m)$.
4. (P, α) realizes $\phi(a_0, \dots, a_n) \rightarrow \psi(b_0, \dots, b_m)$ if and only if, for every R that realizes $\phi(a_0, \dots, a_n)$, $P(\alpha, R, a_0, \dots, a_n)$ computes an R' that realizes $\psi(b_0, \dots, b_m)$.
5. (P, α) realizes $\exists x \phi(x, a_0, \dots, a_n)$ if and only if $P(\alpha, c(a_0), \dots, c(a_n))$ halts with output $(c(b), R)$ where $c(b)$ codes a set b such that R realizes $\phi(b, a_0, \dots, a_n)$.
6. (P, α) realizes $\forall x \phi(x, a_0, \dots, a_n)$ if and only if, for every code $c(a)$ for a set a, $P(\alpha, c(a), c(a_0), \dots, c(a_n))$ halts with output R such that R realizes $\phi(a, a_0, \dots, a_n)$.

16 Carrying this thought further, one might on the other hand be justified in actually being even more restrictive and requiring programs to be "surveyable" or "short." This objection is not critical for the results in this book, though; negative results on program existence show in particular that there is no "surveyable" program, and our positive results were all proved by concrete and surveyable program descriptions. It is only for certain kinds of indirect existence proofs for programs that a conflict may arise.

A statement ϕ is called OTM-realizable if and only if there is some R that realizes R. Below, the prefix "OTM" is usually dropped.

Remark 9.4.2. Note that, by this definition, realizability is only defined for $\in$-formulas where negation is solely applied to quantifier-free subformulas (the same is true for Kleene realizability). This does not seem too bad, since after all, every $\in$-formula can be rewritten in such a form. However, this is done in classical logic, and it is not at all clear that the obtained equivalence can be regarded as "effectively true." We will remain silent about this matter here, as it is irrelevant for our intended applications below.

An axiom frequently criticized from the idealized agent perspective is the power set axiom (abbreviated in the following by "Pot"), i. e., the statement $\forall x \exists y \forall z(z \subseteq x \leftrightarrow z \in y)$. An often attempted justification is that, given a set x, one knows what it means to leave out some of its elements, and thus one can do it "in all possible ways" and collect the results together; see, e. g., H. Wang [177] for such an argument, attributed to K. Gödel:

> To arrive at the totality of integers involves a jump. Overviewing it presupposes an [idealized] infinite intuition. In the second jump, we consider not only the integers as given but also the process of selecting integers as given in intuition. (...) Each selection gives a subset as an object. Taking all possible ways of leaving elements out [of the totality of integers] may be thought of as a *method* for producing these objects. (...) What this idealization—realization of a possibility—means is that we conceive and realize the possibility of a mind which can do it.
> (Wang, [177], Chapter 7, 7.1.19)

However, it is not at all clear that the above actually describes a "method" when it comes to infinite sets. For finite sets, for sure, it is easy to generate all "possible ways of leaving elements out" by, e. g., ordering the elements and enumerating the subsets lexicographically. In the infinite, however, no analogously simple recipe is at hand, not even for a being working in infinite time. Given a subset of the natural numbers, what is the "next" one? Having generated some sets of natural numbers, how does one know that one now has them all and can stop? This apparently presupposes that one knows what the "possible subsets" are, a realm that one is just about to generate!

Based on these arguments, one may thus suspect that the power set axiom should not count as justified from the idealized agent perspective.[17] The following result,

17 The nonconstructive nature of the power set axiom has been noted frequently and early on. For example, in 1927, Fraenkel ([59], p. 113) wrote: "(...) und es ist sogar merkwürdig, daß bei der Kritik gegen die axiomatische Begründung der Mengenlehre es historisch in erster Linie das Auswahlaxiom gewesen ist, das einen wahren Hagel von Angriffen heftigster Art auf sich gezogen hat, und nicht vielmehr das Axiom der Potenzmenge, das von der rein konstruktiven Einstellung ebenso sehr zu beanstanden ist (...)." ("and it is even remarkable that the axiom of choice was the historically first to attract a downright shower of the most vehement attacks in the criticism of the axiomatic foundation of set theory

which was already proved (though with a somewhat different wording) in Chapter 8, Lemma 8.6.4, confirms this suspicion.

Proposition 9.4.3. *There is no realizer (P, α) for the power set axiom.*

Let us now consider Takeuti's criticism of the axiom of comprehension. The idea is that an agent currently building up the constructive universe cannot possibly refer to the end result of his construction while still constructing. We propose the following formal counterpart.

Proposition 9.4.4. *There is an $\in$-formula ϕ such that there is no realizer (P,ρ) for $\forall x \exists y \forall z(z \in y \leftrightarrow (z \in x \land \phi(z)))$.*

Proof. For $\alpha \in \mathrm{On}$, let us write α in the form $\omega\beta_\alpha + i_\alpha$ with $i_\alpha \in \omega$. Now let $\phi(\alpha)$ be an $\in$-formula expressing that $P_{i_\alpha}(\beta_\alpha)\downarrow$. Suppose that comprehension for ϕ was realized by (P,ρ). By computing the next multiplicatively closed ordinal $\gamma > \rho$ and then using (P,ρ) to compute $\{\iota \in \gamma : \phi(\iota)\}$, we would in particular solve the halting problem for OTMs with parameter ρ in the parameter ρ, a contradiction. □

Let ϕ be an $\in$-formula, $\vec{p}$ a parameter, a a nonempty set, $b \in a$ arbitrary. By considering the function $f_{\phi,\vec{p},b,a}$ that maps $x \in a$ to itself when $\phi(x,\vec{p})$ holds and otherwise to b, we see the well-known fact that comprehension for ϕ and $\vec{p}$ is in fact a special case of replacement. It is thus tempting to conclude that there are also instances of the replacement scheme that lack realizers. However, this is not the case; the main reason is that the premise $\forall x \in a \exists y \phi(x,y,u)$ in an instance $\forall u, a[\forall x \in a \exists y \phi(x,y,u) \rightarrow \exists b \forall x \in a \exists y \in b \phi(x,y,u)]$ has a considerably strengthened meaning under the realizability interpretation.

Proposition 9.4.5. *For every $\in$-formula ϕ, the corresponding instance $\mathrm{Rep}_\phi \equiv \forall u, a$ $[\forall x \in a \exists y \phi(x,y,u) \rightarrow \exists b \forall x \in a \exists y \in b \phi(x,y,u)]$ of the replacement axiom is OTM-realizable.*

Proof. Suppose that (codes for) u and a, along with a realizer R for $\forall x \in a \exists y \phi(x,y,u)$ are given; i. e., we have an OTM-program P and some $\alpha \in \mathrm{On}$ such that, for any $x \in a$, P computes (a code for) some y such that $\phi(x,y,u)$ along with a realizer for $\phi(x,y,u)$, given (a code for) x as its input and the parameter α. Thus, all we need to do is to apply $P(\alpha)$ to all elements of a and combine the respective outputs into a set b. □

Given the widespread view that the axiom of choice is highly nonconstructive, the first part of next result may at first be implausible. This immediately changes once one notes that the left-hand side of the implication $\forall x \in F \exists y(y \in x) \rightarrow \exists f : F \rightarrow \bigcup F$

and not rather the power set axiom, which is just as objectionable from a purely constructive perspective (...)" Translation M. C.)

$\forall x \in F(f(x) \in x)$ saying that F is a family of nonempty sets, also receives a constructive reading, which is just the existence of a computable choice-function.[18]

Proposition 9.4.6.
(i) *The statement* $\forall x \in F\exists y(y \in x) \rightarrow \exists f : F \rightarrow \bigcup F\forall x \in F(f(x) \in x)$, *called* AC, *which expresses that every family of nonempty set has a choice function, has a realizer.*
(ii) WO *does not have a realizer.*

Proof. For (i), see the remark preceding the proposition. For (ii), see Chapter 8, and see the proof of Theorem 8.6.25 in Chapter 8. □

With respect to these objections, KP looks much more promising than ZFC: Comprehension is limited to formulas with bounded quantifiers, so that Takeuti's criticism does not apply, and the power set axiom is omitted altogether. Indeed, it seems that the KP axioms are well justifiable on the basis of the idealized agent view. And indeed, we have the following.

Theorem 9.4.7. *Every axiom of KP is realized by some finite tuple R.*

Proof. This is clear for extensionality, pairing, infinity and union.

Given codes $c(a), c(a_0), \ldots, c(a_n)$ for sets and a Δ_0-formula ϕ, a code for $\{x \in a : \phi(x, a_0, \ldots, a_n)\}$ is computable by bounded searches, and thus by an OTM-program in the input $c(a), c(a_0), \ldots, c(a_n)$. Thus Δ_0-separation holds as well.

Let us now turn to Δ_0-collection. For the sake of simplicity, we drop parameters. Let ϕ be a Δ_0-formula, and suppose that (P, α) realizes $\forall x \in a\exists y\phi(x, y)$. We need to compute a realizer (Q, β) for $\exists b\forall x \in a\exists y \in b\phi(x, y)$. But such a b is easily obtained by running $P^x(\alpha)$ for all $x \in a$ and collecting the outputs together, and the respective realizers for $\phi(x, y)$ can also be computed with (P, α).

This leaves us with induction. Again, we drop parameters. Suppose that ϕ is an $\in$-formula and that (P, α) realizes $\forall y(\forall x \in y\phi(x) \rightarrow \phi(y))$. We need to compute a realizer for $\forall z\phi(z)$. This is done recursively: Our goal is thus to compute realizers for $\phi(x)$ for all $x \in z$ and store them on the tape and then to apply (P, α). These realizers can in turn be found by the same strategy, and since $\in$ is well-founded, we will obtain the desired realizer in this way. □

Remark 9.4.8. Those who, like Kitcher, attempt to justify ZFC on the basis of the idealized agent view in spite of the above objections, e. g., against the comprehension axiom sometimes make a distinction between what one may call the "material" and the "cognitive" part of constructing a set: In order for a set x to be constructible, it is only required that all elements of x are "already available," but not that the agent has the cognitive capacity to, e. g., check a formula with unbounded quantifiers (see e.g.

---

18 For the same reason, the axiom of choice is valid in the classical realizability interpretation.

[101], p. 145). One might say that the agent builds the set $\{y \in x : \phi(y)\}$ "accidentally," This would yield the weaker requirement that ZFC holds in the realm of constructible objects. Using OTMs, the formal counterpart of this is of course true, as we saw in Chapter 3 that the class of sets that have a pOTM-computable code is L and $L \models$ ZFC. Whether this can count as a satisfying justification of ZFC on the basis of idealized agency is another matter.

9.4.1 Transfinite creative subjects?

So far, we have restricted our attention to "mechanical" agents that work according to a fixed mechanical rule. In some directions of constructivism, such as Brouwer's intuitionism, the ability of the constructing subject to make free decisions is also taken into account, via the introduction of "choice sequences." At least for the purpose of checking the robustness of the idealized agent view under a variety of formalizations, it seems worthwhile to attempt a transfer of this approach to the infinitary framework as well.

One immediate possibility would be to change Definition 9.4.1 a bit by allowing arbitrary elements of $\{0,1\}^{**}$ as parameters, thus working with Lewis' OTMs with infinite programs already discussed above. It is easy to verify that the above results, along with their proofs, still work in this setting.

We sketch a different way to grant "creativity" to the idealized subject. Let us try to "ordinalize" the creative subject by introducing "ordinal choice sequences." The idea is that the idealized agent builds up a binary sequence of length On of which he has only constructed an initial segment $s_\tau \in \{0,1\}^{**}$ at each moment τ. The sets that can be obtained "mechanically" from the sequence s_τ, i. e., the elements of $L[s_\tau]$, is then the universe that the agent can "see" at time τ.[19] Ordering $\{0,1\}^{**}$ by the initial segment relation, we obtain a class-sized Kripke frame, on which we can now evaluate $\in$-statements according to Kripke's semantics for intuitionistic logic [119].[20]

Definition 9.4.9 (Cf. Kripke, [119]). Let $s \in \{0,1\}^{**}$, let $\phi(x_0,\dots,x_n), \psi(y_0,\dots,y_m)$ be $\in$-formulas and $a_0,\dots,a_n,b_0,\dots,b_m$ sets. We define the relation $s \Vdash \phi$ ("s forces ϕ") by recursion as follows:

19 Here, one might object that building the whole L over s_τ requires a proper class of time after which no time for further "choices" is available. Other reasonable choices of the form $L_\alpha[s_\tau]$ are possible, but they will not influence the results below.

20 Kripke models have been used in the study of constructive set theory in work of Myhill [144], Lubarsky [132], Iemhoff [90] and others. See also Remark 9.4.16 below for the relation of idealized agency to constructive set theory. The idea of considering Kripke frames of ZFC models was also used by Hamkins and Löwe in [78], though with a different motivation and purpose, namely to determine the modal principles valid in this frame.

- If ϕ is Δ_0, then $s \Vdash \phi(a_0, \dots, a_n)$ if and only if $a_0, \dots, a_n \in L[s]$ and $L[s] \models \phi(a_0, \dots, a_n)$.
- $s \Vdash \neg\phi$ if and only if $t \nVdash \phi$ for all $t \supseteq s$.
- $s \Vdash \phi \wedge \psi$ if and only if $s \Vdash \phi$ and $s \Vdash \psi$.
- $s \Vdash \phi \vee \psi$ if and only if $s \Vdash \phi$ or $s \Vdash \psi$.
- $s \Vdash \phi \rightarrow \psi$ if and only if, for all $t \supseteq s$, if $t \Vdash \phi$, then $t \Vdash \psi$.
- $s \Vdash \exists x\phi$ if and only if, $s \Vdash \phi(a)$ for some $a \in L[s]$.
- $s \Vdash \forall x\phi$ if and only if, for all $t \supseteq s$ and all $a \in L[t]$, we have $t \Vdash \phi(a)$.

Suppose that, at time τ, the agent A has constructed $s_\tau \in \{0,1\}^{**}$ and is now looking at $L[s_\tau]$. Let ϕ be an $\in$-formula. In order to say that A "knows" ϕ at this moment, it should (at least) be the case that $L[s_\tau] \models \phi$, and in addition, future modifications should not be able to change this; thus, we should have $L[t] \models \phi$ for all t that extend s_τ. The above definition implements this idea; thus, in order for our agent to know ϕ at time τ, a minimal requirement is that $s_\tau \Vdash \phi$. Let us say that ϕ is "knowable" if and only if there is some s such that $s \Vdash \phi$ and that ϕ is "universally known" if and only if $s \Vdash \phi$ for all $s \in \{0,1\}^{**}$.

Proposition 9.4.10 (Folklore). *The relation $\Vdash$ is monotonic, i. e., if $s \Vdash \phi$ and $t \supseteq s$, then $t \Vdash \phi$. Moreover, $s \Vdash \phi$ implies that $L[t] \models \phi$ for all $t \supseteq s$.*

Proof. An easy induction on ϕ. □

Again, we can use this as a formal approach to the question which axioms of set theory are defensible from an idealized agent perspective by asking what axioms of ZFC are knowable. However, when we do this in the framework of ZFC, we are working in a fixed realm of sequences, which is somewhat opposite to the spirit of "free" choice sequences. In particular, when $V = L$, then all sequences under consideration will be constructible, and all models in our frame will just be L. One way to deal with this is to "simulate" freedom by making extra assumptions implying that the set-theoretical universe is "large." Another would be to work in a multiverse framework as that proposed by J. Hamkins (see, e. g., [75]). We only follow the former path here.

The following is easy to verify.

Theorem 9.4.11. *All axioms of* KP *are universally known.*

On the other hand, if the set-theoretical universe is large enough, the criticism of Takeuti and others is again confirmed.

Proposition 9.4.12. *Suppose that there is a proper class of cardinals κ such that $2^\kappa \neq \kappa^+$. Then we have the following:*

(i) *The power set axiom is not knowable.*
(ii) *There are $\in$-formulas ϕ such that the instance of the comprehension axiom for ϕ is not knowable.*

Proof. (i) Suppose for a contradiction that $s \Vdash$ Pot. Let $\kappa \in L[s]$ be an infinite cardinal greater than the length of s such that $2^\kappa > \kappa^+$, which exists by assumption. The usual argument for GCH in L shows that $(2^\kappa)^{L[s]} = (\kappa^+)^{L[s]}$. It follows that $\mathfrak{P}^{L[s]}(\kappa) \subsetneq \mathfrak{P}(\kappa)$. Pick $x \in \mathfrak{P}(\kappa) \setminus \mathfrak{P}^{L[s]}(\kappa)$, let t be a binary ordinal string that codes x, and let $s \frown t$ be the concatenation of s and t. Then $x \in L[s \frown t]$, so $\mathfrak{P}^{L[s]}(\kappa) \subsetneq \mathfrak{P}(\kappa)^{L[s \frown t]}$. It follows that, for all $a \in L[s]$, $s \Vdash \forall z(z \in a \leftrightarrow z \subseteq \kappa)$ is false. Hence, it is not the case that $s \Vdash \exists a \forall z(z \in a \leftrightarrow z \subseteq \kappa)$, so $s \Vdash$ Pot is false.

(ii) An argument similar to the one given in (i) shows that no s can force comprehension for the formula saying that x is a cardinal. The second claim is then obtained as in the remark following Proposition 9.4.4. □

Exercise 9.4.13. Under the assumptions from Proposition 9.4.12, give an example of an $\in$-formula ϕ such that neither ϕ nor $\neg\phi$ are knowable.

Since the well-ordering principle has no realizer by Proposition 9.4.6 above, the following may come as surprise; however, the surprise ceases when one recalls the highly constructive nature of the objects considered. If all sets are constructed by an idealized cognitive agent working in transfinite time, they are naturally well-ordered by the order in which they were constructed. The picture behind the realizability semantics above is that the universe of sets is "given" and the question is what the agent can effectively know about it; the idea of the Kripke semantics developed here is that the realm of sets is constructed by an agent and the question is what he can know regardless of future developments. The following is in fact easy to prove, and we leave the proof as an exercise.

Proposition 9.4.14. *The well-ordering principle is universally known.*

The same argument actually yields the following result, which however depends much more on the details of our formalization (such as taking models of the form $L[s_\tau]$ as our states rather than models of the form $L_\alpha[s_\tau]$ for reasonably closed α) and, therefore, seems less of an expression of properties of idealized agency:

Exercise 9.4.15. Show that every Π_2-consequence of ZFC + $\exists x V = L[x]$ is universally known.

There is certainly a lot of room for discussion about the approaches mentioned above and their adequacy to the idealized agent views. However, they clearly indicate that there is a certain robustness about the concept of transfinite idealized agency. How far this robustness goes is a matter of future investigation. An ideal result would be that, under appropriate assumptions (multiverse axioms), the universally known sentences coincide with the (intuitionistic)[21] consequences of some extension of KP.

21 Classical logic is ruled out by exercise 9.4.12 as it contains the law of the excluded middle.

Moreover, it would be interesting to "ordinalize" other approaches to the semantics of intuitionistic mathematics.

Remark 9.4.16 (Relation to constructive set theory). The study of the justifiability, or otherwise, of set-theoretical axioms on the basis of idealized agents views of mathematics is clearly related to the study of constructivistic versions of set theory, such as Intuitionistic ZF (IZF), introduced by H. Friedman [60], Myhill's Constructive Set Theory (CST) [144] or Aczel's CZF [4]. We refer to [158] for an overview. One might characterize these approaches as "top-down": Starting with classical set theories like ZFC, one weakens the theory in order to become compatible with basic tenets of constructivism, like failure of the principle of excluded middle. One motivation for doing this is to obtain a suitable background theory for formalising some branches of constructivistic mathematics, such as constructivistic analysis (see [144]). Occasionally, the status of the axioms is discussed from the basic viewpoint of constructivism (see, e. g., Myhills remarks on the power set axiom in [144]), but such considerations do not seem to play a major role constructive set theory.[22]

In contrast, our approach may be characterized as "bottom up"; starting from the basic idea of constructivism that mathematical objects are mental constructions of ideal cognitive subjects, we give a formal meaning to the notion of an "ideal cognitive subject" via infinitary computability and evaluate set theory from this point of view.

It is then natural to ask how close the "top-down" and the "bottom-up" approach are in their resulting theories. This is not the place to address this question in detail, so we restrict ourselves to the following: As we already saw above, basic axioms like extensionality, the existence of the empty set, the existence of an infinite set or the axiom of union or $\in$-induction are OTM-realizable. Both in Myhill's CST and in Aczel's CZF, comprehension is restricted to Δ_0-formulas, for which we saw above that all of its instances are OTM-realizable, while more general instances of comprehension are in general not. Moreover, the "strong collection schema" in Aczel's CZF is OTM-realizable, while neither the "subset collection schema" from CZF nor the "exponentiation axiom" from CST are OTM-realizable. Thus, while OTM-realizability confirms some aspects of constructive set theory, it does not yield any of the usual axiomatic systems used to codify it.

9.5 Infinite time Turing machines and the revision theory of truth

We finally explain another connection between infinitary computability and philosophy, discovered by B. Löwe [128], namely a strong link between infinite time Turing

22 See, e. g., Aczel and Rathjen [5], page 1: "It [constructive set theory, insertion M. C.] is distinctive in that it (...) makes no explicit use of specifically constructive ideas. Of course, its logic is intuitionistic, but there is no special notion of construction or constructive object."

machines and the revision theory of truth, which was introduced independently by H. Herzberger [83] and by A. Gupta and N. Belnap. For a thorough treatment, see Gupta and Belnap [13]. As Löwe's observation allowed it to "import" results from the theory of ITTMs to the revision theory of truth, and thus the solution of some open problems in the field of that theory, this section might as well be considered as belonging to the chapter about "applications." However, we expect philosophically minded people to be attracted more by the title of the current chapter, so we decided to put it here.

9.5.1 The revision theory of truth

We give a brief introduction to the revision theory of truth. Further short introductory accounts are [194] and [174], a comprehensive exposition is Gupta and Belnap [13]. Our account is based on these sources.

Let us have a look at the following two sentences, called "the truth teller" and "the liar," respectively[23]:

1. (The Truth-Teller) sentence (1) is true.
2. (The Liar) sentence (2) is false.

Are these sentences true or false? If we try to find out by experimenting with assumptions about the truth values about (1) and (2), we find that, if we assume that (1) is true/false, then (1) confirms this, while both choices are immediately rejected in the case of (2): If we assume that (2) is false, then what (2) tells us is true, so (2) should be true; if we assume that (2) is true, then it is true that (2) is false, so it follows that (2) is false.

The situation seems to be hopeless if we insist that (1) and (2) have definite, fixed truth-values (many ways of dealing with such sentences have been considered; we will not even try to give a comprehensive discussion). An alternative approach would be to treat the behavior of such sentences as a kind of dynamics: If I make the hypothesis that (2) is true, then I am led to the conclusion that (2) is false. Seen in this way, (2) can be regarded as an "update rule" for the truth value of (2): Starting with an initial hypothesis about the truth value of (2), we get an alternating sequence of successively revised truth values, called the "revision sequence" of (2).

Formally, and considering first-order logic rather than mere propositional logic, we can define revision sequences as follows.

Definition 9.5.1 (Gupta and Belnap, cf. [13]). Let $\mathcal{L}$ be a first-order language the predicate symbols of which include $P_0, \ldots, P_n$, where P_i is of arity n_i. Let $\phi_0, \ldots, \phi_n$ be $\mathcal{L}$-formulas, which may contain occurrences of $P_0, \ldots, P_n$. Then $C := \{P_i(x_0, \ldots, x_{n_i-1}) :=$

23 These standard examples are also, e. g., treated in the web sources mentioned above.

$\phi_i(x_0, \ldots, x_{n_i-1}) : 0 \le i \le n\}$ is called a "revision rule" or a "finite set of circular definitions" in $\mathcal{L}$.

Let X be a set. A "hypothesis" H for C with ground set X is an $(n+1)$-tuple $(P_0^{\mathcal{M}}, \ldots, P_n^{\mathcal{M}})$, where $\mathcal{M}$ is an $\mathcal{L}$-structure $\mathcal{M}$ with domain X. When X and C are clear from the context, they are not mentioned. If $\mathcal{M}$ is an $\mathcal{L}$-structure and H is a hypothesis, then the H-variant $H(\mathcal{M})$ of $\mathcal{M}$ is obtained by replacing the interpretation of $P_0, \ldots, P_n$ in $\mathcal{M}$ in accordance with H. If $\mathcal{M}$ is an $\mathcal{L}$-structure, the corresponding hypothesis is denoted by $H^{\mathcal{M}}$.

Now the C-revision sequence for $(P_0, \ldots, P_n)$ starting with the $\mathcal{L}$-structure $\mathcal{M}$ is a sequence $(H_i : i \in \omega)$ of hypotheses defined as follows: First, let $\mathcal{M}_0 = \mathcal{M}$, $H_0 = H^{\mathcal{M}}$. Then, for $k \in \omega$, $H_k = (P_0^k, \ldots, P_n^k)$, $i \in \{0, \ldots, n\}$ and $\vec{x} \in X^{n_i}$, let $\vec{x} \in P_i^{k+1}$ if and only if $\mathcal{M}_k \models \phi_i(\vec{x})$, let $H_{k+1} = (P_0^{k+}, \ldots, P_n^{k+1})$ and let $\mathcal{M}_{k+1}$ be the H_{k+1}-variant of $\mathcal{M}$. If X is nicely enumerable and the P_i's are unary, revision sequences can nicely be represented by giving the truth values of $P_i(x)$ for all $x \in X$ at the successive stages.

We can now also define revision sequences for arbitrary $\mathcal{M}$-sentences: If ϕ is an $\mathcal{L}$-sentence, then the C-revision sequence of ϕ starting with $\mathcal{M}$ is the sequence $(\phi^{\mathcal{M}_i} : i \in \omega)$ of truth values of ϕ in the sequence of $\mathcal{L}$-structures just defined.

Remark 9.5.2. Note that, if $\phi_0, \ldots, \phi_n$ contain no further predicate letters beyond $P_0, \ldots, P_n$, all that matters about $\mathcal{M}$ are its domain and its interpretations of $P_0, \ldots, P_n$. Thus, the rest of the structures can and will be ignored in this case.

Revision sequences provide an attractive approach to formally modeling circular reasoning figures. As an example, let us model the well-known "surprise examination paradox"[24] (see, e. g., Kritchman and Raz, [121]) in the revision theory of truth.

Example 9.5.3. A teacher tells his class on Friday that there will be a "surprise examination" in the next week. More specifically, there will be an examination at some day (Monday–Friday) of the coming week, and on the morning of the respective day, the students will not know this.

Now, under these conditions, the examination clearly cannot take place on Friday: If the exam has not happened on any earlier day, the students will know on Friday morning that it is due now, since after all they were told that there would be an examination in this week. But now it also cannot happen on Thursday either: For if it has not happened on Monday, Tuesday or Wednesday and given what we know now, namely that it cannot take place on Friday, everyone will know on Thursday morning that it will happen on this day. Iterating this argument, it becomes clear that a surprise examination is impossible on any day. So no one bothered to prepare.

However, the examination took place on Tuesday, to everybody's (unpleasant) surprise.

---

24 Also known by the even more gruesome name "unexpected hanging paradox," see https://en.wikipedia.org/wiki/Unexpected_hanging_paradox

This can be tackled in various ways. With the right approach, one can even get a new proof for Gödel's incompleteness theorem out of this (see [121]). We will try a different approach. One easily gets suspicious that one point of what is going on here is that what counts as "surprising" changes during the course of the argument. This seems like a job for revision theory!

Since it is easier to talk about natural numbers than about days of the week, let us generalize this to a "week" consisting of successive days $d_0, \dots, d_n$ ordered $d_0 < d_1 < \dots < d_n$, $n \in \omega$ for the sake of the discussion. Writing Surprising(d) for "if the exam happens on day d, people will be surprised" and ignoring the punch line for a moment, the student's reasoning can be modeled as follows:

$$\text{Surprising}(d) := \exists d'(d' > d \wedge \text{Surprising}(d')).$$

Indeed, the argumentation above is just this: If no day later than d is surprising and the exam can only happen on a surprising day, then on the morning of day d one would know it must happen on day d, and so it would no longer be surprising. On the other hand, if there is at least one other possibility after d, one cannot know this, and thus will be surprised.

Starting with the hypothesis that every day is surprising (which, if one does not apply any reasoning at all, it is), we obtain the following revision sequence in the case $n = 4$:

Stage	**1**	**2**	**3**	**4**	**5**	**6**	**7**	**8**
Surprising(d_0)	1	1	1	1	1	0	0	...
Surprising(d_1)	1	1	1	1	0	0	0	...
Surprising(d_2)	1	1	1	0	0	0	0	...
Surprising(d_3)	1	1	0	0	0	0	0	...
Surprising(d_4)	1	0	0	0	0	0	0	...

Thus, Surprising(d) is stably false for every day, reflecting the student's conviction that the exam cannot take place.

Let us now account for the obvious reasoning mistake: If the examination is expected to happen at a surprising day and day d is not surprising, then it is indeed surprising if it happens on day d! Indeed, even if the examination took place on Friday, this would come as a surprise to someone who just deduced as above that this cannot be as it would not be surprising (and to the one who did not make such a deduction as well, of course). To account for this extra subtlety, let us change the definition of surprising to:

$$\text{Surprising}(d) := \exists d'(d' > d \wedge \text{Surprising}(d')) \vee \neg\text{Surprising}(d).$$

Thus, d is surprising if some later day is **or if d is not surprising**. What happens to our revision sequences? Let us have a look:

Stage	1	2	3	4	5	6	7	8	9	10	11
Surprising(d_0)	1	1	1	1	1	1	1	1	1	1	...
Surprising(d_1)	1	1	1	1	1	1	1	1	1	1	...
Surprising(d_2)	1	1	1	1	1	1	1	1	1	1	...
Surprising(d_3)	1	1	0	1	0	1	0	1	0	1	...
Surprising(d_4)	1	0	1	0	1	0	1	0	1	0	...

The picture we obtain is rather intriguing: While Friday and Thursday preserve a somewhat special state and alternate at being surprising or non-surprising, all other days remain surprising.

9.5.1.1 Transfinite revision sequences

For many interesting examples such as the liar, the truth teller or the surprise examination paradox, the revision theory as described above works just fine. However, there are cases in which it leads to counterintuitive or at least unwanted results.

As an example, which is inspired by [194] and [174], consider the following argument, which is known as the "interesting number paradox."

Example 9.5.4. "Every natural number is interesting. Indeed, let us suppose otherwise. Then there would be a smallest uninteresting natural number *d*. But *d* would be of tremendous interest! Contradiction."

The idea that "the smallest uninteresting natural number is of tremendous interest" can be captured in the following definition[25]:

$$\text{Interesting}(n) := \forall i < n\text{Interesting}(i) \wedge \neg\text{Interesting}(n).$$

An easy induction now yields that every natural number is interesting. However, starting from the widespread hypothesis that numbers in general are boring, the corresponding revision sequence looks like this (cf. [194, 174]):

Stage	1	2	3	4	5	6	7	8
∀*i*Interesting(*i*)	0	0	0	0	0	0	0	...
Interesting(0)	0	1	1	1	1	1	1	...
Interesting(1)	0	0	1	1	1	1	1	...
Interesting(2)	0	0	0	1	1	1	1	...
Interesting(3)	0	0	0	0	1	1	1	...
Interesting(4)	0	0	0	0	0	1	1	...
				...				

25 Of course, this ignores a great deal of possible reasons why a number might be interesting; however, it captures the aspect relevant for the above argument.

It is easy to see that, from stage i on, Interesting(i) has the value 1. Thus, for every $i \in \omega$, Interesting(i) is stably true. On the other hand, as Interesting($i+1$) is also always false at the ith stage, there is no stage at which $\forall i$Interesting(i) is satisfied, so $\forall i$Interesting(i) is stably false.

Exercise 9.5.5. Show that, for every initial hypothesis H, Interesting(i) is stably true from stage i on.

Clearly, this is not what we would expect to happen with the truth value of $\forall i$Interesting(i). The reason seems to be that the revision sequence is "too short": It is indeed the case that the propositions of the form Interesting(i) all become (and remain) true eventually, but this does not happen at any finite stage and is thus "not noticed." The obvious cure would then be to allow the revision sequence to extend to transfinite stages. And this is what we will do now.[26]

Of course, this forces us to define what should happen at limit stages. This can be done in several reasonable ways [13], but in order to keep our exposition concise, we focus on the following rule, due to Herzberger[27] [83], which appears to be quite natural: At a limit stage δ, exactly those statements are true that became stably true before δ. That is, to determine the truth value of a proposition A at a limit stage δ, consider the sequence $(A_\iota : \iota < \delta)$ or previously assumed truth values and let the value at stage δ be 1 if the sequence is eventually constant with value 1, and otherwise let the value at stage δ be 0.

For our motivating example, this rule clearly works as desired: At stage ω, Interesting(i) has truth value 1 for every $i \in \omega$, hence $\forall i$Interesting(i) has truth value 1 at stage $\omega + 1$, which is never changed back afterwards. Thus, $\forall i$Interesting(i) is now stably true, as it should be.

This hence leads to the following definition.

Definition 9.5.6 (Cf., e. g., [13, 128]). Let $\mathcal{L}, P_0, \ldots, P_n, \phi_0, \ldots, \phi_n, C, X, H$ be as in Definition 9.5.1. Then the transfinite C-revision sequence for $(P_0, \ldots, P_n)$ starting with the $\mathcal{M}$ is the sequence $(H_\iota : \iota \in \mathrm{On})$ of hypotheses defined as follows: First, let $\mathcal{M}_0 = \mathcal{M}$, $H_0 = H^{\mathcal{M}}$. Then, for $\iota \in \mathrm{On}$, $H_\iota = (P_0^\iota, \ldots, P_n^\iota)$, $i \in \{0, \ldots, n\}$ and $\vec{x} \in X^{n_i}$, let $\vec{x} \in P_i^{\iota+1}$ if and only if $\mathcal{M}_\iota \models \phi_i(\vec{x})$, let $H_{\iota+1} = (P_0^{\iota+1}, \ldots, P_n^{\iota+1})$ and let $\mathcal{M}_{\iota+1}$ be the $H_{\iota+1}$-variant of $\mathcal{M}$. Finally, if δ is a limit ordinal, $i \in \omega$ and $\vec{x} \in X^{n_i}$, we let $\vec{x} \in P_i^\delta$ if and only if there is $\alpha < \delta$ such that, for all $\iota \in [\alpha, \delta)$, we let $\vec{x} \in P_i^\iota$, $H_\delta = (P_0^\delta, \ldots, P_n^\delta)$ and let $\mathcal{M}_\delta$ be the H_δ-variant of $\mathcal{M}$.

Exercise 9.5.7. Work out the transfinite revision sequences for the liar and the truth-teller.

---

26 Cf., e. g., [174] for the motivation of extending revision sequences to transfinite stages given here.

27 Actually, we will see below that crucial notions of revision-theoretical truth are quite robust under changes of the limit rule.

Exercise 9.5.8. (a) Consider the definition $A(m,n) := \forall i < m \forall j A(i,j) \wedge \forall j < n A(m,j)$. Work out the corresponding transfinite revision sequences under the starting hypothesis that $A(i,j)$ is false for all $i,j \in \omega$. What happens at times ω, $\omega \cdot 2$, $\omega \cdot 3$, ω^2?

(b) What is the revision sequence for $\forall i \forall j A(i,j)$?

(c) For each $k \in \omega$, find a definition $A(x_0, \ldots, x_n)$ such that $\forall x_0 \ldots \forall x_n A(x_0, \ldots, x_n)$ becomes stably true at time $\omega^k + 1$.

Exercise 9.5.9. (a) Work out the analogue of Example 9.5.4 for countable ordinals instead of natural numbers.

(b) Work out the analogue of Example 9.5.4 for all ordinal numbers. What happens?

But wait a minute. Have not we seen something like Definition 9.5.6 before? We make rather straightforward modifications according to simple rules at successor stages, and at limit stages, we take inferior limits of earlier values. This looks a lot like the definition of an ITTM-computation. Could this be more than a superficial similarity?

9.5.2 Connecting revision theory of truth and ITTMs

Indeed, it is. This was first noticed by B. Löwe in [128], who showed that revision sequences with the Herzberger rule and ITTM-computations are more or less the same thing. This section will take you through both directions of the proof of this surprising connection by some guided exercises.

Exercise 9.5.10 (Löwe, [128]). The goal of this exercise is to show how to simulate a Turing program by a finite set of circular definitions in such a way that the corresponding revision sequence mirrors the course of the Turing computation. To this end, let us fix a Turing program P using n inner states $q_0, \ldots, q_n$. We introduce a binary predicate letter T and unary predicate letters S and H, the intended meanings of which are:

- for $T(i,j)$: "j is the symbol (0 or 1) currently written on tape cell i"
- for $S(j)$: "j is the index of the current inner state"
- for $H(i)$: "the head is currently at position i"

Explain how, based on P, one can obtain (circular) definitions for T, S and H such that, starting with the hypotheses $T(i,0)$ for all $i \in \omega$, $S(0)$ and $H(0)$, the truth values at the kth element of the corresponding revision sequence will correctly describe the snapshot of P at time k. As an example, if P consists of the single rule $(q_0, 0) \to (q_1, 1, +1)$, then the set of definitions should contain $S(j) \leftrightarrow \exists i[j = 1 \wedge S(0) \wedge H(i) \wedge T(i,0)]$.

Exercise 9.5.11 (Löwe, [128]). We now want not extend the simulation of Turing computations by revision sequences just obtained to transfinite revision sequences simulating ITTMs. We start by imitating the "flag" idea with revision sequences:

(a) Consider the set following set of definitions: $\{A := \neg A, B := \neg B, C := (A \leftrightarrow B)\}$. Show that, starting with the hypothesis that A is false, B is true and C is false, the corresponding revision sequence will have C true at the ιth stage ($\iota \in \mathrm{On}$) if and only if ι is a limit ordinal.

The predicate letters used in the preceding exercise need to be modified a bit.

- $T'(i,j)$ now means "the number (0 or 1) currently written on tape cell i is at least j"
- $S'(j)$ now means "the index of the inner state is at least j"
- $H'(j)$ now means "the index of the current head position is at least j, unless we are in a limit state and $H'(j)$ holds for all $j \in \omega$, in which case the index of the head position is 0."

Explain how, based on P, one can obtain (circular) definitions for T', S', H', such that, starting with the hypotheses $T(i,0)$ for all $i \in \omega$, $S(0)$ and $H(0)$, for all $\iota \in \mathrm{On}$, the truth values at the $\iota+1$th element of the corresponding revision sequence will correctly describe the snapshot of P at time ι.

Exercise 9.5.12 (Löwe, [128]). Now suppose that $S = \{D_1, \dots, D_k\}$ is a finite set of definitions of the following form:

- $P_0(x_{i_{00}}, \dots, x_{i_{0n_0}}) := \phi_0(x_{i_{00}}, \dots, x_{i_{0n_0}})$
- ...
- $P_k(x_{i_{k0}}, \dots, x_{i_{kn_k}}) := \phi_k(x_{i_{k0}}, \dots, x_{i_{kn_k}})$

Here, each ϕ_i is a sentence in the first-order language the only nonlogical symbols of which are the predicate letters $P_0, \dots, P_k$.

We want to show that the transfinite revision sequence starting from a hypothesis H can be computed by an ITTM on input H in the sense that the machine has an inner state q such that the sequence of contents of the output tape at times when the program is in state q is the revision sequence of S. To this end, split the working tape into k many portions, where the jth portion is intended to represent the set of n_j-tuples satisfying P_j.

(a) Show that there is an ITTM-program that modifies the tape content correctly at successor levels.

(b) Show that, if the program in (a) makes no further changes on the working tape but replacing each bit with its update at the next stage, then the contents at limit times are automatically correct.

9.5.3 Applying the connection

Given the strong connection between ITTMs and the revision theory of truth established in the last section, one should expect results in one area to have relevance for the other. This is indeed the case. In fact, some results, e. g., about the role of ζ and its model-theoretical characterization that we saw in Chapter 3 were obtained independently in both areas: once by J. Burgess in the context of revision theory, and then by P. Welch for ITTMs (see Section 1.1 of [185]). So let us conclude by seeing some examples of what our knowledge about ITTMs can teach us about the revision theory of truth.

Exercise 9.5.13 (Welch, cf. [185]). Let C be a revision rule and let H be a starting hypothesis with a countable domain. Show that the transfinite revision sequence $(H_\iota : \iota \in \mathrm{On})$ for C with the starting hypothesis H is circular in the following sense: There are $\alpha, \gamma \in \mathrm{On}$ such that, for all $\beta > \alpha$ and all $\iota \in \mathrm{On}$, we have $H_{\beta+\gamma\iota} = H_\beta$. Moreover, show that, if H is coded by a real number h, then $\alpha, \gamma < \Sigma^h$.

Revision theory can be motivated as a theory of circular definitions. This prompts the question what objects can be defined with such definitions when interpreted according to revision theory. Over general domains, this is a bit difficult to answer; so let us focus on sets of natural numbers here, i. e., let us fix our domain as ω.

Let us say that a set $x \subseteq \omega$ is r-definable if and only if there is a finite set of S of definitions including a unary predicate letter A such that, in the revision sequence for S starting with the hypothesis that, for all $k \in \omega$, all k-ary predicate letters P appearing in S and all $(n_0, \ldots, n_{k-1}) \in \omega^k$, $P(n_0, \ldots, n_{k-1})$ is false, $A(i)$ becomes and remains true from some point on if and only if $i \in x$ and otherwise, $A(i)$ becomes and remains false.

The connection between revision theory and ITTM-computability can then be exploited to characterize the subsets of ω that are definable with circular definitions.

Exercise 9.5.14 (Welch, cf. [185]). (a) Show that, if $x \subseteq \omega$ is r-definable, then $x \in L_\zeta$.
(b) Show that, if $x \subseteq \omega$ is eventually writable by an ITTM, then x is r-definable.
Thus, r-definability coincides with eventual writability!

Finally, we mention a problem that played a prominent role in the revision theory of truth, namely the **limit rule problem**: We already mentioned that different proposals have been made concerning the treatment of revision sequences at limit stages. Thus the question arose whether these proposals differed substantially, e. g., with respect to the class of r-definable sets arising from their adoption. A similar question was considered independently for ITTMs, as we mentioned briefly at the end of Chapter 2: One may think of different ways of defining the configuration of an ITTM at a limit time, and thus the question arises whether this changes the class of computable functions.

For ITTMs, this was settled by P. Welch in [180], where it was shown that the liminf limit rule is "complete" among all ways of determining limit configurations via Σ_2-definitions in the sense that the set of writable real numbers arising from the use of such a rule will always be a subset of L_λ, while the corresponding set of eventually writable reals will be a subset of L_ζ. In [63], the question for complete Σ_3, Σ_4 and general Σ_n-rules was solved.

Finally, in Löwe and Welch [131], this was used to solve the limit rule problem for the revision theory of truth: It is shown in ([131], Theorem 4.3) that two kinds of revision-theoretic definability coincide (over the natural numbers) with Π^1_2-definability (and thus with each other), while two kinds of strong revision-theoretic definability coincide with Δ^1_2-definability (and thus with each other).

More specifically, in [131], two formal variants of the limit rule problem of revision theory are treated, both using the connection to ITTM-computability[28]:

1. Is the set of stably true statements independent of the limit rule?
2. Is the set of r-definable subsets of ω independent of the limit rule?

Concerning (1), it is argued in [131] that, since syntax can be encoded in many ways, it is recursive isomorphy rather than strict equality that ought to be considered. (Two sets $x, y \subseteq \omega$ are called recursively isomorphic if and only if there is a recursive bijection $f : \omega \to \omega$ such that $f[x] = y$, i. e., if and only if there is a recursive "translation" between the two sets.) It is then shown that the sets of stably true statements for various limit rules are indeed recursively isomorphic (see [131], Section 4.)

Concerning (2), it is shown that the r-definable sets for various limit rules are exactly the Π^1_2-sets of natural numbers ([131], Corollary 4.4).

Indeed a nice example of the applicability of infinitary computability![29]

28 To be precise, ITTMs are used in the proof of [131], Proposition 4.1, which then leads to the solution to (1) and to Corollary 4.4, which answers (2).

29 As B. Löwe pointed out to us, this is not necessarily the last word about the limit rule problem; see Kühnberger, Löwe, Möllerfeld and Welch [122].

Bibliography

[1] S. Aaronson. NP-complete problems and physical reality. SIGACT News, vol. 36(1), pp. 30–52 (2005).

[2] F. Abramson, G. Sacks. Uncountable Gandy ordinals. Journal of the London Mathematical Society, vol. 14(2), pp. 387–392 (1976).

[3] N. Ackerman, C. Freer, R. Lubarsky. Feedback Turing computability, and Turing computability as feedback. Proceedings of the LICS 2015, Kyoto.

[4] P. Aczel. The type theoretic interpretation of constructive set theory. Studies in Logic and the Foundations of Mathematics, vol. 96, pp. 55–66 (1978).

[5] P. Aczel, M. Rathjen. Notes on Constructive Set Theory. Unpublished Notes, available online: https://pdfs.semanticscholar.org/fffe/b30b80a4201702825b762439b0c892a9925d.pdf (2008).

[6] S. Arora, B. Barak. Computational Complexity. A Modern Approach. Cambridge University Press, Cambridge (2009).

[7] J. Bagaria, N. Castells, P. Larson. An Ω-logic primer. Set Theory, Trends Math. Birkhäuser, Basel, Boston, Berlin, pp. 1–28 (2006).

[8] T. Bartoszynski, H. Judah. Set Theory: On the Structure of the Real Line. A. K. Peters, Wellesley, Mass. (1995).

[9] J. Barwise. Absolute logics and $L_{\infty\omega}$. Ann. Math. Logic, vol. 4, pp. 309–340 (1972).

[10] J. Barwise. Admissible Sets and Structures. An Approach to Definability Theory. Perspectives in Mathematical Logic. Springer, Berlin, New York (1975).

[11] J. Barwise, R. Gandy, Y. Moschovakis. The next admissible set. The Journal of Symbolic Logic, vol. 36(1), pp. 108–120 (1971).

[12] A. Beckmann, S. Buss, S. Friedman. Safe recursive set functions. The Journal of Symbolic Logic, vol. 80(3), pp. 730–762 (2015).

[13] N. Belnap, A. Gupta. The Revision Theory of Truth. MIT Press, Cambridge, MA (1993).

[14] M. Bianchetti. Infinite Time Computation: Strong and Weak Infinite Time Turing Machines. Master's thesis, University of Notre Dame (2017).

[15] M. Bianchetti. Weaker variants of infinite time Turing machines. Preprint.

[16] O. Bournez, S. Ouazzani. Computing to the infinite with Ordinary Differential Equations. Preprint (2018).

[17] N. Bowler, A. Mathias. Rudimentary recursion and provident sets. Notre Dame Journal of Formal Logic, vol. 56(1), pp. 3–60 (2015).

[18] V. Brattka, G. Gherardi and A. Marcone. The Bolzano–Weierstraß Theorem is the Jump of weak König's lemma. Annals of Pure and Applied Logic, vol. 163(6), pp. 623–655 (2012).

[19] M. Carl. The distribution of ITRM-recognizable reals. Annals of Pure and Applied Logic, vol. 165(9), pp. 1403–1417 (2014).

[20] M. Carl. Algorithmic randomness for infinite time register machines. In: A. Beckmann et al. (eds.), Language, Life, Limits. 10th Conference on Computability in Europe, CiE 2014, Budapest, Hungary, 2014. Proceedings. Lecture Notes in Computer Science, vol. 8493. Springer, Berlin, pp. 84–92 (2014).

[21] M. Carl. Optimal results on ITRM-recognizability. The Journal of Symbolic Logic, vol. 80(4), pp. 1116–1130 (2015).

[22] M. Carl. ITRM-recognizability from random oracles. In: A. Beckmann et al. (eds.), Evolving Computability. 11th Conference on Computability in Europe, CiE 2015, Bucharest, Romania, 2015. Proceedings. Lecture Notes in Computer Science, vol. 9136. Springer, pp. 137–144 (2015).

https://doi.org/10.1515/9783110496154-010

[23] M. Carl. Randomness and degree theory for infinite time register machines. Computability, vol. 5(2), pp. 181–196 (2016).

[24] M. Carl. Generalized effective reducibility. In: A. Beckmann et al. (eds.), Pursuit of the Universal. 12th Conference on Computability in Europe, CiE 2016, Paris, France, 2016, Proceedings. Lecture Notes in Computer Science, vol. 9709. Springer, pp. 225–233 (2016).

[25] M. Carl. Infinite time recognizability from random oracles and the recognizable jump operator. Computability, vol. 6(3), pp. 223–247 (2017).

[26] M. Carl. Effectivity and reducibility with ordinal Turing machines. Preprint. arXiv:1811.11630v3 (2018).

[27] M. Carl. Some observations on infinitary complexity. In: F. Manea et al. (eds.), 14th Conference on Computability in Europe, CiE 2018, Kiel, Germany, 2018, Proceedings. Lecture Notes in Computer Science, vol. 10936. Springer, pp. 118–125 (2018).

[28] M. Carl. Space-bounded OTMs and REG^{∞}. Preprint. arXiv:1707.05297v2 (2018).

[29] M. Carl. Towards a Church–Turing-thesis for infinitary computations. Preprint. arXiv:1307.6599v1.

[30] M. Carl. The lost melody phenomenon. In: S. Geschke et al. (eds.), Infinity, Computability and Metamathematics. Festschrift Celebrating the 60th Birthdays of Peter Koepke and Philip Welch. College Publications, London, pp. 49–70 (2014).

[31] M. Carl, B. Durand, G. Lafitte, S. Ouazzani. Admissibles in gaps. In: J. Kari et al. (eds.), Unveiling Dynamics and Complexity. 13th Conference on Computability in Europe, CiE 2017, Turku, Finland, 2017. Proceedings. Lecture Notes in Computer Science, vol. 10307. Springer, pp. 175–186 (2017).

[32] M. Carl, T. Fischbach, P. Koepke, R. Miller, M. Nasfi, G. Weckbecker. The basic theory of infinite time register machines. Archive for Mathematical Logic, vol. 49(2), 249–273 (2010).

[33] M. Carl, B. Löwe, B. Rin. Koepke machines and satisfiability for infinitary propositional languages. In: J. Kari et al. (eds.), Unveiling Dynamics and Complexity. 13th Conference on Computability in Europe, CiE 2017, Turku, Finland, 2017. Proceedings. Lecture Notes in Computer Science, vol. 10307. Springer, pp. 187–197 (2017).

[34] M. Carl, S. Ouazzani, P. Welch. Taming Koepke's Zoo. In: F. Manea et al. (eds.), 14th Conference on Computability in Europe, CiE 2018, Kiel, Germany, 2018, Proceedings. Lecture Notes in Computer Science, vol. 10936. Springer, pp. 126–135 (2018).

[35] M. Carl, B. Rin, P. Schlicht. Writability and reachability for α-tape infinite time Turing machines. Preprint. arXiv:1802.05734v2.

[36] M. Carl, P. Schlicht. Infinite computations with random oracles. Notre Dame Journal of Formal Logic, vol. 58(2), pp. 249–270 (2017).

[37] M. Carl, P. Schlicht. The recognizability strength of cardinal-recognizing infinite time Turing machines. In: J. Kari et al. (eds.), Unveiling Dynamics and Complexity. 13th Conference on Computability in Europe, CiE 2017, Turku, Finland, 2017. Proceedings. Lecture Notes in Computer Science, vol. 10307. Springer, pp. 198–209 (2017).

[38] M. Carl, P. Schlicht. Randomness via infinite computation and effective descriptive set theory. The Journal of Symbolic Logic, vol. 83(2), pp. 766–789 (2018).

[39] M. Carl, P. Schlicht, P. Welch. Recognizable sets and Woodin cardinals. Computation beyond the constructible universe. Annals of Pure and Applied Logic, vol. 169(4), 312–332 (2018).

[40] C. Chong, L. Yu. Recursion Theory. Computational Aspects of Definability. De Gruyter Series in Logic and Its Applications, vol. 8. De Gruyter, Berlin (2015).

[41] P. Cohen. The independence of the continuum hypothesis. Proceedings of the National Academy of Sciences of the United States of America, vol. 50(6), pp. 1143–1148 (1963).

[42] S. Coskey, J. Hamkins. Infinite time decidable equivalence relation theory. Notre Dame Journal of Formal Logic, vol. 52(2), pp. 203–228, 2011.

[43] S. Coskey, J. Hamkins. Infinite time Turing machines and an application to the hierarchy of equivalence relations on the reals. In: Effective Mathematics of the Uncountable, vol. 41. Assoc. Symbol. Logic, La Jolla, CA, pp. 33–49 (2013).

[44] N. Cutland. Computability. Cambridge University Press, Cambridge, New York (1980).

[45] B. Dawson. Ordinal time Turing computation. PhD thesis, Bristol (2005).

[46] V. Deolalikar, J. D. Hamkins, R.-D. Schindler. $P \neq NP \cap co - NP$ for infinite time Turing machines. Journal of Logic and Computation, vol. 15(5), pp. 577–592 (2005).

[47] K. Devlin. Chapter II: The Constructible Universe. In: Constructibility. Springer-Verlag, Berlin, pp. 56–107, (1984), https://projecteuclid.org/euclid.pl/1235419480.

[48] K. Devlin. The Joy of Sets. Fundamentals of Contemporary Set Theory, second edition. Undergraduate Texts in Mathematics. Springer, New York (1993).

[49] R. Downey, D. Hirschfeldt. Algorithmic Randomness and Complexity. Theory and Applications of Computability. Springer, New York (2010).

[50] B. Durand, G. Lafitte. An algorithmic approach to characterizations of admissibles. Proceedings for CiE 2019.

[51] J. Earman, J. Norton. Forever is a day: Supertasks in Pitowsky and Malament–Hogarth spacetimes. Philosophy of Science, vol. 60(1), pp. 22–42 (1993).

[52] Y. Ershov. Forcing in admissible sets. Algebra and Logic, vol. 29(6), pp. 424–430 (1991).

[53] T. Fischbach. The Church–Turing Thesis for Ordinal Computable Functions. Master's thesis, University of Bonn (2010).

[54] T. Fischbach, B. Seyfferth. On λ-definable functions on ordinals. In: P. Bonizzoni et al. (eds.), The Nature of Computation. Logic, Algorithms, Applications. 9th Conference on Computability in Europe, CiE 2013, Milan, Italy, 2013. Proceedings. Lecture Notes in Computer Science, vol. 7921. Springer, Berlin, pp. 135–146 (2013).

[55] T. Fischbach, B. Seyfferth. An ordinal λ-calculus. In: S. Geschke et al. (eds.), Infinity, Computability and Metamathematics. Festschrift Celebrating the 60th Birthdays of Peter Koepke and Philip Welch. College Publications, London, pp. 139–166 (2014).

[56] M. Fitting. Fundamentals of Generalized Recursion Theory. North Holland (1981).

[57] Forcing. Wikipedia article. https://en.wikipedia.org/wiki/Forcing_(mathematics) accessed: 08.08.2018.

[58] L. Fortnow, B. Gasarch. Two proofs of Ladner's theorem. Online lecture notes: http://oldblog.computationalcomplexity.org/media/ladner.pdf.

[59] A. Fraenkel. Vorlesungen über die Grundlegung der Mengenlehre. Wissenschaftlicher Buchgesellschaft, Darmstadt (1972).

[60] H. Friedman. Some applications of Kleene's methods for intuitionistic systems. In: A. Mathias, H. Rogers (eds.), Cambridge Summer School in Mathematical Logic, pp. 113–170 (1973).

[61] H. Friedman, L. Stanley. A Borel reducibility theory for classes of countable structures. The Journal of Symbolic Logic, vol. 54(3), pp. 894–914 (1989).

[62] S. Friedman, P. Welch. Two observations concerning infinite time Turing machines. In: I. Dimitriou (ed.), BIWOC Report. Hausdorff Centre for Mathematics, Bonn, pp. 44–47 (2007).

[63] S. Friedman, P. Welch. Hypermachines. The Journal of Symbolic Logic, vol. 76(2), pp. 620–636 (2011).

[64] L. Galeotti. Computable Analysis Over the Generalized Baire Space. Master's thesis. Universiteit van Amsterdam (2015).

[65] L. Galeotti. Surreal Blum–Shub–Smale machines. LNCS Proceedings of CiE 2019. Preprint available online: https://preprint.math.uni-hamburg.de/public/hbm.html.

[66] V. Gitman, J. Hamkins, T. Johnstone. What is the theory ZFC without power set? Mathematical Logic Quarterly, vol. 62(4–5), pp. 391–406 (2016).

[67] K. Gödel. Russell's mathematical logic. In: S. Feferman et al. (eds.), Journal of Symbolic Logic. Northwestern University Press, pp. 119–141 (1944).
[68] R. Goodstein. Recursive Analysis. Studies in Logic and the Foundations of Mathematics. North-Holland Publishing Company, Amsterdam (1961).
[69] R. Gostanian. The next admissible ordinal. Annals of Mathematical Logic, vol. 17, pp. 171–203 (1979).
[70] C. Gray. Iterated Forcing from the Strategic Point of View. PhD thesis, University of California, Berkeley (1980).
[71] N. Greenberg, D. Hirschfeldt, J. Hamkins, R. Miller (eds.), Effective Mathematics of the Uncountable, vol. 41. Assoc. Symbol. Logic, La Jolla, CA (2013).
[72] M. Habic. Cardinal-recognizing infinite time Turing machines. In: P. Bonizzoni et al. (eds.), The Nature of Computation. Logic, Algorithms, Applications. 9th Conference on Computability in Europe, CiE 2013, Milan, Italy, 2013. Proceedings. Lecture Notes in Computer Science, vol. 7921. Springer, Berlin, pp. 231–240 (2013).
[73] M. Hallett. Cantorian Set Theory and Limitation of Size. Oxford Logic Guides, vol. 10, Clarendon Press, Oxford (1984).
[74] J. Hamkins. A survey of infinite time Turing machines. In: Machines, Computations, and Universality – 5th International Conference MCU 2007, Orleans, France, pp. 62–71 (2007).
[75] J. Hamkins. The set-theoretic multiverse. The Review of Symbolic Logic, vol. 5, 416–449 (2012).
[76] J. Hamkins, C. Leahy. Algebraicity and implicit definability in set theory. Notre Dame Journal of Formal Logic, vol. 57(3), pp. 431–439 (2016).
[77] J. Hamkins, A. Lewis. Post's problem for supertasks has both positive and negative solutions. Archive for Mathematical Logic, vol. 41(6), pp. 507–523 (2002).
[78] J. Hamkins, B. Löwe. The modal logic of forcing. Transactions of the American Mathematical Society, vol. 60(4), pp. 1793–1817 (2008).
[79] J. Hamkins, R. Miller, D. Seabold, S. Warner. Infinite time computable model theory. In: S. Cooper et al. (eds.), New Computational Paradigms. Changing Conceptions of What Is Computable. Springer, New York, pp. 521–557 (2008).
[80] J. Hamkins, D. Seabold. Infinite Time Turing Machines with only one tape. Mathematical Logic Quarterly, vol. 47(2), pp. 271–287 (2001).
[81] J. D. Hamkins, A. Lewis. Infinite time Turing machines. The Journal of Symbolic Logic, vol. 65(2), 567–604 (2000).
[82] V. Harizanov. Pure computable model theory. In: I. Ershov (ed.), Handbook of Recursive Mathematics, vol. 1: Recursive Model Theory. Studies in Logic and the Foundations of Mathematics, vol. 138, North Holland, pp. 3–114 (1998).
[83] H. Herzberger. Notes on naive semantics. Journal of Philosophical Logic, vol. 11, pp. 61–102 (1982).
[84] A. Heyting (ed.). L. Brouwer. Collected Works. Vol. 1: Philosophy and Foundations of Mathematics. North Holland Publishing Company, Amsterdam, Oxford (1975).
[85] D. Hirschfeldt. Slicing the Truth. Lecture Notes Series. Institute for Mathematical Sciences. National University of Singapore, vol. 28. World Scientific, Hackensack, NJ (2014).
[86] D. Hirschfeldt, C. Jockusch. On notions of computability theoretic reduction between Π^1_2-principles. Preprint. Available online: http://www.math.uchicago.edu/~drh/Papers/Papers/upaper.pdf.
[87] W. Hodges. On the effectivity of some field constructions. Proceedings of the London Mathematical Society, vol. 32, pp. 133–162 (1976).
[88] W. Hodges. What is a structure theory? Bulletin of the London Mathematical Society, vol. 19, pp.209–237 (1987).

[89] J. Hromkovic. Theoretische Informatik. Formale Sprachen, Berechenbarkeit, Komplexitätstheorie, Algorithmik, Kommunikation und Kryptographie. Leitfäden der Informatik. Teubner, Wiesbaden (2007).
[90] R. Iemhoff. Kripke models for subtheories of CZF. Archive for Mathematical Logic, vol. 49(2), pp. 147–167 (2010).
[91] G. Igusa, J. Knight. Comparing two versions of the reals. Preprint (2015).
[92] T. Jech. Set Theory. The third Millenium edition. Springer Monographs in Mathematics. Springer, Berlin (2003).
[93] R. Jensen. The fine structure of the constructible hierarchy. Annals of Mathematical Logic, vol. 4, pp. 229–308 (1972).
[94] R. Jensen. Manuscript on fine structure, inner model theory and the core model below one Woodin cardinal. Forthcoming. Available at https://ivv5hpp.uni-muenster.de/u/rds/jensen_script_sept_26_2017.pdf.
[95] R. Jensen, C. Karp. Primitive recursive set functions. In: 1971 Axiomatic Set Theory (Proc. Sympos. Pure Math., vol. 8(1), Univ. California, Los Angeles, Calif.). Amer. Math. Soc., Providence, R.I., pp. 143–176 (1967).
[96] A. Kanamori. The Higher Infinite. Large Cardinals in Set Theory from Their Beginnings. Springer-Verlag, Berlin, Heidelberg (2003).
[97] R. Kaye. Models of Peano Arithmetic. Oxford Logic Guides, vol. 15. Clarendon Press, Oxford (1991).
[98] R. Kaye. The Mathematics of Logic. A Guide to Completeness Theorems and Their Applications. Cambridge University Press (2007).
[99] A. Kechris. Measure and category in effective descriptive set theory. Annals of Mathematical Logic, vol. 5(4), pp. 337–384 (1973).
[100] J. Kennedy, M. Magidor, J. Väänänen. Inner Models from Extended Logics. Isaac Newton Institute Preprint Series. Isaac Newton Institute for Mathematical Sciences, Cambridge, 75 p. (Preprint Series) (2016).
[101] P. Kitcher. The Nature of Mathematical Knowledge. Oxford University Press (1983).
[102] S. Kleene. On the interpretation of intuitionistic number theory. The Journal of Symbolic Logic, vol. 10(4), 109–124 (1945).
[103] J. F. Knight, A. Montalban, N. Schweber. Computable structures in generic extensions. The Journal of Symbolic Logic, vol. 81(3), pp. 814–832 (2016).
[104] P. Koepke. Turing computations on ordinals. The Bulletin of Symbolic Logic, vol. 11, pp. 377–397 (2005).
[105] P. Koepke. Computing a model of set theory. In: S. Cooper et al. (eds.), New Computational Paradigms. Lecture Notes in Computer Science, vol. 3988, pp. 223–232 (2006).
[106] P. Koepke. Infinite time register machines. In: A. Beckmann et al. (eds.), Logical Approaches to Computational Barriers. Second Conference on Computability in Europe, CiE 2006, Swansea, UK. Proceedings. Lecture Notes in Computer Science, vol. 3988. Springer, Berlin, pp. 257–266 (2006).
[107] P. Koepke. Ordinals, computations and models of set theory. In: R. Kahle et al. (eds.), Days in Logic '06. Two Tutorials of the Meeting, Coimbra, Portugal, 2006. Textos de Matematica, Serie B, vol. 38. Universidade de Coimbra, Departamento de Matematica, Coimbra, pp. 43–78 (2006).
[108] P. Koepke. Infinite time register machines. In: A. Beckmann et al. (eds.), Logical Approaches to Computational Barriers. CiE 2006 Proceedings. Lecture Notes in Computer Science, vol. 3988, pp. 257–266 (2006).
[109] P. Koepke. Ordinal computability. In: K. Ambos-Spies et al. (eds.), Mathematical Theory and Computational Practice. Lecture Notes in Computer Science, vol. 5635, pp. 280–289 (2009).

[110] P. Koepke, R. Miller. An enhanced theory of infinite time register machines. In: A. Beckmann et al. (eds.), Logic and Theory of Algorithms. Lecture Notes in Computer Science, vol. 5028, pp. 306–315 (2008).
[111] P. Koepke, A. Morozov. On the computational strength of Infinite Time Blum-Shub-Smale Machines. Algebra and Logic, vol. 56(1) (2017).
[112] P. Koepke, B. Seyfferth. Ordinal machines and admissible recursion theory. Annals of Pure and Applied Logic, vol. 160, pp. 310–318 (2009).
[113] P. Koepke, B. Seyfferth. Towards a theory of infinite time Blum–Shub–Smale machines. In: S. Cooper et al. (eds.), How the World Computes. Turing Centenary Conference and 8th Conference on Computability in Europe, CiE 2012, Cambridge, UK, 2012. Proceedings. Lecture Notes in Computer Science, vol. 7318. Springer, Berlin, pp. 405–415 (2012).
[114] P. Koepke, R. Siders. Computing the recursive truth predicate on ordinal register machines. In: A. Beckmann et al. (eds.), Logical Approaches to Computational Barriers. Computer Science Report Series, vol. 7. Swansea, pp. 160–169 (2006).
[115] P. Koepke, R. Siders. Register computations on ordinals. Archive for Mathematical Logic, vol. 47(6), pp. 529–548 (2008).
[116] P. Koepke, R. Siders. Minimality considerations for ordinal computers modeling constructibility. Theoretical Computer Science, vol. 394, pp. 197–207 (2008).
[117] P. Koepke, P. Welch. A generalised dynamical system, infinite time register machines, and Π^1_1-CA_0. In: B. Löwe et al. (eds.), Models of computation in context. 7th conference on computability in Europe, CiE 2011. Lecture Notes in Computer Science, vol. 6735, pp. 152–159 (2011).
[118] S. Krantz. The Axiom of Choice. Handbook of Logic and Proof Techniques for Computer Science, Birkhäuser, Boston, pp. 121–126 (2002).
[119] S. Kripke., Semantical analysis of intuitionistic logic. Formal Syst. and Recurs. Funct., Proc. 8th Logic Colloquium, Oxford, 1963, pp. 92–130 (1965).
[120] S. Kripke. Naming and Necessity. Basil Blackwell Publisher (1980).
[121] S. Kritchman, R. Raz. The surprise examination paradox and the second incompleteness theorem. Notices of the American Mathematical Society, vol. 57(11), pp. 1454–1458 (2010).
[122] K. Kühnberger, B. Löwe, M. Möllerfeld, P. Welch. Comparing Inductive and Circular Definitions: Parameters, Complexity and Games. Studia Logica, vol. 81. Springer (2005).
[123] K. Kunen. Set Theory. An Introduction to Independence Proofs. Studies in Logic (London), vol. 34. College Publications, London (2011).
[124] J. Lagarias. An elementary problem equivalent to the Riemann hypothesis. The American Mathematical Monthly, vol. 109(6), pp. 534–543 (2002).
[125] A. Lewis. Properties of the jump classes. Journal of Logic and Computation, vol. 22(4), pp. 845–855 (2012).
[126] E. Lewis. Computation with Infinite Programs. Master thesis, ILLC, Amsterdam (2018).
[127] J. Long III. Capabilities and Limitations of Infinite-Time Computation. PhD thesis, Lehigh University (2015).
[128] B. Löwe. Revision sequences and computers with an infinite amount of time, Journal of Logic and Computation, vol. 11, pp. 25–40 (2001).
[129] B. Löwe. Space bounds for infinitary computations. In: A. Beckmann et al. (eds.), Logical Approaches to Computational Barriers. CiE 2006. Lecture Notes in Computer Science, vol. 3988. Springer, Berlin, Heidelberg (2006).
[130] B. Löwe, J. Steel. An introduction to Core model theory. In: Sets and Proofs, Cambridge Univ. Press, Cambridge, pp. 103–157 (1999).
[131] B. Löwe, P. Welch. Set-theoretic absoluteness and the revision theory of truth. Studia Logica, vol. 68(1), pp. 21–41 (2001).

[132] R. Lubarsky. Independence results around constructive ZF. Annals of Pure and Applied Logic, volume 132(2–3), pp. 209–225 (2005).
[133] R. Lubarsky. ITTMs with feedback. Ways of Proof Theory, Ontos Verlag Heusenstamm, pp. 341–353 (2010).
[134] P. Maddy. Book review: The nature of mathematical knowledge by Philip Kitcher. Philosophy of Science, vol. 52(2), p. 312 (1985).
[135] R. Mansfield, G. Weitkamp. Recursive Aspects of Descriptive Set Theory. Oxford Logic Guides, vol. 11, Oxford University Press New York (1985).
[136] D. Marker. Model Theory. An Introduction. Graduate Texts in Mathematics, vol. 217. Springer, New York (2002).
[137] A. Mathias. Provident sets and rudimentary set forcing. Fundamenta Mathematicae, vol. 230(2), pp. 99–148 (2015).
[138] U. Matzner. Ordinalzahltheoretische Berechnungsmöglichkeiten. Bachelor thesis, Bonn (2016).
[139] G. Melles. One cannot show from ZFC that there is an Ulm-type classification of the countable torsion-free abelian groups. In: Set Theory of the Continuum. Math. Sc. Research Inst. Publ., vol. 26, pp. 293–309 (1992).
[140] R. Miller, J. Hamkins. Post's problem for Ordinal Register Machines: An explicit approach. Annals of Pure and Applied Logic, vol. 160(3), pp. 302–309 (2009).
[141] W. Mitchell, J. Steel. Fine Structure and Iteration Trees. Lecture Notes in Logic, vol. 3. Springer, Berlin (1994).
[142] B. Monin, P.-E.t Angles D'Auriac. Genericity and Randomness with ITTMs. Preprint (2018).
[143] Y. N. Moschovakis. Descriptive Set Theory. Studies in Logic and the Foundations of Mathematics, vol. 100, North-Holland Publishing Company, New York (1980).
[144] J. Myhill. Constructive set theory. The Journal of Symbolic Logic, Vol. 40(3), 347–382 (1975).
[145] S. Ouazzani. From algorithmics to logics through infinite time computation. PhD thesis, Montpellier (2016).
[146] G. Pullum. Scooping the Loop Snooper. A proof that the Halting Problem is undecidable. Mathematics Magazine, vol. 73, pp. 319–320 (2000) Correction available at http://www.lel.ed.ac.uk/~gpullum/loopsnoop.html.
[147] M. Rabin. Computable algebra. General theory and theory of computable fields. Transactions of the American Mathematical Society, vol. 95, pp. 341–360 (1960).
[148] F. Ramsey. The foundations of mathematics. In: D. Mellor (ed.), F. P. Ramsey: Philosophical Papers. Cambridge University Press (1990).
[149] B. Rin. The computational strengths of α-tape infinite time Turing machines. Annals of Pure and Applied Logic, vol. 165(9), pp. 1501–1511 (2012).
[150] B. Rin. Mathematical Limits and Philosophical Significance of Transfinite Computation. PhD thesis, Irvine (2014).
[151] B. Rin. Transfinite recursion and computation in the iterative conception of set. Synthese, vol. 192(8), pp. 2437–2462 (2015).
[152] H. Rogers, Theory of Recursive Functions and Effective Computability. MIT Press, Cambridge, MA (1987).
[153] G. Sacks. Higher Recursion Theory. Perspectives in Logic. Cambridge University Press (2016).
[154] E. Schimmerling. A core model toolbox and guide. In: M. Foreman et al. (eds.), Handbook of Set Theory, pp. 1685–1751 (2010).
[155] R. Schindler. $P \neq NP$ for infinite time Turing machines. Monatshefte für Mathematik, vol. 139, pp. 335–340 (2003).
[156] R. Schindler, M. Zeman. Fine structure. In: Foreman, Matthew, Kanamori, Akihiro (eds.), Handbook of Set Theory. Springer, Netherlands, pp. 605–656 (2010).

[157] P. Schlicht, F. Stephan. Automata on ordinals and automaticity of linear orders. Annals of Pure and Applied Logic, vol. 164(5), pp. 523–527 (2013).
[158] Set Theory: Constructive and Intuitionistic ZF. Stanford Encyclopedia of Philosophy. https://plato.stanford.edu/entries/set-theory-constructive/ assessed: 14.01.2019.
[159] B. Seyfferth. Three Models of Ordinal Computability. PhD thesis, Bonn (2012).
[160] B. Seyfferth, P. Schlicht. Tree representations via ordinal machines. Computability, vol. 1(1), pp. 45–57 (2012).
[161] J. Shoenfield. Axioms of set theory. Handbook of Mathematical Logic, Stud. Logic Found. Math., vol. 90, North-Holland, Amsterdam, pp. 321–344 (1977).
[162] R. Shore. Reverse mathematics, countable and uncountable. A computational approach. In: Effective Mathematics of the Uncountable, vol. 41. Assoc. Symbol. Logic, La Jolla, CA, pp. 150–164 (2013).
[163] S. Simpson. Subsystems of Second-Order Arithmetic. Perspectives in Mathematical Logic. Springer, Berlin (1999).
[164] S. Simpson. An extension of the recursively enumerable Turing degrees. Journal of the London Mathematical Society. Second Series, vol. 75, pp. 287–297 (2007).
[165] S. Simpson. Implicit definability in arithmetic. Notre Dame Journal of Formal Logic, vol. 57(3), pp. 329–339 (2016).
[166] S. Simpson, G. Weitkamp. High and low Kleene degrees of coanalytic sets. The Journal of Symbolic Logic, vol. 48(2), pp. 356–368.
[167] R. Soare. Recursively Enumerable Sets and Degrees. A Study of Computable Functions and Computably Generated Sets. Perspectives in Mathematical Logic. Springer, Berlin (1987).
[168] J. Steel. Projectively well-ordered inner models. Annals of Pure and Applied Logic, vol. 74(1), pp. 77–104 (1995).
[169] J. Steel. An outline of inner model theory. In: Foreman, Matthew, Kanamori, Akihiro (Eds.), Handbook of Set Theory. Springer, Netherlands, pp. 1595–1684 (2010).
[170] A. Syropoulos. Hypercomputation. Computing Beyond the Church–Turing Barrier. Springer (2008).
[171] G. Takeuti. Proof Theory. Studies in Logic and the Foundations of Mathematics, vol.81. North-Holland Publishing Co./American Elsevier Publishing Co., Inc., Amsterdam, Oxford/New York (1975).
[172] G. Takeuti. Proof theory and set theory. Synthese, vol. 62(2), The Present State of the Problem of the Foundations of the Mathematics, pp. 255–263 (1985).
[173] S. Terwijn. The mathematical foundations of randomness. In: N. Landsman, E. van Wolde (eds.), The Challenge of Chance, Springer, pp. 49–66 (2016).
[174] The Revision Theory of Truth. Stanford Encyclopedia of Philosophy. https://plato.stanford.edu/entries/truth-revision/ accessed: 20.07.2018.
[175] A. Turing. On computable numbers, with an application to the Entscheidungsproblem. Proceedings of the London Mathematical Society, vol. 2(3), pp. 230–265 (1936).
[176] H. Wang. From Mathematics to Philosophy. Routledge & Kegan Paul Ltd (1974).
[177] H. Wang. A Logical Journey. From Gödel to Philosophy. MIT Press, Cambridge, MA (1996).
[178] K. Weihrauch. Computable Analysis. An introduction. Texts in Theoretical Computer Science. An EATCS Series. Springer, Berlin (2000).
[179] P. Welch. Minimality arguments for infinite time Turing degrees. In: Sets and Proofs. Cambridge Univ. Press, Cambridge, pp. 425–436 (1999).
[180] P. Welch. Eventually Infinite Time Turing Degrees: infinite time decidable reals. The Journal of Symbolic Logic, vol. 65(3), pp 1193–1203 (2000).
[181] P. Welch. The length of Infinite Time Turing Machine computations. Bulletin of the London Mathematical Society, vol. 32(3), pp. 129–136 (2000).

[182] P. Welch. On the possibility, or otherwise, of hypercomputation. British Journal for the Philosophy of Science, vol 55(4) (2004).

[183] P. Welch. Turing Unbound: On the extent of computation in Malament–Hogarth spacetimes. British Journal for the Philosophy of Science, vol. 59(4) (2008).

[184] P. Welch. Characteristics of discrete transfinite time Turing machine models: halting times, stabilization times, and Normal Form theorems. Theoretical Computer Science, vol. 410, pp. 426–442 (2009).

[185] P. Welch. Some observations on Truth hierarchies. The Review of Symbolic Logic, vol. 7(1), pp. 1–30 (2014).

[186] P. Welch. Obtaining Woodin's cardinals. In: A. Caicedo et al. (eds.), Foundations of Mathematics. Logic at Harvard. Essays in Honor of W. Hugh Woodin's 60th Birthday. Proceedings of the Logic at Harvard Conference, Harvard University, Cambridge, MA, USA, 2015. Contemporary Mathematics, vol. 690. American Mathematical Society (AMS), Providence, RI, pp. 161–176 (2017).

[187] P. Welch. $G_{\delta\sigma}$-Games and generalized computation. Preprint.

[188] P. Welch. On the transfinite actions of 1-tape Turing machines. In: S. Cooper et al. (eds.), New Computational Paradigms. CiE 2005 Proceedings. Lecture Notes in Computer Science, vol. 3526. Springer, Berlin, Heidelberg, pp. 532–539 (2005).

[189] P. Welch. Discrete transfinite computation models. In: S. Cooper et al. (eds.), Computability in Context. Computation and Logic in the Real World. Imperial College Press London, pp. 371–410 (2011).

[190] P. Welch. Transfinite machine models. In: R. Downey (ed.), Turing's Legacy. Lecture Notes in Logic Series, vol. 42. Association for Symbolic Logic, Cambridge University Press (2014).

[191] P. Welch. Discrete transfinite computation. In: G. Sommaruga et al. (eds.), Turing's Revolution. The Impact of His Ideas About Computability. Birkhäuser/Springer, pp. 161–185 (2015).

[192] P. Welch, J. Hamkins. $P^f \neq NP^f$ for almost all f. Mathematical Logic Quarterly, vol. 49(5) (2003).

[193] P. Welch. Post's and other problems in higher type supertasks. In: B. Löwe et al. (eds.), Classical and New Paradigms of Computation and their Complexity hierarchies, Papers of the Conference Foundations of the Formal Sciences III. Trends in Logic, vol. 23. Kluwer, Dordrecht, pp. 223–237 (2001).

[194] Revision Theory. Wikipedia article. https://en.wikipedia.org/wiki/Revision_theory accessed: 20.07.2018.

[195] J. Winter. Space complexity in Infinite Time Turing Machines. Master's thesis, Universiteit van Amsterdam (2007).

[196] J. Winter. Is P=PSPACE for infinite time Turing machines? In: M. Archibald, V. Brattka, V. Goranko, B. Löwe (eds.), ILC 2007. Lecture Notes in Computer Science, vol. 5489. Springer, Berlin, pp. 126–137 (2009).

[197] S. Yu. Regular languages. In: G. Rozenberg, A. Salomaa (eds.), Handbook of Formal Languages. Vol. 1: Word, Language, Grammar, pp. 41–110 (1997).

[198] A. Zarach. Unions of ZF^--Models that Are Themselves ZF^--Models. Studies in Logic and the Foundations of Mathematics, vol. 108 (1982).

[199] M. Zeman. Inner Models and Large Cardinals. De Gruyter Series in Logic and Its Applications, vol. 5. Walter de Gruyter Berlin (2002).

Index

De Gruyter Series in Logic and Its Applications

Volume 8
Chi Tat Chong, Liang Hu
Recursion Theory, 2015
ISBN 978-3-11-027555-1, e-ISBN (PDF) 978-3-11-027564-3,
e-ISBN (EPUB) 978-3-11-038129-0, Set-ISBN 978-3-11-027565-0

Volume 7
Itay Neeman
The Determinacy of Long Games, 2008
ISBN 978-3-11-018341-2, e-ISBN 978-3-11-020006-5, Set-ISBN 978-3-11-916429-0

Volume 6
Godehard Link (Ed.)
One Hundred Years of Russell's Paradox, 2008
ISBN 978-3-11-017438-0, e-ISBN 978-3-11-019968-0, Set-ISBN 978-3-11-916554-9

Volume 5
Martin Zeman
Inner Models and Large Cardinals, 2011
ISBN 978-3-11-016368-1, e-ISBN 978-3-11-085781-8, Set-ISBN 978-3-11-179317-7

Volume 4
Rod Downey, Denis R. Hirschfeldt (Eds.)
Aspects of Complexity, 2011
ISBN 978-3-11-016810-5, e-ISBN 978-3-11-088917-8, Set-ISBN 978-3-11-180439-2

Volume 3
Sy. D. Friedman
Fine Structure and Class Forcing, 2011
ISBN 978-3-11-016777-1, e-ISBN 978-3-11-080911-4, Set-ISBN 978-3-11-174922-8

Volume 2
Marat N. Arslanov, Steffen Lempp
Recursion Theory and Complexity, 2014
ISBN 978-3-11-016587-6, e-ISBN 978-3-11-080748-6, Set-ISBN 978-3-11-175948-7

Volume 1
W. Hugh Woodin
The Axiom of Determinacy, Forcing Axioms, and the Nonstationary Ideal,
2nd Edition, 2010
ISBN 978-3-11-019702-0, e-ISBN 978-3-11-021317-1, Set-ISBN 978-3-11-173779-9

Bei Fragen zur Produktsicherheit wenden Sie sich bitte an:
If you have any questions regarding product safety,
please contact:

Walter de Gruyter GmbH
Genthiner Straße 13
10785 Berlin
productsafety@degruyterbrill.com